Progress in Geomathematics

Graeme Bonham-Carter · Qiuming Cheng (Eds.)

Progress in Geomathematics

Graeme Bonham-Carter
Geological Survey of Canada
601 Booth St.
Ottawa, Ontario, Canada, K1A 0E8
gbonhamc@nrcan.gc.ca

Qiuming Cheng
State Key Lab of Geological Processes
and Mineral Resources
China University of Geosciences
Wuhan 430074, Beijing 100083 China
and
Department of Earth and Space Science and
Engineering
York University
4700 Keele Street
Toronto, Ontario, Canada M3J1P3
qiuming@yorku.ca

ISBN: 978-3-540-69495-3   e-ISBN: 978-3-540-69496-0

Library of Congress Control Number: 2008932103

© 2008 Springer-Verlag Berlin Heidelberg

This work is subject to copyright. All rights are reserved, whether the whole or part of the material is concerned, specifically the rights of translation, reprinting, reuse of illustrations, recitation, broadcasting, reproduction on microfilm or in any other way, and storage in data banks. Duplication of this publication or parts thereof is permitted only under the provisions of the German Copyright Law of September 9, 1965, in its current version, and permission for use must always be obtained from Springer. Violations are liable to prosecution under the German Copyright Law.

The use of general descriptive names, registered names, trademarks, etc. in this publication does not imply, even in the absence of a specific statement, that such names are exempt from the relevant protective laws and regulations and therefore free for general use.

Cover design: deblik, Berlin

Printed on acid-free paper

9 8 7 6 5 4 3 2 1

springer.com

A Festschrift for Frits Agterberg

Contents

Introduction 1
Graeme F. Bonham-Carter and Qiuming Cheng

The Role of Frederik Pieter Agterberg in the Development of Geomathematics 5
Graeme F. Bonham-Carter, Václav Němec, Dan F. Merriam, Zhao Pengda and Qiuming Cheng

Another Look at the Chemical Relationships in the Dissolved Phase of Complex River Systems 23
A. Buccianti, J.J. Egozcue and V. Pawlowsky-Glahn

A Critical Approach to Probability Laws in Geochemistry 39
G. Mateu-Figueras and V. Pawlowsky-Glahn

Investigation of the Structure of Geological Process Through Multivariate Statistical Analysis—The Creation of a Coal 53
Lawrence J. Drew, Eric C. Grunsky and John H. Schuenemeyer

Master of the Obscure—Automated Geostatistical Classification in Presence of Complex Geophysical Processes 79
Ute C. Herzfeld

The Rapid Retreat of Jakobshavns Isbræ, West Greenland: Field Observations of 2005 and Structural Analysis of its Evolution 113
Helmut Mayer and Ute C. Herzfeld

Spatiotemporal Continuity of Sequential Rain Suggested by 3-D Variogram 131
Tetsuya Shoji

Anisotropic Scaling Models of Rock Density and the Earth's Surface Gravity Field 151
S. Lovejoy, H. Gaonac'h and D. Schertzer

vii

Non-linear Theory and Power-Law Models for Information Integration and Mineral Resources Quantitative Assessments 195
Qiuming Cheng

Mineral Potential Modelling for the Greater Nahanni Ecosystem Using GIS Based Analytical Methods 227
J.R. Harris, D. Lemkow, C. Jefferson, D. Wright and H. Falck

Map Scale Effects on Estimating the Number of Undiscovered Mineral Deposits 271
Donald A. Singer and W. David Menzie

Are Fractal Dimensions of the Spatial Distribution of Mineral Deposits Meaningful? 285
Gary L. Raines

African Neoproterozoic Mineral Deposits and Pan African Metallogenesis 303
Christien Thiart and Maarten de Wit

On Blind Tests and Spatial Prediction Models 315
Andrea G. Fabbri and Chang-Jo Chung

Strip Transect Sampling to Estimate Object Abundance in Homogeneous and Non-Homogeneous Poisson Fields: A Simulation Study of the Effects of Changing Transect Width and Number 333
Timothy C. Coburn, Sean A. McKenna and Hirotaka Saito

Increasing Resolution in Exploration Biostratigraphy – Part I 353
F.M. Gradstein, A. Bowman, A. Lugowski and O. Hammer

Increasing Resolution in Exploration Biostratigraphy – Part II 369
A. Bowman, F.M. Gradstein, A. Lugowski and O. Hammer

RASC/CASC: Example of Creative Application of Statistics in Geology .. 379
Zhou Di

Euclidean Distances and Singular Value Decomposition: Useful Tools for Geometric Morphometrics in Biology and Paleontology 393
James C. Brower

A Note on Seasonal Variation in Radiolarian Abundance 417
Richard A. Reyment, Isao Motoyama, Miyuki Ota and Yuichiro Tanaka

Application of Markov Mean First-Passage Time Statistics to Sedimentary Successions: A Pennsylvanian Case-Study from the Illinois Basin 435
John H. Doveton

The Beta Distribution for Categorical Variables at Different Support 445
Clayton V. Deutsch and Zhou Lan

Enhancement of Seafloor Maps for Mecklenburg Bay, Baltic Sea, Using Proxy Variables 457
Ricardo A. Olea, Bernd Bobertz, Jan Harff and Rudolf Endler

Statistical Analysis of Physiographic and Structural Directional Data in the U.S. Midcontinent (Kansas) 481
Daniel F. Merriam and John C. Davis

Cross-Wavelet Analysis: A Tool for Detection of Relationships Between Paleoclimate Proxy Records 499
Andreas Prokoph and Hafida El Bilali

On Correlational Properties for Volcanic Earthquakes Associated with Asamayama (Japan), 1983–2005 513
Richard A. Reyment

Crosscorrelation of Sea Levels 529
Joseph E. Robinson

Diversion of Flooding Rivers to Residual Mining Open Pits 535
Ian Lerche

A Christmas Parking Lot Problem 545
Ian Lerche

List of Contributors

Bernd Bobertz
Leibniz Institute for Baltic sea Research Warnemünde (IOW), 18119 Rostock, Germany; Presently at: Institut für Geographie und Geologie, Ernst-Moritz-Arndt-Universität Greifswald, Friedrich-Ludwig-Jahn-Strasse 16, 17487 Greifswald, Germany, bobertz@uni-greifswald.de

Graeme F. Bonham-Carter
Geological Survey of Canada, 601 Booth St., Ottawa, Ontario K1A 0E8, Canada, gbonhamc@nrcan.gc.ca

A. Bowman
Chevron Energy Technology Co, 14141 SW Freeway, Sugarland, TX 77478, USA, abowman2@bigred.unl.edu

James C. Brower
Department of Earth Sciences, Syracuse University, Syracuse, New York, 13244-1070, USA, Karen16@localnet.com

A. Buccianti
Department of Earth Sciences, University di Firenze, Firenze, Italy, antonella.buccianti@unifi.it

Qiuming Cheng
State Key Lab of Geological Processes and Mineral Resources, China University of Geosciences, Wuhan & Beijing, China; Department of Earth and Space Science and Engineering, York University, Toronto, M3J 1P3, Canada, qiuming@yorku.ca

Chang-Jo Chung
Geological Survey of Canada, 601 Booth Street, Ottawa, K1A 0E8 Canada, chung@NRCan.gc.ca

Timothy C. Coburn
Department of Management Science, Abilene Christian University, Box 29315, Abilene, TX 79699, USA, coburnt@acu.edu

W. David Menzie
U.S. Geological Survey, 12201 Sunrise Valley Dr., Reston, Virginia 20192, USA, dmenzie@usgs.gov

John C. Davis
DAVCON, Box 353, Baldwin City, Kansas 66006-0353, USA,
john.davis 5@mchsi.com

Maarten de Wit
AEON – Africa Earth Observatory Network, and Departments of Statistical Sciences, University of Cape Town, Private Bag, Rhodes Gift, Rondebosch 7701, Cape Town, South Africa, maarten.dewit@uct.ac.za

Clayton V. Deutsch
Centre for Computational Geostatistics (CCG), Department of Civil and Environmental Engineering, University of Alberta, Alberta, T6G 2W2, Canada, cdeutsch@ualberta.ca

John H. Doveton
Kansas Geological Survey, University of Kansas, Lawrence, Kansas 66047, USA, doveton@kgs.ku.edu

Lawrence J. Drew
U. S. Geological Survey, Mail Stop 954, 12201 Sunrise Valey Drive, Reston, VA 20192, USA, drew@usgs.gov

J.J. Egozcue
Department of Applied Mathematics III, University Politecnica de Catalunya, Barcelona, Spain, juan.jose.egozcue@upc.edu

Hafida El Bilali
Department of Earth Sciences and Ottawa-Carleton Geoscience Centre, Carleton University, Ottawa, Ontario, K1S 5B6, Canada, hafida.el@sympatico.ca

Rudolf Endler
Baltic Sea Research Institute (IOW), 18119 Rostock, Germany, rudolf.endler@io-warnemuende.de

Andrea G. Fabbri
Università di Milano-Bicocca, Piazza della Scienza, 1, 20126 Milan, Italy and Vrije Universiteit, de Boelelaan 1087, 1081 HV Amsterdam, The Netherlands, andrea.fabbri@unimib.it

H. Falck
Northwest Territories Geoscience Office, 4601-B 52nd Ave, Yellowknife, NT Canada X1A 2R3

H. Gaonac'h
GEOTOP, UQAM, Montreal, Canada, gaonach.helene@uqam.ca

F.M. Gradstein
Geological Museum, University of Oslo, N-O318 Oslo, Norway, felix.gradstein@nhm.uio.no

List of Contributors

Eric C. Grunsky
Geological Survey of Canada, Natural Resources Canada, Rm. 607 615 Booth St.
Ottawa ON K1A 0E9, Canada, egrunsky@nrcan.gc.ca

O. Hammer
Geological Museum, University of Oslo, N-O318 Oslo, Norway,
oyvind.hammer@nhm.uio,no

Jan Harff
Leibniz Institute for Baltic sea Research Warnemünde (IOW), 18119 Rostock,
Germany, jan.harff@io-warnemuende.de

J. R. Harris
Geological Survey of Canada, 615 Booth St., Ottawa, ON K1A 0E9, Canada,
harris@nrcan.gc.ca

Ute C. Herzfeld
CIRES, University of Colorado at Boulder, Boulder, CO 80309-0449, USA,
ute.herzfeld@colorado.edu; Herzfeld@tryfan.colorado.edu

C. Jefferson
Geological Survey of Canada, 615 Booth St., Ottawa, ON K1A 0E9, Canada,
jefferson@nrcan.gc.ca

D. Lemkow
Geological Survey of Canada, 615 Booth St., Ottawa, ON K1A 0E9, Canada,
lemkow@nrcan.gc.ca

Ian Lerche
Institut fuer Geophysik und Geologie, Universitaet Leipzig, Talstrasse 35, 04103
Leipzig, Germany, lercheian@yahoo.com

S. Lovejoy
Physics, McGill, 3600 University st., Montreal, Que. H3A 2T8, Canada,
lovejoy@physics.mcgill.ca

A. Lugowski
Computer Science Department, University of California, Santa Barbara CA
93106-5110, USA,
alugowski@gmail.com

G. Mateu-Figueras
Departament d'Informatica i Matematica Aplicada, Campus Montilivi, Universitat
de Girona, E-17071 Girona, Spain, gloria.mateu@udg.edu

Helmut Mayer
UNAVCO, University of Colorado at Boulder, Boulder, CO, USA,
mayerh@tryfan.colorado.edu

Sean A. McKenna
Geoscience Research and Applications Group, Sandia National Laboratories, Albuquerque, NM 87185, USA, samcken@sandia.gov

Daniel F. Merriam
Kansas Geological Survey, University of Kansas, Lawrence, Kansas 66047, USA, dmerriam@kgs.ku.edu

Isao Motoyama
Department of Earth Evolution Sciences, University of Tsukuba, Tsukuba 305-8572, Japan, isaomoto@sakura.cc.tsukuba.ac.jp

Václav Němec
Krybníčkům 17, CZ-100 00 Praha 10 – Strašnice, Czech Republic, lidmila.nemcova@quick.cz

Ricardo A. Olea
Leibniz Institute for Baltic sea Research Warnemünde (IOW), 18119 Rostock, Germany; Presently at: United States Geological Survey, 12201 Sunrise Valley Drive, Mail Stop 956, Reston, VA, 20192, USA, olea@usgs.com

Miyuki Ota
Department of Earth Evolution Sciences, University of Tsukuba, Tsukuba 305-8572, Japan

V. Pawlowsky-Glahn
Department of Computer Science and Applied Mathematics, Departament d'Informatica i Matematica Aplicada, Campus Montilivi, Universitat de Girona, E-17071 Girona, Spain, vera.pawlowsky@udg.edu

Andreas Prokoph
SPEEDSTAT, 19 Langstrom Crescent, Ottawa, Ontario, K1G 5J5, Canada, aprokocon@aol.com

Gary L. Raines
U.S. Geological Survey (retired), c/o Mackay School of Earth Sciences, UNR, MS 176, Reno, NV 89557, garyraines@earthlink.net

Richard A. Reyment
Section for Palaeozoology, Natural History Museum, Box 50007, 10405, Stockholm, Sweden, richard.reyment@nrm.se

Joseph E. Robinson
Syracuse University, Syracuse, NY 13210, USA, joerobinson1@peoplepc.com

Hirotaka Saito
Department of Ecoregion Science, Tokyo University of Agriculture and Technology, Fuchu, Tokyo 183-8509, Japan, hiros@cc.tuat.ac.jp

List of Contributors

D. Schertzer
Université Paris-Est, ENPC/CEREVE, 77455 Marne-la-Vallee Cedex 2, France,
Daniel.Schertzer@cereve.enpc.fr

John H. Schuenemeyer
Southwest Statistical Consulting, LLC, 960 Sligo St., Cortez, CO 81321 USA,
jackswsc@charter.net

Tetsuya Shoji
School of Frontier Sciences, The University of Tokyo, Kashiwa 277-8583, Japan,
t-t_shoji@jcom.home.ne.jp

Donald A. Singer
U.S. Geological Survey, 345 Middlefield Road, Menlo Park, California 94025,
USA, singer@usgs.gov

Yuichiro Tanaka
Institute of Geology and Geoinformation, Geological Survey of Japan, National
Institute of Advanced Industrial Science and Technology, Higashi 1-1-1, Tsukuba
305-8567, Japan

Christien Thiart
AEON – Africa Earth Observatory Network, and Departments of Statistical
Sciences, University of Cape Town, Private Bag, Rhodes Gift, Rondebosch 7701,
Cape Town, South Africa, christien.thiart@uct.ac.za

D. Wright
Geological Survey of Canada, 615 Booth St., Ottawa, ON K1A 0E9, Canada,
dwright@nrcan.gc.ca

Zhao Pengda
China University of Geosciences, Wuhan 430074, Beijing 100083 China,
pdzhao@cugb.edu.cn

Zhou Di
CAS Key Laboratory of Marginal Sea Geology, South China Sea Institute
of Oceanology, Chinese Academy of Sciences, Guangzhou 510301, China,
zhoudiscs@scsio.ac.cn

Zhou Lan
Centre for Computational Geostatistics (CCG), Department of Civil and
Environmental Engineering, University of Alberta, Alberta, T6G 2W2, Canada

Introduction

Graeme F. Bonham-Carter and Qiuming Cheng

Keywords Festschrift · fractal · quantitative stratigraphy · compositional data · mineral resource assessment · time series · geostatistics

This volume is composed of a series of papers written by colleagues and friends of Frederik (Frits) P. Agterberg – a Festscrift. All the papers were contributed in response to invitations to potential contributors, sent out in late 2006. First drafts were mostly complete by July 2007, and it has taken a further year to complete reviews and revisions for a publication date in time for the International Geological Congress in Oslo, August 2008.

The book consists of 28 (plus 2 introductory) papers by a total of 58 authors, the majority from Canada and USA, but including authors from 12 other countries. The papers cover a diverse range of topics in geomathematics, such as fractal and multifractal modelling, quantitative stratigraphic analysis, compositional data analysis, mineral resource assessment, time series analysis and geostatistics. The diversity of subject matter and the multi-nationality of authors reflect Frits' own broad interests within geomathematics, and the international make-up of his colleagues and friends. Following this introduction, we provide information about Frits Agterberg's career and his many achievements, including his close links with the International Association for Mathematical Geology (IAMG) since its founding in 1968.

Deciding the running order of these papers was not straightforward, because although some papers are naturally linked to others, many are not easily grouped. The subject matter of the book depends only on the current research interests of the contributors and no themes (beyond some geomathematical content) were prescribed in

---

Graeme F. Bonham-Carter
Geological Survey of Canada, 601 Booth St., Ottawa, Ontario K1A 0E8, Canada,
e-mail: gbonhamc@nrcan.gc.ca

Qiuming Cheng
State Key Lab of Geological Processes and Mineral Resources, China University of Geosciences, Wuhan & Beijing, China; Department of Earth and Space Science and Engineering, York University, Toronto, M3J 1P3, Canada, e-mail: qiuming@yorku.ca

advance. The running order of the papers is, therefore, not always logically consistent, being based on a blend of methodology and application.

We begin with three papers about advances and applications of compositional data analysis to geochemical data. Following on the earlier work by Felix Chayes and John Aitchison, Vera Pawlowsky and colleagues have made significant theoretical and applied advances that have raised the awareness of the problem to many. The first 2 papers (Buccianti and Pawlowski; Mateu-Fugueras et al.) provide new material on applications and methodology. Then Drew et al.'s contribution provides an application of logratio transforms to coal data, confirming the importance of evaluating compositional data in coordinates that remove the closure effect.

The next four papers are loosely grouped as being about problems in geophysics. Ute Herzfeld gives an overview of the methodology and application of her geostatistical characterization approach applied to map the texture and morphology of various surfaces, particularly sea floor and ice, both sea ice and glaciers. The following paper by Mayer and Herzfeld is a recent application of the methodology to a Greenland glacier. T. Shoji provides an example of applying 3-D variograms to rainfall data, in order to model the spatial and temporal continuity of this phenomenon. Many will be familiar with the work on multifractals and nonlinear dynamics by Shaun Lovejoy and colleagues. Here Lovejoy et al. extend this approach to rock density and surface gravity. They provide theory and application that allows for vertical anisotropy, clearly important in a stratified Earth.

The following seven papers are clustered under the heading of mineral exploration. Cheng discusses how hydrothermal mineral deposits can be regarded as end products of nonlinear processes that can be modelled as fractals and multifractals. He uses a case study to illustrate how spatial evidence (from geochemistry, structures etc.) can be processed with nonlinear models to reveal singularities in the data that indicate the presence of deposits. Harris et al. provide an application of integration of spatial datasets to evaluate mineral potential in an area of Canada's Northwest Territories proposed as an extension to an already existing National Park. Singer and Menzie explore the problem of estimating the number of undiscovered mineral deposits in an area. Raines demonstrates the bifractal nature of various mineral deposit types, and discusses how this property could be used to estimate the number of undiscovered deposits. Thiart and de Wit show that mineral deposit patterns are distinctly different in older crust that has been remobilized in Pan African belts compared to those in juvenile crust of Neoproterozoic age. Fabbri and Chung argue that mineral (or landslide) potential maps calculated by integrating information from spatial data layers are in a metric that should be regarded as a rank order only, and that blind tests using quantile-type prediction indices should be used to evaluate such spatial prediction models. Coburn et al. use a simulation approach to model strip transect sampling, and although their immediate concern is to find buried ordnance, the methodology has wide application, including its use in oil and mineral exploration.

The next five papers are about the application of mathematical methods in paleontology and stratigraphy. Gradstein et al. discuss the use of the RASC and CASC software (developed over many years by Agterberg, Gradstein and others) to improve the resolution of stratigraphic correlation. Zhou Di illustrates the same methodology applied to data from the Pearl River Mouth Basin, in which a biozonation was developed and applied for correlation of well data. Brower demonstrates how Euclidean distances between landmarks (followed by a singular value decomposition) provide a useful tool for geometric morphometrics in paleontology. Reyment et al. apply a discriminant analysis (on compositional data transformed to overcome closure) to show that some species of radiolarians react differently from season to season, whereas others may be less affected by environmental conditions.

Statistical methods applied to sedimentary and stratigraphic data are exemplified by four papers. Doveton shows how a lithologic transition probability matrix can be transformed to a mean first-passage time matrix that provides a metric useful for characterizing differences between stratigraphic sections. Deutsch discusses the beta distribution for modelling the shape of the scale-dependent multivariate distribution of facies proportions. Olea et al. provide a case study of cokriging to generate a porosity map in an area of the Baltic Sea, using densely sampled bathymetry and grain size measurements as proxy variables. Merriam and Davis use Watson's U^2 statistic to compare structural directions amongst a wide variety of data sources from Kansas.

There are three papers on time series analysis. Prokopf and El Bilali use a cross-wavelet approach to model CO_2 time series from plant cuticle measurements, and temperature from δO^{18} data, for the last 290 my, focussing particularly on the influence of nonstationarities introduced by stratigraphic uncertainties. Reyment examines the volcanic history in an area of Japan using cross-correlations between the magnitudes of A-type earthquakes and depth of events for three periods from 1983 to 2005. Data from harbors on the Baltic Sea illustrate an example of producing sea level measurements from which local variations have been separated, yet with the sea level components in the correct phase at desired locations (Robinson).

The final two papers by Ian Lerche do not seem to cluster with others, so we have put them together at the end of the book—they are last but definitely not least (and we should also mention that these papers were submitted first, at least six months before any others!) They are both good examples of simulation models of an imaginative type, one dealing with strategies to divert flood water to old mine sites, the other about how one can investigate a "Christmas parking lot problem" from the merchant's perspective.

Many of the papers in the book, at the suggestion of the publisher, appeared in recent issues of Mathematical Geosciences and Natural Resources Research, and now are "reprinted" here, although it must be stressed that they were all written for the book, without exception. By appearing also in these journals, the papers involved benefit from a wider circulation and greater availability (particularly online) than would be possible in the book alone. This in no way detracts from the 'specialness' of the book, because the contributors all wrote their papers to honour our friend and colleague Frederik P. Agterberg.

We are very grateful for the time and effort by our many reviewers: Thomas Axel, Geoff Bohling, Fred Bookstein, Antonella Buccianti, John Carranza, Mark Coolbaugh, Roger Cooper, John Davis, Carlos Roberto de Souza Filho, Clayton Deutsch, Juan José Egozcue, Neil Fordyce, Bob Garrett, Michael Goodchild, Felix

Gradstein, Ralf Greve, Pablo Gumiel, John Harbaugh, Jeff Harris, Ute Herzfeld, Mike Hohn, Jerry Jensen, André Journel, Subhash Lele, Gang Liu, Dan Merriam, Mark Mihalasky, Don Myers, Vesa Nykanen, James Ogg, Margaret Oliver, Javier Palarea-Albaladejo, Tim Patterson, Vera Pawlowski-Glahn, Andreas Prokopf, Gary Raines, Joe Robinson, Peter Sadler, Michael Schulz, Bill Sharp, Don Singer, Paul Switzer, John Tipper and Danny Wright.

The Role of Frederik Pieter Agterberg in the Development of Geomathematics

Graeme F. Bonham-Carter, Václav Němec, Dan F. Merriam, Zhao Pengda and Qiuming Cheng

Abstract Frederik Pieter (Frits) Agterberg has written over 250 publications in the field of geomathematics during the past 50 years. He played a major role in the International Association of Mathematical Geology since its inception in 1968, and strongly influenced the development of geomathematics worldwide.

Keywords Geomathematics · geostatistics · mineral prediction · quantitative stratigraphic correlation · fractal · multifractal

1 Introduction

This chapter reviews the part played by Frederik Pieter (Frits) Agterberg on the development and spread of geomathematics. Frits graduated in geology at an exciting time in the geosciences, coinciding with the easy access to computers. Computers catalyzed the quantitative revolution in geology, because they made it possible to apply mathematical and statistical models to large volumes of data (Merriam, 1981). Frits was among a relatively small number of geomathematicians who paved the

---

Graeme F. Bonham-Carter
Geological Survey of Canada, 601 Booth St., Ottawa, Ontario K1A 0E8, Canada,
e-mail: gbonhamc@nrcan.gc.ca

Václav Němec
Krybníčkům 17, CZ-100 00 Praha 10 – Strašnice, Czech Republic,
e-mail: lidmila.nemcova@quick.cz

Dan F. Merriam
Kansas Geological Survey, University of Kansas, Lawrence, Kansas 66047, USA,
e-mail: dmerriam@kgs.ku.edu

Zhao Pengda
China University of Geosciences, Wuhan 430074, Beijing 100083 China,
e-mail: pdzhao@cugb.edu.cn

Qiuming Cheng
York University, 4700 Keele St., Toronto, Ontario M3J 1P3 Canada, e-mail: qiuming@yorku.ca

way for others in the field, by providing examples of theory, associated computational examples, and real-world applications. The breadth and depth of Frits' research activity is best described by providing a thumbnail sketch of his publications, extending over 50 years from 1958 to the present, see a later section of this chapter.

Frits became well known internationally, partly through his publications, but also through his efforts to disseminate geomathematical ideas in many parts of the world. In particular, he has been closely associated with the International Association of Mathematical Geology (IAMG), since its inception in 1968. The history and mission of IAMG is briefly reviewed here, and its impact on the development of East-West contacts that began during a time when the Iron Curtain prevented free exchange of ideas between scientists. He also affected the development of geomathematics in China where his work is widely used and where now more geomathematicians are produced than in any other country.

Frits is a gentle and just man with many friends in the world. He has made an invaluable contribution to the discipline and profession of geomathematics, and it is the wish of all of us that he continue.

2 The Early Years

Frederik Pieter (Frits) Agterberg was born in 1936 in Utrecht, the Netherlands. He was the only child of parents somewhat older than the average, and his father taught at the local school. He started school during World War II, at a time when his country was occupied, and living conditions were spartan. Apparently, he did not

Fig. 1 Frits as a student at University of Utrecht in 1959 when he received a Best Paper Award (photo from Codien Agterberg)

really enjoy the classical school curriculum typical of this period, but his geography teacher in secondary school started his interest in geology. It was not until he went to university at the age of 18 that his intelligence and aptitude, particularly in mathematics, were aroused and recognized. He studied geology and geophysics at Utrecht University obtaining his BSc in 1957. He stayed on at Utrecht for graduate work, where he completed his MSc in 1959, and his PhD in 1961, both *cum laude*. The photograph in Fig. 1 was taken while he was a student in Utrecht.

He has made a major contribution to the geomathematical literature in numerous areas, often as the first to explore new applications and methods, and always with mathematical rigour blended with practical examples. The major research areas can be identified as: statistical frequency distributions applied to geoscience data, mineral-resources appraisal, stratigraphic analysis and time scales, and fractal and multifractal modeling.

Throughout his career, he has been closely associated with the International Association for Mathematical Geology (IAMG), a professional organization that has provided a structure for promoting the use of geomathematics worldwide. Frits Agterberg in his association with the IAMG has been a key figure in its ongoing success.

3 Other Aspects of Frits Agterberg's Life and Kudos

It is said that behind a successful man, there is a supportive woman. Frits met his future wife Codien when he was a student at University of Utrecht, and she was a secretary in the geology department (today she would be termed a "personal assistant"). They eventually married in 1965, and together have raised four sons. She has been a wonderfully supportive wife. The Agterberg household is one that has welcomed many geomathematical visitors over the years, with warm hospitality and excellent cooking! Codien often accompanies Frits on his travels, and their relationship has been a long and mutually supportive one.

Many old friends and colleagues wish to be remembered to Frits. Walther Schwarzacher writes "Frits is an old and extremely good friend. I still remember my first meeting him at the Kansas Time Series Colloquium in 1968. After my talk (Frits, at this stage was smoking a pipe and had no beard) he got up and said: "I was interested in that equation of yours." I have forgotten which equation, but it did strike me at the time that there was at least one real mathematician amongst all us pretenders. Frits kept this reputation and integrity and he has helped and constrained many geologists. We have to be very grateful to him!"

Danie Krige noted "Dr. Agterberg is a highly valued colleague and close personal friend of mine for more than 40 years. With the publication of this book in his honour I would like to add my congratulations and best wishes to him. May he continue to contribute in his field for many years to come for the benefit of us all." Bill Fyson noted the reason why Dutchmen are so tall. "It is an example of Darwinian evolution. Holland is sinking and only those who could keep their heads above water survived."

4 The IAMG and East-West Contacts

4.1 IAMG

IAMG was founded at the ill-fated International Geological Congress (IGC) in Prague in 1968 (Merriam, 2007) and brief histories of the Association have been by Dan Merriam (1978) and Václav Němec (1993a). The Association's charge is to collect, interpret, and disseminate data and information on the application of mathematics and by association, quantitative approaches, and computer applications (Merriam, 2004) to the geological sciences and to bring together those with this common interest. There were fourteen members of an *ad hoc* committee that at the urging of and with the leadership of Richard A. Reyment (Merriam, 2005) formulated the statutes and bylaws for the IAMG. Later, at the Congress twenty geologists took part in the organizational meeting in Prague. Frits was a member of both groups as were Němec and Merriam.

The political situation in Czechoslovakia in the mid-1960s continued to improve with considerable changes leading to the Prague spring as well as the preparations for the IGC and founding meeting of the IAMG. There was an euphoric start of the 23rd IGC with colleagues all over the world reuniting or making new acquaintances. All of this, however, came to an abrupt end with the entry into Czechoslovakia of the armies of the Warsaw Pact changing all plans, but not before the IAMG was born on the day after the invasion.

The need for a society or association stressing geomathematics became apparent from 1961 and the APCOM (Application of Computers in the Mineral Industry) meetings started at the University of Arizona. In 1964 the Kansas Geological Survey (KGS) began publishing computer programs in their *Special Distribution Publication* series and in 1966 the publication *Computer Contributions* was initiated by the Kansas Survey (Merriam, 1999). The publications were so well received that the Survey started a series of meetings termed colloquia on some aspect of the subject of geomathematics and held one every six months from 1966 to 1970. The popular colloquia later were renamed Geochautauquas when hosted by Syracuse University for several years before it became a traveling annual symposium sponsored by other academic organizations in North America. Frits Agterberg was a regular contributor to all of these meetings.

In the meantime, and essentially unknown to those in the West, was the *Mining Příbram Symposium in Technology and Science* organized starting in 1962 (Němec, 1993b). In 1968 Němec and other Czech geologists and geophysicists organized a special section on *Mathematical Methods Geology and Geophysics* at that meeting. The meetings were later cosponsored by the IAMG. These were a long series of regular international geomathematical sessions and was an important and almost unique East-West gate for the IAMG in the period from 1968 to 1989. Němec already had developed many contacts in other eastern countries and was well acquainted with mathematical work by eastern geologists behind the Iron Curtain so the Příbram sessions were well attended.

4.2 The East-West Exchange

For political reasons the series of meetings in the East and in the West continued without much exchange with few exceptions. Most accomplishments of Russian (eastern) geologists were known only through a few western workers and one of those workers was Richard Reyment. In 1965 Němec and two other Czech geologists published their papers in one of the early APCOM meeting proceedings in absentia; they were not allowed to travel to the U.S. From the addresses given in the papers, Merriam sought out Němec in a trip to Prague. This connection started a dialogue, albeit a limited one, between the East and West despite the problems of the Iron Curtain. By 1968 at the IGC in Prague, geologists from both the East and West took part in the organizational meeting of the IAMG. Representing the East were geologists from the Bulgaria, Czechoslovakia (Němec), East Germany, Hungary, and the USSR (Vistelius). The West was represented by geologists from Australia (Watson), Canada (Agterberg), Sweden (Reyment), UK, USA (Merriam), and West Germany. The dialogue was now well underway and important contacts had been made.

Nowadays it is perhaps difficult for many people to imagine how the visits of western specialists behind the Iron Curtain were appreciated by eastern colleagues trying to follow the latest development of the science. The conditions for direct contacts were strongly limited and such events such as regular East-West meetings of mathematical geologists at Pribram was an exception. Many people feared to organize such meetings or to establish any personal contact with western colleagues for the simple reason not to be suspected of illegal activities or even of treason to national interests and the socialist system. Thanks to several Russian colleagues, especially Dimitrii Rodionov, some fundamental western literature on mathematical geology was translated into Russian and published in Moscow. This included Agterberg's book on *Geomathematics*.

Other opportunities for exchange were at the International Geological Congresses (IGC). The Congress after Prague in 1968 was held in Montreal in 1972, Sydney in 1976, Paris in 1980, Moscow in 1984, and Washington D.C. in 1989 and provided opportunity for those from the East to become acquainted with western workers and their work. With the exception of the congress in Moscow, mathematical geology sessions generally were not attended by eastern colleagues. That exception was the 1984 congress in Moscow where many contacts and friendships were made. Sessions on mathematical geology sponsored by the IAMG were attractive to those interested in the subject.

By now it should be obvious that some of the major players in this scenario were Agterberg, Merriam, Němec, Reyment, the Father of the IAMG, and Vistelius (Merriam, 2001). At the organizational meeting of the IAMG the following officers were elected. At the insistence of Reyment, Andrew Vistelius was elected president, then Reyment was elected secretary general, Němec eastern treasurer, Geoff Watson as vice president, Agterberg (amongst others) as a council member, and Merriam editor-in-chief of the IAMG journal yet to be founded – the *Journal of Mathematical Geology*. Vistelius insisted on the use of mathematical geology rather than the more catchy term geomathematics. So, the stage was set for continued and "official" exchange between the East and the West. This exchange was by publications, correspondence, and personal contact. Frits Agterberg spent a month in the Soviet Union in 1977, at the invitation of Vistelius, where he gave a series of lectures in Moscow, Leningrad, Novosibirsk and other cities. He enjoyed the hospitality and survived the vodka!

Personal contacts of course were the best way for exchange. Merriam was able to present a paper at Pribram in 1970, Němec was a visiting scientist in the Mineral Resources Section at the Kansas Geological Survey (KGS) in 1969–1970 during the time Frits Agterberg (Fig. 2) was the Visiting Research Scientist (VRS). The three IAMG functionaries, Agterberg, Němec, and Merriam were working in close proximity and often had brief ad hoc meetings to solve actual problems of the IAMG agenda. Richard Reyment previously had been the first VRS at the KGS in 1966. In those days Lawrence was the hub of activity for the IAMG.

Fig. 2 Frits in field in 1969 while a Visiting Research Scientist at Kansas Geological Survey (photo from Dan Merriam)

4.3 Mathematical Geology in the East After 1989

The political, economic, and social changes in the East occurring especially at the end of 1989 are well known. With the fall of the Berlin Wall, conditions changed for not only the world but in particular for mathematical geology. Direct exchange was now possible on a scale not envisioned prior to this time. The exchange of ideas and scientists could be open and direct, providing the money was available. Němec was able to organize an international section on mathematical geology at the Mining Pribram Symposium in a new atmosphere without any remarks concerning a "too high" audience from the western world.

The Silver Anniversary meeting of IAMG was coming up in 1993 and Richard McCammon, then president, appointed Merriam general chairman of an organizing committee consisting of Agterberg, Peter Dowd, Mike Hohn, Andrea Förster, and Němec to prepare for the meeting in Prague, birthplace of the IAMG. Intensive talks took place in the Němec home in Prague to plan for the combined meeting of IAMG and the Příbram Mining Congress. The Silver Anniversary meeting was well attended and a tremendous success with a few minor glitches by most reports. The cold war which had affected science was now over and there was a free exchange.

The current development of mathematical geology attracts new ideas, new young people, and in general the western colleagues have various occasions to meet their colleagues from the East. To some extent they are not taking into consideration the fact that new political freedom in the East is not identical with the economic freedom of the local population and that therefore any attendance of eastern colleagues in conferences and congresses organized in the West incurs higher financial costs. Fortunately various new technical possibilities have been promoted to make it easier to develop close global contacts.

4.4 Frits P. Agterberg and His Global Contributions

Frits Agterberg (Fig. 3 is a recent photo) has been a major piece on the global chessboard of mathematical geology throughout the forty years of the IAMG existence. Frits is identified with the Association as he formed one of the vital links in the East-West history. He served as an officer of IAMG being elected a Council

Fig. 3 Frits Agterberg in 2006 (photograph from Codien Agterberg)

Member for 1968–1972 and again in 1976–1980; he was vice president 2000–2004 and succeeded to the presidency in 2004. When he turns over the presidency to his successor at the International Geological Congress in Oslo in August 2008, the IAMG will be 40 years old, and Frits will have served the organization in various capacities, not continuously, but as a major player over its whole existence. He has been active in many Association activities and contributed freely to its journals. He has attended many IAMG-sponsored meetings during the years including the colloquia in Lawrence, Geochautauquas in Syracuse and elsewhere, APCOM meetings in several cities, the Mining Příbram Symposium in Czechoslovakia, meetings sponsored by the East bloc organization COMECON (Countries of the Mutual Economic Aid), an *ad hoc* meeting in Novosibirsk, USSR, and several special IAMG sessions at the International Geological Congresses over the years.

Frits has been widely admired for his systematic scientific work as well as for his readiness to serve to promote further his discipline through the IAMG. Frits has particular connections to the geomathematics community in China, and has been particularly influential in fostering IAMG and geomathematical research in that country, as discussed next.

5 Development of Mathematical Geology in China

Frits Agterberg's influence on Chinese mathematical geology can be summarized in five aspects: he was one of the first scholars to visit China; one of the first professors from the West who supervised Chinese students or visiting scholars in the field of mathematical geology; his book *Geomathematics* (translated into Chinese) has been widely used and cited by Chinese authors; he is the first candidate outside China invited to be a distinguished scholar by the Chinese Ministry of Education; and as the Distinguished Lecturer of IAMG he has delivered a series of well-received lectures in China. The subdisciplines of mathematical geology to which he has contributed that have been of particular interest to the Chinese community includes (but are not limited to): mineral-resources quantitative assessments, application of multivariate statistics in geology, quantitative stratigraphy, non-linear theory and fractals, and geographic information system-based decision support modeling.

Starting in the 1980s, academic exchanges between the mathematical geology communities of China and other countries began to increase. In 1980, four American geologists led by Richard McCammon (US Geological Survey), along with Niichi Nishiwaki, a professor at Kyoto University, Japan, went to Beijing and demonstrated the achievements of the International Geological Program Project 98, "Applications of Computer Techniques in Assessment on Mineral Resources." They systematically introduced methods for prediction and assessment of mineral resources. This was followed by a visit by Richard Sinding-Larsen, ex-General Secretary of IUGS, who proposed international cooperation between IAMG and China. As a result, Frits Agterberg invited Zhao Pengda to visit the Geological Survey of Canada in 1982, but he was unable to travel on that occasion. However, in 1984 Frits was the first geomathematical scholar to be invited to visit the China University of Geosciences (CUG). Later, John Davis, Dan Merriam, John Harbaugh, DeVerle Harris, Graeme Bonham-Carter and others visited China and China University of Geosciences, but Frits was the first and the most frequent visitor. In 1990, he attended the International Symposium of Mathematical Geology held in Wuhan, and in 1996, he co-chaired the session on "New Theories, New Methods and Applications of Mathematical Geology" at the 30th IGC in Beijing. In 2001, he spoke on "Development and Future of Mathematical Geology" at the Symposium of Mineralization Diversity and Deposit Spectrum at CUG.

In 1982, Haiqing Zou, a lecturer in the department of Mineral Resources Exploration at CUG was selected as the first visiting scholar to Canada, where he was supervised at the GSC in Ottawa by Frits Agterberg. This was the first example of a Chinese geomathematical scholar studying abroad, and Haiqing remembers well the help he received from Frits in both his daily life and in his studies. In 1984 Yuan Ding, a masters student of Zhao Pengda at CUG, studied with Frits at GSC, later obtaining his doctoral degree at Syracuse University. Then Qiuming Cheng, now a professor of York University (Canada) and currently also the founding director of the State Key Laboratory of Geological Processes and Mineral Resources in China, completed his doctoral degree from University of Ottawa under Frits' supervision in the early 1990s.

In 1974, the Chinese Publishing House of Science and Technology published Agterberg's textbook *Geomathematics*, which was translated by Zhongming Zhang and edited by Zhao Pengda. This was the second important mathematical geology textbook translated into Chinese following Koch and Link's book *Statistical Analysis of Geological Data* (1970, 1971). Both of these texts, with their logical and comprehensive structure and innovative contents, greatly influenced the geomathematical community in China.

Frits Agterberg was invited and sponsored through the special program "Distinguished International Scholars" by the Ministry of Education of China to give lectures in China at CUG in 2002. Besides his lectures on methods of efficient prediction of metallic mineral resources, he also introduced weights-of-evidence, fractal and multifractal models, and digital geological time scales to his audiences. This series of lectures appealed not only to teachers and students at CUG, but also researchers in several other fields.

The IAMG conference held in 2007 in Beijing was co-chaired by Zhao and Agterberg, with Qiuming Cheng as the general secretary. This was a significant event for the Chinese geoscience community. The event showcased Chinese work in mathematical geology, and brought several hundred geomathematicians (especially young researchers) together from all over the world to share their ideas.

These are just some of the ways that Frits Agterberg has fostered and influenced the development of geomathematics in China.

6 Agterberg's Research and Publications

Frits Agterberg's first paper was published in 1958, in his first year of graduate work, receiving that year's Royal Dutch Mining and Geology Association Prize for best paper by a graduate student. The paper appeared in the journal *Geologie en Mijnbouw*, and was entitled "An undulation of the rate of sedimentation in southern Gotland." It described the structure of the Silurian sedimentary basin in Sweden and used a model to interpret the results in terms of shifting axes of maximum deposition (Agterberg, 1958). A geomathematical researcher was born.

For his PhD dissertation under van Bemmelen, Frits undertook field work in the Dolomites, and his dissertation was entitled "Tectonics of the Crystalline Basement of the Dolomites in North Italy." Using his field data on the strikes and dips of Hercynian schistosity planes and attitudes of minor folds, he constructed tectonic maps of mean orientation values in highly deformed terrain, interpreting the results to show major Alpine cross folds in the area. In a paper published in 1959, "On the measuring of strongly dispersed minor folds" (also *Geologie en Mijnbouw*), Frits pointed out that divergent structural trends may belong to a single probability distribution, not represent different structural events (Agterberg, 1959).

In a third paper published while a graduate student, also in *Geologie en Mijnbouw* entitled "The skew frequency distribution of some ore minerals," Frits analyzed data collected by H.J. De Wijs from the Pulacayo deposit in Bolivia. The origin of the skewed distribution were discussed, and the presence of spatial dependence between observations was recognized. Fourier analysis was used to model spatial trends (Agterberg, 1961).

Although a few others may have begun work at this time on spatial statistics in the earth sciences, for a graduate student in a classical geology department, Frits Agterberg showed remarkable innovation and imagination in recognizing the applicability of spatial statistical models to geological data. His ability to pioneer new statistical approaches to analyze data has been a hallmark of his work throughout his career.

After Utrecht, Frits moved to the United States, where he had been awarded a one-year Wisconsin Alumni Research Foundation postdoctorate fellowship at the University of Wisconsin. Here he worked with Lewis Cline, working on the statistical analysis of sedimentary rocks. His first paper in a US journal was "Statistical analysis of ripple marks in Atokan and Desmoinesian rocks in the Arkoma Basin of east-central Oklahoma" in the *Journal of Sedimentary Petrology* (Agterberg and Briggs, 1963). Bonham-Carter remembers reading this paper while a graduate student at University of Toronto in 1963, and learning about the value of a statistical distribution to explain natural variability in orientation directions of sedimentary structures.

In 1962, Frits joined the Geological Survey of Canada (GSC) in Ottawa, initially as a petrological statistician working on the Canadian contribution to the International Upper Mantle Project. GSC and Ottawa were to be his working base and home from 1962 to the present day. His early work at GSC was in providing statistical expertise to other survey members, and developing his ideas on the use of models to describe and interpret spatial and compositional variability of rocks and their characteristics. The 1960s were a time of rapid spread of computers, and Frits was quick to take advantage of computational tools to apply statistical methods.

Examples of his many publications of this period include two papers on spatial statistical applications. In Agterberg (1964) he showed how confidence belts could be calculated for cubic trend surfaces, and how the significance of increasing the degree of the polynomial trend equation could be determined. In Agterberg (1965) he demonstrated the value of serial correlation in establishing the spatial autocorrelation of element concentration values, and in establishing a proper sampling pattern.

In a 1968 paper, he proposed a signal-plus-noise model to account for spatial variation of copper in the Whaleback mine in Newfoundland (Agterberg, 1968). His 1970 paper on autocorrelation functions in geology provided a method to estimate two-dimensional autocorrelation functions from irregularly distributed point data. This paper appeared in a book on *Geostatistics*, edited by Dan Merriam (Agterberg, 1970).

In 1971 he started and led the Geomathematics Section at GSC. One of the main objectives of this group was to develop quantitative methods for mineral-resource assessment. This was a period of increasing availability of digital data from geophysical and geochemical surveys, and Frits was a pioneer in developing new statistical approaches for integrated studies, trying to predict mineral favourability from multivariate spatial data. This clearly was important for mineral exploration, and also for land-use decisions for government planners.

In Agterberg (1971) he published a method to determine a probability index for predicting environments favourable for mineralization. The next year Agterberg et al. (1972) described the method and application of a cell-based approach to mapping mineral favourability over a region, as applied to copper in the Abitibi area. This was the first of several studies carried out by GSC on quantitative estimates – quantitative in terms of statistical probability – of the undiscovered mineral resources of a region based on a statistical appraisal of selected geological and geophysical parameters. Contoured probability maps for copper and zinc were produced at a scale of 1:500,000 for the Abitibi region.

His ground-breaking book *Geomathematics: Mathematical Background and Geoscience Applications* was published by Elsevier in 1974, and sold more than 10,000 copies worldwide. This book (Agterberg, 1974a) established Frits as a leading authority on mathematical and statistical methods in the geosciences.

In a paper in the *Journal of Mathematical Geology* in 1974 (later renamed *Mathematical Geology*, now changed to *Mathematical Geosciences*), entitled "Automatic contouring of geological maps to detect target areas for mineral exploration" Frits introduced the use of logistic regression to estimate mineral favourability (Agterberg, 1974b). This overcame the problem of estimating probability values greater than unity when using ordinary multivariate regression.

In his paper "Statistical methods for regional resource appraisal" (Agterberg, 1977a) Frits provided results of statistical analysis of mineral-deposit data from the Canadian Appalachian region. He recognized the value in fitting gamma and lognormal distributions to areal abundance values for lithological variables, as applied to spatial cell data from the Abitibi area. Their relationship to negative binomial frequencies of pyritic massive sulfide deposits per cell was studied in Agterberg (1977b), a paper presented at the APCOM Symposium held at Pennsylvania State University.

In a 1978 paper with Andrea Fabbri, Frits presented a method to measure the "Spatial correlation of stratigraphic units quantified from geological maps," with measurements obtained from an optical scanner to capture map data (Agterberg and Fabbri, 1978). This novel work was recognized by the Best Paper Award of *Computers & Geosciences* for that year. This was the same year that he was awarded the Krumbein Medal of the International Association for Mathematical Geology— only the third recipient to receive this prestigious award.

In Agterberg (1979), his *Computers & Geosciences* paper "Algorithm to estimate the frequency values of rose diagrams for boundaries of map features" again won Best Paper award. The approach was to measure a two-dimensional autocovariance function of a binary image, which was then converted into a table of intercept values that were used to construct the rose diagram.

In Chung and Agterberg (1980), the methods and applications of ordinary least squares, logistic and Poisson models were summarized for prediction of mineral occurrences, as well as a jackknife method for estimating variances, and a newly proposed "Separate Event Model." With DeVerle Harris in 1981, Frits published a comprehensive review of subjective and mathematical mineral-resource appraisal methods that appeared in the 75th Anniversary Volume of Economic Geology (Harris and Agterberg, 1981).

Starting in the late 1970s, Frits began a collaboration with Felix Gradstein and others on automatic methods of stratigraphic correlation, using fossil data. This problem was important in general, but particularly for offshore well data where the strata were dated using microfossils. Frits brought his statistical expertise to help solve this problem, and developed methods and associated software that have been extensively used by the oil industry.

Frits' paper with Nel in *Computers & Geosciences* entitled "Algorithms for the ranking of stratigraphic events" won the 1981 *Computers & Geosciences* Best Paper award, and was followed in the next year by a paper concerned with the scaling of stratigraphic events (Agterberg and Nel, 1982a,b). These methods were incorporated in their RASC computer program (Ranking and Scaling of stratigraphic events). From 1979 to 1985 he was Leader of the International Geological Correlation Programme's Project on "Quantitative Stratigraphic Correlation Techniques" (IGCP Project 148), as described in Agterberg (1982).

An example of his work during this period was his 1982 paper with Felix Gradstein entitled "Models of Cenozoic foraminiferal stratigraphy – northwestern Atlantic Margin" that appeared in an edited book on Quantitative Stratigraphic Correlation. The RASC method for ranking and scaling of biostratigraphic events was used to erect a zonation for the Cenozoic benthonic and planktonic foraminiferal record (last occurrences of 206 taxa) in 22 wells drilled on the Canadian Atlantic continental margin (Agterberg and Gradstein, 1982).

In 1987, Frits published "Quality of time scales – a statistical appraisal" in another book edited by Dan Merriam on *Current Trends in Geomathematics* (Agterberg, 1987), in which he described a maximum likelihood method for estimating the age of stage boundaries from geological information and radiometric dates. A number of later papers (by Frits and others) used this methodology to improve geological time scales. Agterberg and Gradstein (1988) provided a comprehensive review of quantitative stratigraphic correlation methods. Agterberg (1990) wrote a textbook on the subject, published by Elsevier.

In 1988, Frits published "Spatial analysis of patterns of land-based and ocean-floor ore deposits" (Agterberg, 1988). This applied a Neyman-Scott clustering process-model to massive sulphide deposits in New Brunswick. The spatial cross-covariance function for the association of two rock types was modeled and used to estimate the covariance of rock type composition data for cells of different sizes.

Starting in the late 1980s, Frits worked with Graeme Bonham-Carter, Danny Wright, and others on spatial data measured and displayed with geographical information systems (GIS). The advantage of these systems was that both raster and vector data could be spatially integrated in a single data system, with almost any spatial resolution. Frits proposed the "weights-of-evidence" model, to estimate the probability of occurrence of mineral deposits per unit area. The probabilities could be mapped by the GIS, along with estimates of uncertainty of various types. A series of papers appeared starting with a paper with Graeme and Danny in 1988 in *Photogrammetric Engineering and Remote Sensing* entitled "Integration of geological data sets for gold exploration in Nova Scotia" (Bonham-Carter et al., 1988).

Frits' original work with weights-of-evidence was described in a 1989 paper "Systematic approach to dealing with uncertainty of geoscience information in mineral exploration" presented at an APCOM meeting in Las Vegas that year (Agterberg, 1989). The weights-of-evidence technique was adapted from the medical expert system GLADYS. It was used to incorporate geophysical survey evidence by estimating posterior probabilities for occurrence of massive sulfide deposits in cells. Discussion of this chapter over coffee in Ottawa one day led GB-C to realize the advantage of the approach with GIS data, and the subsequent fruitful collaboration on GIS-based mineral-potential mapping.

In a 1990 paper, "Spatial pattern integration for mineral exploration" that appeared in a book edited by Gaal and Merriam *Computer Applications in Resource Estimation, Prediction and Assessment for Metals and Petroleum*, the theory of weights-of-evidence was developed, and the "contrast" statistic for strength of correlation between a point pattern for discrete events and a binary map pattern was introduced. Equations for variance of weights and variance resulting from missing data were outlined (Agterberg et al., 1990).

During this same period, Frits was supervising Eric Grunsky's PhD dissertation at University of Ottawa. This work involved the development of multivariate spatial factor models, and was published with Grunsky in *Computers & Geosciences* as two papers on computing multivariate spatial autocorrelation, and on spatial factor analysis of multivariate data (Grunsky and Agterberg, 1991a,b).

In the early 1990s, Qiuming Cheng arrived from China, and was supervised by Frits for his PhD at the University of Ottawa. Qiuming and Frits began work on fractal modeling, and in particular the application of multifractal models. The first Agterberg paper to mention fractals was in 1980 "Mineral resource estimation and statistical exploration" in the *Canadian Society of Petroleum Geologists Memoir 6* (Agterberg, 1980). Here, Frits applied fractal analysis to contours on an isopachous map for the Lloydminster Oil Field, Saskatchewan, and to the shapes of oil fields, Bakersfield, California. With Cheng, a number of excellent papers appeared, such as Agterberg, Cheng, and Wright (1993) which examines the multifractal spectrum of gold in bedrock derived for altered Jurassic volcanic rocks in British Columbia (part of Qiuming's dissertation work).

In Agterberg (1994a) Ripley's K-function method was used to show that boundaries of study areas should be considered in the statistical modeling of point patterns. Cheng et al. (1994) showed a new method based on the multifractal property of geochemical patterns for separating anomalous values from geochemical background values. In Agterberg (1994b), Frits showed that a model proposed by H.J. De Wijs in 1948 (previously discussed in Agterberg (1961)) can be regarded as a multifractal. He showed that the semivariogram for this model is approximately semilogarithmic as originally established by Georges Matheron.

Agterberg et al. (1994) "Multifractal modelling of fractures in the Lac du Bonnet Batholith, Manitoba" was presented at the IAMG conference at Mt Tremblant. Cheng and Agterberg (1996) developed an exact equation of the multifractal semivariogram that can be approximated by a simple semilogarithmic form (cf. Agterberg, 1994b). This work was based on Qiuming Cheng's unpublished Ph.D. dissertation that won 1995 best graduate student thesis award of School of Research and Graduate Studies, University of Ottawa.

Prokoph and Agterberg (1999) showed that wavelet analysis is a sensitive method for automatically detecting discontinuities (faults, unconformities), cyclicity and gradual changes in sedimentation rate by transforming depth-related sedimentary signals (e.g. gamma-ray logs) into wavelengths at specific depths. Andreas Prokoph was a post-doctorate fellow at the University of Ottawa.

Agterberg and Bonham-Carter (2005) determined that random cell selection methods recently proposed by others do not provide better mineral-potential maps than those produced by weights-of-evidence and related GIS-based methods. Gold deposits in the Gowganda area of east-central Ontario are used as illustration.

Although he is retired from GSC. Frits continues to be active in geomathematical research and publication, as illustrated by two papers that appeared in 2007. Agterberg (2007a) "New applications of the model of de Wijs in regional geochemistry" shows that Brinck's lognormal approach to frequency distributions can be augmented using multifractal modeling. The model of de Wijs is generalized to a "random cut" model, with practical applications to gold and arsenic in glacial till samples. In Agterberg (2007b), "Mixtures of multiplicative cascade models in geochemistry," Frits showed that mixtures of element concentration frequency distributions can be simulated by adding the results of separate cascade models. Regardless of properties of background levels, an unbiased estimate can be obtained of the parameter of the Pareto distribution characterizing anomalies in the upper tail of element concentration frequency distributions, or the lower tail of the local singularity distribution.

Since 1958, Frits has authored or co-authored over 250 scientific publications, including numerous book reviews. The examples cited here are a small fraction of his output, which has been continuous for 50 years, and illustrate the diversity of his research interests. His publications, at an average rate of 5 publications per year, are a testament to his hard work, enquiring mind, creative imagination, exceptional mathematical ability and underpinned by his education in geology. A search with Google Scholar showed that Agterberg, F.P. papers have been cited more than 2,000 times.

Acknowledgments We thank Codien Agterberg for helping with this paper about Frits, and supplying photographs.

References

Agterberg FP (1958) An undulation of the rate of sedimentation in southern Gotland. Geologie en Mijnbouw, New Series, vol 20, pp 253–260
Agterberg FP (1959) On the measuring of strongly dispersed minor folds. Geologie en Mijnbouw, New Series, vol 21, pp 133–137
Agterberg FP (1961) The skew frequency distribution of some ore minerals. Geologie en Mijnbouw, New Series, vol 23, pp 149–162
Agterberg FP (1964) Methods of trend surface analysis: Quarterly Colorado School Mines 59: 111–130
Agterberg FP (1965) The technique of serial correlation applied to continuous series of element concentration values in homogeneous rocks. J Geol 73:142–154
Agterberg FP (1968) Application of trend analysis in the evaluation of the Whalesback Mine, Newfoundland. Can Inst Mining Metall, special volume 9:77–88
Agterberg FP (1970) Autocorrelation functions in geology. In: Merriam DF (ed), Geostatistics. Plenum Press, New York, pp 113–141
Agterberg FP (1971) A probability index for detecting favourable geological environments. Can Inst Mining Metal, special volume 12 (Decision-Making in the Mineral Industry):82–91
Agterberg FP (1974a) Geomathematics – mathematical background and geo-science applications. Elsevier, Amsterdam, 596 p
Agterberg FP (1974b) Automatic contouring of geological maps to detect target areas for mineral exploration. J Math Geol 6:373–395
Agterberg FP (1977a) Statistical methods for regional resource appraisal. Bulletin, Can Inst Mining Metal 70(778):96–98
Agterberg FP (1977b) Frequency distributions and spatial variability of geological variables. Proceedings, 14th International Symposium on Computers in Mineral Industries, held at Pennsylvania State University, October 1976, American Institute of Mining Engineering, New York, pp 287–298
Agterberg FP (1979) Algorithm to estimate the frequency values of rose diagrams for boundaries of map features. Comput Geosci 5:215–230
Agterberg FP (1980) Mineral resource estimation and statistical exploration. In: Miall AD (ed) Facts and principles of world oil occurrence. Can Soc Petrol Geol Memoir 6:301–318
Agterberg FP (19820 IGCP Project 148: background, objectives and impact. In: Cubitt JM and Reyment RA (eds) Quantitative stratigraphic correlation. Wiley, Chichester, pp 1–4

Agterberg FP (1987) Quality of time scales – a statistical appraisal. In: Merriam DF (ed) Current trends in geomathematics. Plenum, New York, pp 57–103

Agterberg FP (1988) Spatial analysis of patterns of land-based and ocean-floor ore deposits. Proceedings, NATO Advanced Study Institute on "Statistical Analysis in Energy and Mineral Resource Appraisal" held in Il Ciocco, Italy, June 1986, Reidel, Dordrecht, pp 283–299

Agterberg FP (1989) Systematic approach to dealing with uncertainty of geoscience information in mineral exploration. Proceedings, 21st International Symposium on Computers in the Mineral Industry, held in Las Vegas, Nevada, March 1989, Society of Mining Engineers of AIME, Littleton, Colorado, pp 165–178

Agterberg FP (1990) Automated stratigraphic correlation: developments in paleontology and stratigraphy, vol. 13. Elsevier, Amsterdam, 424 p

Agterberg FP (1994a) FORTRAN program for the analysis of point patterns with correction for edge effects: Comput Geosci 20:229–245

Agterberg FP (1994b) Fractals, multifractals, and change of support. In Dimitrakopoulos R. (ed), Geostatistics for the next century. Kluwer, Dordrecht, pp 223–234

Agterberg FP (2007a) New applications of the model of de Wijs in regional geochemistry. Math Geol 39(1):pp 1–26

Agterberg FP (2007b) Mixtures of multiplicative cascade models in geochemistry. In: Cheng Q, Gaonac'h H and Tarquis A (eds) Nonlinear processes in geophysics, special issue on "New developments in nonlinear and scaling processes in Earth's surface and subsurface", pp 201–209

Agterberg FP, Bonham-Carter GF (2005) Measuring the performance of mineral potential maps: Nat Resour Res 14(1):pp 1–18

Agterberg FP, Bonham-Carter GF, Wright DF (1990) Spatial pattern integration for mineral exploration. In Gaal G, Merriam DF (eds) Computer applications in resource estimation, prediction and assessment for metals and petroleum. Pergamon, Oxford, pp 1–21

Agterberg FP, Briggs G (1963) Statistical analysis of ripple marks in Atokan and Desmoinesian rocks in the Arkoma Basin of east-central Oklahoma. J Sediment Petrol 33:393–410

Agterberg FP, Brown A, Cheng Q, Good D (1994) Multifractal modelling of fractures in the Lac du Bonnet Batholith, Manitoba. Proceedings, IAMG'94, Annual meeting of the International Association for Mathematical Geology, held at Mont Tremblant, Québec, October 1994, pp 3–8

Agterberg FP, Cheng Q, Wright DF (1993) Fractal modeling of mineral deposits. Proceedings, 24th International symposium on the application of computers and operations research in the mineral industries, held in Montreal, Quebec, November 1993, Can Inst Mining Metall 1:43–53

Agterberg FP, Chung CF, Fabbri AG, Kelly AM, Springer JS (1972) Geomathematical evaluation of copper and zinc potential in the Abitibi area of the Canadian Shield. Geological Survey of Canada Paper 71–41, 55p

Agterberg FP, Fabbri AG (1978) Spatial correlation of stratigraphic units quantified from geological maps. Comput Geosci 4:284–294

Agterberg FP, Gradstein FM (1982) Models of Cenozoic foraminiferal stratigraphy – northwestern Atlantic Margin. In Cubitt JM, Reyment RA (eds) Quantitative stratigraphic correlation. Wiley, Chicheste, pp 119–173

Agterberg FP, Gradstein FM (1988) Recent developments in stratigraphic correlation. Earth-Sci Rev 25:1–73

Agterberg FP, Nel LD (1982a) Algorithms for the ranking of stratigraphic events. Comput Geosci 8:69–90

Agterberg FP, Nel LD (1982b) Algorithms for the scaling of stratigraphic events. Comput Geosci 8:163–189

Bonham-Carter GF, Agterberg FP, Wright DF (1988) Integration of geological data sets for gold exploration in Nova Scotia. Photogramm Eng Rem Sens 54:1585–1592

Cheng Q, Agterberg FP, Ballantyne SB (1994) The separation of geochemical anomalies from background by fractal methods. J Geochem Explor 51:109–130

Cheng Q, Agterberg FP (1996) Multifractal modeling and spatial statistics Math Geol 28:1–16

Chung CF, Agterberg FP (1980) Regression models for estimating mineral resources from geological map data. Math Geol 12:473–488

Grunsky EC, Agterberg FP (1991a) FUNCORR: a FORTRAN 77 program for computing multivariate spatial autocorrelation. Comput Geosci 17:115–131

Grunsky EC, Agterberg FP (1991b) SPFAC: a FORTRAN 77 program for spatial factor analysis of multivariate data. Comput Geosci 17:133–160

Harris DP, Agterberg FP (1981) The appraisal of mineral resources. Econ Geol (75th anniversary volume):897–938

Koch GS, Jr, Link RF (1970) Statistical analysis of geological data, vol 1. John Wiley & Sons, New York, 375pp

Koch GS, Jr, Link RF (1971) Statistical analysis of geological data, vol 2. John Wiley & Sons, New York, 438pp

Merriam DF (1978) The International Association for Mathematical Geology – a brief history of record and accomplishments. In: Geomathematics: past, present and prospects Syracuse University Geology Contributions 5:1–6

Merriam DF (1981) Roots of quantitative geology, in down-to-earth statistics: solutions looking for geological problems. Syracuse University Geology Contributions 8:1–15

Merriam DF (1999) Reminiscences of the editor of the Kansas geological survey computer contributions, 1966–1970 & a byte. Comput Geosci 25(4):321–334

Merriam DF (2001) Andrei Borisovich Vistelius: a dominant figure in the 20th Century mathematical geology. Nat Resour Res 10(4):297–304

Merriam DF (2004) The quantification of geology. From abacus to pentium: Earth-Sci Rev 67 (1–2):55–89

Merriam DF (2005) Richard Arthur Reyment: father of the International Association for Mathematical Geology. Earth Sci Hist J 23(2):365–373

Merriam DF (2007) Trials and tribulations of international travel: Geol Soc America, Geo-Tales III, 3:24–25

Němec V (1993a) Introduction, In: Davis JC, Herzfeld UC (eds) Computers in geology – 25 years of progress. Oxford Univ. Press, New York, pp 3–11

Němec V (1993b) Fifteen international meetings of mathematical geologists at the Mining Pribram Symposia in Czechoslovakia (1968–1991), In: Nichiwaki N (guest ed) Recent advances in geomathematics and geoinformatics. Geoinformatics 4(3):361–364

Prokoph A, Agterberg FP (1999) Detection of sedimentary cyclicity and stratigraphic completeness by wavelet analysis: application to Late Albian cyclostratigraphy of the Western Canada Sedimentary Basin: J Sediment Res 69:862–875

Another Look at the Chemical Relationships in the Dissolved Phase of Complex River Systems

A. Buccianti, J.J. Egozcue and V. Pawlowsky-Glahn

Reprinted from *Mathematical Geosciences* DOI: 10.1007/s11004-008-9168-2, when citing this article please use the DOI number.

Abstract Large rivers are a major pathway for the erosion products of continents to reach the oceans. The riverine transport of dissolved and particulate materials is generally related to a large number of interactions involving climate, hydrological, physico-chemical and biological aspects. Consequently, the investigation of large rivers allows the erosion processes at a global scale to be addressed, with information about biogeochemical cycles of the elements, weathering rates, physical erosion rates and CO_2 consumption by the acid degradation of continental rocks. Today, good databases exist for the major dissolved ions in the world's largest rivers. Since concentration of ions in river waters has to be considered in a compositional context, it is necessary to study the implications of considering the simplex, with its proper geometry, as the natural sample space. Using the additive (alr) or the isometric (ilr) log-ratio transformations, a composition can be represented as a real vector; but only in the second case can these coordinates be mapped onto orthogonal axes.

Using data related to the dissolved load of some of the most important rivers in the world, the relationships among the major ions frequently used in molar ratio mixing diagrams have been investigated with alternative tools. Following the *balances approach*, an investigation of the properties of aqueous solutions of electrolytes that are often treated in terms of equilibrium constants is undertaken. The role played by the source—rain water, weathering of silic, carbonatic and evaporitic rocks, pollution—from which elements and chemical species can potentially be derived, has been checked through an investigation of a probabilistic model able to describe the relationships among the different components that contribute to the chemical composition of a river water sample.

A. Buccianti
Department of Earth Sciences, Universitat di Firenze, Firenze, Italy,
e-mail: antonella.buccianti@unifi.it

J.J. Egozcue
Department of Applied Mathematics III, Universitat Politecnica de Catalunya, Barcelona, Spain,
e-mail: juan.jose.egozcue@upc.edu

V. Pawlowsky-Glahn
Department of Computer Science and Applied Mathematics, Universitat de Girona, E-17071 Girona, Spain, e-mail: vera.pawlowsky@udg.edu

Keywords Aitchison geometry · simplex · river chemistry · mixing diagrams

1 Introduction

Chemical weathering both shapes the surficial environment and influences soil formations, affecting the global geochemical cycles of the elements and, in particular, that of carbon. Worldwide river chemistry can be consequently used to (1) understand the biogeochemical cycles of major and trace elements, (2) calculate chemical weathering rates, (3) estimate the role of major parameters like relief, climate, lithology, and vegetation that are likely to influence chemical weathering processes, and (4) quantify the effect of rock chemical weathering on the carbon cycle and its potential role on climate change (dissolution of CO_2 in surface waters conveys the protons necessary to weather the rock minerals). Studying river chemistry is fundamental since at the Earth's surface the erosion products are mainly transported by rivers determining, ultimately, the ocean chemistry. In a global approach, solutes in river water are derived from different sources like rain water, silicic, carbonatic and evaporitic rocks weathering (Garrels and MacKenzie, 1971).

Atmospheric inputs to rivers can be evaluated using Cl^- abundance, whose concentration in rocks is very low (Stallard, 1980; Meybeck, 1983). In South America (Stallard and Edmond, 1987), in Western Europe (Meybeck, 1986), and in Central Africa (Négrel et al., 1993), Cl^- originates from the dissolution of atmospheric seasalt particles by rainwater, and shows concentrations which decrease with increasing distance from the coast. Difficulties arise when evaporitic rocks are present in the river drainage, as reported for the Andean rivers (Stallard, 1980). In that situation further investigations are needed to discriminate marine aerosol contributions from those of rocks (Millot et al., 2002). For most large rivers direct rain input is almost insignificant (less than 5%), except for those rivers most influenced by evaporation (Gaillardet et al., 1999).

Waters draining different rock types are characterized by their own chemical and isotopic signatures. These signatures depend on both the chemical composition of the bedrock and on the rate at which it is being eroded. Carbonatic and evaporitic rocks are weathered 12 times and 40–80 times, respectively, more rapidly than granites or gneisses (Meybeck, 1987). As a consequence, evaporites have a major influence on river chemistry even if their outcrops are rather rare. The Sr isotopic ratio and the Ca^{2+}/Na^+, HCO_3^-/Na^+ and Mg^{2+}/Na^+ molar ratios are particularly well suited to distinguish between carbonate and silicate contributions; they also have the important property of being independent of water fluxes, dilution and evaporation effects. Normally, the analysis is based on end–members whose composition is estimated using data for small rivers draining a single lithology (carbonates, silicates and evaporites). Then, the hydrochemistry of a river is considered as a mixture of such end–members. For instance, Négrel et al. (1993) consider that binary relationships between Sr isotopic composition and Ca^{2+}/Sr, as well as between Mg^{2+}/Na^+ and Ca^{2+}/Na^+, or between HCO_3^-/Na^+ and Ca^{2+}/Na^+ molar ratios, can be used to model mixing between end-members. Each river appears as a point in a diagram whose axes are the logarithms of the mentioned ratios and mixing between end–members is represented by straight lines. These diagrams represent a good approximation to the investigation of the phenomena analyzed. However, ratios with the same denominator (Na^+ in this instance) in a logarithmic scale (i.e. additive logratio (alr) coordinates used to represent a composition as a real vector) have some peculiarities as, not being invariant under the permutation of parts, they represent an oblique basis in the Aitchison geometry of the simplex (Egozcue et al., 2003; Egozcue and Pawlowsky-Glahn, 2005, 2006). This fact may lead to misinterpretation and, consequently, a compatible metric is required. Our aim is to describe, using geometrically improved methods, general chemical equilibria among major ions characterizing river composition, their behavior, as well as the role of the different sources, on a global scale. From a general point of view, chemical weathering, highly related to runoff and physical erosion rates, is strongly lithology dependent (Dessert et al., 2001; Chadwick et al., 2003; Oliva et al., 2003; Quade et al., 2003), with basaltic rocks yielding by far the highest chemical weathering rates. As a consequence, basaltic rocks have a disproportionately large effect on CO_2 drawdown, being responsible for as much as 25% of the global carbon flux. However, calculated weathering rates of river basins show only modest dependence on temperature, due to the existence of hot catchment areas in regions of limited denudation, e.g. West Africa (Gaillardet et al., 1999; Anderson et al., 2003).

2 Origin of Data

A data base for 1080 water samples from the most important rivers in the world has been compiled. In it, the concentrations (mg/L) of Na^+, K^+, Mg^{2+}, Ca^{2+}, Cl^-, SO_4^{2-}, HCO_3^- and H_4SiO_4 are reported, together with pH values and, where present, Sr and Rb abundance and $^{86}Sr/^{87}Sr$ isotopic ratios. The analytical procedures for water analysis included ion chromatography for Cl^- and SO_4^{2-}, conventional flame atomic absorption spectrophotometry for Ca^{2+}, Na^+, K^+ and Mg^{2+}, and colorimetry for aqueous silica. On the whole, the uncertainties related to analytical errors are comparable. In most situations samples have been collected taking into account seasonal variability, as well as different positions within the catchment. Catchments were selected to be widely representative of different lithologies, climates and seasons. Contamination, if present, is expected to be revealed by presence of outliers. No correction for the atmospheric contribution has been performed on the original information. Data from the largest rivers of the world are from Gaillardet et al. (1999) and cover a continental area of 53.6×10^6 km^2 or about 54% of the exorheic continental area (open systems in which surface waters ultimately drain to the ocean). This main data base has been expanded with data pertaining to South American (Edmond et al., 1995; Gaillardet et al., 1997; Mortatti and Probst, 2003) and Asiatic regions (Galy and France-Lanord, 1999; Bickle et al., 2005; Tipper et al., 2006). Further data derive from the Deccan volcanic province, located in the western and central parts of India (Dessert et al., 2001) and from the Min Jiang river, a headwater tributary of the Yangtze River in the Chang Jiang area (Qin et al., 2006). African information was supplemented by data from the dissolved load of the Nyong basin rivers (Viers et al., 1997, 2000), the second largest river of Cameroon in terms of length. It has a relatively small drainage area (27,800 km^2), which is characterized by potentially elevated level of chemical weathering (high temperature and precipitation), but strongly restricted due to thick soil and vegetation and lack of tectonic uplift. North America is represented by data from Canada (Slave Province, Northwest Territories and Grenville Province, Québec (Millot et al., 2002); Stikine Province, Western Canadian Cordillera (Millot et al., 2003); Mackenzie River basin, northern Canada (Millot et al., 2003); Fraser, Skeena, Nass river basins, Canadian Cordillera (Spence and Telmer, 2005)). The regions under consideration are characterized by moderate relief and the landscape is typical post-glacial with abundant till deposits. Vegetation is dominated by conifers and population is sparse, making the region remarkably pristine. Further samples derive from White River, Vermont, USA (Douglas, 2006). Data from Europe have been supplemented by considering the Seine (France) (Roy et al., 1999) and Salso (central Sicily, Italy) rivers (Favara et al., 2000). Overall, it has to be kept in mind that the database comprises observations from regions that are not geologically in the same state of evolution. Northern latitude regions are in a transient state and still in the post-glaciation period. There, weathering reactions proceed so slowly that the steady state of soils has not been reached. By contrast, tropical regions, like the Guyana or the African shield, have experienced only small changes during the past million years and have probably reached a steady state for weathering reactions.

3 Methodology

In a classical approach, river geochemistry can be characterized by a number of elemental ratios and the Sr isotopic composition. Several Na^+ normalized molar ratios (Ca^{2+}/Na^+, K^+/Na^+, Mg^{2+}/Na^+, Cl^-/Na^+, SO_4^{2-}/Na^+, HCO_3^-/Na^+) can be used as intensive parameters, independent of dilution and evaporative processes, able to compare rivers draining areas of high runoff with those draining arid basins. Mixing diagrams using Na^+-normalized molar ratios allow a comparison of the chemical composition of each sample with end-member reservoirs estimated using data for small rivers draining one single lithology (i.e. carbonates, silicates and evaporites). However, recall that these scatter diagrams, whose coordinates are given on a log-scale by ratios with the same denominator (e.g. HCO_3^-/Na^+ versus Ca^{2+}/Na^+, Mg^{2+}/Na^+ versus Ca^{2+}/Na^+, Ca^{2+}/Na^+ versus $1000 \times Sr/Na^+$), define a space with non orthogonal axes (Egozcue and Pawlowsky-Glahn, 2005), and that statistical validation of the relationships inside this space (e.g. correlations and their significance, discrimination of groups) requires an appropriate distance due to the specific geometry of compositional data. This standard representation of ratios in fact is equivalent to those of the alr-coordinates (Aitchison, 1986), were $D-1$ of the components of the composition are divided by the remaining component and logarithms taken. The resulting log-ratios are real variables whose coordinates can be mapped onto axes at 60°. In this context it is not possible to use the usual inner product and distances on alr-transformed observations to determine either the angle between two vectors or their distance without modifying the methods of calculation. A simple alternative approach can be used. The methodology is based on the identification of subsets of variables, parts, and their balances (Egozcue and Pawlowsky-Glahn, 2005, 2006). The balances approach was proposed to simplify the analysis of compositional data. It consists of grouping variables into subsets which are interpretable from a geochemical point of view. For example, chemical compositions of river waters include usually a group of anions and another group of cations, leading to a natural first partition of the composition. In this and other similar situations, one may be interested in studying two features of the sample compositions, the relationship or balance between the two groups of parts (inter-group analysis) and the behavior of parts within a group (intra-group analysis). The sequential binary partitioning of parts of a composition (Egozcue and Pawlowsky-Glahn, 2005) represents a tool to design a particular basis in the simplex, such that the corresponding coordinates are directly interpretable as balances between any two groups of parts appearing in some order of the sequential binary partition. Consequently, subcompositional (intra-group) and balance (inter-group) analysis are reduced to orthogonal projections onto subspaces of the simplex, thus guaranteeing consistency of distances and statistical analysis when working in reduced dimensions.

Taking into account the previous considerations, we propose to analyze the chemical relationships among three variables usually used in a scatter diagram with common denominator, e.g. Ca^{2+}, Na^+ and HCO_3^-, to investigate the behavior among samples in the coordinate space. To do so, the coordinates $x = Ca^{2+}/Na^+$ and $y = HCO_3^-/Na^+$, represented on a logarithmic scale, are substituted by the coordinates:

$$x = B_1 = \frac{1}{\sqrt{2}} \ln \frac{Ca^{2+}}{Na^+}, \qquad y = B_2 = \frac{1}{\sqrt{6}} \ln \frac{Ca^{2+} \times Na^+}{(HCO_3^-)^2}, \qquad (1)$$

representing the balances B_1 and B_2 between the part Ca^{2+} versus Na^+, and between the parts Ca^{2+} and Na^+ versus HCO_3^-, with a normalization constant (Egozcue et al., 2003).

¿From a general point of view, consider that Ca^{2+} is contributed to river water almost entirely from rocks weathering and only 9% of it may arise from pollution (Berner and Berner, 1996). The sources of Ca^{2+} mainly consist of $CaCO_3$ bearing sedimentary rocks, with a smaller proportion derived from Ca^{2+}-silicate minerals, chiefly Ca^{2+}-plagioclase, and a minor amount from $CaSO_4$ minerals. Concerning Na^+, recent estimates (Berner and Berner, 1996) indicate that about 28% is derived from contamination, 8% from sea salt, 43% from halite, and the remaining 22% from the weathering of plagioclase. Another possible source of Na^+ in river water is from cation exchange of dissolved Ca^{2+} with Na^+ on detrital clay minerals during marine shale weathering. Finally, all bicarbonate in average river water is derived from rock weathering, pollution accounting for only 2% and cyclic sea salts for far less than 1%. The behavior of this component is affected by the presence of the CO_2 gaseous phase, participating at the carbonate equilibria system. Here acid attack is the principal weathering mechanism with water, the acid being formed by the reaction of carbon dioxide from the atmosphere or from the respiratory cycle of organisms in the soil. With the aim to attribute a geochemical meaning to the balances, two representative weathering reactions given by:

$$CO_2 + H_2O + CaCO_3 \rightarrow Ca^{2+} + 2HCO_3^- \qquad (2)$$

and

$$2CO_2 + 11H_2O + 2NaAlSi_3O_8 \rightarrow 2Na^+ + 2HCO_3^- + Al_2Si_2O_5(OH_4) + 4H_4SiO_4, \qquad (3)$$

are considered. It is reasonable to assume that the two balances may represent their joint contribution, as Ca^{2+} and HCO_3^- are related to the right part of the Eq. (2), whereas Na^+ and HCO_3^- are related to the right part of the Eq. (3). Statistical analysis and investigation of relationships can be performed in the balance space with usual methodologies (e.g. identification of linear or non-linear data fitting, presence of groups, etc.). Moreover, the investigation of the features of balance frequency distributions may give information about equilibrium between the involved parts of the composition, evaluating the nature of the probabilistic model used to describe their behavior. If the distribution is unimodal and the variance is low, an equilibrium situation may be hypothesized; on the other hand, presence of multimodality and/or high variance may be an indication of more complex situations not describable by a single model. The identification of the random variable (normal, skew-normal, bimodal and so on) able to describe a balance represents, finally, a further development in the understanding of how natural phenomena work and their importance.

4 Mixing Versus Balance Diagrams

Mixing diagrams on a logarithmic scale using Na^+-normalized molar ratios in the dissolved phase of rivers are typically used to estimate contribution of end-member reservoirs (i.e. carbonates, silicates and evaporites). When simple binary diagrams have to be constructed for evaluating mixing processes, the balance approach allows us to choose appropriate coordinates for representing the data in a real Euclidean space, in which statistical analysis can be performed. Consider for example the Na^+-normalized diagram in a log-log space, as introduced by Négrel et al. (1993), which is presented in Fig. 1 for Ca^{2+}/Na^+ vs. HCO_3^-/Na^+. In our data base Na^+ ranges from 0.076 (Guayana Shield) to 79 875 mg/L (Salso River, Sicily), Ca^{2+} from

Chemical Relationships in the Dissolved Phase of Complex River Systems

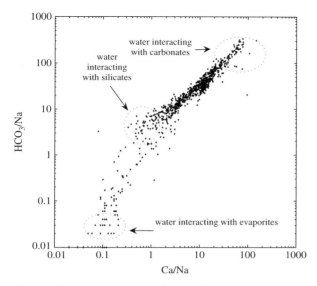

Fig. 1 Mixing diagram using Na^+-normalized ratios in dissolved phase of river water. See text for details

0.040 (Guayana Shield) to 1 070 mg/L (Salso River, Sicily) and HCO_3^- from 0.80 (Amazon river) to 825 mg/L (Ganges-Brahmaputra) with ratios $Ca^{2+}/Na^+ \simeq 0.01$ to 147, $HCO_3^-/Na^+ \approx 0.003$ to 335, $Mg^{2+}/Na^+ \simeq 0.01$ to 40.78. As an alternative, the balances for the three variables, Ca^{2+}, Na^+, and HCO_3^-, are reported in Fig. 2.

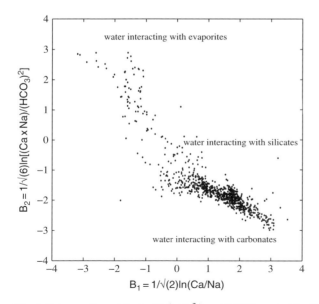

Fig. 2 Balances B_1 and B_2 for Na^+, Ca^{2+} and HCO_3^- content in dissolved phase of river waters

The balances B_1 (x) and B_2 (y) represent the equilibrium among the products, involving, or not, a gaseous phase, of two previous weathering reactions (Eqs. 2, 3).

World average river water composition is dominated by Ca^{2+} and HCO_3^-, both of which are derived predominantly from limestone weathering and, consequently, about 98% of all river water is of the calcium carbonate type, while less than 2% has Na^+ as the principal ion (Meybeck, 1979). From a general point of view, waters draining carbonates show Ca^{2+} and Mg^{2+} dominated reservoirs and Ca^{2+}/Na^+ ratios are often close to 50 (or 3.56 for balance B_1) whereas the average for crustal continental rocks is about 0.6 (or -0.87). Due to the higher solubility of Na^+ relative to Ca^{2+}, lower Ca^{2+}/Na^+ molar ratios are expected in the dissolved load of rivers draining silicates. However, the existence of silicate draining rivers having higher Na^+-normalized ratios can be explained by the presence of rocks with values higher than crustal rocks, with weathering conditions such that Ca^{2+} rich minerals are preferentially dissolved, and with the presence of disseminated calcite within catchment bedrocks.

In order to compare the differences between both representations, Fig. 3 shows the data rescaling of the x-axis of (b) to match exactly the expression of B_1 in (a). Moreover, a set of parallel and not-parallel lines have been superimposed with the goal of visualising the effect of working in a non-orthogonal system. Two features can be observed: (i) there is a reduction of the scale in the y axis of the alr representation (b), thus suggesting a reduction of variability in the y-axis; and (ii) the angle between the intersecting lines has changed, thus suggesting the data are nearer to a linear trend in (b) than in (a). Both facts correspond to the property that alr-axes are not orthogonal, but they form an angle of 60 degrees. The dispersion and curvature of data in Figs. 2 and 3(a), correspond to the Aitchison geometry of the simplex, whereas Figs. 1 and 3(b), exhibit a subtle but important distortion.

In Fig. 2 the scatter of the data tends to increase from carbonate to evaporite end-members due to the higher geochemical variability of those source rocks. The curvature shown by the data indicates that the value of the Ca^{2+}/Na^+ ratio is not proportional to the change of lithology, and small variations in its value have a

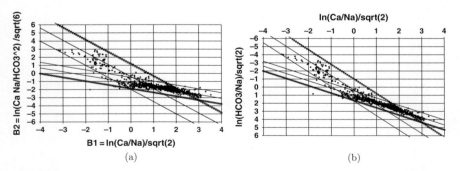

Fig. 3 Parallel and intersecting lines represented in balance-coordinates with chemical composition of river waters (**a**) and in Na^+-normalized alr coordinates (**b**). The x-axis in (**a**) and (**b**) is equally scaled for comparison. In (**b**), y axis has been reversed for easier comparison

large impact on the internal relationships among Ca^{2+}, Na^+ and HCO_3^-, particularly when evaporites are involved. In this framework, the balances diagram appears to take into account the differences among the data in a more sensible way compared to the Na^+-normalized molar ratios.

To investigate the origin of the dispersion in Fig. 2, samples have been divided into subsets by location on a global scale, as reported in Fig. 4. As can be seen, rivers from Europe are characterized by a higher dispersion, with samples moving from carbonate towards evaporite fields. They seem to be well represented by a non-linear model. The shift of the samples upwards is mainly associated with an increase in evaporation (i.e. Mediterranean climate), whereas shift downwards leads to regions characterized by lower evaporation rate and colder temperature. Data from Asia show a different curvilinear pattern, with points moving towards a more accentuated decrease in B_1 compared with those of B_2, the latter showing in the upper part of the diagram fluctuations around a constant value. This situation could be explained by the fact that locally, dependent on bedrock composition, regions have experienced only small changes during the past million years, so that a steady state for weathering reactions could has been achieved. It appear that balance B_2 can describe this situation, representing the joint behaviour of weathering equations (2) and (3). Data from North America contain the most extreme positive values of B_1, moving towards the carbonate end-member.

Data from Africa and South America appear to be located around the zero value for B_1 (Ca^{2+}/Na^+ ratio approximately equal to 1) and between -2 and -1 for B_2,

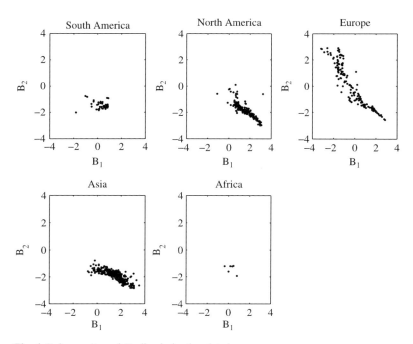

Fig. 4 Balances B_1 and B_2 discriminating data by provenance area

revealing a complex relationship among the three involved variables, Ca^{2+}, Na^+ and HCO_3^-. The balance B_2 behaviour is affected by the different mobility of Ca^{2+} and Na^+ as governed by rainfall, high temperature and intensity of weathering processes, as well as by bedrock lithological nature. In this situation the source of variability can be mainly associated to climate forcing, as recent results suggest that carbonate weathering has a greater sensitivity to runoff because of the faster dissolution kinetics of carbonates relative to silicates (Tipper et al., 2006). Differences in dispersion can be also attributable to less and more organic-dominated environments. When dissolved organic matter and acids are abundant (dissolved humic and fulvic acids play an important part in mineral dissolution), within the drainage, mineral weathering rates tend to increase (Viers et al., 1997, 2000).

5 Identification of Natural Geochemical Laws

In order to check if the balances can represent an equilibrium among the involved parts, histograms of their values are presented in Fig. 5 together with their Gaussian kernel density estimation (Bowman and Azzalini, 1997). In both cases the application of the Kolmogorov-Smirnov, Anderson-Darling (large n) and Shapiro-Wilk (small n) tests for normality indicates that the simple Gaussian model cannot be adopted to describe the data (p-value $\ll 0.05$), and the shape suggests that a unique probabilistic model is not adequate to represent the different processes (we are

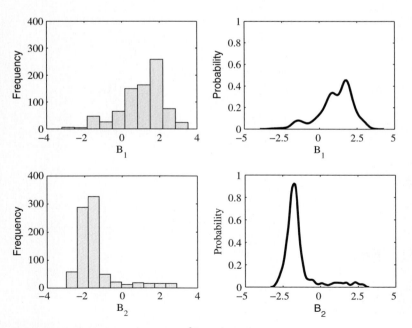

Fig. 5 Histograms of B_1 and B_2 for Ca^{2+}, Na^+ and HCO_3^-

dealing with mixtures) affecting them. The balance B_1, involving the Ca^{2+}/Na^+ ratio, appears to be poly-modal in comparison with B_2, involving the ratios Ca^{2+}/HCO_3^- and Na^+/HCO_3^-. The distribution of B_2 is positively skewed, indicating the likely presence of outliers, i.e. data reflecting rare geochemical processes. In the first B_1 situation, the poly-modality reflects the presence of important variability sources, whereas in the second B_2 situation a more homogeneous condition is achieved, with the exception of a tail in the right part of the histogram. This tail corresponds to most of the samples from Salso river (86), Grenville Province (8 samples with low content of HCO_3^- and pH in the interval 4.63 to 5.96), Rhine, Yukon, Weser, Don, Wisla, Elbe, Murray Darling, Odra, and Ebro, all characterized by high contents of Na^+, Ca^{2+} or HCO_3^-, and often associated with densely populated, industrialized or cultivated areas, or affected by high evaporation rates. If samples with balance values greater than -0.5 are ruled out, normality is achieved (tests p-value > 0.05). In other words, balance B_2 appears to represent an equilibrium condition among the involved cations and anions, and the exceptions are related to rivers affected mainly by anthropogenic contributions or highly affected by evaporation, i.e. outliers due to different geochemical processes. The equilibrium may be attributable to the rapid dissolution reaction of carbonates (involving Ca^{2+} and HCO_3^-) compared to silicates, particularly when larger runoff areas are analyzed and/or to the role played by the carbonate equilibria system. In Fig. 6 box-plots of the balances are reported by continental area of provenance. As can be seen, Europe is characterized by higher variability for both balances, showing both

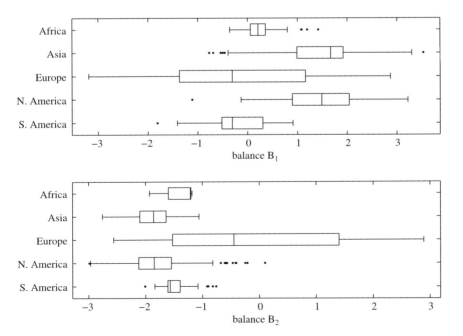

Fig. 6 Box-plots of B_1 and B_2 by continental area of provenance

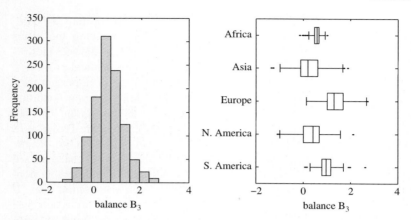

Fig. 7 Histogram of balance B_3 for Ca^{2+}, Na^+ and Mg^{2+} and box-plots by continental area of provenance

negative and positive data; Africa presents values around zero for B_1 (corresponding to a Ca^{2+}/Na^+ ratio equal to 1); Asia and North America are shifted towards positive values, and South America towards negative ones corresponding, respectively, to an increase and to a decrease in the Ca^{2+}/Na^+ ratio. Greater overlap is observed in the box-plots of B_2 where, with the exception of Europe, most of the data are negative; the zero value for B_2 corresponds to the condition in which the geometric mean of Na^+ and Ca^{2+} is equal to that of HCO_3^-.

Since waters draining carbonates have Ca^{2+} and Mg^{2+} reservoirs, the mixing diagram Mg^{2+}/Na^+ versus Ca^{2+}/Na^+ is often considered. The balances to be used to analyze the relationships among the variables Ca^{2+}, Mg^{2+} and Na^+ are the same as those of Eq. (1) with Mg^{2+} substituting for HCO_3^-. Given that the first balance is the same as in the previous example, in Fig. 7 the histogram of the third balance is presented, together with the box-plots by continental areas. Application of the previous normality tests permits the acceptance of the null hypothesis (p-value > 0.05) indicating a degree of equilibrium among these variables. Data from North America and Asia show values more or less around zero, whereas Africa, South America and Europe provide clearly positive values. The mean value of the balance, equal to 0.6 ± 0.64, includes the 0.9 value obtained for small rivers draining only carbonates that have Mg^{2+}/Na^+ ratios close to 10 (Gaillardet et al., 1999). Where Na^+ tends to be important either for silicate, evaporite or contamination contributions, balance data present increasingly positive values.

6 Some Conclusions and Perspectives

The investigation of the fluxes to the ocean of material derived from the chemical and mechanical weathering of the continents is fundamental for the global geochemical budget, and as such, their estimation has attracted attention since the earliest development of earth sciences. In addition, global rates of chemical weathering provide a major source of information concerning CO_2 levels as well as about its consumption. Historically, most attention has been paid to the investigation of the dissolved load of rivers on a global scale by considering data from different parts of the world. The aim of this paper has been to identify the potential source of each element and to hypothesize mixing among end-members to be related to rock weathering or to other natural processes, e.g., evaporation and precipitation, or to anthropogenic contamination phenomena. In this context, molar diagrams based on ratios with a common denominator (alr-coordinates) have been often used. The approach proposed in this paper is based instead on the balance concept as defined by Egozcue and Pawlowsky-Glahn (2005, 2006). Binary mixing diagrams constructed by using balances appear to capture the data variability better than those based on molar ratios with a common denominator. These avoid distortion and reveal details about the complexity of natural phenomena affecting the chemistry of the river water. Since balances are able to represent the joint behavior of different variables they can represent better the development of simultaneous geochemical processes. Further research on this aspect is under progress.

Acknowledgments This research has been financially supported by Italian MIUR (Ministero dell'Istruzione, dell'Università e della Ricerca Scientifica e Tecnologica), PRIN 2004 (through the project GEOBASI, prot. 2004048813–002 and ex 60% funds) and by the Spanish Ministry for Education and Science under projects Ref.: "Ingenio Mathematica (i-MATH)" N. CSD 2006-00032 (Consolider – Ingenio 2010) and Ref.: MTM2006-03040. Graeme F. Bonham-Carter and Robert Garrett are thanked for their invaluable comments.

References

Aitchison J (1986) The statistical analysis of compositional data. Monographs on Statistics and Applied Probability. Chapman & Hall Ltd., London (UK). (Reprinted in 2003 with additional material by The Blackburn Press) p 416

Anderson SP, Lougacre SA, Kraal ER (2003) Patterns of water chemistry and discharge in the glacier-fed Kennicott River, Alaska: evidence for subglacial water storage cycles. Chem Geol 202: 297–312

Berner EK, Berner RA (1996) Global environment: water, air, and geochemical cycles. Prentice-Hall, Upper Saddle River, NJ

Bickle MJ, Chapman HJ, Bunbury J, Harris NB, Fairchild IJ, Ahmad T, Pomiès C (2005) Relative contributions of silicate and carbonate rocks to riverine Sr fluxes in the headwaters of the Ganges. Geochimica et Cosmochimica Acta 69(9): 2221–2240

Bowman AW, Azzalini A (1997) Applied smoothing techniques for data analysis: The kernel approach with S-Plus ilustrations. Clarendon Press, Oxford (GB) p 193

Chadwick OA, Gavenda RT, Kelly EF, Ziegler K, Olson CG, Elliott WC, Hendricks DM (2003) The impact of climate on the biogeochemical functioning of volcanic soils. Chem Geol 202: 195–223

Dessert C, Dupré B, Francois LM, Schott J, Gaillardet J, Chakrapani G, Bajpai S (2001) Erosion of Deccan Traps determined by river geochemistry: impact on the global climate and the $^{87}Sr/^{86}Sr$ ratio of seawater. Earth Planet Sci Lett 188: 459–474

Douglas TA (2006) Seasonality of bedrock weathering chemistry and CO_2 consumption in a small watershed, the White River, Vermont. Chem Geol 231: 236–251

Edmond JM, Palmer MR, Measures CI, Grant B, Stallard RF (1995) The fluvial geochemistry and denudation rate of the Guayana Shield in Venezuela, Colombia, and Brazil. Geochimica et Cosmochimica Acta 59(16): 3301–3325

Egozcue JJ, Pawlowsky-Glahn V (2005) Groups of parts and their balances in compositional data analysis. Math Geol 37(7): 795–828

Egozcue JJ, Pawlowsky-Glahn V (2006) Simplicial geometry for compositional data. volume 264 of Special Publications. Geological Society, London

Egozcue JJ, Pawlowsky-Glahn V, Mateu-Figueras G, Barceló-Vidal C (2003) Isometric logratio transformations for compositional data analysis. Math Geol 35(3): 279–300

Favara R, Grassa F, Valenza M (2000) Hydrochemical evolution and environmental features of Salso River catchment, central Sicily, Italy. Environ Geol 39(11): 1205–1215

Gaillardet J, Dupré B, Allègre CJ, Négrel P (1997) Chemical and physical denudation in the Amazon River Basin. Chem Geol 142: 141–173

Gaillardet J, Dupré B, Louvat P, Allègre CJ (1999) Global silicate weathering and CO_2 consumption rates deduced from the chemistry of large rivers. Chem Geol 159: 3–30

Galy A, France-Lanord C (1999) Weathering processes in the Gange-Brahmaputra basin and the riverine alkalinity budget. Chem Geol 159: 31–60

Garrels RM, MacKenzie FT (1971) Evolution of sedimentary rocks. WW Norton & Co, New York

Meybeck M (1979) Concentrations des eaux fluviales en èlèments majeurs et apports en solution aux ocèans. Rev Gèol Dyn Gèogr Phys 21(3): 215–246

Meybeck M (1983) Atmospheric inputs and river transport of dissolved substances. In: Proceedings of the Hamburg Symposium, August 1983, IAHS, Publ. vol 141

Meybeck M (1986) Composition chimique des ruisseaux non pollués de France. Sci Geol Bull 39(1): 3–77

Meybeck M (1987) Global chemical weathering from surficial rocks estimated from river dissolved loads. Am J Sci 287: 401–428

Millot R, Gaillardet J, Dupré B, Allègre CJ (2002) The global control of silicate weathering rates and the coupling with physical erosion: new insights from rivers of the Canadian Shield. Earth Planet Sci Lett 196: 83–89

Millot R, Gaillardet J, Dupré B, Allègre CJ (2003) Northern latitude chemical weathering rates: clues from the MacKenzie River Basin, Canada. Geochimica et Cosmochimica Acta 67(7): 1305–1329

Mortatti J, Probst JL (2003) Silicate rock weathering and atmospheric/soil CO_2 uptake in the Amazon basin estimated from river water geochemistry: seasonal and spatial variations. Chem Geol 197: 177–196

Négrel P, Allègre CJ, Dupré B, Lewin E (1993) Erosion sources determined by inversion of major and trace element ratios in river water: the Congo Basin case. Earth Planet Sci Lett 120: 59–76

Oliva P, Viers J, Dupré B (2003) Chemical weathering in granitic crystalline environments. Chem Geol 202: 225–256

Qin J, Huh Y, Edmond JM, Du G, Ran J (2006) Chemical and physical weathering in the min jiang, a headwater tributary of the Yangtze River. Chem Geol 227: 53–69

Quade J, English N, DeCelles P (2003) Silicate versus carbonate weathering in the Himalaya: a comparison of the Arun and Seti River watersheds. Chem Geol 202: 275–296

Roy S, Gaillardet J, Allègre CJ (1999) Geochemistry of dissolved and suspended loads of the Seine river, France: anthropogenic impact, carbonate and silicate weathering. Geochimica et Cosmochimica Acta 63(9): 1277–1292

Spence J, Telmer K (2005) The role of sulfur in chemical weathering and atmospheric CO_2 fluxes: evidence from major ions, $\delta^{13}C_{DIC}$, and $\delta^{34}S_{SO_4}$ in rivers of the Canadian Cordillera. Geochimica et Cosmochimica Acta 69(23): 5441–5458

Stallard RF (1980) Major element chemistry of the Amazon River System. PhD thesis, MIT/WHOI

Stallard RF, Edmond JM (1987) Geochemistry of the Amazon, 3. weathering chemistry and limits to dissolved inputs. J Geophys Res 92: 8293–8302

Tipper ET, Bickle MJ, Galy A, West AJ, Pomiès C, Chapman HJ (2006) The short term climatic sensitivity of carbonate and silicate weathering fluxes: insight from seasonal variations in river chemistry. Geochimica et Cosmochimica Acta 40(11): 2737–2754

Viers J, Dupré B, Braun JJ, Deberdt S, Angeletti B, Ngoupayou JN, Michard A (2000) Major and trace element abundances, and strontium isotopes in the Nyong basinrivers (Cameroon): constraints on chemical weathering processes and elements transport mechanisms in humid tropical environments. Chem Geol 169: 211–241

Viers J, Dupré B, Polvé M, Schott J, Dandurand JL, Braun JJ (1997) Chemical weathering in the drainage basin of a tropical watershid (Nsimi-Zoetele site, Cameroon): comparison between organic poor and organic rich waters. Chem Geol 140: 181–206

A Critical Approach to Probability Laws in Geochemistry[1]

G. Mateu-Figueras and V. Pawlowsky-Glahn

Reprinted from *Mathematical Geosciences* DOI: 10.1007/s11004-008-9169-1, when citing this article please use the DOI number.

Abstract Probability laws in have been a major issue of concern over the last decades. The lognormal on the positive real line or the additive logistic normal on the simplex are two classical laws of probability to model geochemical data sets due to their association with a relative measure of difference. This fact is not fully exploited in the classical approach, when viewing both the positive real line and the simplex as subsets of real space with the induced geometry. But it can be taken into account considering them as real linear vector spaces with their own structure. This approach implies using a particular geometry and measure different from the usual one. Therefore we can work with the coordinates with respect to an orthonormal basis. It could be shown that the two mentioned laws are associated with a normal distribution on the coordinates. In this contribution both approaches are compared, and a real data set is used to illustrate similarities and differences.

Keywords Lognormal · additive logistic normal · normal · positive real line · simplex · Aitchison geometry

1 Introduction

Natural processes in geochemistry can arise from the combination of many complex factors. Therefore, suitable probability laws have been a major issue of concern over the last decades (Ahrens, 1953; Agterberg, 2005). Knowledge of the distribution suitable to model geochemical variables is important, as it might be helpful in understanding the generating processes, or to validate the results obtained with some

---

G. Mateu-Figueras
Department d'Informàtica i Matemàtica Aplicada, Campus Montilivi, Universitat de Girona, E-17071 Girona, Spain, e-mail: gloria.mateu@udg.edu

V. Pawlowsky-Glahn
Department d'Informàtica i Matemàtica Aplicada, Campus Montilivi, Universitat de Girona, E-17071 Girona, Spain, e-mail: vera.pawlowsky@udg.edu

[1] Received: ;accepted.

statistical technique. There is a large battery of models in the literature, but finding the adequate model is not an easy task. Two crucial issues, which are usually taken for given, should be considered when selecting a model, namely the sample space and the scale of data.

Most statistical techniques assume data to be realizations of real random vectors, i.e. random vectors with sample space the real space with its usual geometry. Consequently, density functions are expressed with respect to the Lebesgue measure. The Lebesgue measure is the natural measure in real space. It is compatible with its inner vector space structure and, thus, with the absolute measure of difference. One might ask what happens if the structure and the measure in a given sample space is different from the real space structure and the Lebesgue measure. One reasonable possibility is to work with density functions with respect to that measure. This is the approach used by McAlister (1879) when introducing the *law of frequency* as the density of a lognormal variable. But some difficulties arise using it, as the necessary integrals to compute probabilities are not ordinary ones. In order to compute them, McAlister (1879) introduces the *law of facility*, nowadays known as the lognormal density function. This function is the density with respect to the Lebesgue measure. The key issue is to be careful to preserve the coherence with the structure of the sample space; otherwise, interpretation might become tortuous.

The objective in this contribution is to discuss and compare the two strategies on the positive real line and on the D-part simplex as they are two sample spaces that often appear in geochemistry. To illustrate some similarities and differences a real data set corresponding to Skye lavas (Thompson, Esson and Duncan, 1972) is used. Section 2 contains a summary of the theoretical background of the alternative strategy. Section 3 is focused on the positive real line, and Sect. 4 on the simplex as the sample space. Concluding remarks are presented in Sect. 5.

2 Theoretical Background

When a random variable or vector has a constrained sample space, $E \subset \mathbb{R}^D$, the methods and concepts used in real space might lead to absurd results, as is well known from examples like the spurious correlations between proportions stated by Pearson (1897). This problem can be circumvented when E admits a meaningful space structure (Pawlowsky-Glahn and Egozcue, 2001). In fact, if E is a complete inner product vector space, a measure λ_E, compatible with its structure, can be defined (Eaton, 1983). In particular, it can be defined through the Lebesgue measure on orthonormal coordinates (Pawlowsky-Glahn, 2003). A probability density function, f_E, is defined on E as the *Radon-Nikodým derivative* of a probability measure P with respect to λ_E. The measure λ_E has the same properties in E as the Lebesgue measure in real space. Difficulties, arising from the fact that the integral $P(A) = \int_A f_E(\mathbf{x}) d\lambda_E(\mathbf{x})$ is not an ordinary one, are solved working with coordinates, as properties that hold in the space of coordinates transfer directly to the space E. For example, for f_E a density function on E, call f the density function of the coordinates, and then the probability of an event $A \subseteq E$ is computed as $P(A) = \int_V f(\mathbf{v}) d\lambda(\mathbf{v})$, where V and $\mathbf{v}$ are the representation of A and $\mathbf{x}$ in terms of the orthonormal coordinates chosen, and λ is the Lebesgue measure in the space of coordinates. Using f to compute any element of the sample space, e.g. the expected value, the coordinates of this element with respect to the same orthonormal basis are obtained. The corresponding element in E is then given by the representation of the element in the basis chosen.

Every one-to-one transformation between a set E and real space induces a real Euclidean space structure in E, with associated measure λ_E. Particularly interesting are those transformations related to an interpretable measure of difference between observations, which we call *natural* measure of difference. This was evidenced by Galton (1879) when introducing the logarithmic transformation as a means to acknowledge the fact that *sensation = log(stimulus)*.

This simple approach has acquired a growing importance in applications, since it has been recognized that many *constrained sample spaces*, which are subsets of some real space—like $\mathbb{R}_+$ or the simplex—can be structured as Euclidean vector spaces (Pawlowsky-Glahn and Egozcue, 2001). It is important to emphasize that this approach implies using a measure which is different from the usual Lebesgue measure. Its advantage is that it opens the door to alternative statistical models depending not only on the assumed distribution, but also on the measure which is considered as appropriate or natural for the studied phenomenon, thus enhancing interpretation. The idea of using not only the appropriate space structure, but also to change the measure, is a powerful tool because it leads to results coherent with the interpretation of the measure of difference, and because they are mathematically more straightforward.

3 The Positive Real Line

3.1 Space Structure

The real line, with the ordinary sum and product by scalars, has a vector space structure. The ordinary inner product and the Euclidean distance are compatible with these operations. But this geometry is not suitable for the positive real line. Confront, for example, some meteorologists with two pairs of samples taken at two rain gauges, $\{5; 10\}$ and $\{100; 105\}$ in mm, and ask for the difference; quite probably, in the first case they will say there was double the total rain in the second gauge compared to the first, while in the second case they will say it rained a lot but approximately the same. They are assuming a relative measure of difference which is not compatible with the ordinary sum, as it is not invariant under translation. Furthermore, the constrained character of the sample space might lead to problems when shifting a positive number by a positive or negative real number, or when multiplying a positive number by an arbitrary real number, because results can be outside $\mathbb{R}_+$ and would be impossible.

In 1879, Galton realized that the space structure of the real line cannot be adequate on the positive real line when errors are multiplicative and not additive. In those cases, it is more adequate to use the particular vector space structure of $\mathbb{R}_+$, which is summarized as follows (Pawlowsky-Glahn and Egozcue, 2001): Given $x, y \in \mathbb{R}_+$, the internal operation, which plays an analogous role to addition in $\mathbb{R}$, is the usual product $x \oplus y = x \cdot y$ and, for $\alpha \in \mathbb{R}$, the external operation, analogous to the product by scalars in $\mathbb{R}$, is $\alpha \odot x = x^\alpha$. An inner product, compatible with $\oplus$ and $\odot$, is $\langle x, y \rangle_+ = \ln x \cdot \ln y$. It induces a norm, $\|x\|_+ = |\ln x|$, and a *relative measure of difference*, or distance, $d_+(x, y) = |\ln y - \ln x|$. Since $\mathbb{R}_+$ is a 1-dimensional vector space, there are only two orthonormal basis: the unit vector e, and its inverse element with respect to the internal operation, e^{-1}. From now on the first option is considered. Any $x \in \mathbb{R}_+$ can be expressed as $x = \ln x \odot e = e^{\ln x}$, which reveals that $\ln x$ is the coordinate of x with respect to the basis e, and the measure in $\mathbb{R}_+$ can be defined so that, for an interval $(a, b) \subset \mathbb{R}_+$, $\lambda_+(a, b) = \lambda(\ln a, \ln b) = |\ln b - \ln a|$, and then $d\lambda_+/d\lambda = 1/x$ (Mateu-Figueras, 2003; Pawlowsky-Glahn, 2003).

Observations in $\mathbb{R}_+$ are usually expressed in terms of the canonical basis 1 of $\mathbb{R}$. In fact any vector $x \in \mathbb{R}_+$ can be written as $x = x \cdot 1$. The problem is that 1 is a basis in $\mathbb{R}$ but not a basis in $\mathbb{R}_+$. It has zero length (norm) and is orthogonal to any other vector. Using the canonical basis in $\mathbb{R}_+$, we write $x = e^{\ln x}$, and the scalar coefficient of x with respect to this basis is $\ln x$. Working with an orthonormal basis, standard real analysis can be applied to the coordinates. In fact, it is easy to see that the operations defined above are equivalent to the sum and scalar product of the coordinates:

$$x \oplus y = x \cdot y = e^{\ln x} e^{\ln y} = e^{\ln x + \ln y}, \quad \alpha \odot x = x^\alpha = e^{\ln x^\alpha} = e^{\alpha \ln x}.$$

The standard inner product and Euclidean distance in $\mathbb{R}$ can be directly applied to the coordinates with respect to the canonical basis of $\mathbb{R}_+$, as

$$\begin{aligned}\langle x, y \rangle_+ &= \ln x \ln y = \langle \ln x, \ln y \rangle_{eu} \\ d_+(x, y) &= |\ln y - \ln x| = d_{eu}(\ln x, \ln y)\end{aligned} \quad (1)$$

where the subscript '*eu*' refers to standard operations in $\mathbb{R}$. This relationship also holds for the λ_+ measure, i.e. for $(a, b) \in \mathbb{R}_+$, $\lambda_+(a, b) = \lambda(\ln a, \ln b) = |\ln b - \ln a|$. Nevertheless, if results expressed in logarithms are not easy to interpret, they can be expressed in terms of the canonical basis of $\mathbb{R}$, i.e. in terms of the unit vector 1. In conclusion, standard real analysis can be applied to the coordinates, distances are preserved, and results will be coherent with the structure of the sample space.

3.2 The Lognormal Law

The lognormal distribution has long been recognized as a useful model in the evaluation of random phenomena whose distribution is positive and skew, and specially when dealing with measurements in which the random errors are multiplicative rather than additive. The history of this distribution starts in 1879, when Galton observed that the law of frequency of (additive) errors was incorrect in many groups of vital and social phenomena. He based his assertion on Fechner's law which, in its approximate and simplest form, states *sensation* = log(*stimulus*). According to this law, an error of the same magnitude in excess or in deficiency is not equally probable; therefore, he proposed the geometric mean as a measure of the most probable value instead of the arithmetic mean. This remark was followed by the memoir of McAlister (1879), who performed a mathematical investigation concluding with the lognormal distribution and proposed a practical and easy method for the treatment of a data set grouped around its geometric mean: *convert the observations into logarithms and treat the transformed data set as a series around its arithmetic mean.* Thus, the *law of frequency* is introduced as the normal density function applied to the log-transformed variable, i.e. with respect to the measure λ_+. But, in order to compute probabilities in given intervals, he defined the *law of facility*, nowadays known as the lognormal density function, with respect to the usual Lebesgue measure λ. Nowadays, only the *law of facility* is considered for the lognormal distributed variables. It is important to note that using this approach we are considering $\mathbb{R}_+$ as a subset of $\mathbb{R}$ with the induced structure. Summarizing, although the idea behind the definition of the lognormal model was to consider the relative measure of difference, this is not the case when working with the density function with respect to the Lebesgue measure.

A unified treatment of the lognormal theory is presented in Aitchison and Brown (1957) and more recent developments are compiled in Crow and Shimizu (1988). Many authors use the lognormal model from an applied point of view. It is well known that, using standard definitions, the expected value and the variance of a lognormal variable are $\exp(\mu + (1/2)\sigma^2)$, respectively $\exp(2\mu + \sigma^2)\{\exp(\sigma^2) - 1\}$. The median is $\exp(\mu)$ and the mode is $\exp(\mu - \sigma^2)$. One might ask why there is much to say about the lognormal distribution if the data analysis can be referred to the intensively studied normal distribution by taking logarithms. One of the generally accepted reasons is that parameter estimates are biased if obtained from the inverse transformation. We find in the literature an extensive number of procedures to compute consistent estimators and exact confidence intervals for the mean and the variance of a lognormal variable. In most cases, for the expected value, the geometric mean multiplied by a term expressed as an infinite series or tabulated in a set of tables is obtained. For example, Clark and Harper (2000) use Sichel's optimal estimator for the mean obtained as $\exp(\bar{y})\gamma$ where $\bar{y}$ is the arithmetic mean of the log-transformed data (or, equivalently, $\exp(\bar{y})$ is the geometric mean of the original data) and γ is a bias correction factor depending on the variance and on the size of the data set which is tabulated. A similar correction factor is used to obtain confidence intervals (Clark and Harper, 2000). The paradoxical situation is given in practical situations, where the geometric mean of the original data is used to represent the mean and even in some cases to represent the mode of a lognormal distributed variable (Herdan, 1960). But, as Crow and Shimizu (1988) advert, those affirmations cannot be justified using the lognormal theory. In fact, the reason of those paradoxical situations resides in the mixing of two methodologies: the particular vector space structure of the positive real line with a relative measure of difference, and statistical tools defined for the real line with an absolute measure of difference.

Example The Skye lavas is a 3-part compositional data set containing the chemical composition of 23 basalt specimens from the Isle of Skye in the form of percentages of the popular AFM ($A : Na_2O + K_2O$, $F : Fe_2O_3$, $M : MgO$) composition (Thompson, Esson and Duncan, 1972). For illustration purposes consider the ratio F/A. This variable is constrained to the positive real line and a relative scale appears to be sensible. Fitting a lognormal model, the estimates of the parameters obtained from the log-transformed data are $\hat{\mu} = 0.784$ and $\hat{\sigma} = 0.501$. The goodness-of-fit of the lognormal model is checked applying normality tests (Anderson-Darling, Kolmogorov-Smirnov and Ryan-Joiner) to the log-transformed data set. The p-values (greater than 0.05 in all cases) and the visual inspection of the probability plot in Fig. 1 allow us to conclude the adequacy of the lognormal model for the F/A variable. Using tables 7, 8(b) and 8(e) provided in Clark and Harper (2000), the Sichel's correction bias is applied to obtain 2.470 and (2.106, 3.103) as the optimal estimator and 90% confidence interval for the mean.

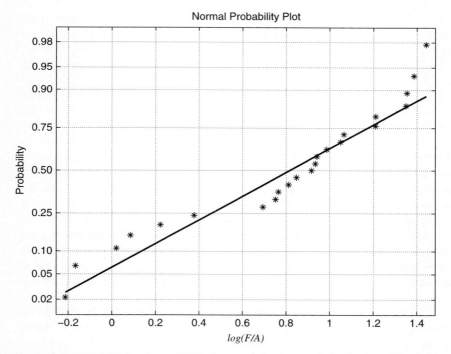

Fig. 1 Probability plot for the $log(F/A)$ data assuming a normal distribution with parameters $\hat{\mu} = 0.784$ and $\hat{\sigma} = 0.501$

3.3 The Normal on $\mathbb{R}_+$ Law

The normal on $\mathbb{R}_+$ model was introduced by Mateu-Figueras, Pawlowsky-Glahn and Martín-Fernández (2002) considering the strategy introduced in Sect. 2. The density function of orthonormal coordinates,

$$\frac{1}{\sqrt{2\pi}\sigma} \exp\left(-\frac{1}{2}\frac{(\ln x - \mu)^2}{\sigma^2}\right), \quad x \in \mathbb{R}_+, \tag{2}$$

is considered. Observe that density (2) is the usual normal density but applied to coordinates $\ln x$, therefore it is the Radon-Nikodým derivative of a probability with respect to the Lebesgue measure in the space of coordinates. Also, it can be viewed as a density in $\mathbb{R}_+$ with respect to the λ_+ measure. This density function corresponds to the *law of frequency* given by McAlister (1879). It exhibits the same properties as the standard normal distribution on the real line (Mateu-Figueras, Pawlowsky-Glahn and Martín-Fernández, 2002; Mateu-Figueras, 2003).

Applying the classical definition of expected value, median and mode to the coordinates using density (2) the parameter μ is obtained in all cases. It is the coordinate with respect to an orthonormal basis of the expected value, median and mode, and taking exponentials they are expressed as elements of $\mathbb{R}_+$,

$$\mathrm{E}_+[x] = \mathrm{Med}_+(x) = \mathrm{Mod}_+(x) = e^{\mu}.$$

The variance of any real random variable x can be understood as the expected value of the squared Euclidean distance around $\mathrm{E}[x]$, i.e. $\mathrm{Var}[x] = \mathrm{E}[d_{eu}^2(x, \mathrm{E}[x])]$. On the positive real line we have the distance d_+ thus $\mathrm{Var}_+[x] = \mathrm{E}[d_+^2(x, \mathrm{E}_+[x])]$. Using (1) we can work with coordinates and apply the classical definition of expected value, thus we obtain

$$\mathrm{Var}_+[x] = \mathrm{E}[d_+^2(x, \mathrm{E}_+[x])] = \mathrm{E}[d_{eu}^2(\ln x, \mathrm{E}[\ln x])] = \sigma^2.$$

This value cannot be interpreted as an element of the support space $\mathbb{R}_+$; thus it is only a numerical value which describes the dispersion of x. But it is a common practice to take the square root of σ^2 as a way to represent intervals centered at the mean and with radius equal to some standard deviations. In our case, to obtain an interval centered at $\mathrm{E}_+[x] = e^{\mu}$ with length $2k\sigma$, we take $(e^{\mu-k\sigma}, e^{\mu+k\sigma})$ as $d_+(e^{\mu-k\sigma}, e^{\mu+k\sigma}) = 2k\sigma$. These kinds of intervals are used in practice, see e.g. Ahrens (1954), who obtains predictive intervals in $\mathbb{R}_+$ taking exponential of predictive intervals computed from the log-transformed data under the hypothesis of normality. It can be shown that it is of minimum length, and it is also an isodensity interval thus, our distribution is *symmetric* around e^{μ}.

An important aspect of this approach is that consistent estimators and exact confidence intervals for the expected value are easily obtained, simply taking exponentials of those derived from standard normal theory and using the log-transformed data, i.e. the coordinates. In this case, given a positive data set $x_1, x_2, \ldots, x_n$, the optimal estimator for the mean of a normal on $\mathbb{R}_+$ population is the geometric mean, $(x_1 x_2 \cdots x_n)^{1/n}$, that equals $\exp(\bar{y})$, where $\bar{y}$ is the arithmetic mean of the log-transformed data. Observe that this result stands in agreement with the remark made by Galton (1879) and provides a justification for using $\exp(\bar{y})$ to represent the mean, the median and the mode of a normal on $\mathbb{R}_+$ distributed random variable.

Example. Applying the normal on $\mathbb{R}_+$ approach to the F/A ratio, the same values for the estimates of the parameters as in the lognormal case are obtained. Nevertheless, the optimal estimator and the exact 90% confidence interval for the mean are 2.191 and $(1.831, 2.622)$, much more conservative than with the classical approach. Observe that in this case the estimator 2.191 is exactly the mid point of the interval considering the d_+ distance. This is not the case in the lognormal approach using either the distance d_+ or d_{eu}.

4 The Simplex

4.1 Space Structure

Compositional data are parts of some whole which give only relative information. Typical examples are parts per unit, percentages, ppm, and the like. Their sample space is the simplex,

$$\mathscr{S}^D = \{(x_1, x_2, \ldots, x_D)' : x_1 > 0, x_2 > 0, \ldots, x_D > 0; \sum_{i=1}^{D} x_i = \kappa\},$$

where the prime stands for transpose and κ is a constant (Aitchison, 1982). For vectors of proportions which do not sum to a constant, always a fill up value can be obtained. The simplex $\mathscr{S}^D$ has a $(D-1)$-dimensional complete inner product space, i.e. Euclidean, structure (Billheimer, Guttorp and Fagan, 2001; Pawlowsky-Glahn and Egozcue, 2001). In fact, let $\mathscr{C}(\cdot)$ denote the closure operation which normalises any vector $\mathbf{x}$ to a constant sum (Aitchison, 1982), and let be $\mathbf{x}, \mathbf{x}^* \in \mathscr{S}^D$, and $\alpha \in \mathbb{R}$. Then, the inner sum, called *perturbation*, is $\mathbf{x} \oplus \mathbf{x}^* = \mathscr{C}(x_1 x_1^*, x_2 x_2^*, \ldots, x_D x_D^*)'$; the outer product, called *powering*, is $\alpha \odot \mathbf{x} = \mathscr{C}(x_1^\alpha, x_2^\alpha, \ldots, x_D^\alpha)'$; and the inner product, with associated norm and distance, is

$$\langle \mathbf{x}, \mathbf{x}^* \rangle_a = \frac{1}{D} \sum_{i<j} \ln \frac{x_i}{x_j} \ln \frac{x_i^*}{x_j^*}, \tag{3}$$

$$\|\mathbf{x}\|_a = \left(\frac{1}{D} \sum_{i<j} \left(\ln \frac{x_i}{x_j} \right)^2 \right)^{1/2}, \tag{4}$$

$$d_a(\mathbf{x}, \mathbf{x}^*) = \left(\frac{1}{D} \sum_{i<j} \left(\ln \frac{x_i}{x_j} - \ln \frac{x_i^*}{x_j^*} \right)^2 \right)^{1/2}. \tag{5}$$

The distance (5) is a *relative measure of difference* and satisfies standard properties of a distance (Martín-Fernández, Barceló-Vidal and Pawlowsky-Glahn, 1998). A natural measure on the simplex, λ_a, compatible with the space structure of $\mathscr{S}^D$, can be defined using orthonormal coordinates (Pawlowsky-Glahn, 2003). This measure is absolutely continuous with respect to the Lebesgue measure on real space, and the relationship between them is $|d\lambda_a/d\lambda| = (\sqrt{D}\, x_1 x_2 \cdots x_D)^{-1}$. The geometry here defined is known as *Aitchison geometry*, and therefore the subindex a is used.

The inner product (3) and its associated norm (4) ensure the existence of an orthonormal basis $\{\mathbf{e}_1, \mathbf{e}_2, \ldots, \mathbf{e}_{D-1}\}$, which leads to a unique expression of a composition $\mathbf{x}$ as a linear combination,

$$\mathbf{x} = (\langle \mathbf{x}, \mathbf{e}_1 \rangle_a \odot \mathbf{e}_1) \oplus (\langle \mathbf{x}, \mathbf{e}_2 \rangle_a \odot \mathbf{e}_2) \oplus \ldots \oplus (\langle \mathbf{x}, \mathbf{e}_{D-1} \rangle_a \odot \mathbf{e}_{D-1}).$$

If $h(\mathbf{x})$ denotes the vector of coordinates, then

$$h(\mathbf{x}^* \oplus (\alpha \odot \mathbf{x})) = h(\mathbf{x}^*) + \alpha \cdot h(\mathbf{x}).$$

Furthermore, the inner product and the distance in $\mathscr{S}^D$ can be computed from the coordinates, as

$$\langle \mathbf{x}, \mathbf{x}^* \rangle_a = \langle h(\mathbf{x}), h(\mathbf{x}^*) \rangle_{eu} \quad \text{and} \quad d_a(\mathbf{x}, \mathbf{x}^*) = d_{eu}(h(\mathbf{x}), h(\mathbf{x}^*)).$$

This relationship also holds for the measure λ_a and the Lebesgue measure λ, i.e. for $A \subset \mathscr{S}^D$, if $h(A)$ denotes the subset of $\mathbb{R}^{D-1}$ characterising A in terms of the coordinates of the elements, then $\lambda_a(A) = \lambda(h(A))$. This means that standard real analysis can be applied to the coordinates.

Like in every inner product space, the orthonormal basis is not unique. It is not straightforward to determine which one is the most appropriate to solve a specific problem, but a promising strategy, based on sequential binary partitions, has been developed by Egozcue and Pawlowsky-Glahn (2005). Another possibility is to consider the specific basis related to the isometric logratio transformation given by Egozcue et al. (2003), whose coordinates are commonly denoted as ilr($\mathbf{x}$).

The particular Euclidean space structure of $\mathscr{S}^D$ is important when working with random compositions, and this is also true for probability laws on $\mathscr{S}^D$. But traditionally the simplex has been considered as a subset of real space and, consequently, some probability laws, like the Dirichlet family, have been defined using the standard approach. In other cases, the inner vector space structure has been considered to define some models, but finally, as in the lognormal case, the expressions of the density with respect to the Lebesgue measure in real space have been considered. This is the case for families of distributions like the logistic normal (Aitchison, 1982, 1986), the logistic skew-normal (Mateu-Figueras, Pawlowsky-Glahn and Barceló-Vidal, 2005), or those defined using the Box-Cox family of transformations (Barceló-Vidal, 1996).

Finally, it is important to note that the Aitchison geometry and the theoretical developments here presented assume no null values for all the components. Nevertheless, in practical modelling, compositional observations can have null values or present values below to detection limits. A detailed description of available methods on how to deal with those situations is beyond the scope of this paper. Martín-Fernández, Barceló-Vidal and Pawlowsky-Glahn (2003) propose a multiplicative zero replacement for rounded zeros coherent with the Aitchison geometry. Also, some new developments based on Markov chain Monte Carlo simulation algorithms have been recently introduced by Palarea-Albaladejo, Martín-Fernández and Gómez-García (2007).

4.2 The Logistic Normal Model

Aitchison (1982) defines the logistic normal law on the simplex. The strategy is standard: transform the random composition from the simplex to the real space, define the density function of the transformed vector and finally go back to the simplex using the change of variable theorem. The result is a density function for the initial random composition with respect to the Lebesgue measure. The logistic normal model was initially defined using the additive logratio transformation (Aitchison, 1986). Using the matrix relationship among the logratio transformations stated by Egozcue et al. (2003), the density function in terms of the isometric logratio transformation can be easily obtained,

$$\frac{(2\pi)^{-(D-1)/2} |\Sigma|^{-1/2}}{\sqrt{D} x_1 x_2 \cdots x_D} \exp\left[-\frac{1}{2} (\mathrm{ilr}(\mathbf{x}) - \mu)' \Sigma^{-1} (\mathrm{ilr}(\mathbf{x}) - \mu)\right]. \qquad (6)$$

The logistic normal model exhibits some interesting and important properties (Aitchison, 1986, Chap. 6). It is a closed family under perturbation, powering, subcompositions and permutation of the parts. The density (6) is a classical density, consequently any moment is computed using the standard definition. Aitchison (1986, p. 116) adverts that the integral expressions of all moments, and in particular of the expected value, are not reducible to any simple form, and that numerical procedures have to be applied to compute them.

Example Fitting the logistic normal model to the Skye lavas gives

$$\hat{\mu} = (0.555, 0.639)' \quad \text{and} \quad \widehat{\Sigma} = \begin{pmatrix} 0.126 & -0.229 \\ -0.229 & 0.456 \end{pmatrix}$$

as the estimates of the parameters obtained from the ilr-transformed data set. Following Aitchison (1986, p. 144), we test the goodness-of-fit of the model applying a battery of 12 tests, based on the Anderson-Darling, Cramér-von Mises and Watson statistics, to the ilr-coordinates of the data. In particular, the tests are applied to the marginal distributions, to the bivariate angle and to the radius. Taking a 1 per cent significance level only one of the marginal tests gives evidence of any departure from normality. Thus, the fit with a logistic normal model seems quite reasonable. In Fig. 2(A) the isodensity curves of the fitted model are represented with the original data.

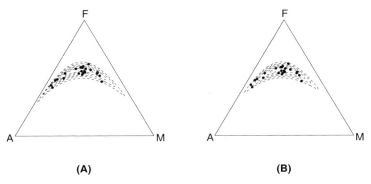

Fig. 2 Skye lavas data set and isodensity curves of the fitted (**A**) logistic normal model and (**B**) normal model on $\mathscr{S}^3$

4.3 The Normal on $\mathscr{S}^D$ Model

Using the algebraic-geometric structure of the simplex and considering the natural measure λ_a, the normal distribution on $\mathscr{S}^D$ is defined through the density function of generic orthonormal coordinates (Mateu-Figueras, 2003). In order to compare it with the logistic normal model, here the ilr-coordinates are considered,

$$(2\pi)^{-(D-1)/2}\left|\sum\right|^{-1/2}\exp\left[-\frac{1}{2}(\mathrm{ilr}(\mathbf{x})-\mu)'\sum\nolimits^{-1}(\mathrm{ilr}(\mathbf{x})-\mu)\right], \quad \mathbf{x}\in\mathscr{S}^D. \quad (7)$$

Observe that Eq. (7) is the usual normal density applied to the ilr-coordinates; therefore, it is a density with respect to the Lebesgue measure in the space of coordinates $\mathbb{R}^{D-1}$. Also, it is a density function on $\mathscr{S}^D$ with respect to the λ_a measure.

As for the logistic normal model, it is closed under perturbation and powering, but in this case it is also invariant under perturbation. This is important, e.g. when centering compositional data (Martín-Fernández et al., 1999), to guarantee that all properties of the original density are preserved.

Applying the classical definition of expected value to the ilr-coordinates using density (7) the parameter μ is obtained. As in the univariate case, it is the vector of ilr-coordinates of the expected value i.e. $\mathrm{E}[\mathrm{ilr}(\mathbf{x})]$. The $\mathrm{E}_a[\mathbf{x}]$ composition is obtained by means of a linear combination. It can be shown that $\mathrm{E}_a[\mathbf{x}] = \mathrm{cen}[\mathbf{x}]$ (Mateu-Figueras, 2003), where $\mathrm{cen}[\mathbf{x}]$ is the center of a random composition, which was introduced as an alternative to the expected value with the argument that it is more representative (Aitchison, 1986). Thus, using the normal on $\mathscr{S}^D$ approach, the expected value with respect to λ_a can be used, and it is not necessary to define another measure of location.

Pawlowsky-Glahn and Egozcue (2002) interpret the variance of a random composition as the expected value of the squared Aitchison distance, d_a, around its expected value and the metric variance, denoted as $\mathrm{Mvar}[\mathbf{x}]$ is obtained. This measure of dispersion is equal to the measure of total variability of a random composition defined by Aitchison (1997). Considering a normal in $\mathscr{S}^D$ model with parameters μ and Σ and using the ilr-coordinates we obtain $\text{Mvar}[\mathbf{x}] = \text{trace}(\Sigma)$. Observe that here we are only interested in a measure of total variability. These kind of measures are often used, e.g., in principal component analysis or in biplots. The matrix Σ is a usual experimental matrix of covariances and can be used for the interpretation of the relationships between logratios following standard rules. A direct interpretation of relations between components is seldom possible. Interested readers are referred to Aitchison (1997) for a discussion about measures of dispersion obtained from different covariance matrices of logratios.

Example Fitting the normal on $\mathscr{S}^3$ model to the Skye lavas data set we obtain the same estimates of the parameters as for the logistic normal model expressed in terms of the ilr-transformed composition. The results in testing the goodness-of-fit are also the same. The isodensity curves of the fitted normal in $\mathscr{S}^3$ are represented in Fig. 2(B). Comparing it with the isodensity curves of the fitted logistic normal in Fig. 2(A) only slight differences are observed.

As both models are closed under perturbation and powering, for illustration purposes a linear transformation in $\mathscr{S}^3$, $\mathbf{a} \oplus (b \odot \mathbf{X})$, is applied to the data, with $\mathbf{a} = g(\mathbf{X})^{-1}$ and $b = \sqrt{3}$. Note that the geometric mean of the resulting data set is the barycenter of the simplex, i.e. $(1/3, 1/3, 1/3)$, because b modifies only the variability, but $\mathbf{a}$ centers the data. This is equivalent to translate the transformed data set, or the coordinates with respect to an orthonormal basis, to the origin of coordinates in real space. For both models the estimates of the parameters are

$$\hat{\mu} = (0.000, 0.000)' \quad \text{and} \quad \widehat{\Sigma} = \begin{pmatrix} 0.377 & -0.688 \\ -0.688 & 1.369 \end{pmatrix}.$$

In Fig. 3(A) and 3(B) the logistic normal and the normal in $\mathscr{S}^3$ fitted models are represented. As can be observed, the same linear transformation leads to a better visualisation of the normal on $\mathscr{S}^3$ fitted model, but in the logistic normal case a completely different model, with two modes, is obtained. In other words, perturbation and powering, which should only move the centre of the density and modify the variability, can generate arbitrary modes, an undesirable property. In Fig. 4 the corresponding normal densities fitted to the ilr-coordinates or, equivalently, to

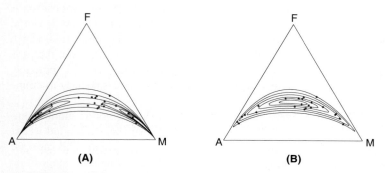

Fig. 3 Linear transformed Skye lavas data set and isodensity curves of the fitted (**A**) logistic normal model and (**B**) normal on $\mathscr{S}^3$

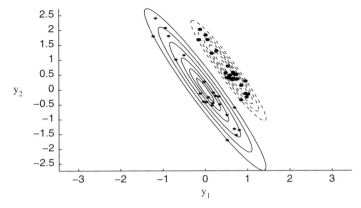

Fig. 4 Ilr coordinates of the Skye lavas data set and fitted normal model to the original data (*dashed line*) and to the linear transformed data (*continuous line*)

the ilr-transformed data set, are represented, because the same graphic is obtained using both methodologies. It is clear that the linear transformation only increases the variability and translates the data to the origin of coordinates.

5 Conclusions

Working with a particular space structure and a compatible measure when dealing with some constrained sample spaces appears not only to be possible, but also desirable. It is mathematically straightforward and, more importantly, it enhances interpretation, as the natural measure of difference can be used to understand variability in data.

Acknowledgments This work has been supported by the Spanish Ministry of Education and Science under project 'Ingenio Mathematica (i-MATH)' No. CSD2006-00032 (Consolider – Ingenio 2010) and under project MTM2006-03040.

References

Agterberg F (2005) High-and low-value tails of frequency distributions in geochemistry and mineral resource evaluation. Geophys Res Abstr vol 7, 03405. SRef-ID: 1607-7962/gra/EGU05-A-03405

Ahrens LH (1953) A fundamental law of geochemistry. Nature 172: 1148

Ahrens LH (1954) The lognormal distribution of the elements. Geochimica et Cosmochimica Acta 5: 49–73

Aitchison J (1982) The statistical analysis of compositional data (with discussion). J R Stat Soc B (Statistical Methodology) 44(2): 139–177

Aitchison J (1986) The statistical analysis of compositional data. Chapman & Hall Ltd., London. Reprinted in 2003 with additional material by The Blackburn Press, p 416

Aitchison J (1997) The one-hour course in compositional data analysis or compositional data analysis is simple. In: Pawlowsky-Glahn V (ed) Proceedings of IAMG'97. International Center for Numerical Methods in Engineering (CIMNE), Barcelona, pp 3–35

Aitchison J, Brown JAC (1957) The lognormal distribution. Cambridge University Press. Cambridge, p 176

Barceló-Vidal C (1996) Mixturas de Datos Composicionales. PhD, Universitat Politècnica de Catalunya, Barcelona, p 261

Billheimer D, Guttorp P, Fagan W (2001) Statistical interpretation of species composition. J Am Stat Assoc 96(456): 1205–1214

Clark I, Harper WV (2000) Practical geostatistics 2000. Ecosse North America Llc., Columbus OH, p 342

Crow EL, Shimizu K (1988) Lognormal distributions. Theory and Applications. Marcel Dekker, Inc. New York, p 387

Eaton ML (1983) Multivariate statistics. A vector space approach. John Wiley & Sons, New York, p 512

Egozcue JJ, Pawlowsky-Glahn V (2005) Groups of parts and their balances in compositional data analysis. Math Geol 37(7): 795–828

Egozcue JJ, Pawlowsky-Glahn V, Mateu-Figueras G, Barceló-Vidal C (2003) Isometric logratio transformations for compositional data analysis. Math Geol 35(3): 279–300

Galton F (1879) The geometric mean, in vital and social statistics. Proc R Soc Lond 29: 365–366

Herdan G (1960) Small particle statistics. Butterwoths, London

Martín-Fernández JA, Barceló-Vidal C, Pawlowsky-Glahn V (1998) A critical approach to non-parametric classification of compositional data. In: Rizzi A, Vichi M, Bock HH (eds) Advances in data science and classification (Proceedings of the IFCS'98). Springer-Verlag, Berlin, pp 49–56

Martín-Fernández JA, Bren M, Barceló-Vidal C, Pawlowsky-Glahn V (1999) A measure of difference for compositional data based on measures of divergence. In: Lippard SJ, Næss A, Sinding-Larsen R (eds) Proceedings of IAMG'99, Tapir Trondheim, pp 211–216

Martín-Fernández JA, Barceló-Vidal C, Pawlowsky-Glahn V (2003) Dealing with zeros and missing values in compositional data sets. Math Geol 35(3): 253–278

Mateu-Figueras G (2003) Models de distribució sobre el símplex: PhD, Universitat Politècnica de Catalunya, Barcelona, p 202

Mateu-Figueras G, Pawlowsky-Glahn V, Martín-Fernández JA (2002) Normal in R+ vs lognormal in R. In: Burger H, Wolfdietrich S (eds) Terra nostra, Proceedings of IAMG'02, vol 3, pp 305–310

Mateu-Figueras G, Pawlowsky-Glahn V, Barceló-Vidal C (2005) The additive logistic skew-normal distribution on the simplex. Stoch Environ Res Risk Assess (SERRA) 19(3): 205–214

McAlister D (1879) The law of geometric mean. Proc R Soc Lond 29: 367–376

Palarea-Albaladejo J, Martín-Fernández JA, Gómez-García JA (2007) Parametric approach for dealing with compositional rounded zeros. Math Geol 39(7): 625–645

Pawlowsky-Glahn V (2003) Statistical modelling on coordinates. In: Thió-Henestrosa S, Martín-Fernández JA (eds) Proceedings of CoDaWork'03. Universitat de Girona, CD-ROM

Pawlowsky-Glahn V, Egozcue JJ (2001) Geometric approach to statistical analysis on the simplex. Stoch Environ Res Risk Assess (SERRA) 15(5): 384–398

Pawlowsky-Glahn V, Egozcue JJ (2002)BLU estimators and compositional data. Math Geol 34(3): 259–274

Pearson K (1897) Mathematical contributions to the theory of evolution. On a form of spurious correlation which may arise when indices are used in the measurement of organs. Proc R Soc Lond LX: 489–502

Thompson RN, Esson J, Duncan AC (1972) Major element chemical variation in the Eocene lavas of the Isle of Skye, Scotland, J. Petrology, 13: 219–253. Cited in Aitchison, 1986

Investigation of the Structure of Geological Process Through Multivariate Statistical Analysis—The Creation of a Coal

Lawrence J. Drew, Eric C. Grunsky and John H. Schuenemeyer

Reprinted from *Mathematical Geosciences* DOI: 10.1007/s11004-008-9176-2, when citing this article please use the DOI number.

Abstract The purpose of this study was to capture the structure of a geological process within a multivariate statistical framework by using geological data generated by that process and, where applicable, by associated processes. It is important to the practitioners of statistical analysis in geology to determine the degree to which the geological process can be captured and explained by multivariate analysis by using sample data (for example, chemical analyses) taken from the geological entity created by that process. The process chosen for study here is the creation of a coal deposit.

In this study, the data are chemical analyses, expressed in weight percentage and parts per million, and therefore are subject to the affects of the constant sum phenomenon. The data array is the chemical composition of the whole coal. This restriction on the data imposed by the constant sum phenomenon was removed by using the centered logratio (clr) transformation. The use of scatter plots and principal component biplots applied to the raw and centered logratio (clr) transformed data arrays affects the interpretation and comprehension of the geological process of coalification.

Keywords Coal · compositional data · multivariate statistics · closure · logratio transform

---

Lawrence J. Drew
U. S. Geological Survey, Mail Stop 954, 12201 Sunrise Valley Drive, Reston, VA 20192, USA, e-mail: drew@usgs.gov

Eric C. Grunsky
Geological Survey of Canada, Natural Resources Canada, Rm. 607, 615 Booth St. Ottawa ON K1A 0E9, Canada, e-mail: egrunsky@nrcan.gc.ca

John H. Schuenemeyer
Southwest Statistical Consulting, LLC, 960 Sligo St., Cortez, CO 81321 USA, e-mail: jackswsc@charter.net

1 Introduction

Coal geologists and geochemists have exerted considerable scientific effort to understand the transformation of peat into coal (Cecil et al., 1985; Schweinfurth, 2003). Compositional data of the Upper Freeport coal, Indiana County, Pennsylvania, are used herein where we attempted to isolate the geological and geochemical processes in the coalification processes by using data and statistical analysis applied to the geochemistry of coal. The environment of the peat that became the Upper Freeport coal is believed have been a planar bog. In this type of bog, the thickness of the peat is controlled by the configuration of the basin and the rate of subsidence. Mineral matter, commonly referred to as ash, usually occurs in the form of silica, kaolinite, and illite. These minerals are often formed by authogenic growth (the source of the SiO_2, K_2O, and Al_2O_3 being the vegetal material that contributed to the formation of the peat) as well as by wind-blown and suspended particles and additionally by ions in solution in ground and surface water (Cecil and others, 1985). The proportion of ash in the Upper Freeport coal ranges from 5.7 to 29.1 percent. The average ash content is 14.9 percent.

Sulfur, ranging from 0.7 to 4.3 percent, occurs in the Upper Freeport coal for the most part in iron pyrite. Pyritic sulfur in coal is most often thought to have been derived from the bacterial reduction of the sulfate radical (SO_4^{+2}) to sulfide and then fixed with iron, which is in abundance in the peat environment. The source of the sulfate is usually believed to have been sea water, and the source of the iron is the shales, which are common in the stratigraphy within the peat bogs. Specific Eh and pH conditions indicate that the iron, as well, has been precipitated from iron-rich groundwater that subsequently circulated throughout the sequence and was deposited. Because there are no marine facies rocks in the stratigraphic sequence, a marine source for sulfate does not seem likely for the sulfur in the Upper Freeport coal. Cecil and others (1985) appealed to variation in the pH within the peat-forming environment to explain the variation in the sulfur content of the Upper Freeport coal. They argued that the abundance of fresh water limestone in the stratigraphic section provided a means for reducing the acidity within the peat-forming environment so that bacterial action could take place. The abundant sulfate in the ground and surface waters in the environment could then be reduced to sulfides. During this process, elements such as arsenic (As) were also precipitated in the iron pyrite.

In addition to isolating the geochemical processes in coal formation by using multivariate statistical analysis, we also were curious to determine if aspects of the consequences of the analytical chemical process could be isolated, the main process being the creation of anhydrite during the ashing of the coal. Before the inorganic chemistry of the coal is determined, it is ashed at 525°C. During this process, a small fraction of the total sulfur in the coal that is vaporized interacts with the available calcium and manganese to form anhydrite. The actual sulfur content of the coal is determined by a wet chemical analysis that was carried out before ashing. The sulfur captured by the available calcium and manganese is only a small fraction of the total sulfur. Another determination made in this study was the advantage gained in understanding geologic processes by interpreting bivariate and multivariate relations in the data, which is a set of proportions or percentages. By their very nature, these data are constrained, and, as such, are generally biased toward producing artificial negative associations (Chayes, 1971). The consequences of such spurious correlations have been known for over 100 years and have been considered a serious problem in geochemistry for more than 50 years (Pearson, 1897; Chayes, 1949; Miesch, 1969; Aitchison, 1986). Here the authors have chosen to replace the older terminology of "closed" and "open" used by Chayes (1971) and Aitchison (1986) with the "raw" data and "centered logratio (clr) transformed" data. The term "clr" means the centered logratio transformed data as it is commonly used by those who study compositional data (i.e. by Aitchison (2003) and (Pawlowsky and Ecozcue, 2006). Although spurious correlations in compositional data analysis has been studied over a long period by statisticians and statistically minded geologists, the significance of this problem often has been overlooked (Miesch, 1969; Rollinson, 1993). Recently, the centered logratio (clr) transformation has been developed to correct this situation (Aitchison, 1986, 1997, 1999). The usefulness of this approach has been confirmed by many other studies (Egozcue et al., 2003; Aitchison and Egozcue, 2005). Buccianti et al. (2006) provided a comprehensive discussion of the theory and practical aspects of compositional data analysis. The authors are aware that the covariance matrix for centered logratio (clr) transformed data is singular. This condition, however, does not preclude the use of appropriate statistical methodologies for the evaluation and interpretation of the data.

In addition, we want to illustrate the changes that occur in statistics such are correlation coefficients when different subcompositions are used. The compositional data for the Upper Freeport coal are presented for four subcompositions of the total coal. The first contains six variables: carbon, hydrogen, nitrogen, oxygen, sulfur, and ash. The second is composed of these six with the addition of three volatile trace elements: arsenic, selenium, and mercury. Both of these subcompositions are determined at room temperature for the organic, sulfur and volatile trace elements where as the concentration of the ash is determine in an ashing process carried out at 525°C. The second subcomposition with nine elements constitute the total coal. Two additional subcompositions using 10 and 25 variables are also analyzed.

Our purpose here is to make a simple statistical-graphical exposition that is easily understandable to the large group of geologists who do not yet understand the problem. The authors want to expose the nature of the spurious correlations in compositional data and the effects caused by changes the number of elements in a subcomposition. We believe that the reader should have this topic illustrated using real data. We will see that the effect of using the centered logratio (clr) transformation can be very large.

2 Geologic Analysis

The geology of the Upper Freeport coal has been studied intensely by Cecil and others (1978, 1979, 1981, 1985) who did detailed sampling. We focus on a subset of their data collected from two adjacent coal mines in Indiana County, Pennsylvania.

Geochemical analysis was carried out in different steps.

1) Analysis for H, C, N, O, sulfur, Sulfate, pyritic sulfur, organic sulfur, ash (defined as the variable StdAsh in the data tables)
2) Individual analysis of Se, Hg, and As
3) Analysis of the StdAsh for 59 elements:, Si, Al, Ca, Mg, Na, K, Fe, Ti, Ag, Au, B, Ba, Be, Bi, Br, Cd, Ce, Co, Cr, Cs, Cu, Dy, Er, Eu, F, Ga, Gd, Ge, Hf, Ho, La, Li, Lu, Mn, Mo, Nb, Nd, Ni, P, Pb, Pd, Pr, Rb, Sb, Sc, Sm, Sn, Sr, Ta, Tb, Th,Tm, U, V, W, Y, Yb, Zn, and Zr

A description of the analytical procedures has been provided by Cecil and others, (1978, 1979, and 1981); ASTM (2001). In this article we refer to the entire composition as constituted by C, H, O, S, N, ash, Se, As and Hg. Note, however, that the ash is also composed of the 59 elements listed above and the term is used to represent them as well. In this study we have chosen to work with various subcompositions of the data—for example, a five element plus ash subcomposition composed of C, H, N, O, S, and ash, as well as the entire composition of C, H, N, O, S, Se, As, Hg and ash. We have also substituted for the ash, 24 of the elements for which the ash was analyzed.

Cecil and others (1978, 1979, 1981, and 1985) illustrated in several diagrams the major elements involved in the formation of peat. The Upper Freeport system is believed to be a planar system with occasional "peat islands." These islands were essential structures in the Upper Freeport bog(s) in that they shed inorganic material, thereby concentrating it laterally from these islands. This fact can be demonstrated by plotting the ash content and the various inorganic elements such as total potassium, silicon, and cerium present in the 50 samples for which a total analysis of the coal is available (Northern and Central Appalachian Basin Coal Regions Assessment Team, 2001). These inorganic substances entered the bog as windblown and suspended particulate material. Also, there was substantial inorganic material in the vegetal mass, silicon being particularly common. An example of a peat island as defined by the spatial distribution of cerium is shown in Fig. 1. This cerium is presumably contained in minerals such as monazite. The location of this peat island is also defined by the spatial distribution of the ash content in the coal (Fig. 2).

The simplest manner by which a coal can be characterized is to list its three main components: (1) a subcomposition of the organic content as represented by its carbon and/or hydrogen content and usually expressed by its BTU content, (2) its ash content, and (3) its sulfur content. We will first examine the sulfur content.

The amount of sulfur in coal is important because of its role as air and water contaminants. Here we are concerned about its statistical properties as an element in the closed array of chemical components in the Upper Freeport coal. The concentration of sulfur in our samples ranged from 0.7 to 4.3 percent, the mean concentration being 2.5 percent. The next question we asked was how much of the total sulfur was contained in iron pyrite and how much was contained in organic sulfur. The sulfur in the iron pyrite is thought to have been precipitated by the bacterial reduction of the sulfate radical (SO_4^{+2}) to sulfide and then fixed with iron, which is abundant in the peat environment. The organic sulfur, however, was originally part of the vegetal mass. Perhaps the best way to answer our question about total sulfur is to inspect the

Investigation of the Structure of Geological Process

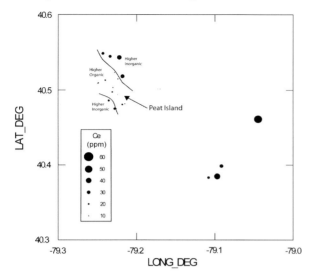

Fig. 1 Map of the the cerium (Ce in ppm) in the Upper Freeport coal in Indiana County, Pennsylvania

scatterplot shown in Fig. 3. There we see a strong linear relation between total sulfur and pyritic sulfur ($r = 0.976$). Furthermore, as this linear relation implies, we can see along the range in total sulfur content a constant difference of about 0.6 percent between the two variables. This relation is important in the subsequent multivariate analyses.

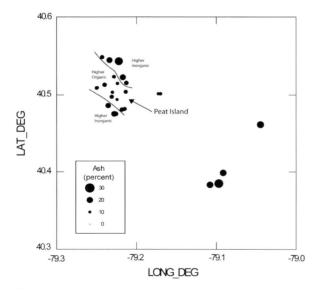

Fig. 2 Map of the the ash (in percent) in the Upper Freeport coal in Indiana County, Pennsylvania

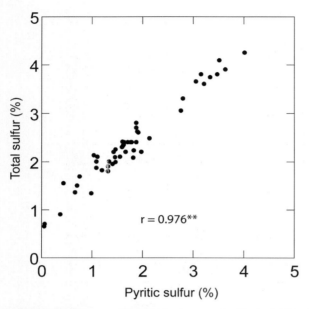

Fig. 3 Scatterdiagram of the pyritic versus total sulfur in the Upper Freeport coal, Indiana County, Pennsylvania. Nonsignificant correlation(ns), significant at the 5-percent level (*), significant at the 1-percent level (**). Sample size, n = 50

The "Scatter**plot** **M**atrix" (SPLOM) is a useful summary display (Fig. 4). Here we can examine in the bivariate relations the six major components of the nine components within the whole coal for the raw data array of analyses for the Upper Freeport coal. The three components not shown are arsenic (As), mercury (Hg), and selenium (Se), which collectively constitute a very small fraction of the coal, but are important elements in the quality of any coal. The equivalent SPLOM for the centered logratio (clr) transformation is shown in Fig. 5. By comparing companion scatterplots in Figs. 4 and 5, we can get a feeling for the concern expressed by Chayes (1971). For example, the relation between the carbon content and the ash content in the Upper Freeport coal, Indiana County, Pennsylvania, is markedly different in the raw data and in the centered logratio (clr) transformed arrays (Figs. 4 and 5). The relation in the raw data shows a nearly perfect linear relation with a correlation coefficient of r = −0.977 (Fig. 6A), whereas this relation in the centered logratio (clr) transformed data, although still linear, has a much smaller correlation coefficient of r = −0.679 (Fig. 6B). We can therefore conclude that about half of the variance explained (49.3 percent) in the relation for the centered logratio (clr) transformed array $[R^2 = (-0.977^2) - (-0.679^2) = 0.493)]$ has been induced by the restraint of the constant sum problem of compositional data. To a large degree, this result could have been anticipated because the sum of the carbon and ash contents is about 90 percent of this coal.

Opening a raw compositional array can produce a variety of changes in relations between variables. For example, in the case of the relation between sulfur and

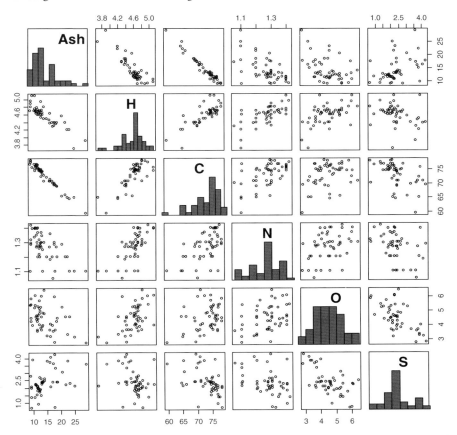

Fig. 4 Scatterplot matrix of the subcompostion of six major components (raw data array) of the Upper Freeport coal, Indiana County, Pennsylvania. Sample size, n = 50

oxygen (Fig. 7), the linear relation in the centered logratio (clr) transformed array is stronger (r = −0.771) than it is in the raw array (r = −0.582). Thus, we get opposing effects from the centered logratio (clr) transformation between the carbon versus ash relation and between the sulfur versus oxygen relation. Miesch (1969) warns that all compositional data are proportions with very complex denominators determined by many relations in the data. Here we have a glimpse into the validity of his warning.

A third bivariate relation—between the carbon content and the nitrogen content—was examined to illustrate another aspect of spurious correlation and the consequence of the use of the centered logratio (clr) transformation (Fig. 8). In this relation for the raw data array, the data plot as a triangle (Fig. 8A). In the centered logratio (clr) transformed data, the data plot in a form that is more linear in appearance (Fig. 8B). The quantized nature of the nitrogen analyses is obvious in the scatterplot shown Fig. 8A and to some degree in the histogram (Fig. 8C). The quantized nature

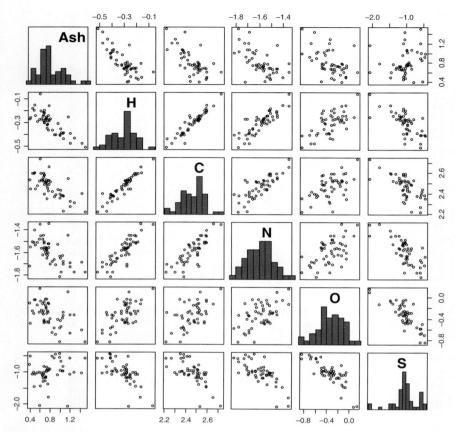

Fig. 5 Scatterplot matrix of the subcomposition of six major components (clr- transformed array) of the Upper Freeport coal, Indiana County, Pennsylvania. Sample size, n = 50

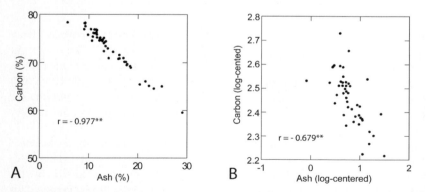

Fig. 6 Comparison of the relation between the carbon and ash content in the Upper Freeport coal. (**A**), raw array, and (**B**), the clr-transformed array. Nonsignificant correlation(ns), significant at the 5-percent level (*), significant at the 1-percent level (**). Sample size, n = 50. Based on a subcomposition of ash, hydrogen, carbon, nitrogen, oxygen, and sulfur

Investigation of the Structure of Geological Process

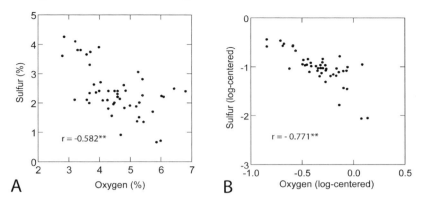

Fig. 7 Comparson of the relation between the sulfur and oxygen content in the Upper Freeport coal. (**A**), raw data array, and (**B**), the clr-transformed array. Nonsignificant correlation(ns), significant at the 5-percent level (*), significant at the 1-percent level (**). Sample size, n = 50. Based on a subcomposition of ash, hydrogen, carbon, nitrogen, oxygen, and sulfur

of the data disappear after log-centering (Figs. 8B, 8D), which is due to the division of the quantized value by the geometric mean for each composition. Therefore, transforming the data array and rescaling into logarithmic space changes the relation between the nitrogen and carbon content. Even though the nitrogen content makes

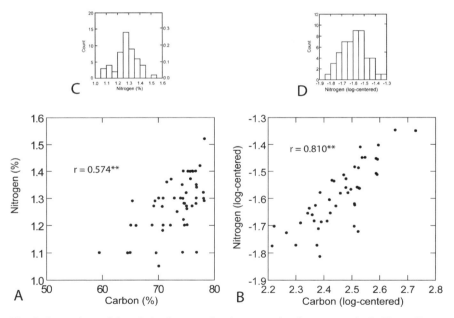

Fig. 8 Comparison of the relation between the nitrogen and carbon content in the Upper Freeport coal. (**A**), raw data array, and (**B**), the clr-transformed array. Nonsignificant correlation(ns), significant at the 5-percent level (*), significant at the 1-percent level (**). Sample size, n = 50. Based on a subcomposition of ash, hydrogen, carbon, nitrogen, oxygen, and sulfur

up at most 1.52 percent of the coal, the effect of the restraints in the compositional data is present in this relation (Fig. 8A) because the carbon content ranges from 59 to 78 percent. Comparing the relations between nitrogen and the other variables in the SPLOM diagrams demonstrates that the restraints in compositional data exists across all relations, not just for variables that make up a large part of the coal (see row 4 in Figs. 4 and 5). We are again reminded of the admonitions of Miesch (1969), Chayes (1971), and Aitchison (1997), who warn us not to ignore the characteristics of the sample space that contains our basic data.

Aitchison (1986) introduced the concept of the variation array where the upper triangle of the array expresses the covariance between any pair of variables on the basis of their log ratios and the lower triangle expresses the mean values for each of these log-ratio pairs. Here, an important aspect of assessing compositions is the calculation of a measure of variability. This calculation is done by creating a variation matrix, composed of a measure of variation of the ratios and the mean values of the ratios.

T is defined by:

$$\tau_{ij} = \text{var}\{\log(x_i/x_j)\}(i=1,\ldots,d; j=i+1,\ldots,D)$$

and the mean, E, is expressed as:

$$\xi_{ij} = E\{\log(x_i/x_j)\}(i=1,\ldots,d; j=i+1,\ldots,D)$$

The variation array T summarizes the contribution that each pair of variables makes in a subcompositional analysis. Table 1a consists of two arrays, an upper and a lower. The upper one shows means and the variation for the set of data used to create Fig. 5. Examination of the upper array of Table a reveals some useful features about the composition which is restricted to the organic, sulfur, and ash elements. The upper triangle of Table 1a shows that ash and organic elements (C, H, N, and O) are more variable, τ values all being 0.13 or greater, whereas, the τ values among the organic elements show very low values (0.04 or less), indicating that the organic elements are much less variable with respect to one another. Similarly, S and the organics and ash pairs, having τ values greater than 0.16, are more variable than

Table 1a Variation matrix for coal geochemistry (organic + sulfur + ash subcomposition)

	Ash	H	C	N	O	S	Upper triangle variability
Ash	0.00	0.13	0.13	0.13	0.20	0.16	0.75
H	1.08	0.00	0.00	0.00	0.03	0.17	0.20
C	−1.68	−2.76	0.00	0.00	0.04	0.18	0.22
N	2.37	1.29	4.05	0.00	0.03	0.18	0.22
O	1.11	0.03	2.79	−1.26	0.00	0.29	0.29
S	1.81	0.73	3.49	−0.56	0.70	0.00	
Total variability							1.67

Upper triangle – variability of the ratios.
Lower triangle – mean values of the ratios.

Investigation of the Structure of Geological Process 63

Table 1b Percent variation matrix for coal geochemistry (organic + sulfur + ash subcomposition)

	H	C	N	O	S	Upper triangle percent variability
Ash	7.63	7.71	7.69	12.00	9.69	44.73
H		0.09	0.19	1.90	9.98	12.16
C			0.27	2.18	10.50	12.96
N				2.06	10.88	12.94
O					17.21	17.21
Total variability						100.00

Upper triangle – variability of the ratios.
Lower triangle – mean values of the ratios.

the ash and organic pairs of variables. The low values of τ for the organic pairs of variables suggest that stoichiometry may play a significant deterministic role in controlling the variability of the organic elements. This topic will be revisited in the discussion of the principal component bi-plots.

The lower triangle of the variation matrix, which contains the mean values (ξ) of the ratios. These values summarize the relative magnitudes of the ratios, which also assist with the interpretation of the variance of the ratios. Positive values of ξ mean that the ratio of the average value of the numerator exceed that of the denominator. Negative values indicate that the average value of the denominator exceeds the average value of the numerator (ratio < 1). Ratios that exceed 1 for ash (ash/H, ash/N, ash/O, ash/S) confirm that the values of ash exceed the values of N, O and S. A large negative value for H/C (-2.76) indicates that the range of values for H is much less than that of C. The ξ value for C/N is 4.05, also indicating a large difference in the magnitudes of the two elements. Ratios that show nearly equal ranges include ash/O, H/S and O/S.

If we wish to know how much variability each ratio contributes to the overall variation of the data, we can express it in percentages, as shown in the upper triangle of Table 1b. Each upper triangular ratio is expressed as a percentage of the sum of all the variances shown in the upper triangular portion of the variation array. From these values, we can see that sulfur paired with the other elements (τ values in percent ranging from 9.69 to 17.21) accounts for more of the variability in the data, followed by the ash pairs (τ values ranging from 7.63 to 12.00). The organic elements (τ values ranging from 0.09 to 2.18) do not vary as much as the S and ash paired with the other elements. The variation matrix will be used below in connection with our analysis of the processes involved in creating a coal.

3 Statistical Behavior of Arsenic and Mercury

The statistical behavior of As and Hg is examined to illustrate two rather different correlation structures in coal that have significant effects on environmental pollution. The correlation structure for Se is not treated for the purposes of brevity. The correlation structure involving As shows a weak yet significant positive correlation with sulfur in the raw data array (Fig. 9F) and small and significant negative correlations with hydrogen, nitrogen, and oxygen (Figs. 9B,D, and E). From these

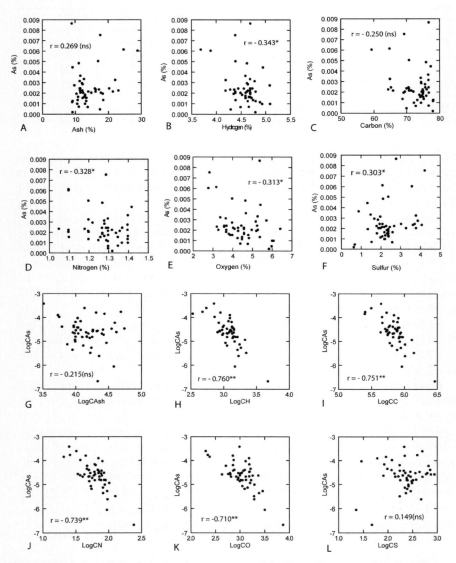

Fig. 9 Scatterplots of arsenic versus six components of the Upper Freeport coal, Indiana County, Pennsylvania. Scatteplots (**A–F**) are for the raw data array and scatterplots (**G–L**) are the respective plots for the clr-transformed array. Nonsignificant correlation(ns), significant at the 5 percent level (*), significant at the 1-percent level (**). Sample size, n = 50. Based on a subcomposition of ash, hydrogen, carbon, nitrogen, oxygen, sulfur, arsenic, mercury and selenium

Table 2a Variation matrix for coal geochemistry (organic + sulfur + ash + selected trace elements subcomposition)

	Ash	H	C	N	O	S	As	Hg	Se	Upper triangle variability
Ash	0.00	0.10	0.10	0.10	0.16	0.14	0.48	0.38	0.10	1.54
H	1.03	0.00	0.00	0.00	0.03	0.17	0.53	0.28	0.16	1.18
C	−1.74	−2.77	0.00	0.00	0.03	0.18	0.53	0.26	0.17	1.18
N	2.32	1.29	4.05	0.00	0.03	0.18	0.54	0.29	0.16	1.22
O	1.07	0.04	2.80	−1.25	0.00	0.29	0.66	0.30	0.22	1.47
S	1.76	0.73	3.49	−0.56	0.69	0.00	0.40	0.47	0.22	1.09
As	8.77	7.74	10.50	6.45	7.70	7.01	0.00	0.54	0.43	0.98
Hg	12.83	11.80	14.57	10.51	11.76	11.07	4.06	0.00	0.41	0.41
Se	11.05	10.02	12.79	8.74	9.99	9.29	2.28	−1.78	0.00	
Total variability										9.07

Upper triangle – variability of the ratios.
Lower triangle – mean values of the ratios.

correlations we might conclude that the abundance of arsenic is negatively associated with the organic content of the coal and slightly positively correlated with the abundance sulfur. These associations are weak because the common variations between As and the other variables range from only 6 to 12 percent. These correlation coefficients should be interpreted with caution because of the uneven scatter of the data. In the variation matrix (Table 2a) we see that the τ values for As ratioed against the organic, sulfur and ash elements are rather large ranging from 0.40 to 0.66 and thus implying a significant variance between As and this group of elements. The corresponding values of ξ range from 6.45 to 10.50 confirming that it is much less abundant relative to the energy elements and ash. The percent variation is shown in Table 2b. The overall variance of the ratios of the energy based elements only account for only 1.2% of the total variability. The variability of the ash and trace element ratios account for 25.8% of the total variability. Again, this confirms the control of stoichiometry on the relationships of the coal as well as a very narrowly defined range of composition of the coal itself.

After the centered logratio (clr) transformation is used, much of the above conclusion is strengthened, although part of it is negated. The part that is strengthened is that between the concentration of arsenic and the concentrations of the organic components (Fig. 9H–K) where the common variance has risen to the 50- to 58- percent range. We can say that these much stronger relations were hidden in the raw data; that is, although there was a hint of their existence in the plots on the raw data, they were not well exposed in the bivariate Figs. 9B–E. The relations between the concentrations of arsenic and ash in the raw and centered logratio (clr) transformed arrays remained nonsignificant (Figs. 9A versus 9G), and the weak positive correlation between arsenic and sulfur in the raw data array was reduced to become non-significant in the centered logratio (clr) transformed array (Figs. 9F versus 9L). The authors suggest caution in interpreting the increase in the correlations (in the

Table 2b Percent variation matrix for coal geochemistry (energy + ash + selectedtraceelements subcomposition)

	H	C	N	O	S	As	Hg	Se	Upper triangle percent variability
Ash	1.06	1.10	1.05	1.71	1.54	5.24	4.22	1.07	16.98
H		0.02	0.04	0.34	1.87	5.87	3.10	1.82	13.05
C			0.05	0.38	1.97	5.84	2.90	1.84	12.98
N				0.37	2.04	5.99	3.22	1.79	13.41
O					3.20	7.31	3.31	2.43	16.26
S						4.43	5.17	2.46	12.06
As							5.98	4.78	10.76
Hg								4.51	4.51
Total variability									100.00

Upper triangle – variability of the ratios.
Lower triangle – mean values of the ratios.

centered logratio (clr) transformed array) between arsenic and the organic elements as being associated with a geochemical process on the basis of element ratios— could this improvement be an artifact of the log-transformation process, if so, what is the mechanism?

The relations between mercury and these same six components are shown for the raw data array in Figs. 10A–F and for the centered logratio (clr) transformed array in Figs. 10G–L. All of the six relations in the raw data array are not significant. In the centered logratio (clr) transformed array (Fig. 10G–L), there are two negative relations that are statistically significant, the other four are also negative, but are not significant at the 5-percent level. Inspecting these Fig. (10G and L) reveals the patterns are not strong and, the correlations may be influenced by outliers. The negative relation with the concentration of sulfur appear to contradict to the findings of Kolker and others (2006), whereas the negative relation with Ash and nitrogen might reflect a meaningful negative relation, similar to those shown in Figs. 9H–K, between the levels of arsenic and the organic constituents of the coal. Certainly, further analysis must be made of the occurrence of mercury in this coal.

The next step in analyzing of the bivariate relations in our samples from the Upper Freeport coal was to include inorganic constituents in the ash. The ash from these Upper Freeport coal samples was analyzed for a total of 60 inorganic elements. Here, we examine the correlation structure of the combination of 8 of the 9 elements examined above, such as carbon and sulfur, along with 16 inorganic elements determined from the ash samples and converted back to a whole-coal basis. These elements include all nine major inorganic elements such as silicon and aluminum and seven inorganic trace elements such as cerium, copper, and manganese. As an aggregate, the remaining major inorganic, trace and other elements sum to a total of between 3 and 13 percent of the whole coal. Problems with the detection limits of many of the trace elements led us to stop the expansion of

Investigation of the Structure of Geological Process

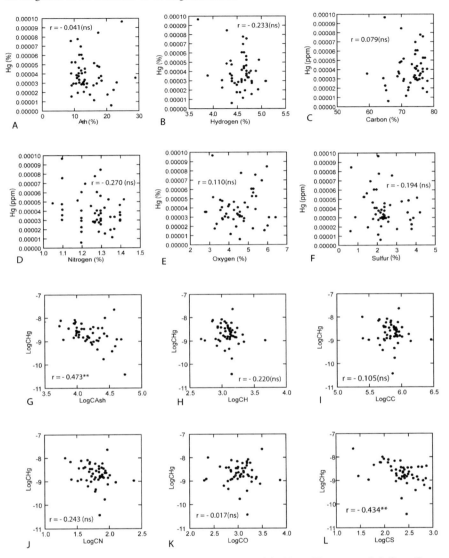

Fig. 10 Scatterplots of mercury versus six components of the Upper Freeport coal, Indiana County, Pennsylvania. Scatteplots (**A–F**) are for the raw data array and scatterplots (**G–L**) are the respective plots for the clr-transformed array. Nonsignificant correlation(ns), significant at the 5-percent level (*), significant at the 1-percent level (**). Sample size, n = 50. Based on a subcomposition of ash, hydrogen, carbon, nitrogen, oxygen, sulfur, arsenic, mercury and selenium

the sample space by partitioning variables from the ash to create a subcomposition of 25 variables. We wish to conserve as many degrees of freedom as possible. The remaining elements in the ash were retained as a variable and labeled ResAsh.

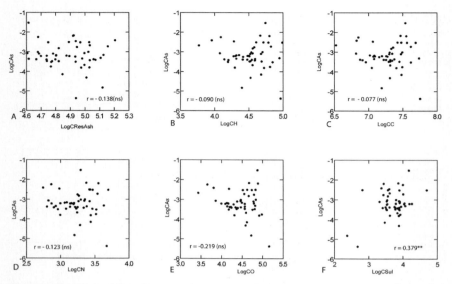

Fig. 11 Scatterplots of Arsenic versus six components in the clr-transformed 25 variable array of the Upper Freeport coal, Indiana County, Pennsylvania. Nonsignificant correlation (ns), significant at the 5 percent level (*), significant at the 1-percent level (**). Sample size, n = 50. Based on a subcomposition of: residual ash, hydrogen, carbon, nitrogen, oxygen, sulfur, Si, Al, Ca, Mg, Na, K, Fe, Ti, S, As, Cd, Ce, Cu, Mn, Pb, Se, Sr, Zn, and Hg

We have learned to anticipate, that when changes are made in the sample space of the raw data array, changes in relations are to be expected in the centered logratio (clr) transformed array. Often, as Fig. 11 shows, the changes are surprising. Here we see that the relations between the concentration of arsenic and the organic components of the coal clearly seen in Figs. 9H–K have vanished. They diminished from having around 50 percent in common variance to having about 1 percent. In contrast, the relation between arsenic and sulfur rose from nonsignificance (Fig. 9L) to significance beyond the 1-percent level (Fig. 11F), this increase in significance may however, be attributed to possible outliers.

These results lead us to ask what changes in the level of correlation (common variance) occur when the sample space is expanded by 1 variable (that is, from 9 to 10)? With the addition of only iron (Fe) to the sample space, the correlations between the organic elements and arsenic decline on average 0.05 across the organic elements (Table 3).

The correlation between sulfur and arsenic also declines with the partitioning of iron out of the ash. This decline is reversed with the addition of more variables and climbs to r = 0.419 for the 25-variable open array. It is beyond the scope of this study to investigate the changes in these correlations as a function of the number of variables taken out of the ash, but it is a study that needs to be done to gain more insight into the problem of the restraints in compositional data analysis.

Table 3 Changes in the correlation between the concentration of arsenic and five constituents in three sample spaces in the Upper Freeport coal, Indiana County, Pennsylvania

Number of variables	Constituents				
	Hydrogen	Carbon	Nitrogen	Oxygen	Sulfur
9	−0.760	−0.751	−0.739	−0.710	0.149
10	−0.718	−0.688	−0.695	−0.656	0.068
24	0.093	0.103	0.066	−0.072	0.419

4 Multivariate Analysis and Geologic Process

Our larger goal was to capture the elements of a geologic process within a multivariate framework by using compositional data without losing sight of the problem created by the restraints in composition. There are at least three processes at work in the creation of a coal: an organic process that reduces the vegetal mass to peat and then to coal, a second organic process whereby bacteria reduces dissolved sulfate to sulfide precipitating iron pyrite, and an inorganic process whereby the elements that form the ash component are introduced and distributed in the peat bog. Further, in preparing the coal for inorganic major and trace element analysis, ashing is used to remove the organic part of the coal. Several processes are involved in the production of the ash. First, the iron pyrite is vaporized; although most of the sulfur escapes, some of it is captured by the available calcium and forms anhydrite ($CaSO_4$). We should see patterns in the principal component bi-plots (Gabriel, 1971) that can be associated with all of these processes.

We began our investigation of the multivariate data space by inspecting the principal component bi-plots for the total coal in the nine-variable sample space (Fig. 12). Here, we identify aspects of the effects of the centered logratio (clr)

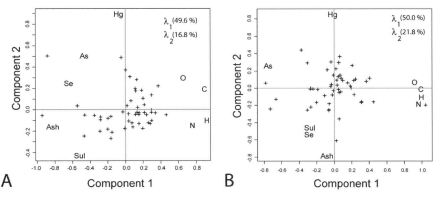

Fig. 12 Principal component bi-plots for the nine components that comprise the whole Upper Freeport coal, Indiana County, Pennsylvania. (**A**), raw data array, and (**B**), clr-transformed array. Based on a subcomposition of ash, hydrogen, carbon, nitrogen, oxygen, sulfur, arsenic, mercury and selenium

transformation as well as the geochemical processes involved in the making of the coal. The organic components (oxygen, carbon, nitrogen, and hydrogen) are more tightly clustered in the centered logratio (clr) transformed array bi-plot (Fig. 12B) than they are in the raw data array bi-plot Fig. 12A. The correlation between oxygen and hydrogen in the raw data array is r = 0.525; it is much higher in the centered logratio (clr) transformed array (r = 0.812). Thus, once we identified a cluster of associated variables such as these organic variables, we could easily identify the effect of centered logratio (clr) transformation and explore the bi-plots for the association of clusters of variables as well as for individual variables, such as the concentration of mercury.

We interpret the positions of the variables in Fig. 12B to suggest that there are at least three principal relations present (interpreted as geochemical processes). In terms of geochemistry, one process is associated with the organic variables; another is associated with the concentration and distribution of the inorganic components (Ash), and a third, which controls the concentration of arsenic (As). It is clear from Fig. 12B that sulfur and selenium have a close association with ash. Mercury shows a strong inverse association with ash, as seen along component 2. Also, mercury appears to have no correlation with respect to the relation of arsenic with the organic variables as seen along component 1.

The variation matrix for this nine variable data array is shown in Table 2a. Several of the conclusions reached from the inspecting of Fig. 12B (about the processes at work in the creation of a coal), are evidenced in this variation matrix. The τ values for the organic element are very low, as in Table 1, implying stoichiometry among the organic variables. The values of ξ in the lower triangle of Table 2a show that As, Hg, and Se occur in significantly smaller abundances than Ash, C, H, N and O. This is obvious as these elements are occur as trace elements only. Other associations are not obvious. For example, one variable can have either a positive or a negative association with other variables depending on the processes from which the associations were derived. Positive and negative correlations between the exogenic, authogenic, and organic elements, may not be observed in the coefficients in the variation matrix derived from the bulk composition. However, these positive and negative correlations are observed in the principal bi-plots. A group of principal component bi-plots using the first three or four components may be required, to describe the processes of coalification, as we think to be the case in the nine variable matrix shown in Table 2a.

The middling position of sulfur (Sul) and selenium (Se) between the positions of Ash and arsenic (As) and the low association between sulfur (Sul) and arsenic (As) in Fig. 12B are explained in the bi-plot shown in Fig. 13. Here the inorganic elements (9 major and 7 trace) determined on the ash were added to create a 25 variable array. Figure 13 shows the bi-plot for the first and second components, which accounts for 49.5 percent of the variance, for the centered logratio (clr) transformed data array. In this array we see, three groups of multi-element associations. It is important to notice the two groups defined along the first component: exogenic (Mg K, Si, Ti Al Cu, Na and Ce) and authogenic (Mn, Fe, Cd, Ca, Zn, Sr, Se, As,

Investigation of the Structure of Geological Process

and S). Mercury appears to have an affinity with the organic elements. This bi-plot is significant because it shows three distinct processes; a distinct difference between the elements that defined exogenic (detrital) sources, a combination of authogenic source (groundwater) correlated with the presence of anhydrite(Ca and S) derived from the ashing of the coal and the organic composition of the coal. It is clear, and logical that the organic elements are distinct from exogenic (detrital) and authogenic (groundwater) elements. We see a grouping of loadings in Fig. 13 that contains Fe and Sul, that reflect the formation of iron pyrite near the loadings of Ca and S that reflect the formation of anhydrite (Ca and S) during the ashing process. These groups merge with the authogenic elements as described above.

Next, we looked into this multivariate data set by returning to the nine-variable data array that was used in Fig. 12B. A bi-plot for principal components 2 and 3 is shown in Fig. 14. These two components account for 34.9 percent of the total variation in this data set. Here we see that the variation of the association between the organic elements is minimal (their vectors plot near the origin) and thereby

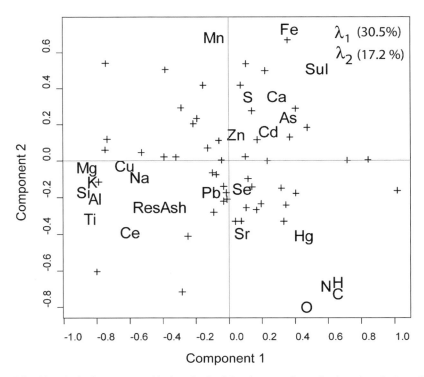

Fig. 13 Principal component bi-plots for 9 of the elements determined on the whole coal and 16 elements determined in the ash of the Upper Freeport coal, Indiana County, Pennsylvania. The data were transformed using the clr transformation. Based on a subcomposition of: residual ash, hydrogen, carbon, nitrogen, oxygen, sulfur, Si, Al, Ca, Mg, Na, K, Fe, Ti, S, As, Cd, Ce, Cu, Mn, Pb, Se, Sr, Zn, and Hg

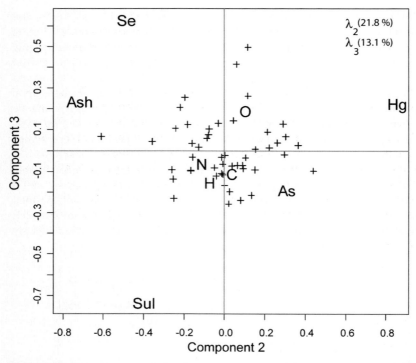

Fig. 14 Principal component bi-plots for components 2 and 3 for the nine constituents that comprise the whole Upper Freeport coal, Indiana County, Pennsylvania. The array is clr-transformed. Based on a subcomposition of ash, hydrogen, carbon, nitrogen, oxygen, sulfur, arsenic, mercury, and selenium

allows us to examine other relationships that appear to be independent of the organic elements. Mercury is inversely associated with sulfur on the basis of their relative compositions. Similarly, both of these variables are inversely associated with selenium, again on the basis of their relative compositions, whereas arsenic and the organic elements contribute little to the relations between the Ash, selenium, sulfur, and mercury elements in this space mapped on principal components 2 and 3.

Before discussing the issues associated with the variation in the concentration of mercury, we will explore the effect of partitioning one variable (Fe) out of the ash and thereby creating a 10-variable data set. The bi-plot for the logratio (clr) transformed data array for this 10-variable system is shown in Fig. 15. Here we see that iron (Fe) plots very close to sulfur (Sul) illustrating the stoichiometry of iron pyrite. We also see that the organic variables form a tight group (positively correlated variables) defined by their stoichiometry in the organic mass. The formation of arsenopyrite in the coal forming environment is seen by the presence of arsenic (As), iron (Fe), and Sulfur (S) along the positive end of component 1 (Fig. 15), which is inversely associated with the organic elements. Pyrite (FeS) shows a close affinity

Investigation of the Structure of Geological Process

with a ash along component 2. The fact that arsenic is negatively associated with Fe and S suggests that arsenopyrite is not present in the ash. Mercury (Hg) appears to be inversely correlated with pyrite along principal component 2. In fact, mercury does not have any positive associations with either the ash or the organic variables (Fig. 15).

We note that the eigenvalue for the third component accounts for 16.2 percent of the total variation. Figure 16 is bi-plot of the first and third components of this 10 variable subcomposition. Sulfur (Sul), iron (Fe), and arsenic (As) are closely associated. These elements represent arsenopyrite which is inversely associated with the organic elements as seen along component 1. Also, the arsenopyrite elements (Fe, As, Sul) are inversely associated with the ash. But the arsenopyrite likely formed during the development of the coal. From these associations we are able to see the three processes at work in the creation of a coal, namely the clustering of the organic elements, the introduction exogenic constituents (ash), and aqueous geochemical reactions during diagenesis involving the creation of pyrite, arsenopyrite, and the concentration of mercury appears to be independent.

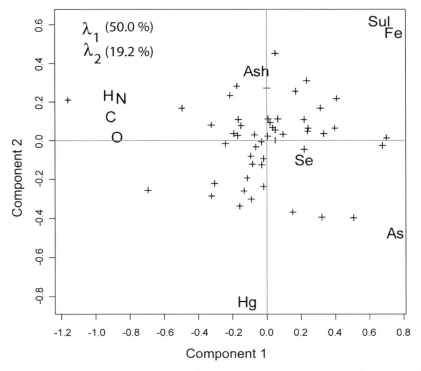

Fig. 15 Principal component bi-plots of the first and second components for the ten chemical components that comprise the whole Upper Freeport coal, Indiana County, Pennsylvania. The data were transformed using the clr transformation. Based on a subcomposition of ash, hydrogen, carbon, nitrogen, oxygen, sulfur, arsenic, mercury, selenium and iron

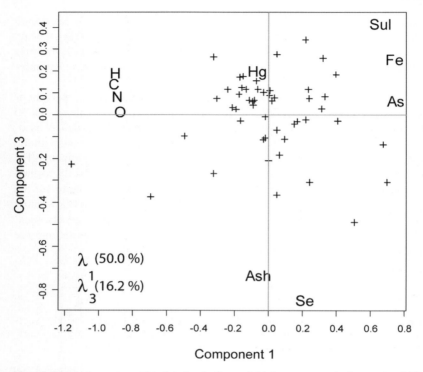

Fig. 16 Principal component bi-plots for the first and third components in the ten chemical components that comprise the whole Upper Freeport coal, Indiana County, Pennsylvania. The data were transformed using the clr transformation. Based on a subcomposition of ash, hydrogen, carbon, nitrogen, oxygen, sulfur, arsenic, mercury, selenium and iron

5 The Correlation of Mercury

The geochemistry of mercury in coal is an important scientific topic because of the metal's effect on air quality and human health. Kolker and others (2006) conclude that mercury is commonly hosted in Fe-sulfides (pyrite and marcasite) in bituminous coals, like the Upper Freeport coal. From our multivariate analysis of the centered logratio (clr) transformed array for the nine-variable data set (Fig. 12B), we interpret the location of the loadings to suggest a negative relation with ash (the points are far apart) and possibly negative relations with the organic elements, ash, selenium (Se), and sulfur (Sul). This interpretation is partially confirmed by the bivariate plots shown in Fig. 10, where we see that the concentration of mercury (Hg) is significantly and negatively correlated with the concentration of Ash (Fig. 10G) and sulfur (Fig. 10L).

Given our analysis of the association of arsenic (As) with the variables in the 9- and 25-variable arrays and the fact that correlations vanished in the larger array, we ask, if mercury could behave in a similar fashion. From Fig. 17 it is clear that

Investigation of the Structure of Geological Process

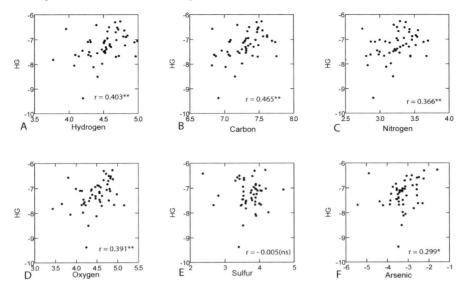

Fig. 17 Scatterplots of mercury versus four organic and two inorganic components in the 25 variable array of the Upper Freeport Coal, Indiana County, Pennsylvania. Scatteplots are for the the clr-transformed array. Nonsignificant correlation(ns), significant at the one percent level (**). Sample size, n = 50. Based on a subcomposition of: residual ash, hydrogen, carbon, nitrogen, oxygen, sulfur, Si, Al, Ca, Mg, Na, K, Fe, Ti, S, As, Cd, Ce, Cu, Mn, Pb, Se, Sr, Zn, and Hg

mercury is poorly correlated with organics, sulfur, and arsenic. From the bi-plots shown in Figs. 13–16) we conclude mercury does not show a strong associations with either the organic or exogenic constituents and appear to occur as an independent variable. That is to say, that mercury behaves in a non-stoichiometric way with the other constituents. This field should be fruitful for further geological and statistical studies.

6 Conclusions

The purpose of this study was to capture the structure of a geological process within a multivariate statistical framework by using compositional geological data generated by that process and, where applicable, by associated processes. The fundamental data is the composition of the rock—in this case a coal (the Upper Freeport coal, Indiana County, Pennsylvania). Therefore, the raw data array sums to a constant. Thus, a technique such as the centered logratio (clr) transformation must be used to remove this restraint on the data. Bivariate correlations and multivariate principal component bi-plots were used to identify the processes that created the coal. During the statistical analysis, we encountered several surprising associations. Relations between variables changed significantly between the choice of a nine and a 25 variable data set. Regardless of the choice of data sets three fundamental processes were identified in the correlation structure of the data. These principal processes that create a coal are authogenic (organic and diagenetic) and exogenic (detrital) processes.

In addition, the correlation of the concentration of arsenic and mercury in the coal was shown to differ depending on the composition of the data sets. In the centered logratio (clr) transformed 10-variable subcomposition, the correlations of arsenic with the concentration of the organic elements were negative and associated with the development of asernopyrite during diagenesis (Fig. 16).

We conclude from the patterns observed that the concentration of mercury does not appear to be associated with the coal formation process, that is it is essentially independent of the concentrations of the other variables.

When using different subcomposition (e.g. 9,10, and 25 variable arrays) the variation matrix can show how the associations change as the number of variables in the subcomposition change. In the case of arsenic expanding the nine variable subcomposition to a 10 variable subcomposition by adding iron (Fe) improved the recognition of asernopyrite in the diagenesis of the coal.

This area of research in statistical analysis of geologic processes deals with a complex and interdependent group of relations induced by the compositional restraints and the number of variables used. Further, we conclude that when we deal with a system of variables such as those involved in the composition of a rock (in the current case, a coal), we must consider both the issue of correlation induced between the variables by compositional restraints and the complex and interacting issue of the number of variables in the data array.

Acknowledgments The authors wish to thank several geologists of the U.S. Geological Survey who made this project possible. First, we thank, Leslie Ruppert for providing the compositional data as well as the metadata on the Upper Freeport coal in Indiana County, Pennsylvania. Second, we thank Frank Dulong and Joseph Hatch for providing basic information on the geology and geochemistry of coal. Third, we thank Curtis Palmer for information on the analytic techniques used to determine the composition of coal. Fourth, we thank Allan Kolker for information on the occurrence of mercury in coal. This manuscript is contribution number 20080234 for the Geological Survey of Canada.

References

Aitchison J (1986) The statistical analysis of compositional data. Methuen, New York, p 416
Aitchison J (1997) The one-hour course in compositional data analysis or compositional data analysis is simple. In: Proceedings of IAMG '97, the Third annual conference of the international association for mathematical geology. In: Pawlowsky-Glahn-Vera RA (ed) Proceedings of the annual conference of the international association for mathematical geology 3, 3–35
Aitchison J (1999) Logratios and natural laws in compositional data analysis. Math Geol 31 (5): 563–580
Aitchison J (2003) The statistical analysis of compositional data, 2nd edn. Blackburn Press, Cardwell, NJ, p 416

Aitchison J., Egozcue JJ (2005) Compositional data analysis: where are we and where should we be heading? Math Geol 37 (7): 829–850
ASTM (2001) Annual book of ASTM standards, section five: petroleum products, lubricants, and fossil fuels, vol 05.06, ASTM, West Conshohocken, PA, p 628
Buccianti A, Mateu-Figueras G, Pawlowsky-Glahn V (2006) Compositional data analysis in the geosciences: from theory to practice. Geological Society, London, Special Publications, vol 264, p 212
Cecil BC, Stanton RW, Allshouse SD, Finkelman RB (1978) Geologic controls on mineral matter in the Upper Freeport coal bed. In: Proceedings: symposium on coal cleaning to achieve energy and environmental coals, U.S. Environ Prot Agency, E.P.A. 60017-79-0998a, vol 1, pp. 110–125
Cecil CB, Renton JJ, Stanton RW, Finkelman RB (1979) Mineral matter in coals of the central Applachian basin. Ninth International congress carboniferous stratigraphy and geology, Abstract paper, p 32
Cecil BC, Stanton SG, Dulong FT (1981) Geology of contaminants in coal: Phase I report of investigations to the U.S. environmental protection agency, U.S. Geol. Surv. Open-file Rep. 81-953-A, 92p
Cecil BC, Stanton RW, Neuzil SG, Dulong FT, Ruppert LF, Pierce BS (1985) Paleoclimate controls on Late Paleozoic sedimentation and peat formation in the Central Applalacian basins (U.S.A.). Int J Coal Geol 5: 195–230
Chayes F (1949) On ratio correlation in petrography. J Geol 57 (3): 239–354
Chayes F (1971) Ratio correlation, The University of Chicago Press, **Chicago, IL**, 99p.
Egozcue JJ, Pawlowsky-Glahn V, Mateu-Figueras G, Barceló-Vidal C (2003) Isometric logratio transformations for compositional data analysis. Math Geol 35 (3): 279–300
Gabriel KR (1971) The bi-plot graphic display of matrices with application to principal component analysis. Biometrika 58 (3): 453–467
Kolker A, Senior CL, Quick JC (2006) Mercury in coal and the impact of coal quality on mercury emissions from combustion systems. Appl Geochem 21 (11): 1821–1836
Miesch AT (1969) The constant sum problem in geochemistry. In Merriam DF (ed) Computer applications in the earth sciences, Plenum Press, New York, pp. 161–176
Northern and Central Appalachian Basin Coal Regions Assessment Team (2001) 2000. Resource assessment of selected coal beds and zones in the Northern and Central Appalachian Basin Coal regions, Professional Paper 1625-C Discs 1 and 2, version 1.0
Pawlowsky V, Ecozcue JJ (2006) Compositional data and their analysis. In: Buccianti A, Mateu-Figueras G, Pawlowsky V (ed) Compositional data analysis in the geosciences, from theory to practice.Geological Society of London, Special Publication, vol 264, pp. 1–9, 212
Pearson K (1897) Mathematical contributions to the theory of evolution: on a form of spurious correlation which may arise when indices are used in the measurement of organs. Proceedings of the Royal Society 60: 489–498
Rollinson HR (1993) Using geochemical data: evaluation, presentation, interpretation, Longman Scientific & Technical, Harrow, England, p 352
Schweinfurth SP (2002) Coal-a complex natural resource; U.S. Geological Survey Circular 1143, p 39

Master of the Obscure—Automated Geostatistical Classification in Presence of Complex Geophysical Processes

Ute C. Herzfeld

Reprinted from *Mathematical Geosciences* DOI: 10.1007/s11004-008-9174-4, when citing this article please use the DOI number.

Abstract The topic of this paper is the retrieval of hidden or secondary information on complex spatial variables from geophysical data. Typical situations of obscured geological or geophysical information are the following: (1) Noise may disturb the signal for a variable for which measurements have been collected. (2) The variable of interest may be obscured by other geophysical processes. (3) The information of interest may formally be captured in a secondary variable, whereas data may have been collected for a primary variable only, that is related to the geophysical process of interest. Examples discussed here include mapping of marine-geologic provinces from bathymetric data, identification of sea-ice properties from snow-depth data, analysis of snow surface data in an Alpine environment and association of deformation types in fast-moving glaciers from airborne video material or satellite imagery. Data types include geophysical profile or trackline data, image data, grid or matrix-type data, and more generally, any two-dimensional or three-dimensional discrete or discretizable data sets.

The framework for a solution is geostatistical characterization and classification, which typically involves the following steps: (1) calculation of vario functions (which may be of higher order or residual type or combinations of both), (2) derivation of classification parameters from vario functions, and (3) characterization, classification or segmentation, depending on the applied problem. In some situations, spatial surface roughness is utilized as an auxiliary variable, for instance, roughness of the seafloor may be derived from bathymetric data and be indicative of geological provinces.

The objective of this paper is to present components of the geostatistical classification method in a summarizing and synoptical manner, motivated by applied examples and integrating principal and generalized concepts, such as hyperparameters and parameters that relate to the same physical processes and work for data in oversampled and undersampled situations, parameters that facilitate comparison among different data types, data sets and across scales, variograms and vario functions of higher order, and deterministic and connectionist classification algorithms.

Ute C. Herzfeld
University of Colorado Boulder, Boulder, CO 80309-0449, USA,
e-mail: Herzfeld@tryfan.colorado.edu

Keywords Spatial structure analysis · spatial surface roughness · vario functions · discrete mathematics · satellite data · marine geophysics · cryospheric sciences · Jakobshavns Isbræ (Greenland) · snow · glaciers · sea ice

Preambula. "Master of the Obscure" was a name given to a rock climber who would always find a way to ascend a rock face, as difficult or strange, demanding or hidden, as *obscure* it may have been, creating a moving solution to any problem presented by geology, if you want to think of it that way. Some 40 years ago, Geomathematics started as a new way of interpreting and utilizing *obscure* information in geologic data, and now comprises many tools to discover patterns and relationships in geology. This paper is dedicated to Frits Agterberg, whose really *clear* descriptions of mathematics in his original book "Geomathematics: Mathematical Background and Geo-Science Applications" (Agterberg, 1974) helped to raise mathematics into *the light* for geologists and geological applications. The topic of this paper is to present, in a summarizing fashion, the geostatistical classification approach that I have developed over several years, attempting to meet the demands of a wide range of geological and geophysical problems and data sets, which all share the problem of *the obscure*.

1 Introduction

Information in geological or geophysical data may be *obscure* for several reasons:

(1) Measurements may have been collected for a given variable of interest, but noise may disturb the signal related to that variable. Examples are: (a) Surface elevation data collected from aircraft, with inaccurate observations of aircraft motion, and (b) snow-depth data collected on sea ice, affected by manual sampling, inaccurate spacing and possibly difficulties of properly determining the boundary between snow and ice.
(2) The variable of interest may be obscured by other geophysical processes. Examples for this situation are: (a) Snow-surface data, collected to study surface morphogenesis and near-surface processes, may actually be influenced by processes at the bottom of the studied snow field and the underlying ground, and (b) ice-surface elevation data, collected from satellite, may be affected by subscale physical features, such as crevasses, which are too small to be measured individually, but still influence the total return signal.
(3) The information of interest may formally be captured in a *secondary variable*, whereas data may have been collected for a *primary variable* only, that is related to the geophysical process of interest. For instance, bathymetric data are usually collected from a ship to map water depth (primary variable) and seafloor topography. Densely-sampled topography contains morphologic information, which in certain situations may be used to infer marine-geologic types and provinces, such as submarine volcanoes, sediment ponds and abyssal hill terrain. Morphology may be considered a secondary variable, and geologic type may be considered a secondary variable in a classification.

Situations (1) and (2) show that "disturbances" of the data may be attributed to measurement errors or to physical processes that were overlooked at the outset of the study, or by a combination of both as in (2b). In a situation of type (3), analysis of one variable may help to understand a second one. The example under (3) may hint at mathematically forming a derivative to access the secondary variable in case of morphology or at applications of artificial intelligence to associate geologic type. However, we will present a simpler approach that is suitable for obscure data situations, including those indicated above; for now, we postpone the methodological discussion.

Looking at the inverse of the relationship between variables of interest and variables of observation opens the door to a wealth of opportunities: Data sets that have already been collected may contain clues to solutions of problems that may be difficult to observe, for instance due to the complexity of the spatial physical processes involved, or to remoteness of the location (in Antarctica, under water, on another planet, in a roadless mountain range), or because a geologic event occurred before anyone noticed. Another hindrance to collection of geophyscial field data may simply be cost. Events such as a volcanic eruption, an earthquake causing deformation of the Earth's surface, or acceleration of an ice stream, are of interest to society because of their immediate (earthquakes) or delayed impact (climate indicators), but observations on the geophysical variables of interest may not be available. However, the event may have been recorded in satellite imagery or other remotely sensed data or simply in videos for media purposes – which brings up the question: How can such information be utilized as geophysical data? Or, how can one use the concept of primary and secondary variables quantitatively?

Such data sets that already exist have been called "data of opportunity". Their analysis may require a geomathematical method that allows hidden, secondary or obscure information to be retrieved. Once that is established, data of opportunity may contain the key for urgent geological problems.

In this paper, we present a family of approaches to the problem of retrieving hidden, *obscure* information from spatial data. The methods are summarized as "Geostatistical characterization and classification" (a bit of a misnomer, as we shall see, lies in geo*statistics*), derived by the author and her group from first mathematical principles (Herzfeld, 1993) to a modular system of tools for a large range of applications and data types. Each aspect, component or generalization of the classification method was developed to meet a challenge posed by an applied problem in the geosciences or by a new type of data encountered in an applied project. Yet together all the methodological components form by now a systematically organized library of tools. The objective of this paper is to present the main components of the geostatistical characterization and classification method in a summarizing and synoptical manner, and to provide the reader with criteria to select suitable aspects of classification for his own applications. Along motivating examples, we will introduce principal and generalized concepts, such as variograms and vario functions of higher order, hyperparameters and parameters that relate to the same physical processes and work for data in oversampled and undersampled situations, parameters that facilitate comparison among different data types, data sets and across scales, and deterministic and connectionist classification algorithms.

Although the geostatistical classification method is self-contained (except for the definition of the (semi-)variogram (Matheron, 1963; Journel and Huijbregts, 1989) and the connectionist part, see Sect. 9), it shares common ground with a wide range of approaches in image classification, neural networks, texture analysis, analysis of remote-sensing data and signal processing. Principles of mathematical morphology are described in Serra (1982). Pattern recognition techniques are also applied in medical pathology, in the analysis of holograms and sonograms, in texture analysis (Weszka et al., 1976; Liu and Jernigan, 1990) and in satellite image analysis as early as the 1980s (e.g. Burger, 1981; Ritter and Hepner, 1990, Franklin, 1991, Franklin and Wilson, 1991). Classification criteria for the analysis of marine bathymetric and sidescan data have been described by Fox (1990) using spectral methods and by Stewart et al. (1992) and Jiang et al. (1994) using a connectionist approach, and by Herzfeld and Overbeck (1999) using spectral, fractal and geostatistical methods. Classification is a widely-used term in the analysis of remote-sensing data (a recent review is given in Lu and Weng, 2007, for treatment in monographs see Tso and Mather, 2001; Landgrebe, 2003). Spectral and statistical methods dominate the field and are often realized in geographic information systems (GIS) (see Bonham-Carter, 1994). Rarely ever is a spatial structure function applied in classification, exceptions are summarized in Atkinson and Lewis (2000)(contributions not including the author's are e.g. Carr, 1999; Carr and Miranda, 1998; Chica-Olmo and Abarca-Hernandez, 2000; Wallace et al., 2000; see also Oliver and Webster,1989).

2 Outline

In the following sections, we will introduce mathematical principles and concepts of the geostatistical characterization and classification method and provide examples of applications.

The geostatistical classification method proceeds by the following steps:

(1) a window is selected from the study area; this may coincide with the entire study area,
(2) statistical or analytical spatial functions are calculated from data in the window (e.g. experimental variograms, directional vario functions of first or higher order, other functions),
(3) spatial functions from (2) may be filtered,
(4) parameters, called *vario parameters* or *geostatistical classification parameters*, are calculated from the functions in (2) or (3),
(5) a feature vector is composed of the parameters,
(6) a deterministic discrimination algorithm or connectionist class association is applied to the feature vector to relate the structures in the window to an object class.

The power of the geostatistical classification method lies in its combination of several methods for the identification and quantification of signals in data and their application at appropriate scales. The structure function allows the capture of properties or features of intermediate scales, relative to the resolution and size of an image or data set. Typical operator windows in traditional spectral or statistical classification are usually smaller (such as 3×3 to 11×11 pixels) or calculations are carried out for the entire data set and then too simple (statistical parameters) or too costly (some Neural Nets). We shall see that the geostatistical classification facilitates manageable operations for larger windows (such as 20×20 or 50×50 cells), while it reduces computational cost compared to analyzing an entire image of thousands by thousands of cells and yet preserves the salient information. In this sense, one may consider structure functions as information-filtering operators. Whether an entire image or data set is analyzed globally or by application of a moving-window operator (in step (1)) depends on the application, the type of data set, and the objective: Characterization, classification, or segmentation of a larger set.

As structure functions, we select experimental variograms or first-order vario functions (step (2)). Other than in interpolation, where the experimental variogram is modeled and thus simplified to facilitate inversion of the kriging matrix, in classification the concepts of spatial surface roughness, local spatial variability and structural homogeneity play a role. We shall see later in the paper how the concepts of surface roughness, homogeneity with respect to surface structure, surface types and surface provinces lead to new ways of utilizing variograms and related functions in a quantitative analysis.

A variogram is the spatial structure function most commonly used in geostatistics, here we use experimental variograms. However, a definition of a vario function in the framework of discrete mathematics is more practical and allows a generalization to higher order, important in situations of obscured data. The reader familiar with geostatistics may think of a first-oder vario function as an experimental variogram (under certain mathematical conditions). In situations of high noise or the presence of other disturbing or overprinting processes, higher-order vario functions may take the place of variograms or first-order vario functions, while in situations of (global) trends of (local) drifts, "residual" vario functions may work best (cf. Herzfeld, 1992, 2002).

The information in the structure function is summarized in geostatistical classification parameters (vario parameters) (step (4)). Depending on the geological application, instrumentation and data quality, different types of parameters are needed to discriminate surface types or provinces. We shall see in Sect. 4 that a one-dimensional approach to describing spatial surface roughness or complexity is insufficient, hence feature vectors, composed of several parameters, are utilized as the basis of classification (step (5)). Each feature vector represents a sample image or line segment of data.

In the last step (6), methods that may be known as classification methods in their own right serve to associate an object to a class. A classifier may be formulated as a deterministic discrimination algorithm (deterministic geostatistical classification) or be represented by a neural network (connectionist-geostatistical classification). Through steps (1–5), the geological object is now given to the classifier as a feature vector, hence the classification is more successful and computationally less costly, and overtraining is less likely. In a colloquial sense, the geostatistical parameters tell the neural net where to look.

Individual components of the classification method are introduced in a sequence that advances from simple to more complex geophyscial situations. First, we will give definitions of characterization, classification and segmentation and of concepts that are essential for classification, such as surface roughness (Sect. 3). With the goal of making the paper more easy to read and more interesting for the applied scientist, the method will be developed here along a series of applied examples from various fields of the geosciences. Each example will motivate a new methodological component or two. The examples, in the order of the paper, are: Sea-floor segmentation: the first automated geostatistical classification (Sect. 4), Glacier roughness surveys of the Greenland Ice Sheet: classification parameters to discriminate simple from complex processes (Sect. 4), Snow-surface roughness in an alpine environment: vario functions of higher order (Sect. 6), Sea ice in the Alaskan Arctic: retrieving information from complex and noisy signals using the hyperparameter concept (Sect. 7), Crevasse patterns as indicators of dynamics in Jakobshanvs Isbræ, West Greenland: mapping of dynamic provinces from ASTER satellite imagery (Sect. 8), and Classification of deformation types during the surge of Bering Glacier: application of the connectionist-geostatistical method Sect. 9).

The following list of theoretical topic sections is provided for the reader interested in the classification method and as a reference:

Components of the Geostatistical Classification Method:

(a) Spatial Surface Roughness, Homogeneity and Surface Provinces, Characterization–Classification–Segmentation: *Theory Topics I, II, III, Sect. 3*
(b) Variograms and (First-Order) Vario Functions: *Theory Topic IV, Sect. 3*
(c) Vario Functions of Higher Order: *Theory Topic VIII, Sect. 6*
(d) On Vario Functions versus Variograms: *Theory Topic IX, Sect. 6*
(e) On Interpolation versus Classification: *Theory Topic IX, Sect. 6*
(f) Vario Parameters (Geostatistical Classification Parameters) I: Parameters for Simple Geologic Situations *pond, mindist, p1, p2*: *Theory Topic V, Sect. 4*
(g) Vario Parameters II: Parameters for Analysis of Complex Geologic and Morphologic Situations (Parameter Types, $p3,\ldots, p12$, *avgspac, deriv*): *Theory Topic VI, Sect. 4*
(h) Vario Parameters III: Hyperparameters: *Theory Topic X, Sect. 7*
(i) Derivation of Roughness Length from Spatial Surface Roughness: *Theory Topic VII, Section 5*
(j) Feature Vectors *Sects. 4 and 7*
(k) Class Association: Deterministic Discrimination and Connectionist Association *Sects. 4, 8 and 9*
(l) Spatial Classification and Scale Dependence *Discussion, Sect. 10*

3 Principles and Concepts

Theory Topic I: Spatial Surface Roughness
Surface roughness is defined as the spatial derivative of surface elevation. Morphologic properties at any scale are not captured by absolute elevation (above an imaginary zero reference, such as sea level) but rather by changes of elevation in space. At larger scale this is topographic relief, at small scale roughness. For instance, spatial ice surface roughness is a morphologic characteristic of ice surfaces. (Note that here we use the convention of "large scale" for low resolution and "small scale" for high resolution, which is opposite to the convention used in cartographic scale.)

Using this definition, surface roughness contains the complete morphologic information of the surface under study. This may sound trivial, but in some disciplines surface roughness is commonly understood as roughness length, a one-dimensional parameter, or root-mean-square roughness, a one-dimensional statistical approximation. For instance, aerodynamic roughness (length) is used in meteorology (e.g. Oke, 1987). However, roughness length can be derived from (spatial) surface roughness (see section (5.) and Herzfeld et al., 2006a).

Theory Topic II: Characterization – Classification – Segmentation
The objective of (mathematical) *characterization* is to determine a set of parameters that uniquely describe an object. The objective of *classification* is to associate a given object to one of a number of classes; so that the class association can be carried out for each case automatically, a *rest class* is required to collect all objects that do not belong to any of the characteristic classes. In each new application, the characterization problem must be mastered before the classification problem. By moving a classification operator over a large spatial data set, such as an image, each window is associated to a class, and a segmentation may be achieved. A segmentation into reasonably coherent subsets or segments may require smoothing; however, in a good classification this is not necessary (see Herzfeld and Higginson (1996) for examples and details). The sea-floor study (Sect. 4) is an example of a classification and segmentation of an area into geologic provinces.

Theory Topic III: Concepts of Homogeneity and Surface Province
Extraction of parameters requires the concept of homogeneity: An area is called *homogeneous with respect to surface structure* if the area has the same surface structure throughout, but the structure may be complex and resulting from several morphogenetic events, phases, or processes, as long as all those processes have affected the entire area in the same way. The term *heterogeneous* is reserved to mean "not homogeneous".

The concept of homogeneity matches the definition of a *surface province*, which is an area of maximal extent that is homogeneous with respect to *surface type*, seen in the field as a collection of morphologic features (e.g., for snow and ice surfaces: sastrugi, crevasses, ablation ridges, melt ponds, ablation runnels, fields of meltwater streams, new snow surface, or complex combinations of any of the above). In surveys, the size of the area needs to be selected such that the largest feature occurs several times.

A *complex surface* is one that has experienced several morphogenetic processes. The resultant forms may be superimposed on each other. The selection of parameters in the geostatistical classification need to reflect and capture the complexity of the surface.

Theory Topic IV: How do we Analyze Spatial Surface Roughness? – Variograms and (First-Order) Vario Functions

Spatial surface roughness may be analyzed as the derivative in analytical mathematics, or by investigating a spatial structure function, such as the variogram or the vario function.

Recalling that morphology is related to the spatial derivative rather than to the absolute value of elevation, we need to calculate the spatial derivative from a data set. One way to do this is by building differences of measured values over given distances. Forming the (mathematical) limit of such expressions, for distance increments approaching zero, yields the value of the surface slope in a given location:

$$lim_{x \to x_0} \frac{z(x_0) - z(x)}{x_0 - x} \tag{1}$$

For elevation measurements, this approach gives a local slope value, the slope at point x_0, for snow-depth measurements, the local change in snow-layer thickness at x_0.

If we are interested in characterizing the surface structure in a given area, then the actual value at each location is not as relevant as parameters that describe the morphology more generally for the whole study area. Therefore, we form differences of measurement values again, but average over all points that have the same common distance (or, distance and direction), creating *vario functions*.

$$v_1(h) = \frac{1}{2n} \sum_{i=1}^{n} [z(x_i) - z(x_i + h)]^2 \tag{2}$$

for pairs of points $(x_i, z(x_i))$, $(x_i + h, z(x_i + h)) \in \mathscr{D}$, where $\mathscr{D}$ is a region in $\mathscr{R}^2$ (case of survey profiles) or $\mathscr{R}^3$ (case of survey areas) and n is the number of pairs separated by h; the distance value h is also termed "lag" ($\in$-element of; $\mathscr{R}^2, \mathscr{R}^3$- two- and three-dimensional space of real numbers (coordinates), respectively). The function $v_1(h)$ is called the *first-order vario function*. This function exists always and has a finite value, because only finitely many data points enter the calculation.

Geostatisticians note that the first-order vario function is formally equivalent to the variogram, however, the variogram is defined for a data set that may be considered a realization of a spatial random function satisfying the intrinsic hypothesis (see Matheron, 1963). A discussion of the theoretical and practical differences between vario functions and variograms is given in Sect. 6.

Vario functions are more robust analysis tools than the covariance function and the power spectrum (power spectral density) and exist under conditions where those functions do not exist. This makes them particularly useful in geoscience applications. In the geostatistical classification method, the vario function is used akin to the spectrum, but in contrast to the spectrum, it is not easily perturbed by disturbances over small scales, which is an important aspect in the study of roughness characteristics.

4 Sea-Floor Segmentation: The First Automated Geostatistical Classification

The geostatistical classification method was originally developed for the classification of sea-floor provinces, based on bathymetric data, as described in Herzfeld and Higginson (1996), based on mathematical principles derived in Herzfeld (1993). The objective of this project is to identify and map geologic provinces and thus to derive a segmentation of an area of the sea floor. Automated mapping from bathymetric data is a useful approach to geological mapping, since naturally the sea floor is, other than terrestrial areas, only accessible via remote sensing, with punctual sampling through drilling or dredging. The study area is located on the western flank of the Mid-Atlantic Ridge near 26° North, between the Kane Fracture Zone in the south and the Atlantis Fracture zone in the north. Dominant geological processes in the area are generation of crust at the ridge axis, submarine volcanism, sea-floor spreading and sinking of the crust with increasing age and distance from the ridge axis, and sedimentation (see Macdonald et al., 1991; Tucholke and Lin, 1994). Each of these processes leads to specific morphological forms: the ridge, submarine volcanoes, abyssal hill terrain (fields of near parallel ridges and valleys) and sediment ponds. Spacing of abyssal hills and occurrence and sizes of sediment ponds are indicative of age of the crust.

During a seven-week marine geophyscial experiment, the bathymetry of a 150 by 100 km large area near 26° North was surveyed using the HYDROSWEEP instrument. Bathymetric data were collected along dense profiles to almost cover the study area, and the classification is based on gridded data with gaps (see Fig. 1). Bathymetry and azimuth maps give a visual impression of the area, however, the goal is now to map geologic units automatically. Following steps (1–6) of the geostatistical classification method, as outlined in Sect. 2, classification maps – geological maps – were derived. Examples are given in Fig. 1. We used parameters as given in the next section and a deterministic classification algorithm. The classes in step (6) are simply color-coded, which results in a thematic map (here, a geological map).

Theory Topic V: Vario Parameters (Geostatistical Classification Parameters) I: Parameters for Simple Geologic Situations
Maximum vario-function value (*pond* parameter) is the simplest parameter, used to distinguish flat areas from areas with topographic relief. For instance, the vario function of an evenly-spaced hill-and-valley profile in topography is a sinusoidal wave. Then the lag to the first minimum in the sine curve is the characteristic spacing of hills and valleys in the topographic relief. The *mindist* parameter is defined as the lag of the first minimum after the first maximum in the vario function. *mindist* gives the spacing of parallel ridges, for instance, in the abyssal hill terrain. We further define the significance parameters $p1$ and $p2$:

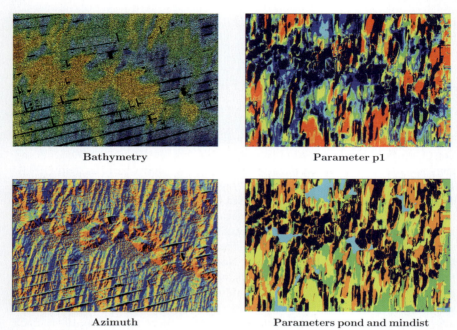

Fig. 1 Geostatistical seafloor classification, based on bathymetric data of an area on the western flank of the Mid-Atlantic Ridge. *Left top*: **Bathymetry** of study area. Depth increasing on continuous color scale with *blue* 2,300 m, *light blue* 3,600 m, *green* 3,900 m, *yellow* 4,000 m, *orange* 4,300 m, *dark orange* 4,500 m. Black stripes are survey gaps. *Left bottom*: **Azimuth** of seafloor gradient. Directions (with 0°N, increasing clockwise) on continuous color scale with *blue* 145°, *light blue* 180°, *green* 215°, *orange* 252°, *dark orange* 288°, *red* 360°. *Right top*: **Classification based on parameter** $p1$, **significance of abyssal hill terrain (slope)**. *Dark blue*: $p1 < 0.5$, *light blue*: $0.5 \leq p1 < 1.0$, *green*: $1.0 \leq p1 < 2.0$, *yellow*: $2.0 \leq p1 < 4.0$, *orange-red*: $4.0 \leq p1 < 6.0$, *red*: $6.0 \leq p1$, *black*: $p1$ not defined. *Right bottom*: **Classification based on combination of parameters** *pond* **and** *mindist*. *Light blue*: *pond* for residual variogram smaller than 5000 m^2. For all other colors, the pond criterion does not apply, and *mindist* has the following values (in meters): *Dark blue*: $mindist < 2000$, *green*: $2000 \leq mindist < 4000$, *yellow*: $4000 \leq mindist < 6000$, *orange*: $6000 \leq mindist < 7600$, *black*: parameters not defined. Study area located at easting 300,000–450,000, northing 2,850,000–2,950,000 UTM zone 23 (central meridian 45°W), 47°–45°30′W/25°45′–26°40′N. Depth approx. 2300–4500 m. (after Herzfeld and Higginson, 1996, Figs. 8 and 10, reprinted with permission from Computers & Geosciences)

$$p1 = \frac{\gamma_{max_1} - \gamma_{min_1}}{h_{min_1} - h_{max_1}} \qquad (3)$$

$$p2 = \frac{\gamma_{max_1} - \gamma_{min_1}}{\gamma_{max_1}} \qquad (4)$$

$p1$ is the slope parameter and $p2$ the relative significance of the first minimum min_1 after the first maximum max_1, and h_x and γ_x denote lag and vario-function value of x, respectively (Herzfeld and Higginson, 1996). In this notation,

Master of the Obscure—Automated Geostatistical Classification

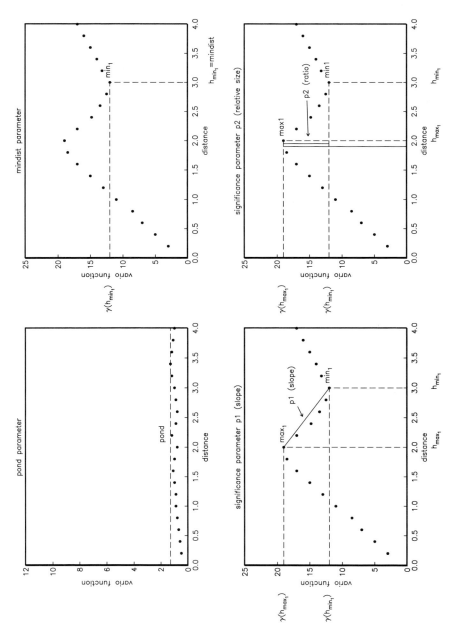

Fig. 2 Geostatistical classification parameters

$$mindist = h_{min_1} \qquad (5)$$

(for illustration, see Fig. 2). A large parameter $p1$ together with a large *mindist* value characterizes structures that are absolutely big in size and also significant. A large value of $p2$ indicates an area of structures that are relatively significant for their size, but may be absolutely small, for instance, small, pronounced features may have a large $p2$ value. (Note that $p2$ is always positive and absolutely less or equal than 1). The definition of parameter $p1$ includes absolute size, but the definition of $p2$ does not, this property makes parameter $p2$ and its generalizations (see Theory Topics VI (Sect. 4 and X (Sect. 7) useful for comparison of features among different data sets and across scales.

To study anisotropy, one investigates variograms of different directions (directional variograms) and compares parameters, for example, *dirmin* is the direction in which *mindist* is smallest. In the seafloor study, *dirmin* identifies the across-strike direction of the abyssal hills.

Results of Application: By application of the automated geostatistical classification method, it is possible to identify and discriminate, characterize and classify marine geological prototypes: submarine volcanoes, Mid-Atlantic Ridge, abyssal hill terrain, and sediment ponds (see Fig. 1). Differences in the distribution and sizes of sediment ponds in one ridge segment indicate that this segment was older than the other segments, and the orientation of the ridges, as characterized by the parameters *dirmin* and *mindist*, shows that the segment was also rotated, which together with a smaller significance parameter $p2$ pointed at a more complicated geological history (see also Herzfeld and Higginson 1996, and Tucholke and Lin 1994). The key to success of a relatively simple classification approach lies in two facts: (1) the geological situation is relatively simple, and (2) the data situation is good. (1) The situation is that only few morphological types exist, and these are directly related to certain geologic processes: Volcanoes are near-symmetrical, near-round cones or hills, abyssal hill terrain consists of fields of near-parallel ridges and valleys, their spacing being related to age, and sedimentation occurs in those valleys over time. Sediment ponds have a smooth surface, compared to other geological provinces. (2) The data accuracy is in the decimeter to meter range, and the geological features are tens to hundreds of meters in relief. Hence a solution is possible with only five parameters and one maximum-minimum sequence.

5 Glacier Roughness Surveys of the Greenland Ice Sheet: Classification Parameters to Discriminate Simple from Complex Processes

In a project to study ice-surface morphogenetic processes in the Greenland Ice Sheet and to provide high-resolution field information for ERS Synthetic Aperture Radar (SAR) data, micro-topographic and surface roughness data were collected with the Glacier Roughness Sensor (GRS), and instrument specifically designed for this purpose (Herzfeld et al., 1997, 1999, 2000a,b). The GRS is a field instrument that is pulled by a skier across the ice surface, recording surface elevation in eight channels with mechanical measurements and electronic transmission, instrument position using differential GPS and instrument attitude using clinometers, all stored in an on-board computer (Fig. 3). GRS data have about 10-cm along-track spacing, 20-cm across-track spacing and sub-centimeter accuracy (Fig. 4).

Compared to the sea-floor example, the ice surface is more complex, several processes including snowfall, wind deposition and erosion of snow, melting and distant crevassing all affect the ice surface repeatedly and interact with each other. Consequently, more parameters are required. To give an example, parameters are needed that can discriminate simple processes from overprinted ones. Their definition

Fig. 3 Surveying with the Glacier Roughness Sensor in the Jakobshavns Isbræ Drainage Basin south of the South Ice Stream. *Top*: Spring 1997 (MICROTOP97) (from Herzfeld et al., Zeitschrift für Gletscherkunde und Glazialgeologie, vol. 35, 1999, reprinted with permission). *Bottom*: Summer 1999 (MICROTOP99). Note differences in ice-surface roughness: The sping surface is formed by snowfall, melting, wind and distant crevassing. The ice surface is generally smoother in spring than in summer, at the scale of decimeters to several meters, due to formation of pronounced ablation ridges, melt runnels, cryoconite holes and other ablation features (Fig. 4 and text)

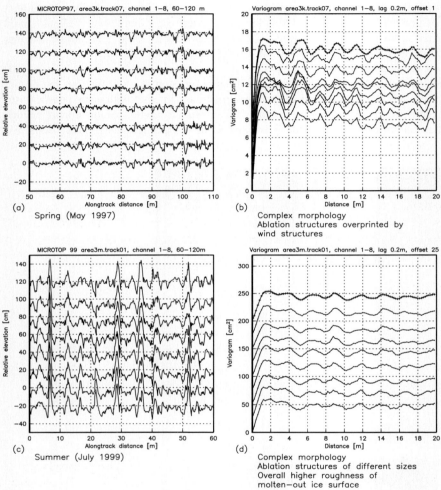

Fig. 4 Comparison of spatial surface roughness in area "above ice camp" (cf. Fig. 3). *Left* panels: GRS data, *right* panels: variograms, one per channel, and global variograms (*top*). For explanations, see text

involves automatically examining more sequences of maxima and minima in the variogram or vario function.

Theory Topic VI: Vario Parameters II: Parameters for Analysis of Complex Geologic and Morphologic Situations

We define $p2$-type parameters $pt2(\cdot,\cdot)$ as

$$pt2(max_i, min_j) = \frac{\gamma_{max_i} - \gamma_{min_j}}{\gamma_{max_i}} \tag{6}$$

for $i \leq j$. These are useful if $\gamma_{max_i} \geq \gamma_{min_j}$, which is easily checked because it is the case if and only if $pt2 \leq 0$. With this notation, we define significance parameters for the first few minima in the vario function:

$$p3 = pt2(max_1, min_2) \tag{7}$$

$$p4 = pt2(max_2, min_2) \tag{8}$$

$$p5 = pt2(max_s, min_2) \tag{9}$$

Parameter $p3$ is the significance of the second minimum min_2 relative to the first maximum, $p4$ is the significance of the second minimum min_2 relative to the second maximum max_2, $p5$ is the significance of the second minimum min_2 relative to the highest previous maximum max_s, where γ_{max_s} is the maximum of γ_{max_1} and γ_{max_2}. So if the first maximum is higher than the second, $p5$ equals $p3$, otherwise it equals $p4$. Using $p5$ rather than $p3$ or $p4$ serves to better characterize morphologies with smaller features superimposed on larger ridge structures. Note that $\gamma_{max_i} \geq \gamma_{min_j}$ is the case if and only if $pt2 \leq 0$.

Similarly, slope parameters, or $p1$-type parameters $pt1(\cdot,\cdot)$ are defined by generalizing the definition of $p1$ as

$$pt1(max_i, min_j) = \frac{\gamma_{max_i} - \gamma_{min_j}}{h_{min_j} - h_{max_i}} \tag{10}$$

for $i \leq j$. As noted for $p2$-type parameters, $p1$-type parameters are useful if $\gamma_{max_i} \geq \gamma_{min_j}$, which is easily checked because it is the case if and only if $pt1 \leq 0$. Useful $p1$-type parameters are

$$p6 = pt1(max_1, min_2) \tag{11}$$

$$p7 = pt1(max_2, min_2) \tag{12}$$

$$p8 = pt1(max_s, min_2), \tag{13}$$

for instance $p8$ is the slope parameter analogous to significance parameter $p5$. In case the resolution of the vario function is high relative to the feature size, checking more minima is useful:

$$p9 = pt2(max_3, min_3) \tag{14}$$

$$p10 = pt2(max_1, min_3) \tag{15}$$

$$p11 = pt2(max_2, min_3) \tag{16}$$

$$p12 = pt2(max_4, min_4) \tag{17}$$

The *avgspac* parameter, defined as the weighted average of lags to all minima in a vario function, also measures the spacing of dominant structures. It is of importance if structures in a vario function repeat regularly, but do not start clearly at the first minimum:

$$avgspac = \frac{1}{n}\sum_{i=1}^{n}\frac{1}{i}h_{min_i} \qquad (18)$$

calculated typically for $n = 3$ or $n = 4$. We also use *conc*, the concavity of the vario function before the first maximum, and *deriv*, the derivative or slope of the vario function at its origin, approximated by

$$deriv = \frac{\gamma(3h_0) - \gamma(h_0)}{2h_0} \qquad (19)$$

All these parameters are employed in analyses of snow and ice data. As a result, we are able to characterize a variety of surface forms that occurr in the ablation zone of the Greenland Inland Ice, including ridges, formed by an interaction of snow accumulation and melting, small melt runnels, melt ponds, cryoconite holes (formed by dust particles with primitive plants). Some forms remain characteristically the same for several years, while other forms depend on season, and others yet on weather (Fig. 4). For instance, a dominant ablation structure of about 4 m spacing with a subordinate structure of 2 m characteristic scale remains for at least several seasons, but the overall roughness, indicated by the maximal vario value or *pond* parameter, is much higher in summer than in spring (Figs. 3 and 4). In the photographs (Fig. 3) one can see that roughness, on a scale of few decimeters to meters, is larger in summer (bottom panel) than in spring. However, a larger parameter *p2* for the spring surface than for the summer surface indicates that spring features are more pronounced (Herzfeld et al., 2000a,b). More generally, this example demonstrates that several parameters are necessary to characterize the spatial roughness of the ice surface at several scales (see also, Discussion and Outlook, Sect. 10).

6 Derivation of Roughness Length z_0 from Spatial Surface Roughness (Theory Topic VII)

Geostatistical classification parameters can be applied to relate spatial surface roughness measurements to aerodynamic roughness length z_0. Empirical relationships between geometrical surface properties and z_0 can be employed (see also Lettau, 1969; Kondo and Yamazawa, 1986) to derive the following approximation formulas (Herzfeld et al., 2006a): In case the vario function fluctuates around a *sill* (for larger distances), the *pond* parameter is only slightly higher than the *sill* and can be used for a first approximation:

$$z_0 = z_{0_{f1'}} = \frac{1}{2}\sqrt{2pond} \qquad (F1') \qquad (20)$$

However, formula (F1) likely overestimates z_0, because the surface elements in the geometric derivation are implicitly assumed to be rectangular. A formula using vario-function components resultant from surface features of different sizes is

$$z_0 = z_{0_{f3}} = \frac{1}{2} \frac{\sum_{k=1}^{n_k} h_{min_k} \sqrt{2\gamma_{max_k}}}{\sum_{k=1}^{n_k} h_{min_k}} \quad (F3) \tag{21}$$

Summation is over significant features of different scale and involves classification parameters h_{min_k} (distance to minimum k) and γ_{max_k} (vario-function value of maximum k). (F3) is a scale-dependent equation.

7 Snow-Surface Roughness in an Alpine Environment: Vario Functions of Higher Order

One might think that snow-surface observations are similar to ice-surface observations. However, a new problem arose in a project to study snow-surface changes related to environmental conditions throughout a winter and summer, in an Alpine snowfield (Niwot Ridge Saddle at 3250 m above sea level, Indian Peaks, Colorado Front Range, see Caine, 1995a,b; Losleben, 1990 for hydrology and climate, Herzfeld et al., 2003). Snow surface features are much smaller in size than the ice-surface features observed in the Greenland Inland Ice (amplitudes to 2 meters for ice-surface features versus amplitude of a few centimeters to ca. 30 cm for snow-surface features on Niwot Ridge, cf. Figs. 4 and 5a), so the temporal and spatial variability in a snow-surface GRS data set comes close to the limit of accuracy of the GRS—and hence one is now looking at a mathematically ill-posed problem.

We recall that the derivation of geostatistical classification parameters depends on the ability of an automated routine to detect sequences of minima and maxima in the vario function. For snow-surface data, minima cannot be determined easily in a variogram or (first-order) vario function (Fig. 5). This motivates the introduction and investigation of a second-order vario function (see Theory Topic VIII). In Fig. 5, the second-order vario function of the same data set clearly shows a minimum at 7 m. The minimum corresponds in size to the ground patterns underlying the snowpack, which indicates that melting and compaction of snow is influenced by the ground patterns on a scale of several meters. In contrast, at a smaller scale self-organisational processes dominate (Herzfeld et al., 2003).

Here, using variograms of variograms provided a practical solution—and with it a mathematical problem, really a reincarnation of the old problem of geostatistics, that of the "ensemble" and the "population" in geostatistics, in case only one data set is recorded. Whereas for the variogram that problem has been discussed (e.g. Matheron, 1973; Armstrong, 1984; Journel, 1985; a summary of solutions in other disciplines is given in Herzfeld 1992, e.g. Kaula, 1967; Grafarend, 1975; Lauritzen, 1977; Tarantola and Valette, 1984), reviewers of the vario-of-the-vario idea were critical, although the approach had already proven itself useful in identifying snow hydrological processes (Herzfeld et al., 2003). So what was wrong? The method or the calculation? The same numerical calculation rewritten in the framework of discrete mathematics rather than stochastics leads to *vario functions* rather than variograms, and the vario functions can be taken to higher order without much

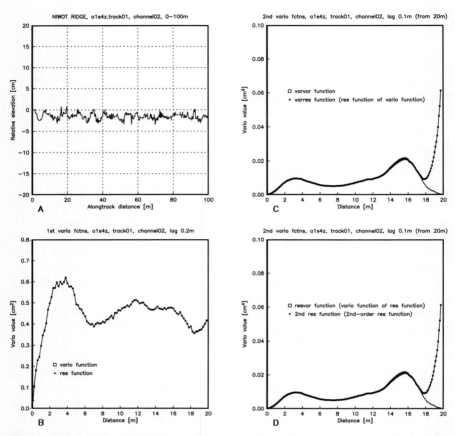

Fig. 5 GRS data of snow-surface roughness on Niwot Ridge Saddle, vario function and 2nd-order functions resvar and varres. For explanations, see Eq. (22) and text

complication. Vario functions exist for every discrete data set, but their theoretical introduction requires a bit of a framework, such as criteria for homogeneity of a data set (see Herzfeld, 2002).

Theory Topic VIII: Definition of Vario Functions of Higher Order
Vario functions, defined for discrete spatial data sets with a domain in 2 or 3 dimensions in Theory Topic IV (Section 3), can be generalized to higher order. If we let the values of the first-order vario function take the place of the data in Eq. (2), we can calculate the *second-order vario function* v_2, termed *varvar function* for short and defined as

$$v_2(k) = \frac{1}{2s} \sum_{j=1}^{s} [v_1(h_j) - v_1(h_j + k)]^2 \qquad (22)$$

with $((h_j, v_1(h_j)), (h_j + k, v_1(h_j + k))) \in \mathscr{V}_1 = (h, v_1(h))$, the first-order vario-function set. Higher-order vario functions are defined recursively, following an analogy to Eq. (22). The varvar functions allow the detection of shallow large-scale features in the presence of locally dominating small-scale features, and the detection of morphologic features in presence of strong sampling artefacts.

Matching the concept of the residual variogram in geostatistics, we have also introduced residual vario functions of first and higher order, through a *res* operator. The *res* operator compensates for a drift component, hence *res* functions facilitate modeling of spatial structures in presence of a trend or drift. Noting that vario functions are calculated for sampling neighborhoods, it is possible to analyze phenomena with non-linear trends, since in effect the "residual" operator takes out a piecewise linear drift. At each order, the *res* operator and the *var* operator can be applied, creating *resvar* and *varres* functions. A more formal treatment and discussion of properties of higher-order vario functions is given in Herzfeld (2002).

Theory Topic IX: On Vario Functions and Variograms. On Interpolation Versus Classification.
Although the equation of the first-order vario function (Eq. 2) has the same form as the equation of the variogram (or semi-variogram), the functions are not the same. The vario-function theory is set in a discrete-mathematics framework, whereas the variogram is defined in the stochastic framework typically employed in geostatistics, the theory of regionalized variables (Matheron, 1963; Journel and Huijbregts, 1989). There are practical advantages persuant to these theoretical differences which manifest themselves in numerous applications (for a theoretical discussion, see Herzfeld, 2002).

In geostatistical estimation the role of the variogram is to quantify the transitive behavior of the regionalized variable from typically good covariability of closely neighboring samples to poor covariability of samples spaced farther apart, and to use the resultant quantitative model to solve kriging equations. The experimental variograms, calculated from the data, are usually reduced to variogram models, which are selected from a small and useful zoo of prototypes, including the spherical model, the linear model with sill and the Gaussian model. Variogram models from the zoo share the properties that (1) any ordinary or universal kriging system built with such a model has a unique solution, and (2) the variogram models may be characterized by their function type and a small number of parameters: sill, range, and nugget effect. However, if solving kriging equations is not the objective, then the zoo of models is no longer helpful but rather limiting, because much of the structural information is potentially lost in the reduction to such a model (cf. Herzfeld, 1999).

For lags exceeding a certain distance, the variogram "fluctuates" around a sill value, indicative of the total variance. In the stochastic framework of classic geostatistics, these fluctuations are attributed to the random nature of the variogram for lags exceeding a certain range, which is also the range of covariation. In geostatistical classification, these are the source of information, and their cause is by no means random, but related to small-scale geophysical processes. Different parameters are needed for classification. (Simulations of superimposed almost sinusoidal processes and their cova functions given in Herzfeld (1990) may serve to demonstrate basic relationships between processes of different scales).

Casting out probability renders the geostatistical classification method not geo*statistical* in the strict sense any more, however, as "The deterministic side of geostatistics" (Journel, 1985) has tradition in geostatistics as well, the vario function is only a logical step for a useful tool away from probabilistic concepts, so we kept the name. Notably, the new definition was motivated by a practical application and is not just an exercise in mathematical theory. The change in concept does, however, open the door to a practically oriented philosophy: If a tool yields useful results, then it is a good tool. This view is more at home in connectionism—if a Neural Net works, we can use it, and do not need to understand the processes involved (which is actually also possible, but a black-box attitude of "it just does it, like the human brain" is more well-known. Neural Networks are related to mathematical optimization, see Herzfeld and Zahner (2001)). The open view at tools motivated to try a few of those—see Sect. 8.

8 Sea Ice in the Alaskan Arctic: Retrieving Information from Complex and Noisy Signals using the Hyperparameter Concept

As part of a validation campaign for satellite passive microwave data from AQUA AMSR-E over Arctic sea ice, we analyze snow-depth data collected on the sea ice in several offshore provinces in Arctic Alaska (Herzfeld et al., 2006b; Maslanik et al., 2006). In the neighborhood of Point Barrow, Alaska, the Chukchi and Beaufort Seas meet, and together with ice types encountered in near-by Elson Lagoon, ice provinces in the area represent a good subset of Arctic sea-ice regimes. Not much is known about sea-ice properties, spatial characteristics of older versus younger ice, and processes that lead to sea-ice formation. An example of snow-depth data over sea-ice is given in Fig. 6. The complexity of sea ice, together with the properties of manually collected snow-depth data, provided new challenges for the classification method:

The *variable* complexity of sea-ice morphogenetic processes finds an expression in the variable number of max-min sequences that needs to be studied to find relevant parameters, i.e. geostatistical classification parameters that tell the "story" of sea-ice formation. For marine geophysical studies, the first application of the classification method, the needed number of max-min sequences is 1, for Jakobshavns area Glacier Roughness Sensor (GRS) data, 2, for SAR data of the same region, 4 (Herzfeld and Higginson, 1996; Herzfeld et al., 1999; Herzfeld et al., 2000b); but for sea-ice snow-depth data now it is unknown and variable (Fig. 7). As an answer, we are developing a robust search algorithm for newly defined hyperparameters in complex geo-data sets, which automatically determines relevant "bigger" max-min sequences and associated generalized parameters, and also the optimal group size for the data set at hand. In this work, we apply software modules of a new classification library that realize the mathematical algorithms introduced in this paper and accompanying data-handling functions (Geostatistical Classification Software

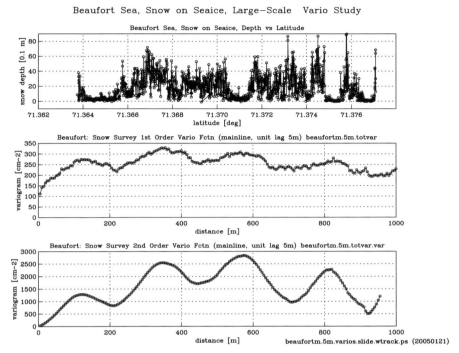

Fig. 6 Snow-depth data on sea ice (*top*), 1st-order vario function (*center*) and 2nd-order vario function (*bottom*), Beaufort Sea. From 2nd-order vario function, a characteristic length of 210 m is easily determined automatically

Library "geoclass", by U.C. Herzfeld and S. Williams; for information on the *geoclass* system, write to ute.herzfeld@colorado.edu).

Theory Topic X: Hyperparameters
Again, the classification starts by determining sequences of minima and maxima in a vario function, on which parameter calculations are based. We determine *bigmax*, the largest maximum in a group of g maxima, and then *bigmin*, the smallest minimum in a group of g minima following *bigmax*. For a fixed group size g, several big minima and maxima can be determined, starting the search for the first one, $bigmin_1$ at the first $max - min$ sequence of the vario function, and continuing to find $bigmin_2$, $bigmin_3$ after that. As for parameters, we distinguish $p1$-type and $p2$-type hyperparameters, in generalization of $p1$-type and $p2$-type parameters.

$$pt1(bigmax_i, bigmin_j) = \frac{v_{bigmax_i} - v_{bigmin_j}}{h_{bigmin_i} - h_{bigmax_j}} \tag{23}$$

and

$$pt2(bigmax_i, bigmin_j) = \frac{v_{bigmax_i} - v_{bigmin_j}}{v_{bigmax_i}} \tag{24}$$

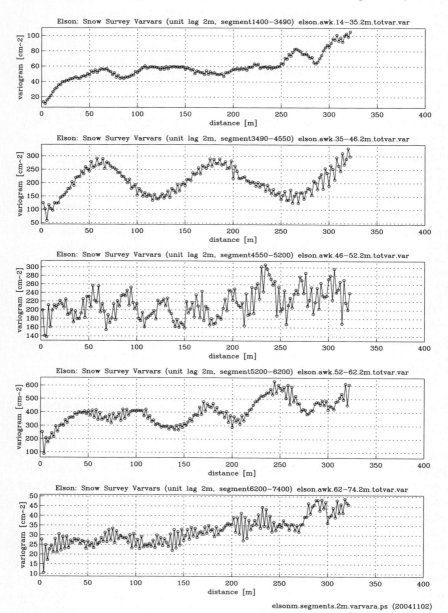

Fig. 7 A segmentation of a profile across Elson Lagoon, Alaska: Parameters and hyperparameters derived from second-order vario functions quantify specific characteristics for each segment

In the characterization of Point-Barrow sea-ice provinces, we will most frequently use the first hyperparameters, written $p1big$ and $p2big$ for simplicity, where

$$p1big = pt1(bigmax_1, bigmin_1) \qquad (25)$$

and

$$p2big = pt2(bigmax_1, bigmin_1) \qquad (26)$$

Similar to the original parameters, $p1big$ and $p2big$ can now be used to quantify the relative size of small-scale features on top of large-scale features, for those large features whose characteristic spacing is identified by $h_{bigmin_1} = mindist_{big}$. The spacing of the small-scale features may be derived automatically, by counting the intermediate minima in the varvar function plots; the intermediate minima are the minima between two consecutive *bigmins* and lead to a varvar function that is reminiscent of a sawtooth pattern.

The optimum groupsize g can also be determined automatically, here we have applied a criterion to find stable groupsizes, such that the *bigmax-bigmin* pair stays the same for 3 consecutive groupsizes, and in addition, such that *bigmax-bigmin* sequences that are multiples of previous ones are avoided. This rule leaves one with the first one or two (or more) maxima and minima in a sequence, even if it is overprinted with variations caused by smaller surface features or sampling noise.

Application continued. Good examples are given in the Elson Lagoon profiles, where subordinate features are significant relative to the largest features (Fig. 7). Application of classification with hyperparameters composed into feature vectors leads to a segmentation of Elson Lagoon into sea-ice provinces (Fig. 8).

Feature vectors of the following form

$$v = (z_0, z_0(SLT), pond, mindist_{big}, p1_{big}, p2_{big}) \qquad (27)$$

were utilized, where z_0 denotes surface roughness, $z_0(SLT)$ roughness associated with the thickness of the snow layer on top of the sea ice, *pond* a vario parameter and $mindist_{big}, p1_{big}, p2_{big}$ hyperparameters as defined above. We note that linear relationships $(z_0, z_0(SLT))$ yield different outcomes in many clustering and optimization routines than quadratic ones $(z_0, pond)$, whereas a neural network can be trained to recognize that $pond, z_0, z_0(SLT)$ are always related.

In the Beaufort profiles, where structures are chaotic but large features dominant, an analysis of second-order vario functions with the usual parameters $(p1, p2)$ leads to a classification (Fig. 6). One may think of this as hyperparameters defaulting to usual parameters when the vario functions are not overprinted by "sawtooth" patterns. Based on the geostatistical classification from feature vectors that consist of parameters and hyperparameters, applied to snow-depth data from Beaufort Sea, Chukchi Sea and Elson Lagoon, a first morphogenetic model for development of sea-ice types is derived (Herzfeld et al., 2006b).

The hyperparameter concept is still in development. It provides tools for signal processing that fit in the context of geostatistical structure analysis and as such is generally applicable to spatial data analysis, without the context of classification. As

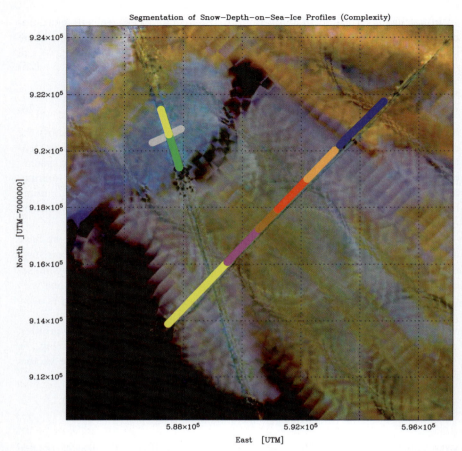

Fig. 8 Segmentation of snow-depth-on-sea-ice profiles, derived using geostatistical classification with 6-component feature vector (Eq. 27), superimposed on color-composition of aerial Polarimetric Scanning Radiometer (PSR) data. Color-coding of snow-depth segments according to increasing complexity of snow-depth structures (*green-yellow-orange-brown-red-purple-blue*), indicating increasing complexity of sea-ice types. Segments are (on Elson-Beaufort profiles, trending SW-NE): Elson segment 1 (*yellow*), Elson segment 2 (*purple*), Elson segment 3 (*brown*), Elson segment 4 (*red*), Elson segment 5 (*orange*), Beaufort (*blue*), on Chukchi profiles: Chukchi main profile segment mod1 (*green*), Chukchi main profile segment mod2(12-15) (*yellow*), Chukchi main trends SE-NW; Chukchi cross profile (*grey*). Underlying PSR data are shown as composition of data from three channels, 10.7 GHz H-Polarization (*red*), 19 GHz H-Polarization (*green*), 37 GHz H-Polarization (*blue*). Note that snow-layer segments match areas that appear homogeneous in PSR-data composition, and that increasing complexity matches a progression of colors in homogeneous areas, indicating a relationship between snow-depth provinces, and hence sea-ice provinces, and PSR-data (Herzfeld et al., 2006b, reprinted with permission from IEEE Transactions on Geoscience and Remote Sensing)

far as passive microwave data are concerned, application of geostatistical characterization with hyperparameters has demonstrated that complexity of spatial snow depth structures as captured in multi-dimensional feature vectors directly influences the passive microwave signal received from sea-ice with snow cover, and, less directly, snow-depth and surface roughness length (Herzfeld et al., 2006b). These results indicate that passive microwave data in general may be affected by spatial snow-depth and surface roughness, with a dependence on scale and quantified by geostatistical classification.

9 Crevasse Patterns as Indicators of Dynamics in Jakobshavns Isbræ, West Greenland: Mapping of Dynamic Provinces from ASTER Satellite Imagery

In satellite imagery, the problem is typically one of undersampling rather than oversampling, because the resolution of any satellite image is limited. Hence we return to using the primary parameters *pond*, *mindist*, $p1$ and $p2$, calculated in several direction and distance classes. If property mapping or segmentation is the goal for a large image, the entire classification is performed as a moving-window operation (as in the Jakobshavns example in this section). If the goal is to associate the entire image to a class, then every step from vario-function calculation to parameter derivation and class association is undertaken for one entire image at a time (see the Bering Glacier example in Sect. 9. The optimal window size depends on the problem and the data set.

Our next example stems from a mapping of deformation structures in Jakobshavns Isbræ, West Greenland, the Earth's constantly fastest-moving glacier, which a few years ago (in 1999) has entered a state of rapid retreat (for literature on Jakobshavns Isbræ see Echelmeyer and Harrison, 1990; Echelmeyer et al., 1991; Bennike et al., 2004; Podlech and Weidick, 2004; Krabill et al., 1999; Thomas et al., 1993; Mayer and Herzfeld, 2000, 2001).

Figure 9 shows the retreating ice front, the North Ice Stream and part of the longer South Ice Stream, whose crevasses can be followed 80 kilometers up into the Greenland Ice Sheet. The calving front is located near ($69°N$, $50°W$).

Crevasses are the manifestation of deformation on the surface and in the upper layer of a glacier. Although they are conspicuous, little research has attended to their interpretation, largely because of the connection of physical glaciology to the principles of mechanics of a continuous medium (which in itself does not negate structural failure); for exceptions, see Hambrey and Milnes (1977), Mayer et al. (2002), and Herzfeld et al. (2004). Our approach is based on the principles of structural geology, where deformation history is deduced from inspection of the image or the material itself and the patterns left by the deformation processes (for a modern treatment of the topic, see Ramsay and Lisle (2000), for applications to glaciology see Mayer and Herzfeld, 2000, 2001; Herzfeld et al., 1997, 2003).

The structural glaciology approach can be combined with the classification method, (1) by mathematically exploiting the geometry of movement (Herzfeld et al., 2004), (2) by mapping vario parameters and relating them to deformation, and (3) by associating structure and deformation as in the next Sect. 9. Following avenue (2), we apply the classification to ASTER data in Fig. 9. Here, we use the new software library "geoclass", which in addition to parameter and hyperparameter algorithms also contains utility functions for satellite data conversion.

Fig. 9 Jakobshavns Isbræ North Icestream. May 28, 2003 (ASTER Data)

Parameter maps may already be considered classification maps, when displayed with a colour-coded contouring algorithm as in Fig. 10. (Parameter maps are simply given as matrices, which may be input into an image analysis software or a contouring program). In the seafloor example (Sect. 4), we went a step further, mapping provinces based on a deterministic class-association algorithm, based on a combination of parameters and criteria. Fig. 10a which shows the *pond* parameter identifies areas of increased roughness, which generally coincide with crevassed areas, while low *pond*-parameter areas coincide with non-crevassed areas (compare Figs. 9

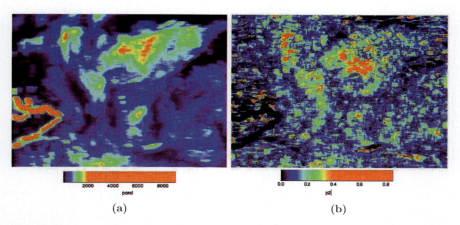

Fig. 10 Parameter maps, created with geostatistical classification library from ASTER data of Jakobshavns Isbræ, North Icestream in Fig. 9. (**a**) Parameter *pond* (directly related to roughness length). (**b**) Significance parameter $p2$ (relative size and significance). Note that different structural provinces (in Fig. 9) have different parameter values

and 10). This tells us *where* deformation occurred, but not *what type of deformation*. A step to identification of deformation type is given by the example of the map showing parameter $p2$ values: A large $p2$ indicates dominance of structures that are significant relative to their size, such as fields of clear-cut crevasses developed under rapid movement. Combinations with criteria derived from other parameters narrow down a characterization. This process can be pursued until a full characterization is achieved, or a neural network may be built: see next Sect. 9.

10 Classification of Deformation Types During the Surge of Bering Glacier: Application of the Connectionist-Geostatistical Method

Rather than further explore and develop the relationships of parameters to structural or geological processes in a deterministic way, one may resort to the connectionist approach and train a neural network to recognize deformation types from surface imagery.

A glacier surge is a sudden acceleration of a glacier to a multiple of its normal velocity (typically 100 times its speed, for Bering Glacier, see Herzfeld, 1998). The ice surface breaks up into fields of square-cut fresh crevasses. Surges occur quasi-periodically (for Bering Glacier, about every 25 years). It is not known why some glaciers surge and others do not, in fact, Bering Glacier surges but neighbouring Steller Glacier only pulses (Post, 1972), and other glaciers which flow out of Bagley Ice Field, the source of Bering and Steller glaciers, do not surge at all. On the scale of mountain ranges, surge glaciers are geographically clustered; there are surge glaciers in Alaska, in Greenland, and in Antarctica and there were some in the Alps in the last century. For background on the surge phenomenon, see Meier and Post (1969), Weertman (1964), Post and LaChapelle (1971), Dolgushin and Osipova (1975), Raymond (1987), Clarke (1987), Kamb et al. (1985), Herzfeld (1998). We detected the onset of the 1993–1995 Bering Glacier surge suddenly (Lingle et al., 1993) and by good fortune (a surge of a huge glacier is an impressive and beautiful sight), and with no geophysical survey program in place that could possibly handle the complexity of the surge. Instead, video data, collected for media reports, became our data of opportunity at the start of the surge in 1993. In overflights in 1994 and 1995, we collected more aerial video data and GPS observations, having by then realized that video data may be treated and analyzed as a source of geophysical information.

"*How can video data of crevassed ice surfaces be used to extract information on ice deformation and kinematics?*" became the more sophisticated form of the question "What does a geophysicist do if there are no geophysical data?" During flights over Bering glacier we observed that crevasse fields exhibit the same surface type for large, connected deformation provinces, inside which the patterns repeat (see Fig. 11), and with sharp boundaries to neighboring fields of a different crevasse type. This observation suggests using vario functions and classification parameters

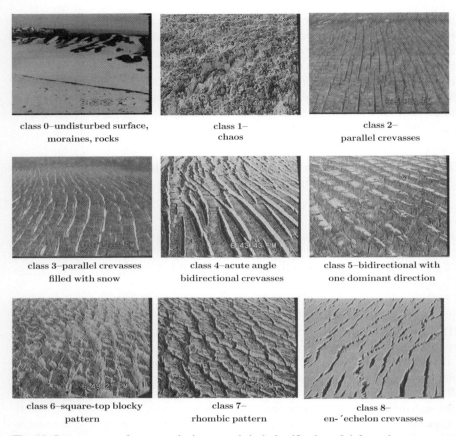

Fig. 11 Crevasse types for connectionist-geostatistical classification of deformation structures. Bering Glacier surge 1993–1995. Video images

for a basic analysis. On the other hand, the multitude of data (video frames) and the complexity of surface patterns motivates us to try the connectionist approach (i.e., to build a neural net). Using geostatistical processing as a first step in the classification, the data for each video-frame subset are reduced to three-directional vario functions. This creates a manageable input for network training, having efficiently captured the salient information in the image data (Herzfeld and Zahner, 2001).

Using a neural network rather than tediously working out deterministic rules sounds advantageous at first, but a neural net requires training, and this in turn requires a thorough understanding of the geological situation and background. In this instance, expert knowledge on how to deduce the type of deformation that a given piece of material (rock or ice) experienced from a single picture of the deformed material is needed to create a near-exhaustive collection of possible deformation types (unless one wants to end up with a large rest class of cases that prospective NN cannot assign). We used the crevasse types shown in Fig. 11.

The selection of the type of neural net and other technical and mathematical information on neural net training, as well as on the geostatistical processing is given in Herzfeld and Zahner (2001). The resultant network is 95% correct in associating images of the ice surface to the deformation type, for video frames not previously known to the neural net. A notable disadvantage of the connectionist approach is, however, that a new net needs to be trained for eachapplication, whereas parameter maps (as in Sect. 8) can be calculated from any image data set and often provide a good "first cut" at segmentation into structural provinces.

11 Summary, Discussion and Outlook

Summary
The objective of this paper has been to exploit hidden information in data, as a means to learn about geophysical properties and processes that are secondary, or hard to observe, or not in the variable that was originally measured, or measured directly but overprinted by too much noise or other effects. In this paper, we have outlined the classification approach using examples from different geophysical fields as well as from different types of data sets. The geostatistical characterization and classification method utilizes statistical and analytical spatial functions and proceeds by extraction of parameters from those functions, composition of feature vectors, and association of classes by application of deterministic and connectionist algorithms.

Discussion and Outlook
Generalizations and Applications
The examples presented in this paper stem from marine geology, cryospheric sciences and climate research. However, the geostatistical characterization and classification method may be generally applied to any problem of image analysis, signal processing or classification based on regularly or irregularly distributed data. Potential fields for applications may be in forestry, medical tomography, any discipline of human and physical geography, oceanography, archaeology, planetary research, biology and ecology.

Limitations of the method lie in the symmetry of vario functions, and a resolution here may lead to inclusion of higher-order moments. More generally, new types of generalized functions in the spatial and spectral domain may be utilized. As one direction of future research, we discuss the topic of spatial classification and scale dependence.

Spatial Classification and Scale Dependence
The problem of spatial classification is related to the problem of scale dependence, as geophysical phenomena, data sets, spatial resolution, and characteristic lengths are each given at a specific scale, or range of scales. Two examples are compared in Table 1. At small scale, we may study spatial surface roughness of the Greenland ice sheet, resulting from morphogenetic processes forming melt runnels, sastrugi, ridges, etc, as described in Sect. 4. At large scale, spatial structure resulting from

Table 1 Spatial classification problems

Small-Scale	Large-Scale
GRS measurements	SAR data, ASTER data
complex classification parameters	simple classification parameters
($pond$, $mindist$, $p1, \ldots, p12$, $conc$, $avgspac$)	($pond$, $mindist$, $p1$, $p2$)
spatial surface roughness	spatial structure analysis
surface forms	deformation state
(ridges, melt runnels)	(crevasse patterns)
morphogenetic processes	kinematic and dynamic processes

deformation processes in glaciers may be analyzed, based on satellite rather than field data. Whereas the field data analysis typically yields a situation of oversampling, satellite data analysis usually concerns undersampled situations, because the energy in a signal is measured and averaged over the footprint of the instrument, which is usually larger than some surface features. In the geostatistical classification, complex parameters are used for extraction of information in oversampled situations, as described for field data of sea ice, Greenland inland ice and snow surfaces in Sect. 4, 6 and 7. In undersampled data situations, simple parameters are better utilized. As indicated in the table, we may study different types of processes at different scales. An interesting question in geophysical investigations is that of scaling versus scale-dependence, and parameters of type $pt2$ facilitate direct comparison of significance of surface features across scales. The examples in this paper already demonstrate that different processes reign at different scales. Ongoing research concerns the definition and application of transfer functions that relate spatial characteristics at the field scale to those at the regional scale and at continental scale. Results are relevant for climatic and atmospheric modeling and to study the interaction between the ice surface and the atmosphere.

Acknowledgments I like to thank Graeme F. Bonham-Carter and Qiuming Cheng for inviting me to contribute to this volume and Frits Agterberg for countless interesting discussions over the years. Thanks are due to Andre Journel, Phaedon Kyriakidis and Helmut Mayer for information on some of the references, to Axel Thomas and Graeme F. Bonham-Carter for comments on the manuscript, and to my collaborators and students without whom the projects mentioned in this paper would not have been possible (for brevity, I refer to the acknowledgements of the quoted papers). Support through Deutsche Forschungsgemeinschaft (DFG He1547/4, He1547/7, He1547/8, He1547/9), National Science Foundation (NSF EAR-0001514), NASA Cryospheric Sciences Program (NASA IPY NNX07AR21G) and a CIRES Visitor Fellowship is thankfully acknowledged.

References

Agterberg FP (1974) Geomathematics: mathematical background and geo-science applications. Elsevier, Amsterdam, p 596

Atkinson PM, Lewis P (2000) Geostatistical classification for remote sensing: an introduction. Comput Geosci 26 (4): 361–371

Armstrong M (1984) Problems with universal kriging. Math Geol 16 (1): 101–108

Bonham-Carter GF (1994) Geographic information systems for geoscientists. Pergamon, Oxford, p 398
Bennike O, Mikkelsen N, Klinge Pedersen H, Weidick A (eds) (2004) Ilulissat Icefjord. Geological Survey of Denmark and Greenland (GEUS), Copenhagen, p 116
Burger H (1981) Untersuchungen zur Klassifizierung von Gesteinsoberflächen auf Landsat-Aufnahmen mit Hilfe von Signatur- und Texturparametern. Berliner geowissenschaftliche Abhandlungen A 35
Caine N (1995a) Snowpack influences on geomorphic processes in Green Lakes Valley, Colorado Front Range. Geogr J 161: 55–68
Caine N (1995b) Temporal trends in the quality of streamwater in an alpine environment: Green Lakes Valley, Colorado Front Range, U.S.A. Geografiska Annaler 78A: 207–220
Carr JR (1999) Classification of digital image texture using variograms. In: Atkinson PM, Tate NJ (eds) Advances in remote sensing and GIS analysis. John Wiley and Sons, New York, p 135–146
Carr JR, Miranda FP (1998) The semivariogram in comparison to the co-occurrence matrix for classification of image texture. IEEE Trans Geosci Remote Sens 36 (6): 1945–1952
Chica-Olmo M, Abarca-Hernandez F (2000) Computing geostatistical image texture for remotely sensed data classification. Comput Geosci 26 (4): 373–384
Clarke GKC (1987) Fast glacier flow: ice streams, surging, and tidewater glaciers. J Geophys Res 92 (B9): 8835–8841
Dolgushin LD, Osipova GB (1975) Glacier surges and the problem of their forecast. Symposium on snow and ice in mountain regions, IUGG, IAHS Commission of Snow and Ice, XVth General Assembly, Moscow, 1971. IAHS Publication 104: 292–304
Echelmeyer K, Clarke TS, Harrison WD (1991) Surficial glaciology of Jakobshavns Isbræ, West Greenland: Part I. Surface morphology. J Glaciol 37 (127): 368–382
Echelmeyer K, Harrison WD (1990) Jakobshavns Isbræ, West Greenland: seasonal variations in velocity — or lack thereof. J Glaciol 36: 82–88
Fox CG (1990) Objective classification of ridge crest terrain using two-dimensional spectral models of bathymetry. Eos Trans Amer Geophys Union 71 (43): 1571
Franklin SE (1991) Image transformations in mountainous terrain and the relationship to surface patterns. Comput Geosci 17 (8): 1137–1149
Franklin SE, Wilson BA (1991) Spatial and spectral classification of remote-sensing imagery. Comput Geosci 17 (8): 1151–1172
Grafarend E (1975) Geodetic stochastic processes. Method Verf Math Phys 14: 1–27
Hambrey MJ, Milnes AG (1977) Structural geology of an Alpine glacier (Griesgletscher, Valais, Switzerland). Eclogae geologicae Helvetiae 70 (3): 667–684
Herzfeld UC (1990) Cova functions for unevenly and noncorrespondingly spaced processes. Comput Geosci 16 (5): 733–749
Herzfeld UC (1992) Least squares collocation, geophysical inverse theory, and geostatistics: A bird's eye view. Geophys J Internat 111 (2): 237–249
Herzfeld UC (1993) A method for seafloor classification using directional variograms, demonstrated for data from the western flank of the Mid-Atlantic ridge. Math Geol 25 (7): 901–924
Herzfeld UC (1998) The 1993–1995 surge of Bering Glacier (Alaska)–a photographic documentation of crevasse patterns and environmental changes. Trierer Geograph Studien 17: 211 p, Geograph Gesellschaft Trier and Fachbereich VI – Geographie/Geowissenschaften, Universität Trier, Trier, p 211
Herzfeld UC (1999) Geostatistical interpolation and classification of remote-sensing data from ice surfaces. Int J Remote Sens 20 (2): 307–327
Herzfeld UC (2002) Vario-functions of higher order–definition and application to characterization of snow surface roughness. Comput Geosci 28 (5): 641–660
Herzfeld UC, Higginson CA (1996) Automated geostatistical seafloor classification – principles, parameters, feature vectors, and discrimination criteria. Comput Geosci 22 (1) 35–52
Herzfeld UC, Mayer H (1997) Surge of Bering Glacier and Bagley Ice Field, Alaska: an update to August 1995 and an interpretation of brittle-deformation patterns. J Glaciol 43 (145): 427–434

Herzfeld UC, Mayer H (2003) Seasonal comparison of ice-surface structures in the ablation area of Jakobshavn Isbræ drainage system, West Greenland. Ann Glaci 37: 199–206

Herzfeld UC, Overbeck C (1999) Analysis and simulation of scale-dependent fractal surfaces with application to seafloor morphology. Comput Geosci 25 (9) 979–1007

Herzfeld UC, Zahner O (2001) A connectionist-geostatistical approach to automated image classification, applied to the analysis of crevasse patterns in surging ice. Comput Geosci 27: 499–512

Herzfeld UC, Kausch B, Stauber M, Thomas A (1997) Analysis of subscale ice-surface roughness from ultrasound measurements and its relevance for monitoring environmental changes from satellites. Trierer Geograph Studien 16: 203–228

Herzfeld UC, Mayer H, Feller W, Mimler M (1999) Glacier roughness surveys of Jakobshavns Isbrae Drainage Basin, West Greenland, and morphological characterization. Zeitschrift für Gletscherkunde und Glazialgeologie 35 (2): 117–146

Herzfeld UC, Mayer H, Feller W, Mimler M (2000a) Geostatistical analysis of glacier-roughness data. Annals Glac 30: 235–242

Herzfeld UC, Stauber M, Stahl N (2000b) Geostatistical characterization of ice surfaces from ERS-1 and ERS-2 SAR data, Jakobshavn Isbræ, Greenland. Annals Glac 30: 224–234

Herzfeld UC, Mayer H, Caine N, Losleben M, Erbrecht T (2003) Morphogenesis of typical winter and summer snow surface patterns in a continental alpine environment. Hydrol Process 17: 619–649

Herzfeld UC, Clarke GKC, Mayer H, Greve R (2004) Derivation of deformation characteristics in fast-moving glaciers. Comput Geosci 30: 291–302

Herzfeld UC, Box JE, Steffen K, Mayer H, Caine N, Losleben MV (2006a) A case study on the influence of snow and ice surface roughness on melt energy. Zeitschrift Gletscherkunde Glazialgeol 39 (2003/2004, printed 2006): 1–42

Herzfeld UC, Maslanik J, Sturm M (2006b) Geostatistical characterization of snow-depth structures on sea ice near Point Barrow, Alaska–A contribution to the AMSR-ICE03 Field Validation Campaign. IEEE Transactions on Geoscience and Remote Sensing (Special Issue on the March 2003 EOS AQUA AMSR-E Arctic Sea Ice Field Campaign) 44 (11): 3038–3056, DOI 10.1109/TGRS.2006.883349

Jiang M, Stewart WK, Marra M (1994) A neural network approach to classification of sidescan sonar imagery from a Mid-Ocean Ridge area. IEEE J Oceanic Eng 19 (2): 214–224

Journel A (1985) The deterministic side of geostatistics. Math Geol 17 (1): 1–15

Journel AG, Huijbregts C (1989) Mining geostatistics, 2nd edn. Academic Press, London, p 600

Kamb WB, Raymond CF, Harrison WD, Engelhardt H, Echelmeyer KA, Humphrey N, Brugman MM, Pfeffer T (1985) Glacier surge mechanism: 1982–1983 surge of Variegated Glacier, Alaska. Sciemce 227 (4686): 469–479

Kaula WM (1967) Theory of statistical analysis of data distributed over the sphere. Rev Geophys Space Phys 5: 83–107

Kondo J, Yamazawa H (1986) Aerodynamic roughness over an inhomogeneous ground surface. Bound-Layer Meteorol 35: 331–348

Krabill W, Frederick E, Manizade S, Martin C, Sonntag J, Swift R, Thomas R, Wright W, Yungel J (1999) Rapid thinning of parts of the southern Greenland ice sheet. Science 283: 1522–1524

Lauritzen S (1977) The probabilistic background of some statistical methods in physical geodesy. Publ 48, Dan Geod Inst, Copenhagen, p 96

Landgrebe DA (2003) Signal theory methods in multispectral remote sensing. Wiley and Sons, Hoboken, NJ

Lettau H (1969) Note on aerodynamic roughness parameter estimation on the basis of roughness element description. J Appl Meteorol 8: 828–832

Liu S, Jernigan ME (1990) Texture analysis and discrimination in additive noise. Proc Comput Vis Graph Image Process 49: 52–67

Lingle CS, Post A, Herzfeld UC, Molnia BF, Krimmel RM, Roush JJ (1993) Bering Glacier surge and iceberg-calving mechanism at Vitus Lake, Alaska, U.S.A. J Glaciol 39 (133): 722–727

Losleben MV (1990) Climatological data from Niwot Ridge, East Slope, Front Range, Colorado – 1989, Institute of Arctic and Alpine Research, University of Colorado, Boulder, Long-Term Ecological Research Data Report DR-90/1, p 108

Lu D, Weng Q (2007) A survey of image classification methods and techniques for improving classification performance. Int J Rem Sensing 28 (5): 823–870

Macdonald KC, Scheirer DS, Carbotte SM (1991) Mid-ocean ridges: discontinuities, segments, and giant cracks. Science 253 (5023): 986–994

Maslanik J, Sturm M, Belmonte Rivas M, Cavalieri D, Gasiewski AJ, Heinrichs JF, Herzfeld UC, Holmgren J, Klein M, Markus T, Perovich DK, Sonntag JG, Stroeve JC, Tape K (2006) Spatial variability of Barrow-area shore-fast sea ice and its relationships to passive microwave emissivity. IEEE Trans Geosci Rem Sens (Special Issue on the March 2003 EOS AQUA AMSR-E Arctic Sea Ice Field Campaign) 44 (11): 3021–3031, DOI 10.1109/TGRS.2006.879557

Matheron G (1963) Principles of geostatistics. Econ Geol 58: 1246–1266

Matheron G (1973) The intrinsic random functions and their applications. Adv Appl Prob 5: 439–468

Mayer H, Herzfeld UC (2000) Structural glaciology of the fast-moving Jakobshavn Isbræ, Greenland, compared to the surging Bering Glacier, Alaska, U.S.A. Annals Glac 30: 243–249

Mayer H, Herzfeld UC (2001) A structural segmentation, kinematic analysis and dynamic interpretation of Jakobshavns Isbræ, West Greenland. Zeitschrift für Gletscherkunde und Glazialgeologie 37(2)(2001, printed 2002): 107–123

Mayer H, Herzfeld UC, Clarke GKC (2002) Analysis of deformation types in fast-moving glaciers. Terra Nostra 4: 273–278

Meier MF, Post A (1969) What are glacier surges? Can J Earth Sci 6: 807–817

Oke TR (1987) Boundary layer climates, 2nd edn. Methuen & Co., London (reprinted 1995, Routledge, London, New York), p 435

Oliver MA, Webster R (1989) A geostatistical basis for spatial weighting in multivariate classification. Math Geol 21 (1): 15–35

Podlech S, Weidick A (2004) A catastrophic break-up of the front of Jakobshavn Isbræ, West Greenland, 2002/03. J Glaciol 50 (168): 153–154

Post A (1972) Periodic surge origin of folded medial moraines on Bering piedmont glacier, Alaska. J Glac 11 (62): 219–226

Post A, LaChapelle ER (1971) Glacier Ice. University of Washington Press, Seattle, WA

Ramsay JG, Lisle RJ (2000) The techniques of modern structural geology, vol 3: Applications of continuum mechanics in structural geology. Academic press, San Diego: 701–1061

Raymond CF (1987) How do glaciers surge? A review. J Geophys Res 92 (B9): 9121–9134

Ritter ND, Hepner GF (1990) Application of an artificial neural network to land-cover classification of Thematic Mapper imagery. Comput Geosci 16 (6): 873–880

Serra JP (1982) Image analysis and mathematical morphology. Academic Press, London

Stewart WK, Marra M, Jiang M (1992) A hierarchical approach to seafloor classification using neural networks. Proceedings of the IEEE Oceans 92 conference, Honolulu, Hawaii, 109–113

Tarantola A, Valette B (1984) Generalized nonlinear inverse problems solved using the least squares criterion. In: Grafarend EW, Rapp RH (eds) Advances in geodesy. American Geophys Union, Washington DC, 69–82 (from: Rev Geophys Space Phys 20 (2)[1982]: 219–232)

Thomas RH, Abdalati W, Frederick E, Krabill WB, Manizade S, Steffen K (2003) Investigation of surface melting and dynamic thinning on Jakobshavn Isbræ, Greenland. J Glaciol 49 (165): 231–239

Tucholke BE, Lin J (1994) A geological model for the structure of ridge segments in slow spreading ocean crust. J Geophys Res 99 (B6): 11937–11958

Tso B, Mather PM (2001) Classification methods for remotely sensed data. Taylor and Francis, New York.

Wallace C, Watts JM, Yool S (2000) Characterizing the spatial structure of vegetation communities in the Mojave desert using geostatistical techniques. Comput Geosci 26 (4): 397–410

Weertman J (1964) The theory of glacier sliding. J Glaciol 5: 287–393

Weszka JS, Dyer C, Rosenfeld A (1976) A comparative study of texture measures for terrain classification. IEEE Trans Syst Manage Cybern 6 (4): 2269–2285

The Rapid Retreat of Jakobshavns Isbræ, West Greenland: Field Observations of 2005 and Structural Analysis of its Evolution

Helmut Mayer and Ute C. Herzfeld

Reprinted from *Natural Resources Research* DOI: 10.1007/s11053-008-9076-7, when citing this article please use the DOI number.

Abstract Jakobshavns Isbræ in West Greenland (terminus at $\approx 69°\,10'\text{N}/50°\,\text{W}$), a major outlet glacier of the Greenland Ice Sheet and a continuously fast-moving ice stream, has long been the fastest moving and one of the most productive glaciers on Earth. It had been moving continuously at speeds of over 20 m/day with a stable front position throughout most of the latter half of the 20th century, except for relatively small seasonal changes. In 2002 the ice stream apparently suddenly entered a phase of rapid retreat. The ice front started to break up, the floating tongue disintegrated, and the production of icebergs increased.

In July 2005, we conducted an extensive aerial survey of Jakobshavns Isbræ to measure and document the present state of retreat compared to our previous field observations since 1996. We use an approach that combines structural analysis of deformation features with continuum mechanics to assess the kinematics and dynamics of glaciers, based on aerial imagery, satellite data and GPS measurements. Results from interpretation of ERS-SAR and ASTER data from 1995 to 2005 in combination with aerial imagery from 1996 to 2005 shed light on the question of changes versus stability and their causes in the Jakobshavns Isbræ dynamical system. The recently observed retreat of Jakobshavns Isbræ is attributed to climatic warming, rather than to an inherent change in the glaciodynamic system. Close to the retreating front, deformation structures are characteristic of extension and disintegration. Deformation provinces that do not border the retreating front have had the same deformation characteristics throughout the past decade (1996–2005).

Keywords Glacier · glacial retreat · global warming · deformation features

---

Helmut Mayer
UNAVCO, University of Colorado at Boulder, Boulder, Co, USA,
e-mail: mayerh@tryfan.colorado.edu

Ute C. Herzfeld
CIRES, University of Colorado at Boulder, Boulder, CO 80309-0449, USA,
e-mail: ute.herzfeld@colorado.edu

1 Introduction

Jakobshavns Isbræ (Sermeq Kujalleq) calves into Jakobshavns Isfjord (Ilulissat Isfjord, Kangia) at 69° 10′N/50° W in West Greenland. Jakobshavns Isbræ is known as the continuously fastest-moving glacier on Earth, moving at speeds of 20.6 m/d (7.5 km/yr, Pelto et al., 1989; 8.360 km/yr according to Echelmeyer and Harrison, 1990) at the calving front, and one of the most productive with 36.6 km^3/yr discharge, which corresponds in volume to 6% of the total accumulation of the Greenland Ice Sheet. The heavily-crevassed fast-moving south ice stream can be identified in the slow-moving Greenland inland ice up to about 80 km from its calving front. Most conspicuous is the wide shear margin that forms between the fast-moving ice and the slow-moving inland ice (approximately 0.3 m/d). The shorter northern branch (North Ice Stream) joins the south ice stream below a bulge termed "rumple" and continues side-by-side with the south ice stream, with a linear feature termed "the zipper" as the delimiter (both terms by Echelmeyer and Harrison, 1990). Jakobshavns Isbræ follows a geologic trough that exists in the bed of the Greenland Inland Ice and hence draws ice from a large drainage basin surrounding the trough into fast motion. The drainage basin covers 6.5% of the inland ice (Echelmeyer et al., 1991). The trough is 1200 m deep (below sea level) at the calving front and continues in the ice fjord. The terminus of the ice stream is about 800 m thick at its front and floating. Drift of icebergs into Disko Bay is obstructed by a large sea-floor shoal. Work on Jakobshavns Isbræ is summarized in Bennike et al. (2004). Ilulissat, a town of 5000 inhabitants and the commercial center of northwest Greenland with a harbour, an airport, a hospital, 6 elementary schools and a high school-boarding school, is situated two kilometers north of the mouth of the ice fjord on the shore of Disko Bay.

In terms of its glacial dynamics, Jakobshavns Isbræ has been considered the prototype of an autonomous dynamic system, which does not share the seasonal velocity variations of other glaciers (Echelmeyer and Harrison, 1990). An autonomous dynamic system is a dynamic system whose dynamic behaviour is time-independent ($A(x(t)) = A(x)$ for all times t in the time domain under consideration). Notably, an autonomous system can exhibit complex dynamics and, in case of a glacier, the position of the front can change. For Jakobshavns Isbræ changes in the front position have been documented by Weidick (in Bennike et al. 2004). In contrast, velocity variations have been noted in recent observations (Joughin et al., 2004 report velocities of 6700 m/yr in 1985, 5700 m/yr in 1992, 9400 m/yr in 2000 and 12600 m/yr in 2003, see also Zwally et al., 2002). In recent years, likely starting in 1999 according to our observations and culminating in 2005, Jakobshavns Isbræ has entered a state of drastic retreat, accompanied by considerable surface lowering (Zwally et al., 2002; Thomas et al., 2003; Abdalati et al., 2003; Joughin et al., 2004; Podlech and Weidick, 2004; see also Krabill et al., 1999).

More generally, fast-flowing ice streams are important, because they play a key role in the problem of possible instability of the Earth's ice sheets (Hollin, 1962; Alley and Whillans, 1991; Thomas, 1977; van der Veen, 1985; Muszynski and

Birchfield, 1987; Thomas and Bentley, 1978; Clarke, 1987; Schubert and Yuen, 1982). Under the influence of climatic warming, whether attributed to the usual variations in insolation (Milankovich cycles) or caused by increased atmospheric CO_2, the Antarctic ice sheets will probably not melt away slowly, but break up catastrophically, starting from the fast-moving ice streams (calculation: Huybrechts, 1993; Mercer, 1978; Hughes, 1973). That catastrophic disintegration of ice does indeed occur, has been documented by the recent break-up of Wordie Ice Shelf, Antarctica (Vaughan, 1993), Larsen Ice Shelf (in spring 2003) and other ice shelves around the Antarctic Peninsula.

Now the question is, where in relationship to these existing models of disintegration does Jakobshavns Isbræ fall? Is the retreat of Jakosbhavns Isbræ a dynamically or climatically forced event? Will it reverse? Or, at the other extreme, is it the start of a disintegration of the entire Greenland ice sheet?

For the Greenland ice sheet, meteorologic observations indicate that the annual mean air temperature in central Greenland increased by $2°C$ for the time period 1995–1999 compared to the standard decade 1951–1960 (Steffen and Box, 2001), while the mean temperature change decreased to approximately $1°C$ for the elevation 1000–2000 m. Changes in temperature lead to changes in small-scale surface structures, which may be summarized as spatial surface roughness. Melt energy depends with a factor of two on surface roughness for the range of measured roughness values (Herzfeld et al., 2006), and hence a self-enhancing process may be perpetuated involving warming, roughness and ablation in the Greenland ice sheet.

Surveys of high-resolution surface features in the area south of Jakobshavns Isbræ were undertaken during MICROTOP 1997 and 1999 expeditions with the Glacier Roughness Sensor, to provide subscale information for ERS SAR data and to study the morhpogenesis of the ice surface in the ablation area throughout several seasons (see Herzfeld et al., 1999, 2000a,b, Herzfeld and Mayer, 2003). While surface features in the largely non-crevassed area yield information on environmental processes, large-scale crevasse fields provide insight in the glacier's dynamic behavior (Mayer and Herzfeld, 2000, 2001; Herzfeld et al., 2004).

In July 2005, we undertook systematic aerial photographic, videographic and GPS surveys of Jakobshavns Isbræ with the objectives (1) to observe and document the extent of recent changes in the ice stream, and (2) to investigate whether the front retreat is accompanied by a change in deformation structures in the interior of the glacier, which may be attributed to an internally caused change in the glacier's dynamic system, or whether the retreat may be attributed primarily to climatic warming. The focus of this paper is on the presentation of aerial photographic, videographic and GPS data collected in summer 2005 and structural-glaciological analysis of this data material in comparison with satellite and photographic data from 1995–2003, covering both the changes in the location and appearance of the ice front, of the South and North Ice streams, and the surface structure of deformation provinces in central and upstream areas.

2 Calving Front

1995–1999. As is documented by a time-series of Synthetic Aperture Radar (SAR) data sets from ERS-1 and ERS-2, the position of the calving front changed only locally about a decade ago (Fig. 1c). In the three years 1995–1997, spring positions were generally more downstream (May/June 1995–1997) than autumn positions (September 1995–1997). For comparison with 2003 ASTER data (Fig. 2) and 2005 aerial photographs of the front position, it is worth noticing the calving-front position relative to the southern and northern headlands that mark the termination of Jakobshavns Isfjord and the start of the Greenland Inland Ice. The ice fjord is 7 km wide, measured from the westernmost end of the northern land (east of a small bay that is bordered by Nunatarsuaq in the west) to the north shore of Qeqertaarsunnguit Nunataat in the south (which is west of the embayment visible in the SAR data). The extension of the northern land segment is 12.9 km from the aforementioned small bay to the tip of the northern headland (Vandrekort Nordgrønland, Ilulissat, 1995/1996). The front position is about 10.1 km west of the tip of the northern headland in the May 1995 ERS-2 SAR (Fig. 1c) and at a similar position in spring 1996 and 1997, with about a 3.5 km seasonal fluctuation (the calving front is located farther east in September 1995–1999 than in spring of the same years). At the southern edge of the ice fjord, the floating ice front appears to be tied to a point at the northwestern head of Qeqertaarsunnguit Nunataat, while the retreating front forms an almost rectangular concave edge part-way across the northern limit of the bay between Qeqertaarsunnguit Nunataat, Isua, the southern headland, and the trough of the ice fjord that continues under the floating tongue. On 19 July 1999, at the end of the MICROTOP 1999 expedition to the inland ice, we noticed that the front in the area of the bay had retreated much farther (a few kilometers) than in previous years, and strong waterfalls had formed along the southern edge. The rapid retreat, reported by others as starting in 2002 or 2003 (Abdalati et al., 2003; Joughin et al., 2004) may have begun in the summer of 1999.

2003. A satellite image derived from ASTER data of 28 May 2003 shows the front of Jakobshavns Isbræ with several characteristics of a rapid retreat and disintegration (Fig. 2). The Advanced Spaceborne Thermal Emission and Reflection Radiometer (ASTER) aboard NASA's TERRA satellite (http://terra.nasa.gov/About/ASTER/about-aster.html; http://asterwebjpl.nasa.gov) is the result of a cooperative effort between NASA and Japan's Ministry of Economy and Trade. ASTER data, collected in 14 different wavelengths of the electromagnetic spectrum from visible to thermal infrared light with three telescopes (VNIR, SWIR, and TIR) at high resolution (15–90 m^2 per pixel) may be considered the next generation in remote-sensing imaging, compared to LANDSAT Thematic Mapper and the JERS-1 OPS scanner. ASTER data from the three high-resolution bands of the scene have been processed and color-enhanced using ENVI (Research Systems Incorporated, Boulder, Colorado, USA), such that the contrast created by crevasses and edges is emphasized, in order to facilitate a visual interpretation of deformation structures as

(a)

(b)

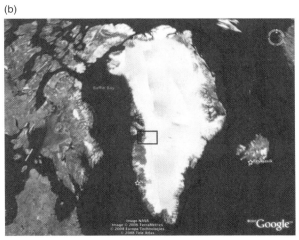

(c)

Fig. 1 (continued)

Fig. 2 Jakobshavns Isbræ Front Position. May 28, 2003 (ASTER Data, processed by U.C. Herzfeld)

captured by crevasse patterns. The result is a false-color composite, parts of which are shown in the figures in this paper.

In May 2003, the calving front meets the northern shore at 2.2 km from the tip of the headland, which is 7.9 km upstream from the May 1995 position. The calving front forms an arc, with the farthest-upstream-located point at the zipper, in a position that is 2.4 km upstream of the connecting line from the south shore to the north shore, and only 2.5 km downstream from the lower end of the "rumple". The arc reaches a point closest to the south shore that coincides with the eastern tip of the southern headland and extends 3.5 km westward across the bay in a small stripe of floating ice (the entire bay is 12 km wide in east-west, measured along the edge of the main ice fjord). This indicates that once the floating ice had disconnected from the tie point at the western edge of the bay, it retreated more rapidly from the bay than from the main ice fjord, which is an effect related to the presence of the bay rather than the retreat of Jakobshavns Isbræ.

Fig. 1 Jakobshavns Isbræ(Sermeq Kujalleq, Ilulissat Icestream), West Greenland. (**a**) Overview of Jakobshavns Isbræ and its structural glaciologic provinces, based on a structural segmentation of an ERS SAR-data composition (from Mayer and Herzfeld, Zeitschrift für Gletscherkunde und Glazialgeologie, vol. 37, 2001). (**b**) Location of Jakobshavns Isbræ in West Greenland (modified after Google Earth). (**c**) Lower Jakobshavns Isbræ and position of the calving front, May 1995. Based on ERS-2 SAR data (European Space Agency)

Large floes of crevassed ice that appear to have just recently calved off are floating in front of the south ice stream. The fjord is filled with dense brash ice for a few kilometers within the front, as was also the case in 1995–1997 (see the SAR images in Fig. 1). Large transverse crevasses indicative of increasingly fast extensional movement are visible in the lower North Ice Stream, in particular near the rumple where they extend 3.5 km upstream. Pieces in the North Ice Stream within the lowest kilometer may form the next icebergs. Strong extensional crevassing affects the lowest 1.5 km of the South Ice Stream, and much weaker extensional crevassing can be seen up to 7 km upstream of the calving front. This means that the lowest part of the North Ice Stream is affected most drastically by disintegration, whereas the South Ice Stream appears stiffer and more connected, which allows the extensional forces to be propagated higher up this branch of the ice stream (visibly to 7 km). We defined this as the areas neighboring the calving front and define the interior areas as those not apparently affected by the rapid extension that accompanies the retreat (deformation provinces that do not border the calving front; see Sect. 3).

2005, compared to previous decade. On 16 July 2005, we undertook an extended aerial survey of Jakobshavns Isbræ North and South Ice streams, to study front retreat and deformation changes. Photographic and videographic and GPS data were collected from a helicopter during several longitudinal and transverse profile flights. Calving-front-GPS positions obtained during overflights were $69°\,13.07'\text{N}/49°\,43.76'\text{W}$ and $69°\,13.148'\text{N}/49°\,43.683'\text{W}$ (Pt 55) for the North Ice Stream and $69°\,9.75'\text{N}/49°\,39.87'\text{W}$ and $69°\,11.084'\text{N}/49°\,40.738'\text{W}$ (Pt 58) for the South Ice Stream. Positions were obtained with the onboard GPS of the aircraft and with a handheld GPS, flying directly over the front in direction normal to the front position and looking straight down, and interpolating/averaging multiple measurements.

In Fig. 3, we present photographs from aerial observations undertaken in August 1996 and July 2005 that capture the essential deformation features of Jakobshavns Isbræ near the calving front, front position, morphology and calving activity. The 1996 image (Fig. 3a) clearly exhibits the autonomous behavior: A very regular surface structure is created by continuous movement throughout time, the same types of forces break up crevasses and move them along downstream, which leads to the formations of braided features and grouping of crevasses in longitudinal bands (see also Sect. 3). The "zipper" is seen in the upper center of the image, starting in a triangular area at the foot of the "rumple" (in center of image), at this time about 8 km from the front; the South Ice Stream is on the left and the North Ice Stream on the right. This area is near the calving front in 2005 (see below). The ice fjord is filled with large icebergs and brash ice; the open waters of Disko Bay are visible in the background and the ice-capped mountains of Disko Island are on the horizon on the right. View-direction is almost due west.

Figure 3b (Photograph gro2005-6-11) shows the location of the calving front in July 2005, looking southeast. The North Ice Stream, in the foreground, is retreating drastically, and ice near the headland is melting. Compared to 2003 (Fig. 2), the calving front still forms an arc, but has retreated farther, the contact point with the northern headland is now almost at the tip (a retreat of approximately 2 km since 2003, 6 km since September 1996, and 10 km compared to 1995–1997 spring positions. The upstream maximum of the arc has almost reached the bottom of the

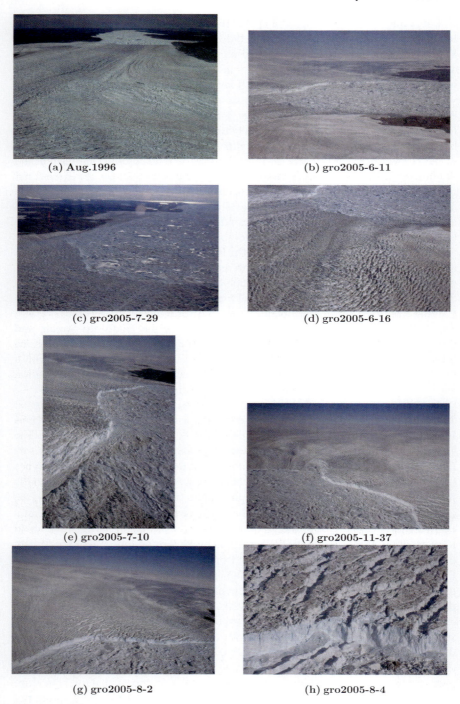

Fig. 3 (continued)

"rumple" (seen in Fig. 3b in the left and in the center of Fig. 3f), which indicates an 8 km retreat along the joint of the North and South Ice streams since summer 1996. The contact point of the calving front on the southern shore is best seen in Fig. 3c (gro2005-7-29), near a transform fault in the center of the headland (approximately 2.6 km from the position in May 2005).

Figure 3d shows extensional crevasses in longitudinal bands in the North Ice Stream. While the crevasse patterns in this province still bear the characteristics of their formation, a large gash has formed along the base of the "rumple" (Fig. 3f), detaching the North Ice stream from the "rumple" and the South Ice Stream. Water and ice brine stream out along a corridor following this gash, at a high velocity indicated by the linear patterns in the photograph and observed from the air during the survey flight. This is the same area that showed largest extensional crevassing in the 2003 ASTER image. The gash is also visible left of the center and left of the "rumple" in Fig. 3f (gro2005-11-37). The nature of the calving and the structure of the front area of the South Ice Stream are essentially the same as in 2003, the near-front ice bears the characteristics of deformation provinces in 1996, but more large ice floes have broken off. Extensional crevasses open up wider, as the fast-moving ice nears the front and the backholding force of lower parts of the ice tongue is disappearing with the calving events and with decreasing distance from the front. Extensional crevassing reaches less far upstream in 2005 than in May 2003 (the retreat may not be accelerating any more?). Widening of extensional crevasses affects in particular the southern half of the south ice stream, as seen in Fig. 3g (gro2005-8-2), reaching upstream for a few kilometers (about 4 km). A close-up view of the calving front in this area (Fig. 3h, gro2005-8-4) shows that huge pieces calve off the front, seracs are collapsing, leaving ice-debris cones at the base of the front, because the calved-off pieces cannot float off fast enough to clear the area of new-calve-ice production before more calving occurs. The black piece in the front must stem from a medial moraine or other rocky debris enclosed in the glacier farther upstream. In summary, the entire scene is one of rapid and very active disintegration.

Fig. 3 Aerial photographs of the calving front of Jakobshavns Isbræ (all photographs by H. Mayer and U.C. Herzfeld). (**a**) Lower Jakobshavns Isbræ looking downstream from above the "rumple" (junction of north and south ice streams) along the "zipper" and out over Jakobshavns Isfjord (covered with icebergs and brash ice). Open water of Disko Bay in distance. Disko Island on *right*. August 1996. (**b**) Front position in July 2005, looking southeast. North Ice Stream in foreground, rumple at left center, South Ice Stream in upper half of image (gro2005-6-11). (**c**) Front position in July 2005, looking westsouthwest over the ice fjord (covered with icebergs and brash ice). Open water of Disko Bay in distance (gro2005-7-29). (**d**) Crevasse fields of lower North Ice Stream neighboring calving front, July 2005 (gro2005-6-16). (**e**) Disintegrating front of North Ice Stream in foreground, South Ice Stream front in left half, looking south to southern headland. July 2005 (gro2005-7-10). (**f**) Front position has receded to "rumple" (in center), North Ice Stream on *left*, South Ice Stream on *right*, looking northeast (gro2005-11-37). (**g**) Active calving at front of South Ice Stream and extensional crevasse fields, looking eastsoutheast. July 2005 (gro2005-8-2). (**h**) Close-up of calving front of South Ice Stream (detailed view of area in center of (**g**)). July 2005 (gro2005-8-4)

3 Interior Areas

Motivated by the observation that structural provinces that border the calving front are affected by increased extensional crevassing, caused by lack of backholding force, we define the interior area of Jakobshavns Isbræ as the union of the areas of any provinces that are not bordering the calving front (this matches the mathematical definition of the interior of a set in topology, if a point with a neighborhood is translated into a structural province). As is seen on most photographs and in the ASTER data, structural provinces are clearly delineated. Inside a structural province, or deformation province, the same forces reign, creating a homogeneous crevasse pattern. We derive deformation characteristics from structural patterns, in this case, crevasse patterns, and deduce information on the kinematics and dynamics of the ice in a given province, and in the system of provinces (see Mayer and Herzfeld, 2001). This structural glaciological approach utilizes principles of structural geology and continuum mechanics in a semi-quantitative way, i.e. relative velocities and sizes of forces can be deduced, but not absolute values. The advantage is that more complex situations can be analyzed than, for instance, by calculating the velocity field or using other one-parameter approaches.

A structural segmentation of Jakobshavns Isbræ was undertaken based on 1995 ERS-SAR data (Mayer and Herzfeld, 2001, see also Mayer and Herzfeld, 2000). We compare to this in the sequel, to investigate the question whether deformation provinces of Jakobshavns Isbræ have changed their characteristics, location and extension in a principal way, and ultimately, whether Jakobshavns Isbræ has undergone a change as a glacio-dynamical system.

Figure 4 shows the South Ice Stream of Jakobshavns Isbræ near the most prominent and southernmost bend in a subset of the 28-May-2003 ASTER data set. The South Ice Stream has the characteristics of a continuously fast-moving ice stream with a wide shear margin on its north and south sides. It is evident that ice streaming in from the side joins the central core of fast flow in the main trough (see an area in the bottom center of the image). Localized bulges, as are particularly visible in the lower left of the image, indicate where the glacier overcomes obstructions and the ice is slowed down, and accelerates below, as compressional features form on the upstream side of an obstacle and extensional crevasse fields on the downstream side. The fast-flowing core of the center appears smooth, compared to the heavily crevassed wide shear margin; the core is actually segmented into five longitudinal bands per side, as sketched in Fig. 5; the definition of the bands is given in the figure caption. The center is smoothest and has a braided pattern (Mayer and Herzfeld, 2001).

Our 1996 aerial surveys extended to about 48° W (about 62 km from the tip of the northern headland near 49° 49′W), and from a survey altitude of 3000 m a.s.l. one can see much further, to beyond the highest crevasses. In 2005 we flew to an easternmost position of 69° 10.012′N/49° 13.699′W about 18 km East of the present calving front at the South-Ice-Stream center and 23 km east of the headland's tip. At the easternmost location, we gained altitude with the helicopter, to get an overview

The Rapid Retreat of Jakobshavns Isbræ, West Greenland 123

Fig. 4 Jakobshavns Isbræ South Ice Stream. May 28, 2003 (ASTER Data, acquired through U.S. Geol.Survey/ EOS, processed by U.C. Herzfeld)

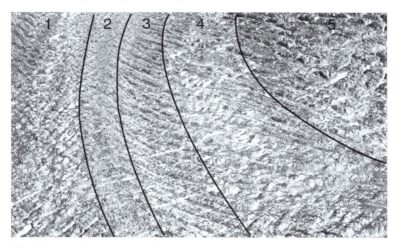

Fig. 5 Longitudinal zonation of the focused core of Jakobshavns Isbræ South Ice Stream, based on 1996 aerial photograph. From center (*left*) to margin (*right*), view in flow direction: (1) closed compressional and conjugate shear faults, (2) very smooth surface with pervasive shear deformation, (3) closed conjugate shear faults with one dominant direction curving into flow direction towards center, (4) closed conjugate shear faults, (5) slightly open extensional crevasses in multiple directions. Zones repeat symmetrically on other side of center. (from Mayer and Herzfeld, Zeitschrift für Gletscherkunde und Glazialgeologie, vol. 37, 2001)

over a much larger area. Crevasse patterns in the upper part of the glacier appeared unchanged from those of 1996 in photographic material and in the field.

In Fig. 6, we analyze and compare surface structures observed in 2005 with those observed in earlier years (1996, 1997, 1999). For more photographs and their analysis from the last decade, the reader is referred to Mayer and Herzfeld (2000, 2001). Figure 6a shows the longitudinal bands of Jakobshavns Isbræ in August 1996 and the wide shear margin on the north and south sides (south is on the right-hand side of the image and north is on the left, the view is upstream). Figure 6b shows the same central area of Jakobshavns Isbræ (also looking upstream). The succession of longitudinal bands is the same as in 1996, indicating that the dynamics in the core center of the ice stream has remained unchanged. Surficial water is indicative of a warm glacier, as the ice is melting on the surface during summer and not connected to subglacial channels (else it would drain). Surficial lakes existed in August 1996 also, for example, the lake in Mayer and Herzfeld (2001, Fig. 11) is in the same location as the central lake in Fig. 6b (gro2005-9-2), and another melt pond is forming on the outside of the bend where a larger one was located in August 1996.

Notably, surface structures in both the fast-moving part of Jakobshavns Isbræ and in the adjacent slow-moving parts of its drainage basin remain essentially unchanged throughout at least several seasons, modulo changes due to short-term weather, snowfall, and wind, albeit for different reasons (see Herzfeld and Mayer, 2003). The edges of the crevasses in some photographs may appear somewhat less sharp, which may be a consequence of warmer surface temperatures, or in the noise of image accuracy. A count of the distribution, size and time of occurrence of melt ponds on the surface of Jakobshavns Isbræ and its drainage basin may be a worthwhile number to derive from satellite imagery; one assumes that with warmer temperatures as reported in Steffen and Box (2001) the melt ponds would increase, which appears to be the case (J. Box, pers. comm).

Figure 6c (gro2005-8-17) provides a cross-section of the longitudinal zones in the South Ice Stream, this aerial view also serves to demonstrate that the concept of provinces is an appropriate one, as the zones are clearly delineated. Figure 6d (gro2005-10-21) gives a close-up view of zones 1–5 ((1) closed compressional and conjugate shear faults, (2) very smooth surface with pervasive shear deformation, (3) closed conjugate shear faults with one dominant direction curving into flow direction towards center, (4) closed conjugate shear faults, (5) slightly open extensional crevasses in multiple directions). The symmetry of the longitudinal zones in the center of the core is still prevailing, as is documented by Figure 6e (gro2005-9-3). Figure 6f demonstrates that clear borders exist between structural provinces, also in the shear margin. If the dynamics of the ice stream had undergone a significant change, such as a surge or another type of large acceleration, then at least some provinces would show signs of another type or generation of deformation overprinting the patterns observed in previous years (see Herzfeld and Mayer, 1997, Herzfeld, 1998, Mayer and Herzfeld, 2000). However, the surface patterns in all interior provinces are essentially the same as throughout the last decade (since 1995).

We conclude that Jakobshavns Isbræ is still a continuously fast-moving glaciodynamic system with a distribution of structural provinces as observed in 1996 (see

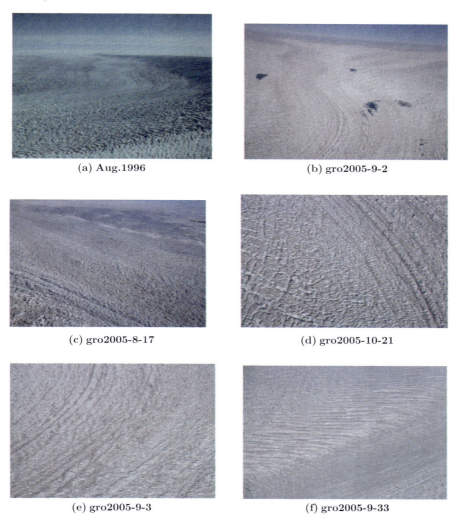

Fig. 6 Aerial photographs of ice-surface structures in interior areas of Jakobshavns Isbræ (all photographs by H. Mayer and U.C. Herzfeld). (**a**) Central and upper Jakobshavns Isbræ South Ice Stream, looking upstream (east). Heavily-crevassed marginal areas and comparatively smooth central core of fast-moving ice stream. August 1996. (**b**) Central and upper Jakobshavns Isbræ South Ice Stream, looking upstream (east). Photographer's position is farther upstream than in (**a**). Heavily-crevassed marginal areas and comparatively smooth central core of fast-moving ice stream. Surficial melt lakes. July 2005 (gro2005-9-2). (**c**) Detailed view of longitudinal structural zones of southern half of South Ice Stream (cf. Fig. 5), looking southsoutheast. July 2005 (gro2005-8-17). (**d**) Close-up of longitudinal structural zones of South Ice Stream (cf. Fig. 5). July 2005 (gro2005-10-21). (**e**) Symmetry of longitudinal structural zones around the central longitudinal axis, South Ice Stream. July 2005 (gro2005-9-3). (**f**) Clear borders between structural provinces, shear margin, South Ice Stream. July 2005 (gro2005-9-33)

Fig. 2 in Mayer and Herzfeld, 2001). For an interpretation of the recently observed changes in the front, the actual distribution of deformation provinces is less important than the conclusion that the glacio-dynamic system in the interior has not changed significantly. Hence the retreat of the ice front is not caused by an internal forcing in the dynamics, but should primarily be attributed to climatic warming and related melting.

Discussion. We distinguish between the observed changes and the potential causes of such changes. Changes in velocity, in particular near the ice front, as deduced from analysis of satellite data (Joughin et al., 2004, Luckman and Murray, 2005) are naturally associated with the retreat and advance of the ice front, both during the retreat and during seasonal changes in earlier years. In the sense of the definition of an autonomous dynamic system, the velocity field of Jakobshavns Isbræ is not time-independent any more, i.e. the glacier system does not constitute an autonomous system at present. There are divergences in the literature concerning the question whether it is correct that Jakobshavns Isbræ was an autonomous system, as stated by Echelmeyer and Harrison (1990).

Two important questions have been investigated: The first one concerns the extent of the area that is affected by the observed present changes. The second one concerns the physical explanation of the observed changes – are they caused by an internal change in the glacio-dynamic system, or are the changes caused by external forces, which may include environmental causes? A key to answering the second question lies in the answer to the first one, hence we have divided our investigations into those of the front and the interior areas. Large velocity variations are observed in the floating part, and are clearly associated with front-position changes, as have been seasonal variations of smaller magnitude in earlier years. Velocity variations farther upglacier are much smaller.

Luckman and Murray (2005) derive velocity fields from feature tracking applied to ERS-SAR data of 1998–2000, which is possible for an area near the front and areas of inflowing ice around the fast-moving part of Jakobshavns Isbræ. Large velocity changes occurred within a few kilometers of the front and rapidly increased from ≈ 16 m/d in 1992, 1995–1999 to ≈ 25 m/d in 2000 at a location "just downstream of the grounding line" (Luckman and Murray, 2005), whereas velocity at a point 30 km upstream varied only within the error bar of the analysis and increased slightly in 2000 (from 4.2 to 5.1 m/d). According to our observations, the glacier has now (by July 2005) retreated to the location of the previous grounding line. The decrease in velocity gradients with distance from the front matches our observations of drastic changes in the deformation state of structural provinces bordering the retreating front, and essentially stable deformation types in interior areas. With the front retreat, the extensional forces travel back up the glacier. Since the South Ice Stream has a lower gradient than the North Ice Stream, backpropagation of the relaxation reaches farther upstream in distance. Relatively small and gradually occurring velocity changes do not change the deformation type significantly, whereas deformation changes occurring during a glacier surge lead to overprinting of crevasses, indicating a different deformation state (Herzfeld and Mayer, 1997; Herzfeld, 1998; Mayer and Herzfeld, 2000; Herzfeld et al., 2004).

We explain the changes in velocity and front position as a consequence of climatic warming, which has been observed in Greenland (Steffen and Box, 2001). Melt-induced acceleration has also been given as an explanation in Zwally et al. (2002). Warming may have led to increased surface ablation and that to increased run-off on Jakosbhavns Isbræ. In addition, the contributions of ice that flows into the fast-moving part from slow-moving parts of the drainage basin may have varied with local ablation, surface topography and bed topography, causing some fluctuations with an overall positive trend. The reason that the deformation provinces in the interior are largely unchanged may lie in the existence of the large subglacial trough, which maintains its geological position and is the cause for the continued fast flow of the central ice stream. At present, changes in the mass of flowing ice are small in comparison, hence the flow and deformation patterns in the interior remain the same. However, if the warming trend continues and the temperature of the ice rises, its hydrological properties and viscosity may change, which may in turn affect the dynamic behavior and enhance or counteract variability.

4 Calve-ice in Jakobshavns Isfjord and Disko Bay

The increased production of calve-ice caused by the rapid retreat of Jakobshavns Isbræ brings changes to the fjord and the ocean areas in its neighborhood. In this context, we briefly study the ice cover of the fjord. In the interpretation of Fig. 3g,h it was already noted that the drift of calve-ice away from the calving front is overwhelmed by the increasing production. The entire fjord is filled with icebergs and brash ice, more densely in a several-kilometer-wide area near the front. The largest icebergs are more than a kilometer in diameter.

The entrance of Jakobshavns Isfjord is protected by a seafloor shoal, which forms a barrier to large icebergs, drifting out of the ice fjord. The large icebergs stranded

(a) gro2005-13-6   (b) gro2005-13-25

Fig. 7 Drift of icebergs of Jakobshavns Isbræ (**a**) Large icebergs and brash ice in Ilulissat Ice Fjord (Jakobshavns Isfjord), open water in Disko Bay. Drift of icebergs out of the fjord is impeded by large seafloor shoal. July 2005 (gro2005-13-6). (**b**) Comparative sizes of icebergs from Jakobshavns Isbræ in Disko Bay and houses (Ilulissat). July 2005 (gro2005-13-25)

on the shoal hold the smaller bergs and brine back, to some extent. Whenever the pressure of following ice exceeds a threshold, the large icebergs burst out into Disko Bay, across the shoal. This has occurred each spring and been related to seasonal front retreat. Figure 7a (gro2005-13-6) demonstrates that the shoal is still in existence in July 2005 and has not been eroded by the passage of large icebergs, which increasingly cross the shoal now during break-outs. The size of icebergs floating in Disko Bay and stranded near the shore is illustrated in Fig. 7b, in comparison to houses on the coast in Ilulissat.

5 Summary and Conclusions

Aerial observations of the surface structure and front position of Jakobshavns Isbræ led to the following results: The glacier is at present (July 2005) retreating rapidly, likely since 1999. The retreat of the front is 8–10 km, compared to 1995–1997, depending on season of comparison and location along the calving front. Ice areas near the front show characteristics of very active calving and extensional crevassing. The lower parts of the North Ice Stream are more heavily affected by disintegration, whereas the extensional forcing that is related to increasing lack of back-pressure propagates farther up the South Ice Stream than the North Ice Stream. The joint of the two ice-stream branches is replaced by a corridor in the rumple area, along which ice from the North Ice Stream flows out beyond the calving front into the fjord.

While glacial structural provinces near the front are affected by the retreat, all interior areas of Jakobshavns Isbræ exhibit the same deformation characteristics as in the past decade (compared to 1995–1999 and 2003 data). In conclusion, the recently observed retreat of Jakobshavns Isbræ is attributed to climatic warming, rather than to a change in the glaciodynamic system caused by internal forcing.

The seafloor shoal at the entrance of Jakobshavns Isfjord is still in existence and holds back large icebergs, until a threshold is reached and a break-out of bergs into Disko Bay occurs, as has been observed throughout past decades. As a result of the increased production of Jakobshavns Isbræ more icebergs are present in Disko Bay near Ilulissat than in previous years.

Acknowledgments This paper is dedicated to Frits Agterberg. We would like to thank our pilots Dr. med. Thomas Rose (private), Jan Wilken and Egon Dietz (Grønlandsfly) and Bo Isaaksen (Air Alpha) for excellent survey flights, Monika Stauber, Oliver Zahner and Marion Stellmes, Geomathematik Universität Trier, and Steven Sucht, CIRES, University of Colorado, for assistance with acquisition and processing of the 1995–1999 ERS SAR data, Koni Steffen, CIRES, University of Colorado, for acquisition of the raw ASTER data from 2003, to Graeme F. Bonham-Carter and Qiuming Cheng for inviting us to contribute to this issue and to Ralf Greve for helpful suggestions on the manuscript. Support provided by Deutsche Forschungsgemeinschaft (grants He 1547/4, He1547/8 and Ma2486/1) and through a CIRES Visitor Fellowship (UCH) is gratefully acknowledged.

References

Abdalati W, Manizade S, Golder J, Thomas RH, Krabill W, Csatho B (2003) Recent increase in flow rates of the Jakobshavn Isbræ, Greenland. Eos Trans Am Geophys Union 84 (46 Suppl.), F370.
Alley RB , Whillans IM (1991) Changes in the West Antarctic ice sheet. Science, 254, 950–963
Bennike O, Mikkelsen N, Klinge Pedersen H, Weidick A (eds) (2004) Ilulissat Icefjord. Geological Survey of Denmark and Greenland (GEUS), Copenhagen, p 116
Clarke GKC (1987) Fast glacier flow: ice streams, surging, and tidewater glaciers. J Geophys Res 92 (B9): 8835–8841
Echelmeyer K, Clarke TS, Harrison WD (1991) Surficial glaciology of Jakobshavns Isbræ, West Greenland: Part I. Surface morphology. J Glaciol 37 (127): 368–382
Echelmeyer K, Harrison WD (1990) Jakobshavns Isbræ, West Greenland: seasonal variations in velocity – or lack thereof. J Glaciol 36: 82–88
Herzfeld UC (1998) The 1993–1995 surge of Bering Glacier (Alaska) – a photographic documentation of crevasse patterns and environmental changes. Trierer Geograph Studien 17: 211 p, Geograph Gesellschaft Trier and Fachbereich VI – Geographie/Geowissenschaften, Universität Trier, Trier, p 211
Herzfeld UC, Mayer H (1997) Surge of Bering Glacier and Bagley Ice Field, Alaska: an update to August 1995 and an interpretation of brittle-deformation patterns. J Glaciol 43 (145): 427–434
Herzfeld UC, Mayer H (2003) Seasonal comparison of ice-surface structures in the ablation area of Jakobshavn Isbræ drainage system, West Greenland. Ann Glaciol 37: 199–206
Herzfeld UC, Mayer H, Feller W, Mimler M (1999) Glacier roughness surveys of Jakobshavns Isbrae Drainage Basin, West Greenland, and morphological characterization. Zeitschrift für Gletscherkunde und Glazialgeologie 35 (2): 117–146
Herzfeld UC, Mayer H, Feller W, Mimler M (2000a) Geostatistical analysis of glacier-roughness data. Ann Glaciol 30: 235–242
Herzfeld UC, Stauber M, Stahl N (2000b) Geostatistical characterization of ice surfaces from ERS-1 and ERS-2 SAR data, Jakobshavn Isbræ, Greenland. Ann Glaciol 30: 224–234
Herzfeld UC, Clarke GKC, Mayer H, Greve R (2004) Derivation of deformation characteristics in fast-moving glaciers. Comput Geosci 30: 291–302
Herzfeld UC, Box JE, Steffen K, Mayer H, Caine N, Losleben MV (2006) A case study on the influence of snow and ice surface roughness on melt energy. Zeitschrift Gletscherkunde Glazialgeol 39 (2003/2004, printed 2006): 1–42
Hollin J T (1962) On the glacial history of Antarctica. J Glaciol 4 (32): 173–195
Hughes T J (1973) Is the West Antarctic Ice Sheet disintegrating? J Geophys Res 78: 7884–7910
Huybrechts P (1993) Glaciological modelling of the late cenozoic East Antarctic ice sheet: stability or dynamism?, Gegrafiska Annaler 75 A: 221–238
Joughin I, Abdalati W, Fahnestock M (2004) Large fluctuations in speed on Greenland's Jakobshavn Isbræ glacier. Nature 432 (7017): 608–609
Krabill W, Frederick E, Manizade S, Martin C, Sonntag J, Swift R, Thomas R, Wright W, Yungel J (1999) Rapid thinning of parts of the southern Greenland ice sheet. Science 283: 1522–1524
Luckman A, Murray T (2005) Seasonal variation in velocity before retreat of Jakobshavn Isbræ, Greenland. Geophys Res Lett 32, L08501, doi:10.1029/2005GL022519
Mayer H, Herzfeld UC (2000) Structural glaciology of the fast-moving Jakobshavn Isbræ, Greenland, compared to the surging Bering Glacier, Alaska, U.S.A. Annals Glaciol 30: 243–249
Mayer H, Herzfeld UC (2001) A structural segmentation, kinematic analysis and dynamic interpretation of Jakobshavns Isbræ, West Greenland. Zeitschrift für Gletscherkunde und Glazialgeologie 37(2)(2001, printed 2002): 107–123
Mercer JH (1978) West Antarctic ice sheet and CO2 greenhouse effect: a threat of disaster. Nature 271: 321–325

Muszynski I, Birchfield GE (1987) A coupled marine ice stream – ice shelf model. J Glaciol 33: 3–15

Pelto MS, Hughes TJ , Brecher HH (1989) Equilibrium state of Jakobshavns Isbræ, West Greenland. Ann Glaciol 12: 127–131

Podlech S, Weidick A (2004) A catastrophic break-up of the front of Jakobshavn Isbræ, West Greenland, 2002/03. J Glaciol 50 (168): 153–154

Schubert G, Yuen DA (1982) Initiation of ice ages by creep instability and surging of the East Antarctic ice sheet. Nature 292: 127–130

Steffen K, Box JE (2001) Surface climatology of the Greenland ice sheet: Greenland climate network 1995–1999. J Geophys Res 106 (D24): 33951–33964

Thomas RH (1977) Calving bay dynamics and ice sheet retreat up the St. Lawrence valley system. Geogr Phys Quat 31: 167–177

Thomas RH, Bentley CR (1978) A model for Holocene retreat fo the West Antarctic ice sheet. Quat Res 10:150–170

Thomas RH, Abdalati W, Frederick E, Krabill WB, Manizade S, Steffen K (2003) Investigation of surface melting and dynamic thinning on Jakobshavn Isbræ, Greenland. J Glaciol 49 (165): 231–239

Vandrekort Nordgrønland, Ilulissat, Scale 1:100000, contour interval 25 m, 1995/96, Compukort, Denmark

van der Veen CJ (1985) Response of a marine ice sheet to changes at the grounding line. Quat Res 24: 257–267

Vaughan DG (1993) Implications of break-up of Wordie Ice Shelf, Antarctica for sea level. Antarctic Sci 5 (4): 403–408

Zwally HJ, Abdalati W, Herring T, Larson K, Saba J, Steffen K (2002) Surface melt-induced acceleration of Greenland ice-sheet flow. Science, 297 (5579): 218–222

Spatiotemporal Continuity of Sequential Rain Suggested by 3-D Variogram

Tetsuya Shoji

Reprinted from *Natural Resources Research* DOI: 10.1007/s11053-008-9074-9, when citing this article please use the DOI number.

Abstract A series of rainfalls observed in central Japan from noon on the 13th to midnight on the 14th, August 1999 (36 h) has been analyzed by spatiotemporal variograms in order to reveal the continuity of rain precipitation in a 3-D space defined by geographic coordinates and time. All instances of zero precipitation are considered, but have been treated as four different cases: Case 0 excludes all zero data, Case 1 includes a zero datum neighboring to each finite value, Case 2 includes a zero neighboring to each finite value and the next neighboring zero, and a fourth case (termed Case A) includes all zeros. Hourly precipitation has a statistical distribution best approximated by a Weibull model, and somewhat less well by a normal distribution, in all four cases. A rectangular variogram of measured values of total precipitation shows that the best continuity appears approximately along the N-S direction (the ranges given by directional variograms are 500 and 80 km in the N-S and W-E directions, respectively). In contrast, temporally stacked rectangular variograms of hourly precipitation shows that the best continuity direction is W-E in all cases (the ranges in Case A are 50 and 100 km along the N-S and W-E directions, respectively). A spatial variogram gives a spatial range independently of time, whereas a temporal variogram gives a temporal range. When geographic coordinates are normalized by the spatial range (here 80 km given by the temporally stacked omnidirectional variogram in Case A), and time is normalized by the temporal range (here 7 h given by the spatially stacked temporal variogram), geographic coordinates and time can be treated as equivalent variables. Consequently, a spatiotemporal variogram can be calculated along a given direction in 3-D space using the normalized coordinates. The continuity direction of a series of rainfalls can be best understood by display on a Wulff net, where each range value is written at a point corresponding to the direction. The direction of the best continuity is N0°W + 20° in the normalized space. A rectangular variogram in the normalized space, in which the horizontal and vertical axes represent N-S direction and time, respectively, suggests that the series of heavy rainfalls examined here had a continuity pattern that was elongated from west to east (the range values are 20–30 km and

Tetsuya Shoji
School of Frontier Sciences, The University of Tokyo, Kashiwa 277-8583, Japan,
e-mail: t-t_shoji@jcom.home.ne.jp

100 km along N-S and W-E, respectively), and that migrated from south to north with a speed of 30 km/h.

Keywords Rain · hazard · spatiotemporal · variogram · Wulff net · stacking · Japan

1 Introduction

Topographically Japan is characterized by many mountains and active volcanoes, because it is tectonically located on or near boundaries between continental and oceanic plates such as the European (continental), the North American (continental), the Philippine Sea (oceanic) and the Pacific (oceanic) plates. The mountainous topography means that the land is steep in many places. Meteorologically, Japan is characterized by high rainfall, with frequent heavy rain, because it is climatologically located in the western Pacific monsoon zone. Because of the topographical and meteorological character, therefore, Japan has many natural hazards caused by heavy rain such as floods, landslides and other problems. Understanding the spatial and temporal continuity of a series of heavy rainfalls may be an important contribution in predicting such kinds of natural hazards.

The purpose of this paper is to obtain fundamental knowledge for predicting natural hazards caused by heavy rain on the basis of statistical and geostatistical analysis of rainfall data. Rainfall data are recorded as functions of space and time. Therefore, analysis of rainfall is one of the most interesting themes in the spatiotemporal research field, and hence many papers have been already published. For example, Cox and Isham (1988) proposed a simple spatiotemporal model of rainfall in which storms arrive in a Poisson process, each storm consisting of a circular region of rain which moves with random velocity for a random time. They did not apply, however, the model to real data. Wheater et al. (2000) examined the spatial and temporal structure of rainfall in SW England by cross correlations and autocorrelations. The structure is shown, however, only in a geographic space and a temporal space independently, but not in a spatiotemporal space. Although the research field has many objectives, this paper focuses mainly on understanding spatiotemporal continuity of rainfall. A continuity structure in a 3-D geometric space can be recognized easily by stereographic projection using a Wulff net, on which range values of directional variograms are shown (e.g. Shoji et al., 2000). In contrast to a geometric space, a spatiotemporal space has a time axis which is very different from geometric axes. In order to use a Wulff net, consequently, time and geometric coordinates have to be transformed as they are equivalent. This paper tries to solve this problem by temporal and spatial stacking as discussed later. The objective is to obtain a result in which the spatiotemporal continuity of rainfall is shown visually.

This paper, parts of which were already presented orally (Shoji, 2005, 2006a), uses rain data in the period from noon (1200 h) on the 13th to midnight (2400 h) on the 14th of August 1999 (36 h). This rain event caused a severe flood in the Kanto district, which consists of a wide plain, the widest plain in Japan. The data were obtained from AMeDAS (Automated Meteorological Data Acquisition System) published by the Japan Meteorological Agency. The data are the same as those used in Shoji and Kitaura (2001, 2006), in which the rain data in the Kanto district and mountainous Chubu district (respectively located east and west in central Japan), were analyzed statistically and geostatistically. The main results are as follows: (1) hourly, daily and annual precipitations show lognormal distributions independently of the districts, whereas only monthly precipitation shows exponential distributions; (2) spatial variograms of annual precipitation has ranges of 130 km in both districts; (3) temporal variograms of hourly precipitation through a year have ranges of 8 h independently of the district; (4) if it rains heavily in a wide area with a series of rainfalls, spatial variograms of hourly precipitation in Kanto show clear ranges, which are generally 50–70 km; (5) ranges of variograms increase with increasing accumulation time, and become constant at 120–150 km over 3–5 h; and (6) ranges of temporal variograms are about 16 h in the situation when it rained for 30 h.

Sample stations at elevations lower than 200 m were selected, in order to remove the influence of topography. The district is about $2.1 \cdot 10^4$ km^2 in area, and includes 81 stations. Of these, 78 stations have complete rainfall records for the period under study (Fig. 1). The spatial density of stations is, accordingly, about 1/270

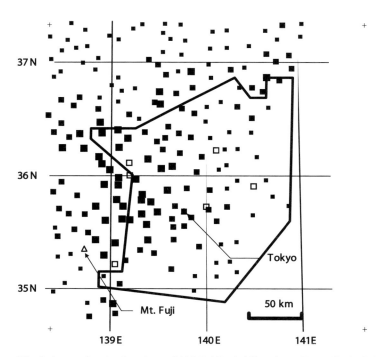

Fig. 1 A map showing locations of AMeDAS rainfall stations. Kanto district is bounded by broken line. Precipitation values (mm/36h) accumulated from 1200 h August 13 to 2400 h August 14, 1999 are coded by size: *large* > 200, *middle* = 200–100, *small* < 100, and *open* = unrecorded

station/km^2. Thus the nearest interstation distance is about 16 km in a tetragonal grid or 18 km in a trigonal grid.

2 Statistical Distribution Models

First, basic statistics were obtained from hourly and total precipitation data in the study period. Normal (Fig. 2), lognormal (Fig. 3), exponential (Fig. 4), and Weibull (Fig. 5) distribution functions were examined. Total precipitation values are finite at all stations, but hourly precipitation values are zero at many stations and intervals (ca. 50% ≈ 1416/2811). For the rain precipitation data, "zero" is interpreted in two ways: "absolutely zero" and "practically zero" (Shoji and Kitaura, 2006). "Absolutely zero" is for a sunny day, and "practically zero" represents almost immeasurable accumulation rain precipitation. Since these data show a series of rainfall events over the studied interval, most of "zero" data values can be treated as "practically zero". In this paper zero data were treated in four ways, termed here as "cases":

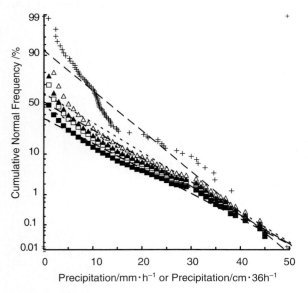

Fig. 2 Statistical plot of total and hourly precipitation values on a normal probability diagram. Horizontal and vertical axes represent precipitation rates (cm/36 h for total precipitation, and mm/h for hourly precipitation) and cumulative normal frequency, respectively. Symbols: *plus* = total precipitation, *open triangle* = hourly precipitation excluding all zero data (Case 0), *solid triangle* = hourly precipitation including a zero datum neighboring to each finite value (Case 1), *open circle* = hourly precipitation including zero data neighboring to each finite value and next neighboring zero (Case 2), and *solid circle* = hourly precipitation including all zero data (Case A). Each dashed line shows a fitted probability model calculated by least-square method. See sample coefficients of determination summarized in Table 1 for fitness of each model

Spatiotemporal Continuity of Sequential Rain Suggested by 3-D Variogram

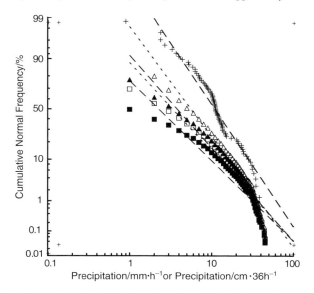

Fig. 3 Statistical plot of total and hourly precipitation values on a lognormal probability diagram. Axes and symbols are same as those in Fig. 2

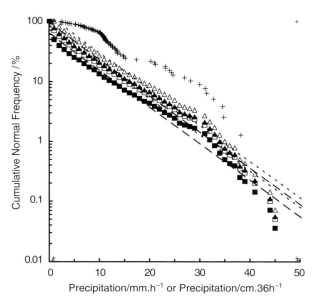

Fig. 4 Statistical plot of total and hourly precipitation values on a semi-log diagram. Axes and symbols are same as those in Fig. 2

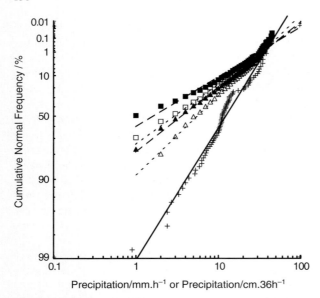

Fig. 5 Statistical plot of total and hourly precipitation values on a Weibull probability diagram. Axes and symbols are same as those in Fig. 2

(i) all zero data were excluded (Case 0), (ii) any zero data value occurring next to a positive value in time at a particular station is included (Case 1), (iii) any zero data value observed within the second neighbor (in time and at the same station) of a positive value is included (Case 2), and (iv) all zero data values are included (Case A).

Table 1 summarizes sample coefficients of determination (square of correlation coefficient) between precipitation values and the corresponding values of the inverse function for the four distributions. In the hourly precipitation data, the highest correlation is seen for the combination of Case 0 (i.e. excluding all zeros) and the Weibull distribution, and the coefficient of determination is 0.993 ($r = 0.996$). The

Table 1 Sample coefficients of determination (square of correlation coefficient) between precipitation and frequency cumulated from high-value side in some statistical distribution models

			Statistical Model			
Data	Case*	No. of Data	Normal	Lognorm.	Exponen.	Weibull
Hourly Precipitation	0	1395	0.966	0.951	0.972	0.993
	1	1805	0.965	0.926	0.974	0.987
	2	2006	0.972	0.916	0.974	0.980
	A	2811	0.982	0.892	0.973	0.958
Total Precipitation		78	0.884	0.943	0.959	0.968

* Case 0 excludes all zero data, Case 1 includes a zero datum neighboring to each finite value, Case 2 includes a zero neighboring to each finite value and the next neighboring zero, and Case A includes all zeros.

fitness of the Weibull distribution models deteriorates gradually with an increasing number of zero data taken into account. The second highest coefficient of determination is for the combination of Case A and the normal distribution ($r^2 = 0.982$ and $r = -0.991$). Contrary to the Weibull distribution, however, the fitness of the normal distribution deteriorates when a *decreasing* number of zero data are taken into account.

The empirical values of hourly precipitation show a bend at about 30 mm/h (e.g. Fig. 4). The slope is steep on the right side compared with the left side in any of the empirical distributions. The bending seems to suggest that the hourly precipitation data may come from two populations. The cause for the bending is not clear at present. For this reason, the present analysis does not take this point into account.

In contrast to the hourly precipitation values, total precipitation values do not show such large coefficients (Table 1). The Weibull distribution shows the largest coefficient of determination value (0.984). The exponential distribution has the next largest coefficient, followed by the lognormal distribution.

It seems to be satisfactory to calculate sample variograms using measured values of hourly precipitation, because a normal model fits these values well. However, hourly precipitation values studied previously over a longer time period of a year show a lognormal distribution (Shoji and Kitaura, 2006). For this reason, sample variograms were also calculated using logarithmic values of hourly precipitation, and compared with variograms of untransformed values. In this paper, the logarithmic hourly precipitation values are referred to as Case L (zero data are excluded in this case).

3 Spatial Variograms of Total Precipitation Values

Three types of sample variograms were calculated. The first and second types are omnidirectional variogram and directional variogram, respectively. The third type is rectangular variogram, in which distance vector $\boldsymbol{h}$ is given as (ξ, η) in rectangular coordinates. All variograms were calculated by a program written in MS-Excel/VBA, see Appendix, (Shoji, 2002a,b) under the condition that the lag interval and lag tolerance were 5 km and ±10 km (a circle whose radius is 10 km in the 2-D variogram calculation), respectively. A spherical model for every sample directional variogram was fitted as the following procedure and condition: (1) the spherical model of each omnidirectional variogram was calculated, and (2) a spherical model of each directional variogram was calculated under the condition that the nugget and sill values were equal to those of the omnidirectional variogram. This means that anisotropy is assumed to be geometric.

Omnidirectional and directional sample variograms of measured and logarithmic values of total precipitation values were calculated, and spherical models for the directional variograms were fitted. When measured values are used, the range of the omnidirectional variogram is 110 km, and the ranges of the directional variograms vary from 80 to 510 km. When logarithmic values are used, on

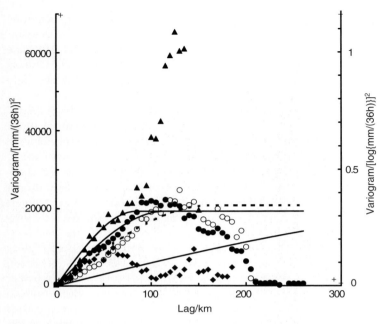

Fig. 6 Omnidirectional variograms (*circle*), N-S directional variogram (*rhomb*), and W-E directional variogram (*triangle*) of measured (*solid symbol*) and logarithmic (*open symbol*) values of total precipitation values. Solid lines (untransformed value) and dashed lined (logarithmic value) show fitted spherical models. Models of N-S and W-E directional variograms are calculated under condition that their nugget and sill values are equal to those of omnidirectional variogram

the other hand, the range of the omnidirectional variogram is 160 km, and the ranges of the directional variograms vary from 110 to 680 km. Figure 6 shows omnidirectional variograms together with N-S and W-E directional variograms of measured and logarithmic values of total precipitation. When the variograms are compared between measured and logarithmic values, the omnidirectional variograms and the W-E directional variograms are similar each other, whereas the N-S directional variograms are different. This difference is clearly shown by correlation coefficients between variogram values calculated from measured values and logarithmic values, where a data pair consists of variogram values calculated from measured values and logarithmic values at a given lag: they are 0.93, 0.07 and 0.98 in omnidirection, N-S direction and W-E direction, respectively. Figure 6 also shows spherical models, each of which is fitted for the corresponding sample variogram. The calculated ranges are 480 and 84 km along the N-S and W-E directions in the case of measured values, respectively, and 580 and 120 km along the N-S and W-E directions in the case of logarithmic values, respectively. (If two sample variograms are perfectly similar in form to each other, the correlation coefficient between variogram values of the variograms is 1, where two values consisting of a data pair for calculating the coefficient are connected with a lag vector. It is expected, therefore,

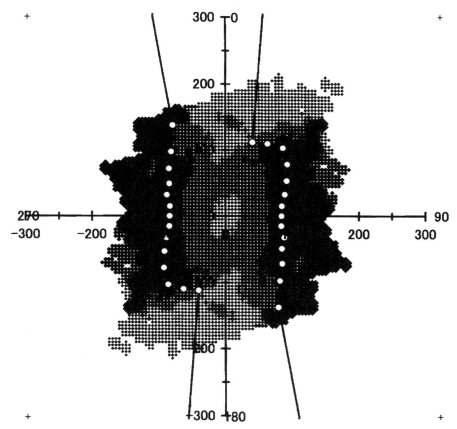

Fig. 7 A rectangular variogram of measured values of total precipitation. Degree of darkness represents values of variogram in (mm/36 h)^2: *dark* > 20000, *moderate* = 20000–5000, *light* < 5000. Broken lines (parts of a rose diagram) with *open circles* show ranges of directional variograms

that a high correlation coefficient of two sample variograms indicates the similarity between both variograms is also high.)

Figures 7 and 8 show rectangular variograms of measured and logarithmic values of total precipitation, respectively. The rectangular variograms are similar. In practice, the correlation coefficient between the two variograms is 0.80. Figures 7 and 8 show also ranges of directional variograms as rose diagrams (note that a rectangular variogram is point-symmetric), although their north and south parts are drawn out of the figures. Both rectangular variograms indicate that the best continuity direction is N-S, and that the worst one is W-E. Since the continuity along the N-S direction is good, the directional variograms along N-S do not reach the sills (Fig. 6), and the rose diagrams are open at north and south (Figs. 7 and 8).

The N-S directional sample variograms gradually increase with increasing lag, and do not reach the sills within the drawn range: that is, the variograms increase

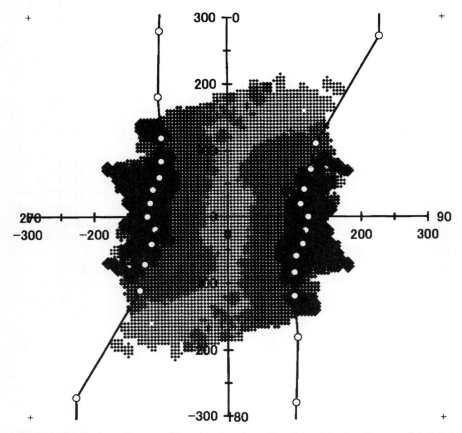

Fig. 8 A rectangular variogram of logarithmic values of total precipitation. Degree of darkness represents values of variogram in $\log^2$(mm/36 h): *dark* > 0.5, *moderate* $= 0.5$–0.1, *light* < 0.1. A rose diagram with *open circles* shows ranges of directional variograms

linearly. A variogram of the linearly increasing type is frequently obtained when a variable increases or decreases monotonously. According to Fig. 1, precipitation values are greatest in the west and least in the east. The variogram might thus be expected to increase linearly along the W-E direction. In practice, however, the ranges of the E-W variograms are clear. In contrast, the ranges of the N-S variograms are too long in the diagram, although precipitation values do not appear to increase or decrease monotonicly in the N-S direction. It is concluded, therefore, that the long range of the N-S variogram does not result from such a precipitation pattern, with precipitation being much greater on one side than the opposite side. The omnidirectional variograms increase with increasing lag, but decrease for lags greater than 150 km. This trend may be caused by occurrence of large precipitation values observed in the western side of the study area.

4 Spatial and Temporal Variograms of Hourly Precipitations

An omnidirectional sample variogram of hourly precipitation values was calculated at every hour with a lag tolerance of ±10 km in all cases, and was approximated individually by a spherical model. The ranges vary from 10 to 550 km except variograms of the linearly increasing type. An averaged omnidirectional variogram is obtained by the following equations:

$$\gamma(h_s) = \sum_t \gamma_t(h_s) n_t(h_s) \bigg/ \sum_t n_t(h_s) \tag{1}$$

where h_s is distance, and $\gamma_t(h_s)$ and $n_t(h_s)$ represent variogram value and pair number at time t, respectively. Let us use "temporally stacked" as the modifier representing a weighted average given by Eq. (1) (i.e. Eq. (1) gives a temporally stacked spatial variogram). The ranges of the temporally stacked omnidirectional variograms are 78, 66, 60, 50 and 88 h in Cases A, 2, 1, 0 and L, respectively. This indicates that the range values decrease with a decreasing number of zero data values taken into account, except for Case L.

Directional sample variograms of the temporally stacked type were also calculated by the following equation:

$$\gamma(h_s, \theta_s) = \sum_t \gamma_t(h_s, \theta_s) n_t(h_s, \theta_s) \bigg/ \sum_t n_t(h_s, \theta_s) \tag{2}$$

where θ_s denotes azimuth given at every 10 degrees. Spherical models for the directional variograms were fitted. Figure 9 shows the omnidirectional variogram and some directional variograms together with their fitted spherical models in Case A as examples of temporally stacked variograms. The ranges of the directional variograms of measured values vary from 40 to 100 km, and those of logarithmic values are from 70 to 120 km. Figure 10 shows ranges in Cases A, 2, 1, 0 and L as rose diagrams, which show ranges obtained from spherical models fitted for temporally stacked directional variograms. Figure 10 shows that the ranges of directional variograms increase with increasing number of zero data as with the omnidirectional variograms. All rose diagrams shown in Fig. 10 are elongated along the W-E direction. This implies that the best continuity direction of hourly precipitation values is W-E, whereas the worst continuity direction is N-S. This continuity pattern is contrary to the continuity pattern of total precipitation, where the best continuity direction is N-S.

Rectangular sample variograms were also stacked temporally by the following equation:

$$\gamma(\xi_s, \eta_s) = \sum_t \gamma_t(\xi_s, \eta_s) n_t(\xi_s, \eta_s) \bigg/ \sum_t n_t(\xi_s, \eta_s) \tag{3}$$

where (ξ_s, η_s) denotes lag vector $\boldsymbol{h}$, and $\gamma_t(\xi_s, \eta_s)$ and $n_t(\xi_s, \eta_s)$ represents variogram value and pair number at time t, respectively. Figure 11 shows a temporally stacked rectangular variogram calculated using all measured values (Case A) as an

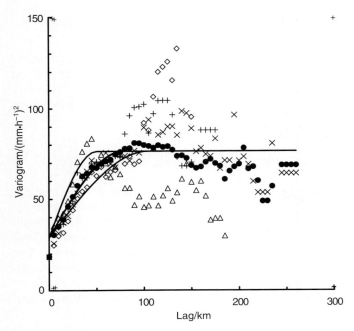

Fig. 9 A temporally stacked omnidirectional variogram (*solid circles*) and some temporally stacked directional variograms calculated by Eqs. (1) and (2), respectively (*open triangles* = N0°E, *crosses* = N40°E, *open rhombi* = N90°E, and *pluses* = S40°E) of hourly precipitations, and fitted spherical models (*solid lines*). Applied tolerance is a circle whose radius is 10 km.

example. Temporally stacked rectangular variograms in the other cases (i.e. Cases 2, 1, 0 and L) are similar to the pattern shown in Fig. 11. The spherical models are shown on the rose diagram by a broken line with open circles in Fig. 11 (the rose diagram is the same as shown in Fig. 10).

Shoji and Kitaura (2006) also calculated temporal variograms using the same data in order to reveal spatial and temporal continuities in short term. Figure 12 shows some of the temporal variograms at the stations where rain was recorded for longer than 26 h. Figure 13 shows the variogram summarized in the district, which is obtained by the following equations:

$$\gamma(h_t) = \sum_s \gamma_s(h_t) n_s(h_t) \bigg/ \sum_s n_s(h_t) \qquad (4)$$

where h_t is time lag, and $\gamma_s(h_t)$ and $n_s(h_t)$ represents variogram value and pair number at station s, respectively. Let us modify an average weighted by Eq. (4) with "spatially stacked" (i.e. spatially stacked temporal variogram). The ranges of the spatially stacked temporal variograms are 7, 9, 10, 19 and 19 h in Cases A, 2, 1, 0 and L, respectively. The ranges decrease with increasing number of zero data values.

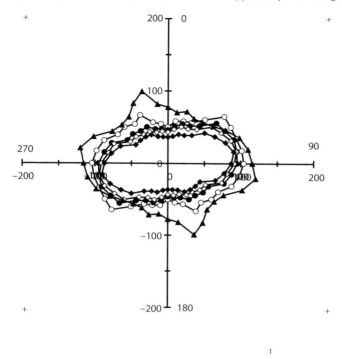

Fig. 10 Rose diagrams showing ranges of spherical models fitted for directional sample variograms stacked temporally by Eq. (2) under condition where nugget and sill values equal to those of omnidirectional variogram. From center outwards, roses with *solid rhombi, open rhombi, solid circles, open circles* and *solid triangles* show Cases 0, 1, 2, A and L, respectively

Cases 0 and L, which exclude all zero data, show extremely long ranges. It may be suggested, therefore, that the long ranges result from excluding zeros.

5 Spatiotemporal Variogram of Hourly Precipitations

In order to calculate a spatiotemporal sample variogram, geographic coordinates (i.e. northing and easting) and time data were normalized by the range (80 km) of the temporally stacked omnidirectional variogram, and the range (7 h) of the spatially stacked temporal variogram in Case A, respectively. This normalization makes every coordinate dimensionless, and hence makes all coordinates equivalent. Directional variograms were calculated with a lag tolerance of 0.2 (i.e. 16 km and 1.4 h) in the normalized spatiotemporal space. A spherical model was fitted for every sample variogram, where nugget and sill values were equal to those of the omnidirectional variogram. Figure 14 shows range values of directional variograms in the normalized spatiotemporal space on an upper hemisphere of a Wulff net, where left-right

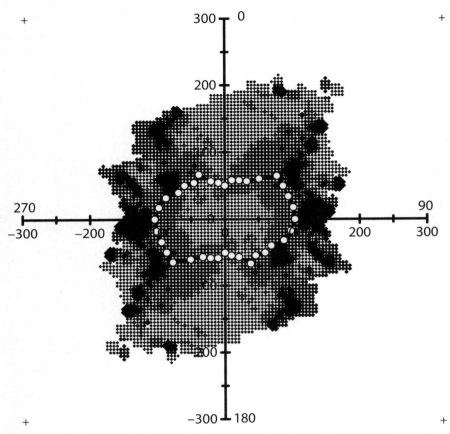

Fig. 11 A temporally stacked rectangular variogram of hourly precipitation values including all zero data, which was calculated by Eq. (3) (Case A). Degree of darkness represents variogram values in (mm/h)^2: *dark* > 0.4, *moderate* = 0.4–0.3, *light* <0.3. Applied tolerance is a circle whose radius is 10 km. A rose diagram with *open circles* shows ranges of spherical models fitted for directional sample variograms under condition where nugget and sill values are equal to those of omnidirectional variogram

and top-bottom correspond to west-east and north-south in the geographic coordinates, respectively, and the center corresponds to future in the temporal coordinate (past if a lower hemisphere is used). When the ranges are approximated by an ellipsoid in the normalized space, the directions of the longest and shortest ranges are N3°W + 18° and S55°E + 62° on a Wulff net, respectively.

The direction of the best continuity is N3°W + 18° (i.e. ca. N0°E + 20°) as stated previously. Considering this direction, Fig. 15 shows a 2-D variogram on the plane including the N-S azimuth (looking westwards), where the horizontal and vertical axes represent normalized N–S distance and normalized time, respectively. The best continuity direction runs from lower left (south-past) to upper right (north-future)

Spatiotemporal Continuity of Sequential Rain Suggested by 3-D Variogram 145

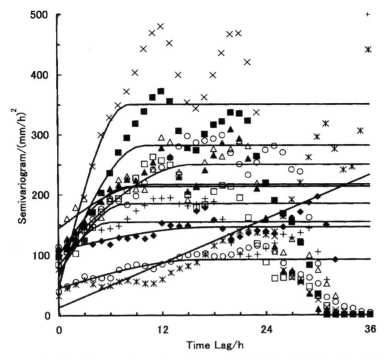

Fig. 12 Selected temporal sample variograms and fitted spherical models of hourly precipitations including all zero data (Case A) at stations which recorded rain more than 26 h. Stations: Kamisatomi = *star*, Fujioka = *open rhomb*, Kumagaya = *plus*, Hatoyama = *solid rhomb*, Hanno = *open triangle*, Oume = *solid triangle*, Hachioji = *cross*, Sagamiko = *solid square*, Sagamihara = *open circle*, and total area = *solid circle*

in the diagram, and the slope is about 20°. A unit on the horizontal axis is 80 km, whereas a unit on the vertical axis is 7 h. Therefore, the slope is about 0.03 h/km ($\approx$ 7 h/80 km × tan20°). The slope value suggests that the rain pattern migrated from south to north with a speed of about 30 km/h ($\approx$ 1/0.03 h/km). In other words, the zone of heavy rain was expected to have been observed at about 30 km north of the present position after 1 h, and at 60 km after 2 h.

The exact definition of "hourly precipitation" is the precipitation accumulated for one hour. This means that the range value of temporal variograms of hourly precipitation is determined by the combination of continuity of momentary rainfall and migration of rain areas. The range value along the N-S direction is 50 km as stated previously. On the other hand, the rain is estimated to have migrated from south to north with a speed of 30 km/h. These two values indicate, therefore, that the range value of momentary precipitation values along the N-S direction was 20 or 30 km ($\approx$ 50 km − 30 km/h × 1 h). Contrary to the N-S direction, the range values of total precipitation and hourly precipitation are similar as respectively 110 and 80 km along the W-E direction. This similarity suggests that the rain scarcely migrated

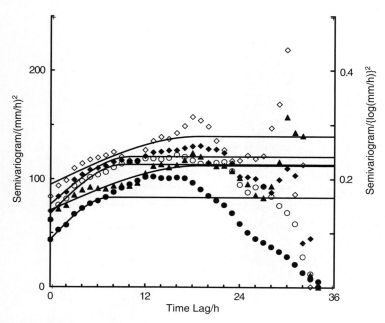

Fig. 13 Spatially stacked temporal variograms and fitted spherical models of hourly precipitation values in total area in five cases: Case 0 = *open rhomb*, Case 1 = *solid rhomb*, Case 2 = *open circle*, Case A = *solid circle*, and Case L = *solid triangles* (*right scale*). Stacking was carried out by Eq. (4)

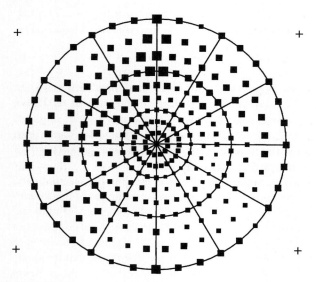

Fig. 14 Range values represented on an upper hemisphere of a Wulff net where *left-right* and *top-bottom* correspond to W-E and N-S, respectively, and center corresponds to time. Unit of geographic coordinates is normalized by range of temporally stacked spatial omnidirectional variogram (80 km), and that of time by range of spatially stacked temporal variogram (7 h). Size of symbols represents values of ranges: *large* > 2, *moderate* = 2–1, *small* < 1

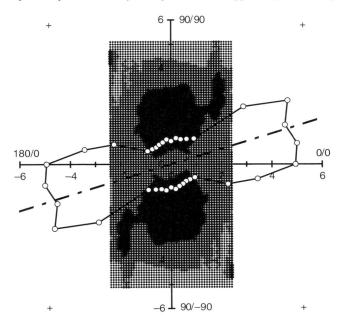

Fig. 15 A 2-D variogram where horizontal and vertical axes represent geographic coordinates (*left* = S and *right* = N) and time (*bottom* = past and *top* = future), respectively. Unit of horizontal axis is 80 km, and that of vertical axis is 7 h

along the direction. It is concluded, therefore, that the series of heavy rainfalls studied here had a continuity pattern elongated from west to east, and migrated from south to north.

The variogram range of momentary precipitation along the N-S direction is estimated to be 20–30 km. In contrast, the interstation distance is 16–18 km. This value is almost equal to the variogram range, and therefore seems to suggest that the observation system of rainfall should be denser. The temporal accumulation is always necessary in order to obtain precipitation data. Viewing from the rain, it is considered that a station migrated 30 km in an hour (i.e. during an accumulation interval of the measurement). This migration distance is 1.5 times the interstation distance. Each recorded precipitation is regarded, accordingly, to have been an average of values measured at 1.5 stations. In order to avoid such averaging, therefore, it is concluded that the accumulation interval should be shorter.

6 Conclusions and Remarks

Spatiotemporal variograms of the heavy rain events observed on August 13–14, 1999, in central Japan indicate the following features: (1) hourly precipitation values show statistical distributions approximated by the Weibull model and somewhat less well by the normal distribution; (2) a rectangular variogram of total precipitation shows that the continuity is the best approximately along the N-S direction (the ranges given by directional variograms are 500 and 80 km along the N-S and W-E directions, respectively); (3) rectangular variograms of hourly precipitation values show that the best continuity direction is W-E (the ranges are 50 and 100 km along the N-S and W-E directions, respectively); (4) the direction of the best continuity is $N0°W + 20°$ in the normalized space (i.e. on a Wulff net), where geographic coordinates are normalized by the spatial range (here 80 km given by the temporally stacked omnidirectional variogram in Case A), and time is normalized by the temporal range (here 7 h given by the spatially stacked temporal variogram); (5) the series of heavy rainfalls treated here had a continuity pattern elongated from west to east (the range values are 20–30 km and 100 km along N-S and W-E, respectively), and migrated from south to north with a speed of 30 km/h.

The present conclusions indicate that geostatistical analysis is one of powerful tools for predicting natural hazards caused by heavy rainfall. They are not enough, however, at many points. For instance, treating the data non-parametrically (e.g. a normal score transform) may give another result. The treatment seems to have to be taken into account, especially when kriging is applied for estimating rainfall amounts in spatiotemporal space. In order to predict natural hazards caused by heavy rain, rainfall amounts in a spatiotemporal space have to be simultaneously forecast. Often, actual forecasts are carried out using fewer data than in the present analysis, but are restricted by the time and circumstances of the forecast. This means the accuracy of many forecasts is often low compared with the present result. In contrast, however, the present results indicate the need for a denser station network and a shorter accumulation time. One of the most important research topics as a followup to this paper is, accordingly, to reveal forecast accuracies as functions of station densities and accumulation intervals. This paper did not compare the results obtained by geostatistical analysis with weather charts, because the main objective was to reveal the spatiotemporal continuity of rainfall visually. However, comparison between geostatistical results and meteorological charts is an important future research topic.

Acknowledgments I would like to thank Dr. Graeme F. Bonham-Carter of the Geological Survey of Canada, and Prof. Ryuji Kimura of The University of the Air for their critical reading of the manuscript and valuable suggestions. Acknowledgements are also given anonymous reviewers whose comments helped me to present this topic in greater depth.

Appendix

The software used in this paper is written in MS-Excel/VBA, and can give three types of sample variograms: omnidirectional and directional variograms, and rectangular variograms (Shoji, 2002). In a 2-D space, directional variogram values are given as a function of distance h and azimuth θ of lag vector $\boldsymbol{h}$ (only h for omnidirectional variogram) in polar coordinates, whereas rectangular variogram values are given as a function of W-E component ξ and N-S component η of h in rectangular coordinates. Either a fan or a circle is applied for tolerance in the calculation of a 2-D variogram. Fan-shaped tolerance can fill up a 2-D space exclusively and thoroughly, but the shape changes depending on h. On the other hand, circle tolerance has a shape independent of h, but is partly overlapped when it fills up a 2-D space thoroughly. The interval between neighboring lag classes is given independently of tolerance, and hence neighboring lag classes are overlapped when the lag class interval is shorter than the width (given by tolerance) of each lag class. Prepared variogram models are linear, spherical, exponential and Gaussian. Each of them is fitted by the following algorithm: (1) assume a range value; (2) obtain nugget and sill values of a variogram model approximating sample variogram values by least squares or weighted least squares (for a linearly transformed function if possible, otherwise of successive approximation), where weight is number of pairs at each lag value, and lags longer than the range are considered to be the range value for linear and spherical models; (3) calculate the deviation, which is sum of squared differences between the sample variogram and the variogram model; (4) return to Step 1, if the deviation is larger than the given permissible value. This fitting gives three parameters (nugget, sill and range) of a variogram model. If nugget and sill values are given, then only range is calculated (if either nugget or sill is given, range and another parameter are calculated). When data are 3-dimensional, the software can give 247 directional sample variograms automatically. A range value (nugget or sill value, too) of each directional variogram model obtained in a 3-D space can be represented by a figure or a symbol on a point of a Wulff net corresponding to the direction (e.g. Fig. 14), where positions of squares show 247 directions along which directional sample variograms are calculated. The software can be downloaded from the homepage of Japan Society of Geoinformatics (http://www.jsgi.org/).

References

Cox DR, Isham V (1988) A simple spatial-temporal model of rainfall. Proc R Soc A, Math Phys Sci 415: 317–328

Shoji T (2002a) Calculation of variograms by MS-Excel/VBA. Geoinformatics 13: 9–21 [in Japanese with English abstract]

Shoji T (2002b) Stereographic projection and variogram calculation by MS-Excel/VBA. In Proceedings annual conference of international association for mathematical geology (IAMG 2002), Berlin, Germany, 15–20 September, 2002 (Terra Nostra, Schriften der Alfred-Wegener-Stiftung 03/2002), pp. 461–464

Shoji T (2005) Continuity of a series of rain suggested by spatio-temporal variogram. In Proceedings annual conference of international association for mathematical geology (IAMG'05) Toronto, Canada, 21–26 August, 2005, vol. 2, pp. 687–691 (CD-ROM SN1213-101o#0687-0691_shoji)

Shoji T (2006) Spatiotemporal variograms: analysis of continuity of a series of rain, Preprints of Spring Meeting of MMIJ (Mining and Materials Processing Institute of Japan), Mar. 27–29, 2006, Kikaku 55–58. [in Japanese]

Shoji T, Kitaura H (2001) Statistical and geostatistical analysis of rain fall in central Japan. In Proceedings annual conference of international association for mathematical geology (IAMG'01), Cancun, Mexico, September 6–12, 2001, CD-ROM (Session D)

Shoji T, Kitaura H (2006) Statistical and geostatistical analysis of rainfall in central Japan. Comput Geosci 32: 1007–1024

Shoji T, Sasaki M, Nakamura T (2000) Geostatistical analysis of the stockwork gold mineralization in the Nurukawa kuroko-type deposits, northeastern Japan. In Proceedings of international Symposium geostatistical simulation in mining, 28th–29th October 1999, Perth, Australia

Wheater HS, Isham VS, Cox DR, Chandler RE, Kakou A, Northrop PJ, Oh L, Onof C, Rodriquez-Iturbe I (2000) Spatial-temporal rainfall fields: modeling and statistical aspects. Hydrol Earth Sys Sci 4: 581–601

Anisotropic Scaling Models of Rock Density and the Earth's Surface Gravity Field

S. Lovejoy, H. Gaonac'h and D. Schertzer

Reprinted from *Mathematical Geosciences* DOI: 10.1007/s11004-008-9171-7, when citing this article please use the DOI number.

Abstract In this paper we consider a anisotropic scaling approach to understanding rock density and surface gravity which naturally accounts for wide range variability and anomalies at all scales. This approach is empirically justified by the growing body of evidence that geophysical fields including topography and density are scaling over wide range ranges. Theoretically it is justified since scale invariance is a (geo)dynamical symmetry principle which is expected to hold in the absence of symmetry breaking mechanisms. Unfortunately to date most scaling approaches have been self-similiar, i.e. they have assumed not only scale invariant but also isotropic dynamics. In contrast, most nonscaling approaches recognize the anisotropy (e.g. the strata), but implicitly assume that the latter is independent of scale. In this paper, we argue that the dynamics are scaling but highly anisotropic i.e. with scale dependent differential anisotropy.

By using empirical density statistics in the crust; and a statistical theory of high Prandtl number convection in the mantle we argue that $P(\underline{K}, k_z) \approx (|K/k_s|^{H_z} + |k_z/k_s|)^{-s/H_z}$ is a reasonable model for the 3-D spectrum ($\underline{K}$ is the horizontal wavevector and K is its modulus, k_z a vertical wavenumber), (s, H_z) are fundamental exponents which we estimate as (5.3, 3), (3,3) in the crust and mantle respectively. We theoretically derive expressions for the corresponding surface gravity spectrum. For scales smaller than ≈ 100 km, the anisotropic crust model of the density (with flat top and bottom) using empirically determined vertical and horizontal density spectra is sufficient to explain the (Bouguer) g_z spectra. However, the crust thickness is highly variable and the crust-mantle density contrast is very large. By considering isostatic equilibrium, and using global gravity and topography data, we show that this thickness variability is the dominant contribution to the surface g_z spectrum

---

S. Lovejoy
Physics, McGill, 3600 University st., Montreal, Que. H3A 2T8, Canada,
e-mail: lovejoy@physics.mcgill.ca

H. Gaonac'h
GEOTOP, UQAM, Montreal, Canada, e-mail: gaonach.helene@uqam.ca

D. Schertzer
Université Paris-Est, ENPC/CEREVE, 77455 Marne-la-Vallee Cedex 2, France,
e-mail: Daniel.Schertzer@cereve.enpc.fr

over the range ≈ 100–≈ 1000 km. Using estimates of mantle properties (viscosity, thermal conductivity, thermal expansion coefficient etc.), we show that the mantle contribution to the mean spectrum is strongest at ≈ 1000 km, and is comparable to the variable crust thickness contribution. Overall, we produce a model which is compatible with both the observed (horizontal and vertical) density heterogeneity and surface gravity anomaly statistics over a range of meters to several thousand kilometers.

Keywords Geogravity · geopotential theory · fractals · multifractals · scaling

1 Introduction

1.1 Gravity as a Probe of the Earth's Interior: Gravity Anomalies and Depths to Sources

The Earth's gravity field is highly variable over a very wide range of spatial scales. There are two approaches which have been used to understand this. The most common has been to seek one to one (deterministic) relations between the fluctuations in surface gravity at a given scale and density anomalies at corresponding depths. In local or regional studies, this idea is commonly used to infer the depth to the source of gravity anomalies from the spectral peaks of surface gravity [e.g. Bullard and Cooper (1948), Spector and Grant (1970), Maus and Dimri (1996); the methods of wavelet analysis represent the most recent development in this type of application (e.g. Fedi et al. (2005))]. The second (neglected) approach has aimed at understanding and explaining the overall scale dependence of the statistics and the relations between the rock density and surface gravity statistics. Both approaches exploit a basic result of potential theory which shows that the contribution to the surface gravity at horizontal wavenumber K falls off exponentially with the depth of the layer.

Globally, the deterministic approach has attempted to interpret the separation of the density heterogeneities from different rheological layers – the lithosphere, asthenosphere, lower mantle, and core in order to understand the relationship between geodynamic processes and planetary gravity fields (see Bowin (2000) for a recent review). Figure 1 shows the earth geoid up to 360th order the EGM96 model, Lemoine et al. 1998, indeed it is plausible that the breaks at scales corresponding to ≈ 3000 km and to ≈ 100 − 200 km can be associated with the depths of core-mantle and mantle-crust boundaries. That the break at 100 km–200 km is indeed the reflection of the crust-mantle boundary with the mechanism of isostatic compensation can be confirmed by the comparison of the gravity and topography power spectra (Fig. 1). For uncompensated topography, the two spectra must be parallel (as they appear to be at high wave numbers). If the topography is completely compensated, the low wavenumber gravity power spectrum will be attenuated depending on the depth of compensation. The exact wavenumber where the break occurs depends on the flexural rigidity which varies locally presumably in a scaling manner.

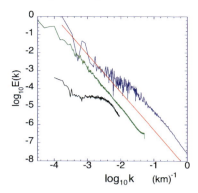

Fig. 1 Comparison of the surface gravity spectrum (*bottom*, shifted vertically for ease of comparison) with the ETOPO5 topography spectrum (*middle*) and continental US (*top*, two strips, each 512 × 65526 pixels long, each pixel, 90 m, also shifted vertically for clarity). The red reference line has slope -2.1. Wave number units: km^{-1}. Black, lower left shows the (isotropic) (spherical harmonic) global surface gravity spectrum calculated from the data discussed in Lemoine et al. 1998. The modes 1, 2 have been excluded since they go far off-scale. The extreme high frequency ($k = 360$) corresponds to ≈ 100 km

The topography for wavenumbers $< (200 \text{ km})^{-1}$ (see Sect. 3.5) is apparently fully compensated whereas for wavenumbers $> (100 \text{ km})^{-1} - (200 \text{ km})^{-1}$, it is not. Finally, the break at ≈ 3000 km in Fig. 1 could be a manifestation of the mantle core discontinuity.

A final note before continuing: we use the terms "crust" and "mantle" somewhat loosely; we recognize that in many cases the terms "lithosphere" and "asthenosphere" might be more technically exact.

1.2 Geophysical Scaling and Surface Gravity

The deterministic approaches have been most successful in determining characteristic scales – either of rheological transitions, or of the depths of anomalies. They give no information about – nor understanding of – the statistics as functions of scale between the break points, nor of the statistics of strong anomalies at fixed scales. In order to understand the observed wide range variability, some scaling (scale invariant) type assumptions are virtually mandatory since otherwise a (largely ad hoc) hierarchy of individual (nonscaling) sources of variability would have to be invoked every factor of 2 or 3 in scale.

Indeed there have been scaling models in solid earth geophysics ever since Vennig-Meinesz (1951) suggested that the spectral density[1] at horizontal wave number K in the topography spectrum follows:

---

[1] This is the (horizontal) angle integrated spectrum not the angle averaged spectrum; see discussion below.

$$E(K) \approx K^{-\beta_{top}} \tag{1}$$

with $\beta_{top} = 2$. Figure 1 shows that even with modern data, Vennig-Meinesz's spectrum is still an excellent approximation even down to scales of below 1 km (although $\beta_{top} \approx 2.1$, see Gagnon et al. 2003, 2006). If we perform isotropic scale reductions by a factor λ such that horizontal vectors are transformed as $\underline{X} \to \lambda^{-1}\underline{X}$ then the corresponding wave vectors are transformed as $\underline{K} \to \lambda\underline{K}$; we see that the power law form of $E(K)$ is conserved (it is "scaling"); the exponent β is "scale invariant". Spectra of this form are therefore expected if the underlying dynamical processes are also symmetric with respect to isotropic scaling transformations (systems symmetric with respect to such isotropic scaling transformations are called "self-similar").

The implications of the scaling of the topography for the gravity field have also been considered for some time. Kaula (1963) noted that the spectrum of the earth's geoid follows a power law of the type (1) but with $\beta_{geoid} \approx 3$. Since $\beta_{geoid} = \beta_g + 2$, and over the range 3000–200 km, Fig. 1 shows a flat gravity spectrum ($\beta_g \approx 0$) a value $\beta_{geoid} \approx 2$ is more realistic. Kaula already noted that the power spectrum of the gravity potential due to uncompensated surface topography should have $\beta_{geoid} = 4$ corresponding to a much more rapid fall-off than that observed. The discrepancy reflects the fact that the surface topography is by and large isostatically compensated.

Although studies of the scaling properties of rock density do not cover the same range of scales as those of gravity or topography, they have also tended to support the idea that various rock properties are scaling over wide ranges. For example, several recent (1-D) studies Leary (1997); Pilkington and Todoeschuck (1993), Shiomi et al. (1997) and Marsan and Bean (1999) have shown the rock density in boreholes to be scaling over the range ≈ 2 m to ≈ 1 km. The discovery of such empirical scalings have encouraged and Maus and Dimri (1995, 1996), Maus (1999) to explore the consequences for the surface gravity implied by assuming self-similar (isotropic, unstratified) scaling rock density fields; their basic result is $\beta_g = \beta_\rho + 2$ where β_ρ is the exponent of the 3-D isotropic rock density spectrum.

1.3 Anisotropic Scaling, Geomagnetism, Geogravity

The assumption of isotropic rock density statistics is quite unrealistic if only because it contradicts the obvious fact of geological stratification. This anisotropy has been noted and quantified for the magnetic susceptibility by Pilkington and Todoeschuck (1995), and – more extensively – for various different physical properties (including density) by Leary (1997, 2003) who compared the spectra of horizontal and vertical boreholes. These authors (and recently Tchiguirinskaia (2002) for hydraulic conductivity in both the vertical and horizontal) made the important point that the scaling is obeyed in both horizontal and vertical directions, but with different exponents in the different directions. In Lovejoy and Schertzer (2007), we review these and other scaling properties of intensive quantities such as rock density, magnetic susceptibility etc.

If the statistics in the horizontal and vertical directions are both scaling but different, then the overall system will be symmetric with respect to a scale change more general than isotropic reductions. The general formalism for handling scaling transformations is Generalized Scale Invariance (GSI; Schertzer and Lovejoy (1985a,b)). GSI shows that scale invariance is a (nonclassical) dynamical symmetry principle; as usual with symmetries; they are expected to hold in the absence of symmetry breaking mechanisms. A generic consequence of wide range GSI scaling are the existence of fractal structures with multifractal statistics; these features have now been reported in many areas of geophysics (including the topography analyzed in Fig. 1, Gagnon et al. (2003, 2006) and scaling has been proposed as a unifying paradigm for geodynamics Schertzer and Lovejoy (1991), Lovejoy and Schertzer (1998). These papers also argued that the development of scaling models in geophysics has been held back because of the all too frequent reduction of scaling to the isotropic special case of self-similarity.

In a pair of papers [Lovejoy et al. (2001), Pecknold et al. (2001)], we explicitly proposed that the earth's magnetization (M) respects such an anisotropic scaling symmetry. Using potential theory and with the help of multifractal simulations of M and the associated surface magnetic field (B) we explored the consequences for the B field anomalies and the relationships between the M and B statistics. Not only were we able to reconcile stratified, anisotropic M scaling with the surface B scaling, we also showed that the anisotropy leads to a qualitatively new scaling regime which could explain the intermediate scale (100–2000 km scale) surface B statistics.

Our aim in this paper is to extend these results to gravity which – also being a potential field – has several similarities. For example, as with magnetism, an important success of anisotropic scaling models is evident at scales smaller than that of the crust thickness. This is because the isotropic (self-similar) relation between gravity and rock density spectral scaling exponents ($\beta_g = \beta_\rho + 2$) is untenable since regional Bouguer gravity surveys have $\beta_g \approx 5$ (see Sect. 3.2.2) whereas empirically, $\beta_\rho \approx 1$ (see Sect. 3.2.1). However, since we show theoretically that[2] $\beta_g = \beta_\rho + 1 + H_z$ and the rock spectra indicate the anisotropy exponent $H_z \approx 3$, we will see that the small scale gravity exponent is correctly predicted by the theory and the rock density exponents.

Beyond these regional scales, the magnetism and gravity problems have important differences. For example, below the Curie depth (which is above the crust-mantle boundary), the magnetization vanishes whereas on the contrary, the corresponding gravity field has a source in the convective mantle. Although our information on the mantle density fluctuations is quite limited and indirect, a recent anisotropic scaling theory of high Prandtl number convection (Sect. 3.3) predicts that over the range ≈ 20–≈ 3000 km the density should indeed respect anisotropic scaling (also with $H_z = 3$); we calculate the corresponding surface gravity statistics and compare them to the global gravity spectra. An additional difference between magnetism and gravity is that M does not appear to have a sharp discontinuity at

---

[2] This formula is valid if β_ρ is the horizontal density exponent and $H_z > 1$; see Sect. 3.

the Curie depth so that the horizontal variation in the cut-off depth does not seem to generate a strong surface B anomaly. On the contrary, the density contrast at the crust-mantle boundary is large and is best modeled as a discontinuity across a (multi) fractal surface. We do not consider scales larger than ≈ 3000 km, so that we ignore core and possible core-mantle boundary contributions. Although several of the present calculations are similar to those in the magnetism problem, by making a small change in our scale function ansatz, we are able to obtain many exact results and therefore can make precise comparisons between the gravity theoretically predicted from the density observations/models, and both global and regional surface gravity surveys. Although a full statistical characterization of the fields requires statistics of all orders for relative simplicity, we here limit ourselves to spectra (which are 2nd order). While for quasi Gaussian (e.g. monofractal) models these spectral results fully characterize the statistics, for multifractal models they only give a partial characterization.

This paper is structured as follows. In Sect. 2 we review the basic theory of anisotropic scaling and relevant results of potential theory and derive the connections between the second order rock density and surface gravity statistics (in both real and fourier space). In Sect. 3, we apply the results to the crust, we develop a scaling model of mantle convection and apply the result to the mantle and estimate the contribution to surface gravity of a scaling mantle-crust interface/topography model. In Sect. 4 we conclude.

2 Symmetries and the Relation Between the Density and Gravity Fields

2.1 The Standard Density – Gravity Relations

In order to show how anisotropic scaling of the rock density field $(\rho(\underline{r}))$ can lead to a scale break in the surface gravity (g_z) or gravitational potential (ϕ), first recall the solution of the Poisson equation:

$$\underline{g} = -\nabla \phi; \quad \phi(\underline{r}) = G \int \rho(\underline{r}') \frac{1}{|\underline{r}-\underline{r}'|} d\underline{r}' \tag{2}$$

where G is the universal gravitational constant. The convolution in the above can be regarded as a fractional integration of order 1, hence if the problem (including the surface boundary conditions) were isotropic, the relative orders of singularity of the two fields (ρ, ϕ) would be simply shifted by one, leading to simple relations between the multifractal statistics of the two fields. However, the boundary conditions are clearly not isotropic, the classical assumption being that the rock is distributed over a half-volume bounded at $z = 0$. This amounts to ignoring the topography[3] and sphericity of the earth.

---

[3] or, to assuming that its effects can be removed/"corrected".

With this half-volume boundary condition, we obtain (e.g. Naidu (1968), Blakely (1995)) the following particularly simple expression for the horizontal fourier transform of the surface gravity:

$$\tilde{g}_z(\underline{K}) = 2\pi G \int_0^\infty \tilde{\rho}(\underline{K},z) e^{-zK} dz; \quad z \geq 0 \quad (3)$$

the horizontal wavevector is $\underline{K} = (k_x, k_y)$, $K^2 = |\underline{K}|^2 = k_x^2 + k_y^2$, and $\tilde{\rho}(\underline{K}, z)$ is the horizontal fourier transform of the density at depth z. We use the following convention for the D-dimensional fourier transform pair $\tilde{f}(\underline{k}), f(\underline{x})$:

$$\tilde{f}(\underline{k}) = \int_{-\infty}^\infty f(\underline{x}) e^{-i\underline{k}\cdot\underline{x}} d^D\underline{x}; \quad f(\underline{x}) = \frac{1}{(2\pi)^D} \int_{-\infty}^\infty \tilde{f}(\underline{k}) e^{i\underline{k}\cdot\underline{x}} d^D\underline{k} \quad (4)$$

and also the convention that $z > 0$ downward. Equation (3) shows that the contributions of deep layers are exponentially attenuated. Defining the three-vector $\underline{k} = (\underline{K}, k_z)$, a more convenient equivalent expression is obtained in terms of the 3-D Fourier transforms $\tilde{\rho}(\underline{K}, k_z)$:

$$\tilde{g}_z(\underline{K}) = 2\pi G \int_{-\infty}^\infty \tilde{\rho}(\underline{K}, k_z) \frac{dk_z}{-K + ik_z} \quad (5)$$

If we now assume statistical translational invariance, then the various fourier modes are statistically independent (Eq. 7) and the horizontal spectral density is easily obtained by multiplying the above by the complex conjugate and ensemble averaging:

$$P_g(\underline{K}) = 2(2\pi G)^2 \int_0^\infty \frac{P_\rho(\underline{K}, k_z) dk_z}{(K^2 + k_z^2)} \quad (6)$$

(the additional factor of 2 comes from the contributions for $k_z < 0$), and P_ρ, P_γ are the spectral densities of ρ, g_z and (from statistical translational invariance) we have used:

$$\langle \tilde{\rho}(\underline{k})\tilde{\rho}(\underline{k}') \rangle = P_\rho(\underline{k})\delta(\underline{k}+\underline{k}'); \quad \langle \tilde{g}_z(\underline{k})\tilde{g}_z(\underline{k}') \rangle = P_g(\underline{k})\delta(\underline{k}+\underline{k}') \quad (7)$$

Note that here and below, the symbol "<>" denotes ensemble (statistical) averaging.

We have used the symmetry $\tilde{f}(\underline{k}) = \tilde{f}^*(-\underline{k})$ (complex conjugate) valid when $f(\underline{x})$ is real.

If we now assume horizontal statistical isotropy, then the horizontal spectral density is a function only of K, and we define the isotropic spectrum $(E(K))$ by:

$$E(K) = 2\pi K P_g(\underline{K}); \quad K = |\underline{K}| \quad (8)$$

The isotropic spectrum E is usually used in the turbulence literature; in isotropic systems it has the advantage that (contrary to P), it is independent of the dimension of space (e.g. 1-D cross-sections will have the same E as for the full three dimensional system; this is not true for P).

2.2 Anisotropic Scaling

Up until the 1980s, scaling was restricted to isotropic systems with unique fractal dimensions. Since then two generalizations have been important for geophysical applications. First, the treatment of statistics of all orders (not only second order): multiscaling/multifractality, second the extension to anisotropic differentially stratified, and/or rotating systems: Generalized Scale Invariance (GSI, Schertzer and Lovejoy 1985a,b). In the following, for simplicity we pursue the second order statistics. Our results will be valid for both anisotropic fractal and multifractal rock density models although in the latter, they will only provide a rather limited characterization of the statistics.

In order to understand GSI it is helpful to introduce the dimensionless "scale function" $\|(\underline{X},z)\|$ which is the physically relevant notion of scale. The scale function satisfies the functional scale equation:

$$\|T_\lambda(\underline{X},z)\| = \lambda^{-1}\|(\underline{X},z)\| \tag{9}$$

where T_λ is the scale changing operator:

$$T_\lambda = \lambda^{-G} \tag{10}$$

and G is the generator. The scale function defines the physically relevant notion of size, scale, it is analogous to a norm, but need not respect the triangle inequality[4].

In the special case where the statistics of the anisotropy are independent of location (but not on scale), G is a matrix (linear GSI) and there exists conjugate fourier space scale functions which satisfy:

$$\| T_\lambda^T(\underline{K},k_z) \| = \lambda^{-1} \| (\underline{K},k_z) \| \tag{11}$$

where the "T" indicates the transpose (note the scale function in Eq. (10) is not generally the same as the real space counterpart which satisfies Eq. (9)).

If we have pure (scaling) stratification in the z direction, we may take the generator to be diagonal (this leads to self-affine statistics):

$$G = G^T = \begin{pmatrix} 1 & 0 & 0 \\ 0 & 1 & 0 \\ 0 & 0 & H_z \end{pmatrix} \tag{12}$$

---

[4] It need only define a series of decreasing balls: i.e. if $B_\lambda = T_\lambda B_1$ then $\lambda' > \lambda \Rightarrow B_{\lambda} \subset B_{\lambda'}$.

(the first two rows/columns refer to k_x, k_y, the last to k_z), and an anisotropic spectral density may be written:

$$P_\rho(\underline{K}, k_z) = P_0 \|(\underline{K}, k_z)\|^{-s} \tag{13}$$

where s is the spectral density exponent and P_0 is a constant determining the amplitude of the spectrum; if ρ is in Kg m^{-3}, then P_0 is in Kg^2 m^{-3}. A convenient, but not unique, choice of $\|(\underline{K}, k_z)\|$ is:

$$\|(\underline{K}, k_z)\| = \left(\left(\frac{K}{k_s}\right)^{H_z} + \frac{k_z}{k_s}\right)^{1/H_z} \tag{14}$$

where we have introduced a "sphero-wave number" k_s at corresponding spheroscale $l_s = 2\pi/k_s$ (note that K, k_s, $k_z > 0$). At this scale, $P_\rho(k_s, 0, 0) = P_\rho(0, k_s, 0) = P_\rho(0, 0, k_s)$, i.e. P_ρ is roughly constant over a sphere, since $\|(k_s, 0, 0)\| = \|(0, k_s, 0)\| = \|(0, 0, k_s)\| = 1$ horizontal and vertical fluctuations have the same variance. Indeed, here and in the following, any scale function satisfying linear GSI (i.e. including those in which G has off-diagonal elements, as long as its eigenvalues are real[5]) will give essentially the same qualitative results (including for the gravity spectrum) as those discussed here.

Using the sphero-scale as a reference scale, dimensional analysis gives:

$$P_0 = C\rho_s^2 l_s^3 = C\rho_s^2 (2\pi)^3 k_s^{-3} \tag{15}$$

where ρ_s^2 is the density variance at the sphero-scale and C is a dimensionless constant which depends on the exact definition of ρ_s and of the unit ball.

2.3 Second Order Horizontal and Vertical Density Statistics; the Crust

The above choice of P_ρ (Eqs. 16, 17) determines the second order horizontal and vertical density statistics. The horizontal spectrum is:

$$E_\rho(K) = 2\pi K \int_0^\infty P_\rho(K, k_z) dk_z = A_{\rho x} \rho_s^2 k_s^{-1} \left(\frac{K}{k_s}\right)^{-\beta_x};$$
$$\beta_x = (s - H_z - 1); \quad (\text{if } s > H_z) \tag{16}$$

(if $s < H_z$, then there is a high wavenumber divergence; if we prevent the divergence by using a finite high frequency cut-off, then $\beta_x = -1$). Here and below, the dimensionless constants will be denoted by A (spectral), B (real space), C (other)

---

[5] The case of complex eigenvalues involves an infinite number of rotations of structures as the scale is varied from 0 to ∞; it is probably not relevant to the vertical stratification problem. See Pecknold et al. (2001) for applications in surface magnetic anomaly mapping.

Table 1 A comparison of various density and gravity formulae assuming $s > H_z > 1$ (applicable to the crust), see Appendix 1

	Crust Statistics	
	Spectra	Structure Functions
Density, horizontal	$E_\rho(K) =$ $C_c 2(2\pi)^4 \frac{H_z}{s-H_z} \rho_s^2 k_s^{-1} \left(\frac{K}{k_s}\right)^{1+H_z-s}$; $s > H_z$	$S_\rho(\Delta X, 0) = \rho_s^2 \left(\frac{\Delta X}{l_s}\right)^{s-H_z-2}$
Density, vertical	$E_\rho(k_z) =$ $\frac{C_c(2\pi)^4 \Gamma\left(\frac{2}{H_z}\right)\Gamma\left(\frac{s-2}{H_z}\right)}{H_z \Gamma\left(\frac{s}{H_z}\right)} \rho_s^2 k_s^{-1} \left(\frac{k_z}{k_s}\right)^{(2-s)/H_z}$; $s > 2$	$S_\rho(0, 0, \Delta z) =$ $B_{\rho z} \rho_s^2 \left(\frac{\Delta z}{l_s}\right)^{\left(\frac{s-2}{H_z}\right)-1}$; $s - 2 > H_z$
Column integrated density fluctuation	$E_{I\rho}(K) = 2C_c(2\pi)^4 \rho_s^2 k_s^{-3} \left(\frac{k_s}{k_c}\right)^{1+s/H_z} \left(\frac{K}{k_s}\right)$; $K \ll k_s \left(\frac{k_c}{k_s}\right)^{1/H_z}$ $E_{I\rho}(K) =$ $2C_c(2\pi)^4 \rho_s^2 k_s^{-3} \frac{H_z+s}{H_z} \left(\frac{k_s}{k_c}\right) \left(\frac{K}{k_s}\right)^{1-s}$; $K \gg k_s \left(\frac{k_c}{k_s}\right)^{1/H_z}$	$S_{I\rho}(\infty) =$ $B_{I\rho}(2\pi)^2 \rho_s^2 k_s^{-2} \left(\frac{k_c}{k_s}\right)^{-\left(\frac{s-2}{H_z}\right)-1}$
Surface Gravity	$E_g(K) = A_{gch} \rho_s^2 k_s^{-3} G^2 \left(\frac{K}{k_s}\right)^{-s}$; $K \gg k_s$; $H_z > 1$ $E_g(K) = A_{gcl<} \rho_s^2 k_s^{-3} G^2 \left(\frac{K}{k_s}\right)^{-s/H_z}$; $s < H_z$; $K \ll k_s$ $E_g(K) = A_{gcl>} \rho_s^2 k_s^{-3} G^2 \left(\frac{K}{k_s}\right)^{-s-1+H_z}$; $s > H_z$	$S_g(\Delta X) \approx \frac{A_{gch}}{8\pi^2} G^2 \rho_s^2 \frac{\Delta X^2}{s-3}$
Geoid	$E_{geoid}(K) = \frac{E_g(K)}{g_z^2 K^2}$	$S_{geoid}(\Delta X) \approx \frac{A_{gch}}{8\pi^2} \frac{G^2}{g_z^2} \rho_s^2 \frac{\Delta X^2}{s-1}$; $\Delta X < l_s$

$$B_{I\rho} = C_c \frac{4\pi \Gamma\left(\frac{2}{H_z}\right)\Gamma\left(\frac{s-2}{H_z}\right)}{(s-2+H_z)\Gamma\left(\frac{s}{H_z}\right)}; \quad A_{gch} = C_c \frac{(2\pi)^7}{2}; \quad A_{gcl<} = C_c(2\pi)^7 \frac{\cos\left(\frac{\pi}{2}\left(1-\frac{s}{H_z}\right)\right)}{\cos\left(\frac{\pi}{2}\left(1-\frac{2s}{H_z}\right)\right)}; \quad s < H_z$$

$$A_{gcl>} = C_c 2(2\pi)^6 \frac{H_z}{s-H_z}; \quad s > H_z$$

and can when necessary be found by comparing the exact results in Tables 1, 2 with the corresponding formulae in the text (see appendices 1, 2 respectively). The corresponding vertical spectrum is:

$$E_\rho(k_z) = 2\pi \int_0^\infty P_\rho(K, k_z) K dK = A_{\rho z} \rho_s^2 k_s^{-1} \left(\frac{k_z}{k_s}\right)^{-\beta_z};$$

$$\beta_z = (s-2)/H_z \quad ; \quad (s > 2) \tag{17}$$

Anisotropic Scaling Models of Rock Density and the Earth's Surface Gravity Field 161

Table 2 A comparison of various formulae for $s = H_z$ with an exponential high wave number cut-off at k_s and ρ_s as defined such that $R_{\rho m}(0,0,0) = \rho_s^2$ corresponding to the mantle convection model, see Appendix 2. For the mantle model, take $H_z = 3$

	Mantle Statistics	
	Spectra	Structure Functions
Density, horizontal	$E_\rho(K) \approx$ $2C_m(2\pi)^4 \rho_s^2 k_s^{-2} K \log\left(\frac{k_s}{K}\right)^{H_z}$; $K << k_s$	$R_\rho(\Delta X, 0) \approx$ $2H_z C_m \rho_s^2 \left(\frac{l_s}{\Delta X}\right)^2$; $\Delta x >> l_s$
Density, vertical	$E_{\rho z}(k_z) = C_m(2\pi)^4 \left(\frac{2-H_z}{H_z^2}\right) \Gamma\left(\frac{2}{H_z} - 1\right) \times$ $\Gamma\left(1 - \frac{2}{H_z}\right) \rho_s^2 k_s^{-1} \left(\frac{k_z}{k_s}\right)^{2/H_z - 1}$	$R_{\rho z}(0, 0, \Delta z) =$ $B_{\rho z} \rho_s^2 \left(\frac{l_s}{\Delta z}\right)^{2/H_z}$; $\Delta z >> l_s$
Column integrated density fluctuation	$E_{I\rho}(K) =$ $2C_m(2\pi)^4 \rho_s^2 k_s^{-3} \left(\frac{k_s}{K}\right)^2 \left(\frac{k_s}{k_m} - 1\right)$; $K > k_s \left(\frac{k_m}{k_s}\right)^{1/3}$ $C_m(2\pi)^4 \rho_s^2 k_s^{-4} K \left(\left(\frac{k_s}{k_m}\right)^2 - 1\right)$; $K < k_s \left(\frac{k_m}{k_s}\right)^{1/3}$	
Surface Gravity	$E_g(K) \approx \frac{4C_m(2\pi)^6 G^2 \rho_s^2}{k_s^2 K} \log\left(\frac{k_s}{K}\right)$; $K << k_s$ $E_g(K) = 2C_m(2\pi)^6 G^2 k_s \rho_s^2 K^{-4}$; $K >> k_s$ downward continuation distance z_c: $E_{gd}(K) = E_g(K) e^{-2Kz_c}$	$S_g(\Delta X, 0) \approx$ $2C_m \left((2\pi)^2 G \, k_s^{-1} \rho_s \left(\log\left(\frac{k_s \Delta X}{2}\right) + \gamma_E\right)\right)^2 \Delta X >> l_s$
Geoid	$E_{geoid}(K) = \frac{E_g(K)}{g_z^2 K^2}$	$S_{geoid}(\Delta X, 0) \approx$ $C_m \left(4\pi^2 G \rho_s k_s^{-1} \Delta X \log \frac{k_s}{k_m}\right)^2$ $\Delta X >> l_s$

$$H_z = 3; \quad C_m = \frac{1}{(2\pi)^2 \Gamma\left(\frac{2}{3}\right)}; \quad B_{\rho z} = C_m 2(2\pi)^{3-2/H_z} H_z^{-1} \cos\left(\frac{\pi}{H_z}\right) \Gamma\left(1 - \frac{2}{H_z}\right) \Gamma\left(\frac{2}{H_z}\right)^2$$

(if $s < 2$, then there is a high wave number divergence; using a finite high frequency cut-off, we obtain $\beta_z = 0$).

Although k_s is the sphero-wave number as defined by the spectrum P_ρ, we note that:

$$\frac{E_\rho(k_z = k_s)}{E_\rho(K = k_s)} = \frac{\Gamma\left(\frac{2}{H_z}\right) \Gamma\left(\frac{s-2}{H_z}\right) (s - H_z)}{2H_z^2 \Gamma\left(\frac{s}{H_z}\right)} \tag{18}$$

which is not exactly unity (for the empirical crust exponents, $s = 5.3$, $H_z = 3$, we obtain a ratio 0.18; Γ is the usual gamma function). This fact points to the inherent inaccuracy of estimates of the sphero-scale obtained from 1-D spectra E (rather than from the spectral density P). We also note that here, the elliptical dimension characterizing the rate of increase in volumes of typical structures[6] is $d_{el} = 2 + H_z$.

---

[6] This type of spectrum was first proposed in the atmosphere Schertzer and Lovejoy (1985a) where the values $\beta_x \approx 5/3$, $\beta_z \approx 11/5$ (hence $d_{el} = 23/9 = 2.555...$) were derived from dimensional analysis and confirmed by observation.

When $H_z = 1$, we obtain the isotropic value $d_{el} = 3$, with the corresponding isotropic relation between exponents: $\beta_z = \beta_x = s - 2$.

It will also be convenient to express the statistics in real space via the correlation function (R) and structure functions (S). For statistically horizontally homogeneous systems these are defined by:

$$R(\underline{\Delta x}) = \langle f(\underline{x}) f(\underline{x} + \underline{\Delta x}) \rangle \tag{19}$$
$$S(\underline{\Delta x}) = \langle (f(\underline{x}) - f(\underline{x} + \underline{\Delta x}))^2 \rangle = 2(R(0) - R(\underline{\Delta x}))$$

¿From the Wiener-Khintchin theorem we have:

$$R(\underline{\Delta x}) = \frac{1}{(2\pi)^D} \int_{-\infty}^{\infty} P(\underline{k}) e^{i\underline{k} \cdot \underline{\Delta x}} d^D \underline{k}; \quad S(\underline{\Delta x}) = \frac{2}{(2\pi)^D} \int_{-\infty}^{\infty} P(\underline{k})(1 - e^{i\underline{k} \cdot \underline{\Delta x}}) d^D \underline{k} \tag{20}$$

For the models discussed here which are anisotropic in the vertical plane, but isotropic in the horizontal, we have:

$$R(0, 0, \Delta z) = \frac{1}{\pi} \int_0^\infty \cos(k_z \Delta z) E(k_z) dk_z;$$

$$S(0, 0, \Delta z) = \frac{2}{\pi} \int_0^\infty (1 - \cos(k_z \Delta z)) E(k_z) dk_z \tag{21a}$$

$$R(\underline{\Delta X}, 0) = \frac{1}{(2\pi)^2} \int_0^\infty E(K) J_0(K \Delta X) dK;$$

$$S(\underline{\Delta X}, 0) = 2 \int_0^\infty (1 - J_0(K \Delta X)) E(K) dK \tag{21b}$$

where $\underline{\Delta X} = (\Delta x, \Delta y)$ is a horizontal vector; $\Delta X = |\underline{\Delta X}|$ is the 2-D modulus and J_0 is the 0th order Bessel function.

Equations (16, 17) have been derived by assuming that the scaling of P is respected for all K, k_z; see Appendix 1 for the effect of finite cut-offs (necessary in particular to account for the finite crust thickness). The constants $A_{\rho x}$, $A_{\rho z}$ have been chosen so that ρ_s^2 is the (horizontal) sphero-scale density fluctuation variance (structure function):

$$S_\rho(\underline{\Delta X}, 0) = \rho_s^2 \left(\frac{\Delta X}{l_s}\right)^{\beta_x - 1} \quad s > H_z$$
$$S_\rho(0, 0, \Delta z) = B_{\rho z} \rho_s^2 \left(\frac{\Delta z}{l_s}\right)^{\beta_z - 1}; \quad s > H_z + 2 \tag{22}$$

i.e. by definition of ρ_s, $S_\rho(l_s,0,0) = \rho_s^2$. Note that rather than defining the sphero-scale via the fourier space k_s using $l_s = 2\pi/k_s$, one could define the sphero-scale in real space (l_{rs}) using for example $S_\rho(l_{rs},0,0) = S_\rho(0,l_{rs},0) = S_\rho(0,0,l_{rs})$. Since $B_{\rho z}$ is of order unity, the difference will generally not be large. However, if the β's are close enough to one (as is apparently the case in the crust), the difference can be large, see Appendix 1. Here and throughout, we use the fourier space definition $l_s = 2\pi/k_s$.

2.4 Second Order Horizontal and Vertical Density Statistics; the Mantle

The mantle model is discussed in Sect. 3.3, the spectrum is of the same general form as that discussed for the crust, hence it is appropriate to discuss the corresponding gravity formulae here. Although the mantle and the crust formulae share the same basic anisotropic scaling form, there are nevertheless significant differences. For example, for the mantle $s = H_z (= 3)$ whereas for the crust, $s > H_z$. This is significant since when $s = H_z$ there must be a high wavenumber cut-off at the sphero-scale to assure convergence of the horizontal spectra, i.e. formula 16 is only valid for K, $k_z < k_s$. Physically, the convection model upon which the density scaling law is based breaks down for these scales, the corresponding Peclet number is less than one, convection becomes ineffective. The necessity of a large wavenumber cut-off poses a technical problem: what is the most realistic/and or mathematically tractable cut-off? A related problem is the definition of the sphero-scale fluctuation variance ρ_s^2. The model choices made in dealing with these issues are considered in Appendix 2; they will alter the constants in the following by a factor of order unity (comparison of various models indicates that the factors may be as large as 4).

In the special case $s = H_z$, we have:

$$E_\rho(K) \approx A_{\rho x}\rho_s^2 k_s^{-2} K \log\left(\frac{k_s}{K}\right)^{H_z} \quad K \ll k_s \quad (23a)$$

$$E_\rho(k_z) \approx A_{\rho z}\rho_s^2 k_s^{-1}\left(\frac{k_z}{k_s}\right)^{2/H_z - 1} ; \quad k_z \ll k_s \quad (23b)$$

Similarly, in real space:

$$R_\rho(\underline{\Delta X},0) \approx B_{\rho x}\rho_s^2 \left(\frac{l_s}{\Delta X}\right)^2 \quad \Delta x \gg l_s \quad (24a)$$

$$R_\rho(0,0,\Delta z) = B_{\rho z}\rho_s^2 \left(\frac{l_s}{\Delta z}\right)^{2/H_z} \quad \Delta z \gg l_s \quad (24b)$$

(for the mantle, put $H_z = 3$ in the above). The correlation (rather than structure) function is used since the corresponding spectrum is an increasing function of horizontal wave number up to the cut-off so that R rather than S is a pure power law (see Eq. 20 for the relation between them). Following the discussion in Appendix 2, the optimum choice is the exponential cut-off model with the definition of ρ_s such that $R_\rho(0,0,0) = \rho_s^2$; these choices were used in determining the theoretical constants in Table 2.

2.5 Symmetries, Symmetry Breaking and the Gravity Statistics

We have seen that the gravitational potential ϕ is the convolution denoted "*" of density ρ with the Green's function $|\underline{r}|^{-1}$:

$$\phi \propto \rho * \frac{1}{|\underline{r}|} \tag{25}$$

(Eq. 2; $|\underline{r}|$ is the usual norm/distance) however, the Green's function is symmetric with respect to scale changes with isotropic generator $\boldsymbol{G} = \boldsymbol{1}$:

$$\left|\lambda^{-G}\frac{1}{\underline{r}}\right| = \lambda\frac{1}{|\underline{r}|}; \quad \boldsymbol{G} = \begin{pmatrix} 1 & 0 & 0 \\ 0 & 1 & 0 \\ 0 & 0 & 1 \end{pmatrix} \tag{26}$$

The result is that ϕ has broken symmetry. A direct calculation of the horizontal spectrum of the vertical component of gravity (with low frequency cut-off; k_c) gives:

$$E_g(K) = 2(2\pi)^3 G^2 K \int_0^\infty \frac{P_\rho(K,k_z)dk_z}{|(\underline{K},k_z)|^2} = 2(2\pi)^3 G^2 K \int_0^\infty \frac{dk_z}{|(\underline{K},k_z)|^2 \|(\underline{K},k_z)\|^s}$$

$$= 2(2\pi)^3 G^2 K \int_0^\infty \frac{dk_z}{(K^2 + k_z^2)[(K/k_s)^{H_z} + (k_z/k_s)]^{s/H_z}} \tag{27}$$

We now consider in turn the two cases $s > H_z$, $s = H_z$.
i) $s > H_z$:
For the crust, ($s > H_z$, no high frequency cut-off), this yields:

$$E_g(K) = A_{gch}\rho_s^2 k_s^{-3} G^2 \left(\frac{K}{k_s}\right)^{-\beta_h} \quad K \gg k_s$$

$$E_g(K) = A_{gcl}\rho_s^2 k_s^{-3} G^2 \left(\frac{K}{k_s}\right)^{-\beta_l} \quad K \ll k_s \tag{28}$$

i.e. there are two distinct regimes with high and low wavenumber exponents β_h, β_l given by:

$$\beta_l = s + 1 - H_z; \quad s > H_z$$
$$\beta_l = s/H_z; \quad H_z > s > 1 \quad H_z > 1$$
$$\beta_h = s \tag{29a}$$

This result shows that the incompatibility of the anisotropic scaling of the density with the isotropic scaling of the gravitational Green's function produces a break at the sphero-scale.

The corresponding formulae for $H_z < 1$ are:

$$\beta_l = s - 2;$$
$$\beta_h = s - 3 + H_z; \quad s > H_z; \quad H_z < 1$$
$$\beta_h = 2/H_z - 2; \quad s < H_z \tag{29b}$$

¿From Eqs. (33a,b) we see that if $s > H_z$, then for any H_z: $\beta_h - \beta_l = H_z - 1$. However, we shall see that the empirical rock density spectra constrain $s > 1$ and from Fig. 1 we see that all of the transitions have $\beta_h - \beta_l > 1$ so that anisotropic scaling with $H_z < 1$ cannot explain them. In addition, we will see that the empirical evidence is fairly clear that $\beta_x > \beta_z$ for various rock properties including density, implying $H_z > 1$ (see also the survey Lovejoy and Schertzer, 2007). Final evidence that $H_z > 1$ is that $\beta_h \approx 5$ so that $H_z < 1$ would imply (Eq. 39b) that $s \approx 8$, β_x, $\beta_z > 6$ which are much too large. In what follows, we shall concentrate on the parameter range $H_z > 1$ (in particular, the values $s = 5.3$, $H_z = 3$ give a reasonable fit to the high wavenumber rock and gravity spectra). Note that for $s > H_z > 1$, we have $\beta_l = \beta_{\rho x} + 2$ which provides a strong constraint on models since the mantle regime $((\approx 3000 \text{ km})^{-1} < k < (\approx 200 \text{ km})^{-1})$, has $\beta_g \approx 0$, and $\beta_{\rho x} \approx 1$. This rules out a simple linear GSI model for the crust/mantle transition. Finally, when $H_z = 1$ (isotropy), we recover the standard result $\beta_l = \beta_h = s = \beta_{\rho x} + 2 = \beta_{\rho z} + 2$.

ii) $H_z = s$:

For the mantle ($s = H_z = 3$), we obtain:

$$E_g(K) \approx A_{gmx,l} \frac{G^2 \rho_s^2}{k_s^2 K} \log\left(\frac{k_s}{K}\right); \quad K << k_s \tag{30a}$$

$$E_g(K) = A_{gmx,h} G^2 k_s \rho_s^2 K^{-4}; \quad K >> k_s \tag{30b}$$

Note that this formula ignores the downward continuation factor e^{-2Kz_c} necessary to take into account the fact that the mantle is at a depth z_c below the crust. The corresponding real space results are given in Table 2 and in Appendix 2.

It is also of interest to calculate the corresponding formulae for the geoid. The relation of the geoid and gravity spectra is:

$$E_{geoid}(K) = \frac{E_g(K)}{g_z^2 K^2} \tag{31}$$

Corresponding formulae are given in Tables 1, 2 and the Appendices 1, 2.

We have already noted that breaks in the gravity spectra introduced by the anisotropy of the rock density scaling, cannot in themselves explain the shape of the gravity spectrum (Fig. 1) if only because the latter has two breaks. Since the effect of low wave number cut-off is not trivial; and an understanding is helpful in evaluating this and other (more realistic) models discussed in Sect. 3, we give details of the effect of a cut-off in Appendix 1.

3 Scaling Models of the Density of the Crust, Mantle, Topography and Interface

3.1 Discussion

In modeling the density of the crust-mantle system, we will need hypotheses about the topography, the crust-mantle spatial correlations and the nature of their interface; the latter being important because of the large (typically ≈ 400 Kgm^{-3}) crust-mantle density contrast. Because of isostatic equilibrium the crust-mantle interface and the topography contributions are intimately connected; (see Sect. 3.4). The spatial correlations between the crust and the mantle are most simply dealt with by considering them to be statistically independent systems. Physically, the most unrealistic consequence of this neglect of mantle "roots" of crustal structures is that it implies a strong statistical discontinuity in structure at the interface; however since the interface will be treated as a (statistically independent) fractal discontinuity surface, this lack of statistical crust-mantle continuity may be less significant.

This model leads to the following equation (c.f. Eq. 3) for the surface Fourier transform:

$$\tilde{g}_z(\underline{K}) = \int_0^{z_c} \tilde{\rho}_c(\underline{K},z) e^{-zK} dz + e^{-z_c K} \int_0^{z_m - z_c} \tilde{\rho}_m(\underline{K},z) e^{-zK} dz; z \geq 0 \qquad (32)$$

where crust and mantle parts are indicated with indices "c", "m" and the crustal region is down to depth z_c, and the mantle between z_c and z_m. With the assumption of statistical independence of the crust and mantle (but also of fourier components, Eq. 7), we obtain

$$P_g(\underline{K}) = \left\langle |\tilde{g}_z|^2 \right\rangle \approx \int_{k_c}^{\infty} \frac{[P_c(\underline{K},k_z) + e^{-2Kz_c} P_m(\underline{K},k_z)]}{K^2 + k_z^2} dk_z$$

$$+ e^{-2Kz_c} \int_{k_m}^{k_c} \frac{P_m(\underline{K},k_z)}{K^2 + k_z^2} dk_z \qquad (33)$$

Where the factor e^{-Kz_c} takes into account the fact that the mantle layer starts at a depth z_c, not at $z = 0$ and where we have used the more convenient step-function fourier space cut-offs: $k_c \approx 1/|z_c|, k_m \approx 1/|z_m - z_c| \approx 1/|z_m|$ (i.e. take $z_m \gg z_c$). Equation (33) shows how the crust and mantle contributions to the surface gravity may be combined.

3.2 The Crust

3.2.1 Empirical Estimates of Model Parameters

Unfortunately, very few data exist on spectral exponents for the rock density. Leary, (1997) has probably the most extensive analyses with both horizontal and vertical spectra from similar regions. Due to strong (presumably multifractal) intermittency/ variability, (see Marsan and Bean (1999), Pecknold et al. (2001) individual boreholes have a fair amount of spectral variability (recall that the spectrum is an ensemble averaged quantity; the scaling is almost surely violated on every individual realization).

Before proceeding, it is useful to invert the relations (16–17) to obtain:

$$H_z = \frac{(\beta_x - 1)}{(\beta_z - 1)} \qquad (34)$$

which is a convenient formula for estimating H_z from spectra.

The difficulty in estimating H_z (and the sphero-scale) is that Leary's results give roughly $\beta_z \approx 1, \beta_x \approx 1$; his precise analysis of 45 spectra (30 vertical, 15 horizontal) yields $\beta_z \approx 1.1 \pm 0.12, \beta_x \approx 1.34 \pm 0.12$, yielding $H_z \approx 3$ (the nearest integer). A comparable value ($H_z \approx 2-3$) was obtained for the magnetization (M) (Lovejoy et al. (2001), Pecknold et al. (2001)); in obvious notation, if $H_{zM} = H_{z\rho}$ and $s_M = s_\rho$, then a statistical version of Poisson's relation may hold[7]. The spectrum from the much longer KTB borehole yields: $\beta_z \approx 1.2$ Lovejoy and Schertzer (2007); similarly Shiomi et al. (1997) obtains $\beta_z \approx 1.1 - 1.3$ for sedimentary, $\beta_z \approx 1.3 - 1.6$ for volcanic rock. Finally, we should note that Leary also gives nearly identical values for the exponents for gamma decay and sound velocity; this supports the idea that the value of H_z (and hence d_{el}) may be the same for different physical properties and hence supports the notion that it may be a fundamental characteristic of the geological stratification.

The poor estimates of H_z (due to their small horizontal/vertical difference) leads to great uncertainty in estimating the sphero-scale. It can be roughly estimated using Leary's spectra (which are over the range $\approx 1 - 10^3$ m), by extrapolating the horizontal and vertical spectra to their crossing point (although he gives exponents for 45 spectra, he only shows a single horizontal and a single vertical density spectrum). For the above exponents, this gives a crude estimate of the sphero-scale

---

[7] Poisson's relation is between magnetic and pseudo-gravity potentials and should not be confused with Poisson's equation. More precisely, if M has a constant direction and is everywhere proportional to ρ then both Poisson's relation and $H_{zM} = H_{z\rho}$ and $s_M = s_\rho$ follow. However, the latter does not necessarily imply the former.

to be[8] ≈ 100 km, but this value is very sensitive to the exact values of β_x, β_z. In order to improve the reliability of this estimate and to use Shiomi's (vertical only) density spectra, we first graphically estimated the prefactors in the formulae $E_\rho(k_z) \approx E_{0z} k_z^{-1.1}$, $E_\rho(K) \approx E_{0x} K^{-1.3}$; these are shown in the table below where the units are rad m^{-1} for k, Kg^{2}m^{-5} for E (Shiomi obtained an exponent 1.27, but this is not too different from the 1 value from Leary). Shiomi normalized his densities by an unknown mean; from the graph of his borehole data, we estimated a mean density of $\rho_0 = 2.5 \times 10^3$ Kgm^{-3} and we used his graph to estimate the $E_{0\rho z}$ in Table 3.

The second step in obtaining a reliable estimate of ρ_s, k_s, was to use the DNAG Bouguer gravity data (Fig. 2). These anomalies were from 8 continental regions in North America; the compilations were made for the Decade of North American Geology (DNAG); resolution ≈ 5 km, 1024×1024 pixels. We note that the high wave number regime, down to 10^{-4} rad m^{-1} or so is fairly linear on a log-log plot with slope $s \approx 5.3$ as predicted by the high frequency gravity (approximately given by $E_g(K) = C_h \rho_s^2 k_s^{-3} G^2 \left(\frac{K}{k_s}\right)^{-s}$). In this power law regime, we estimate $E_g(K) \approx E_{0g} K^{-5.3}$ with $E_{0g} = 3.0 \times 10^{-25}$. Using this high wave number gravity formula, leads to $\rho_s^2 k_s^{2.3} = 2.0 \times 10^{-7}$ which can then be used as a constraint in the density spectrum (which also depends on k_s, ρ_s, see Eqs. 16, 17). Unfortunately, due to the low wavenumber cut-off, these theoretical formulae are not too precise. However, numerics (assuming the crust thickness in the range 40–160 km, see Fig. 2 for the limited dependence on k_c) give the solutions in Table 3 for k_s, ρ_s the overall "best" values being $\rho_s = 215$ Kgm^{-3}, $l_s = 2\pi/k_s = 250$ km, $l_s \rho_s = 5.4 \times 10^7$ Kgm^{-2}. The fact that $k_s < k_c$ (the crust cut-off) means that ρ_s cannot be interpreted as the actual sphero-scale variance; it is simply a dimensional parameter. In passing, we may note that the assumption of self-similar rock scaling ($H_z = 1$) is untenable since the DNAG estimate $s \approx 5$ for the surface gravity exponent would imply $\beta_x = \beta_z = 5 - 2 = 3$ which is much to steep to be compatible with the borehole data; indeed, the difference is so large that we can probably safely rule out the use of self-similar models in explaining the high wave number surface gravity variability.

Table 3 A comparison of various parameters estimated for the density field using the constraint from the DNAG gravity that $E_g(K = 10^{-4} \text{rad m}^{-1}) = 5 \times 10^{-4}$ m^3 (and crust thickness = 80 km, but the result is not too sensitive to this, see Fig. 2)

	E_0	k_s (rad m^{-1})	l_s (km)	ρ_s (Kgm^{-3})	$\rho_s l_s$ (Kgm^{-2})
Shiomi et al. (1997)* (vertical)	1.17×10^4	$10^{-4.5}$	250	233	5.8×10^7
Leary 1997 (vertical)	1.92×10^4	$10^{-4.7}$	310	300	9.5×10^7
Leary (horizontal)	2.3×10^3	$10^{-4.3}$	125	113	1.4×10^7
Overall		**$10^{-4.5}$**	**250**	**215**	**5.4×10^7**

*(For the Shiomi relative density fluctuations, we assumed a mean density $\rho_0 = 2.5 \times 10^3$ Kg m^{-3}).
+These values assume $E = E_0 k^{-\beta}$ with $\beta = \beta_x = 1.3$ (horizontal), $\beta = \beta_z = 1.1$ (vertical), and units of k in rad m^{-1}, units of E in Kg^2 m^{-5}.

---

[8] For comparison for gamma emission, we obtain ≈ 1 m, whereas for the velocity we find ≈ 1 km, but these are all quite inaccurate. In addition, for magnetic susceptibility, Lovejoy et al. 2001 estimate a sphero-scale at $\approx 10^4 - 10^5$ km.

Anisotropic Scaling Models of Rock Density and the Earth's Surface Gravity Field   169

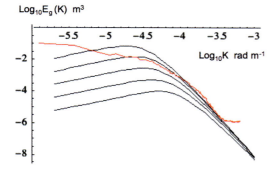

Fig. 2 A comparison of the crust model thickness 10, 20, 40, 80, 160 km with the DNAG spectrum (from North American continental Bouguer data over an area ≈ 5000 km across). The parameters $\rho_s = 215$ Kgm^{-3}, $l_s = 250$ km have been adjusted to fit the function at $K = 10^{-4}$ rad m^{-1}, 80 km thick, and the Shiomi and Leary borehole density data as above. The latter curve agrees well with the gravity data up to $\log_{10} K \approx -4.5$ i.e. up to about 200 km. The model low wave number slope is $+1$, the high wave number slope $-s = -\beta_h = -s = -5.3$ (the intermediate wave number regime discussed in the text is not visible since $k_c > k_s$)

3.2.2 Crust Density and Gravity Spectra, Structure Functions (<300 km)

Figure 2 shows that with these parameters the measured horizontal and vertical rock density statistics up to scales of a kilometer can be extrapolated up to vertical scales comparable to the crust thickness and horizontal scales of at least the order of several hundred kilometers without contradicting the surface gravity spectra. We argue in Sect. 5 that the breakdown at the larger horizontal scales is due to the large contribution from the fractal crust-mantle boundary which dominates for scales > 100–300 km rather than because of a break in the horizontal scaling of the rock densities. Indeed, since the crust contribution to surface gravity falls off at low wave numbers with $\beta_l = -1$ (see Fig. 2), the crust contribution to the spectrum rapidly becomes smaller than the contribution from the crust-mantle interface or mantle. From the gravity spectrum alone, we cannot rule out the possibility that the horizontal crust density scaling continues up to planetary scales.

Using these parameters, we can numerically calculate the crust statistics; these are shown in Fig. 3a–d.

3.3 The Mantle

3.3.1 Theoretical Statistics Far from Boundaries

The basic starting point is the consideration of very large (most often taken as infinite) Prandtl number convection (Pr $= \nu/\kappa =$ viscosity/diffusivity; typical values for the mantle yield 10^{24}). This implies that inertial terms are totally negligible

Fig. 3a Horizontal density spectrum with crust thickness $2\pi/k_c = 10, 20, 40, 80, 160$ km (bottom to top), $l_s = 2.5 \times 10^5$ m, $\rho_s = 215$ Kgm^{-3}. The maximum is proportional to $(k_c/k_s)^{(1-s/Hz)}$; see Appendix 1

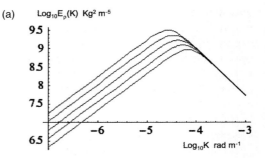

Fig. 3b Horizontal density structure function $S_\rho(\Delta x)$ showing the variance of the density fluctuations as a function of separation distance Δx. The curves are for crust thickness 10, 20, 40, 80, 160 km, $l_s = 2.5 \times 10^5$ m, $\rho_s = 250$ Kgm^{-3}. The maximum is $\approx (k_c/k_s)^{(1+(2-s)/Hz)}$, see Appendix 1: at 80 km it is about $(160\,\text{Kg m}^{-3})^2$

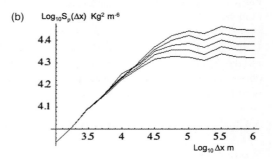

Fig. 3c The surface gravity structure function corresponding to 3b for $l_s = 2.5 \times 10^5$ m, $\rho_s = 215$ Kgm^{-3} and lithosphere thickness 10, 20, 40, 80, 160 km. $S_g = 10^{-10}$ m^{2}s^{-4} corresponds to 1(mGal)^2

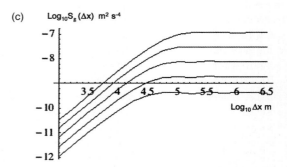

Fig. 3d The horizontal gravity spectra corresponding to $l_s = 2.5 \times 10^5$ m, $\rho_s = 215$ Kgm^{-3} and lithosphere thickness 10, 20, 40, 80, 160 km

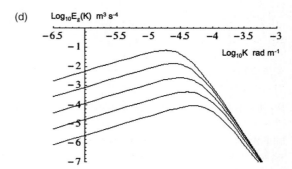

(e.g. the Reynolds number $\text{Re} = vL/\nu \approx 10^{-19}$ for typical values $L = 3 \times 10^6$ m, $v = 10^{-8}$ m/s, $\nu = 3 \times 10^{17}$ m^{2}s^{-1}). The standard approach to mantle convection concentrates on either a) the boundary layer where most of the temperature drop occurs; one nondimensionalizes the equations with typical external lengths, temperature gradients etc. or b) the linearized nondimensional equations which are used to estimate the critical Rayleigh number (Ra) for the onset of convection (the latter is typically estimated at 1000–2000, however the Ra for the entire Mantle is probably $> 10^6$ so that chaotic behaviour (as found in high resolution numerical models) is expected.

Lovejoy et al. (2005) describe a turbulence-type approach which is expected to be valid far from boundaries within high Prandtl number convection with quasi-constant heat flux. The basic argument is that if we are interested in the statistics in the interior mantle region far from boundaries, then the type of statistics should not depend on the outer boundaries; our approach is analogous to that used to obtain the Kolmogorov spectrum in fully developed turbulence (the latter is also expected to be insensitive to the nature of the forcing and boundaries). The most satisfying way to derive the Mantle convection scaling laws is to start from the basic convection equations for the fluctuations around the conductive solutions (see e.g. Busse (1989)):

$$\nabla \cdot \underline{u} = 0 \quad \text{incompressibility} \tag{35}$$

$$\nu^{-1}\left(\frac{\partial \underline{u}}{\partial t} + \underline{u} \cdot \nabla \underline{u}\right) = -\frac{\nabla p}{\nu \rho_0} + \nabla^2 \underline{u} - \underline{g}\alpha \nu^{-1} T \quad \text{velocity equation}$$

(Boussinesq approx.) (37)

$$\frac{\partial T}{\partial t} + \underline{u} \cdot \nabla T = -\frac{\nabla \cdot \underline{H}}{c_p \rho_0} + \frac{J}{c_p \rho_0} + \frac{v_z Q}{\kappa c_p \rho_0} \quad \text{Temperature equation} \tag{36}$$

$$\underline{H} = -\kappa \rho_0 c_p \nabla T + Q\hat{z} \quad \text{Heat diffusion equation} \tag{38}$$

T, p are respectively the temperature and pressure differences with respect to a reference temperature and pressure (the solutions of the static equations), $\underline{H}$ is the heat flux, J is a volume heat source. We now ignore J with respect to the heat originating in the core, and take a typical value of $(\underline{H})_z = Q$, the vertical heat flux imposed by the bottom heating, top cooling. The v_z term in the temperature equation arises because of the use of fluctuating T; as does the $Q\hat{z}$ term in the heat diffusion equation ($\hat{z}$ is the vertical unit vector).

Consider first the velocity Eqs. (35, 36). Due to the very low Reynold's number we take $D\underline{u}/Dt \approx 0$ ("D/Dt" is the advective derivative). In addition, as usual the role of the pressure term is simply to maintain the incompressibility condition. Therefore Eq. (36) depends only on the dimensionless combination $\frac{g\alpha}{\nu}$.

Considering the temperature equation, various arguments show that with the assumption about the imposed vertical heat flux boundary condition, that the main variations are in the z direction, i.e.:

$$v_z \frac{\partial T}{\partial z} \approx \frac{1}{c_p \rho_0} \frac{\partial \langle H \rangle_z}{\partial z} \tag{39}$$

Integrating over a layer and using a typical value of $\langle H \rangle_z = Q$, this implies that only the combination of variables $\frac{Q}{c_p \rho_0}$ enters into the problem. Finally, we can note that the heat conductivity equation only contains the dimensionless combination $\kappa c_p \rho_0$.

We thus see that the dynamics depend on the three combinations: $\frac{g\alpha}{\nu}$, $\frac{Q}{c_p \rho_0}$, $\kappa c_p \rho_0$; since there are also three fundamental dimensions (temperature, distance, time), we obtain the unique dimensional quantities:

$$l_s \approx \left(\frac{\rho_0 c_p \nu \kappa^2}{g \alpha Q} \right)^{1/4} \tag{40}$$

$$\tau_s \approx \left(\frac{c_p \rho_0 \nu}{Q g \alpha} \right)^{1/2}$$

$$T_s \approx \left(\frac{Q^3 \nu}{g \alpha \rho_0^3 c_p^3 \kappa^2} \right)^{1/4}$$

¿From these, we may derive a characteristic density and velocity:

$$\rho_s = \rho_0 \alpha T_s \approx \left(\frac{\alpha^3 \rho_0 Q^3 \nu}{g c_p^3 \kappa^2} \right)^{1/4} \tag{41}$$

$$v_s = l_s / \tau_s$$

The significance of these numbers can be seen by considering the fluctuation Peclet number $Pe = \frac{l_s v_s}{\kappa}$ for fluctuations at scale l_s, velocity v_s. This dimensionless group characterizes the typical ratio of the dynamic heat transport terms to the heat diffusion terms. Using the above dimensional quantities, we obtain:

$$Pe = \frac{l_s v_s}{\kappa} = 1 \tag{42}$$

i.e. for scales smaller than l_s, the heat transport is dominated by conduction, convection can be neglected, l_s is therefore the inner scale of the convection regime. We have used the subscripts "s" in anticipation of the fact that the inner scale is also a sphero-scale (see below).

Before continuing, we can note that using standard empirical estimates for the various parameters, we obtain quite reasonable values for l_s, v_s, ρ_s. In the final column, we give the combination $\rho_s l_s = \left(\frac{\alpha \rho_0 Q \nu}{g c_p} \right)^{1/2}$ since according to Eq. (30a) (ignoring log corrections) this is the quantity that determines the mantle contribution to the surface gravity:

To obtain the behaviour of the statistics, we must perform a more detailed analysis of the equations. This may be done by considering the horizontal and vertical extent of convective plumes. In particular, it is possible to obtain two fundamental empirical laws relating the horizontal and vertical extent of laboratory generated plumes (C. Jaupard private communication). If these laws are applied to an ensemble of plumes, the following anisotropic scaling results:

$$\Delta T(\Delta x) = T_s \left(\frac{\Delta x}{l_s}\right)^{-1} \quad \Delta T(\Delta z) = T_s \left(\frac{\Delta z}{l_s}\right)^{-1/3}$$

$$\Delta v(\Delta x) = v_s \left(\frac{\Delta x}{l_s}\right)^{1} \quad \Delta v(\Delta z) = v_s \left(\frac{\Delta z}{l_s}\right)^{1/3} \quad (43)$$

the density fluctuations may be obtained by multiplying the temperature equations by $\alpha \rho_0$. these equations justify the interpretation of l_s as the sphero-scale of the convection. Comparing this with Eq. (24) we see that $s = H_z = 3$. Physically, the decrease of temperature differences for points increasingly seperated (the negative exponents) seems reasonable since it reflects the ability of the convection to better uniformize the temperatures over larger distances.

3.3.2 Mantle Parameters: Density, Gravity, Spectra, Structure Functions (> 100 km)

We may see that Eq. (43) predicts reasonable typical external velocities, temperatures. Taking the following values from Table 4, $T_s = 375$ K, $\rho_s = 30$ Kgm^{-3} and $l_s = 20$ km, $v_s = 2$ mm/yr, and defining λ as the scale ratio $\lambda = \frac{L}{l_s}$ where $L \approx 3000$ km is the thickness of the mantle and $l_s = 20$ km, we obtain $\lambda = 150$, so that the typical temperature, velocity, density horizontal and vertical fluctuations with $\Delta x = \Delta z = 3000$ km are:

$$\Delta T(\Delta x) = T_s \lambda^{-1} \approx 2.5 K; \quad \Delta T(\Delta z) = T_s \lambda^{-1/3} \approx 70 K$$

$$\Delta v(\Delta x) = v_s \lambda \approx 45 cm/yr; \quad \Delta v(\Delta z) = v_s \lambda^{1/3} \approx 1.5 cm/yr \quad (44)$$

$$\Delta \rho(\Delta x) = \rho_s \lambda^{-1} \approx 0.5 K; \quad \Delta \rho(\Delta z) = \rho_s \lambda^{-1/3} \approx 6 K$$

This shows that at large enough scales, the free convection zone far from boundaries is indeed nearly isothermal (these values are fluctuations with respect to the static diffusive solutions of the equations). The typical vertical velocity horizontal shear of 45 cm/yr is also a rough estimate of the horizontal advection velocity at the top. Finally, we can consider the Rayleigh number (Ra) which must be high for convection. We obtain:

$$Ra = \frac{g \alpha \Delta T \Delta z^3}{\nu \kappa} = \left(\frac{\Delta z}{l_s}\right)^{8/3} \quad (45)$$

Using the largest scale $\Delta z = 3000$ km, $l_s = 20$ km, we obtain $Ra \approx 10^6$ which is comparable to but a little smaller than other estimates (see e.g. the review in Jarvis and Pelletier (1989) where values 7×10^6–6×10^7 are suggested depending on the exact specification of the boundary conditions).

We can now use these values to calculate the second order surface gravity statistics; the main additional assumptions that are needed concern the details of the high wave number convection cut-off (these details are discussed in Appendix 2 and imply uncertainties of a factor of four or so). All the following mantle calculations use an exponential cut-off at the sphero-scale defined in real space at $l_s = 20$ km as the value for which $R(l_s, 0, 0) = \rho_s^2$. The upper bounds of the mantle are considered to be flat, lying directly underneath the crust (only the mantle contribution is shown); downward continuation to this depth is used; see Fig. 4a,b for the effect of varying depths to the top of the mantle. An additional assumption affecting the low wave number behaviour is necessary at the lower bounds of the mantle. Since there is additional variability in the core, putting a drastic truncation at the wave number corresponding to the bottom of the mantle k_m would underestimate the true variability; hence a cutoff corresponding to 6000 km rather than 3000 km was used. As seen in Fig. 6, this difference is only noticeable at the lowest wave numbers.

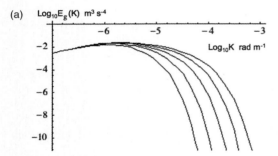

Fig. 4a Mantle gravity spectrum the curves correspond to downward continuations of 10, 20, 40, 80, 160 km (*right* to *left*), $l_s = 20$ km, $\rho_s = 30$ Kgm^{-3}. Mantle thickness is 6000 km so as to partially account for the core

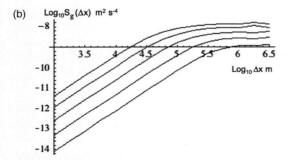

Fig. 4b Mantle gravity structure functions, $l_s = 20$ km, $\rho_s = 30$ Kg m^{-3}, the curves correspond to with downward continuations of 10, 20, 40, 80, 160 km (*top* to *bottom*)

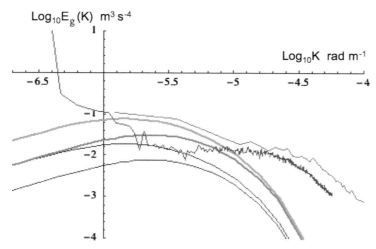

Fig. 5 Global and DNAG gravity (the *bottom, top* empirical curves respectively), the thin theoretical curves are for the optimum estimates $\rho_s l_s = 6 \times 10^5$; the thick continuous curves are the corresponding curves for double this: $\rho_s l_s = 1.2 \times 10^6$. In each case, the thick upper curves are for mantle thickness 6000 km (to avoid an artificial drastic low wave number truncation), the lower thin curves for 3000 km. All four model curves assume a downward continuation of 80 km

The surface gravity field provides one of few ways of empirically testing the model, we therefore compared the theoretical predictions with both the global and DNAG spectra (Fig. 5). With the optimum parameters (Table 4), the figure shows that the contribution of the mantle to the surface gravity spectrum is barely discernable. However, in the next section we see that out to about $\log_{10} K \approx -5.7$ (i.e. 3000 km) that the surface gravity can be explained quite adequately by a fractal mantle-crust density discontinuity, so that this result is not surprising. Indeed, had the model predicted an effect larger even for a factor of only 4 or so – then the absence of a clear signature would have been difficult to explain. We should note that these conclusions are for ensemble averaged effects only; we may expect local regions to have somewhat larger mantle contributions to surface gravity, in these regions, a mantle gravity signature may be visible.

3.4 Topography and the Crust/Mantle Interface Regime (≈ 300–3000 KM)

Up until now, we have considered the earth's surface as well as the mantle-crust boundary to be flat. However, Fig. 1 showed clearly that the topography is on the contrary scaling up to planetary scales, in addition processes of isostatic equilibrium imply that the variations in high wave number surface topography are associated with particularly deep "roots" (mantle-crust boundary depths). This suggests that we can use the observed surface topography as a statistical surrogate for the overall

Table 4 The mean of the four estimates of $\rho_s l_s$ is 6×10^5 Kg m^{-2}; which is the result that we use below. We also took $l_s = 20$ km, $\rho_s = 30$ Kg m^{-3}, but as long as l_s is small enough, only the product is important at low frequencies

Quantity	α expansion coeff.	ρ_0 density	κ thermal cond.	c_p heat capacity	Q thermal flux	ν kinematic viscosity	l_s spheroscale	v_s typical vertical velocity l_s	T_s typical temp. fluct.	ρ_s typical density fluct.	$\rho_s l_s$ product
Units	K^{-1}	Kgm^{-3}	m^{2}s^{-1}	m^{2}s^{-2}K^{-1}	Kgs^{-3}	m^{2}s^{-1}	km	mm/yr	K	Kgm^{-3}	Kgm^{-2}
Poirier 1991	3×10^{-5}	4×10^3	10^{-6}	10^3	8×10^{-2}	3×10^{17}	15.0	2.1	300	36	5.4×10^5
Jarvis, Pelltier (1989) (upper mantle)	2×10^{-5}	3.7×10^3	1.5×10^{-6}	1.26×10^3	0.099	2.7×10^{17}	19.5	2.4	276	20.5	4.0×10^5
Jarvis, Pelltier (1989) lower mantle	1.4×10^{-5}	4.7×10^3	2.5×10^{-6}	1.26×10^3	0.099	4.3×10^{17}	32.9	2.3	219	14.4	4.7×10^5
Overall	2×10^{-5}	4×10^3	1.5×10^{-6}	1.2×10^3	0.09	3.1×10^{17}	24	2	340	26	6×10^5

crust thickness if we assume that on average, the two are related by a numerical factor χ. The following derivation corresponds thus to the Airy model of isostatic equilibrium.

In order to determine the implications of varying crust thickness for the surface gravity, consider a crust with thickness varying as $h(\underline{R})$ where $\underline{R} = (x,y)$ is a surface position vector. This thickness takes into account the entire thickness of the crust (including the topography), except that all the corresponding mass is considered to reside in a column of uniform density $\rho(\underline{R}',z)$, and the top of the column is $z = 0$. The typical crust/mantle density contrast $\Delta\rho_m = \rho_m - \rho_c$ is about 400 Kgm^{-3} ($= 3300 - 2900$). This uniform density approximation should not be too bad for scales comparable to or larger than the thickness.

If we assume that the "roots" of the topography are χ times larger, then we have:

$$h'(\underline{R}) = \chi h_t(\underline{R}) \qquad (46)$$

where h_t is the altitude of the topography. Using this model (see Appendix 4), we obtain:

$$E_g(K) \approx G^2 \Delta\rho_m^2 h_0^2 \chi^2 E_{h_t}(K) K^2; \quad K < 1/H \qquad (47)$$

i.e. in this range, $E_{h_t}(K) \propto E_{geoid}(K)$ (see Eq. 31). h_0 represents a mean crust thickness about which the topography with "roots" represents a fluctuating part and $H > h_0$ is the thickest part of the crust encountered. Since $K \approx (1 - \exp(h_0 K))$ (see eq. 3), this formula is essentially the small K limit of a layer thickness h_0; we have ignored the flexural rigidity of the crust which – if fixed – would break the scaling; presumably a scaling rigidity model is required which is beyond our scope.

We can test out the implications of the above by comparing E_g with $K^2 E_{h_t}$. If the latter is multiplied by the factor $2.1 \times 10^{17} G^2$ we obtain the excellent agreement indicated below over the range 300–3000 km. If $h_0 = 100$ km, this implies $\chi = 11$ which seems reasonable (see Fig. 6). We therefore conclude that such a fractal crust-mantle discontinuity surface can reasonably account for the surface gravity field all the way up to several thousand kilometers in scale.

3.5 Buoyancy Forces

One of the interesting properties of scaling models is the very long range of the implied correlations. In particular, fluctuations of column-integrated densities can be much larger than one would expect from classical (non scaling, Markov process type) statistical models. Horizontal variations in column integrated density will give rise to buoyancy forces; if these are large enough they could play a significant dynamical role. In Appendix 3, we also obtain an analytic approximation to the maximum variance of the difference in column integrated densities. Using the empirical values of the constants (with $z_c = 80$ km), we estimate that for the crust the maximum standard deviation is the equivalent of about 100 m of rock (this occurs for horizontal distances of about $\Delta x = 170$ km). The corresponding numerically

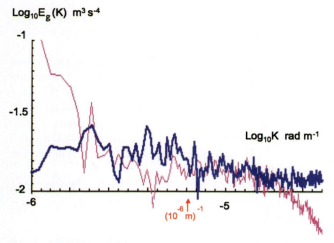

Fig. 6 Above: comparison of the global gravity spectrum (*magenta*), with that simulated from the global topography (*blue*) using the parameter $\Delta\rho_m^2 h_0^2 \chi^2 = 2.1 \times 10^{17}\,\mathrm{Kg\,m^{-2}}$. Putting in $\Delta\rho_m = 4 \times 10^2\,\mathrm{Kg\,m^{-3}}$, $h_0 = 100\,\mathrm{km}$, we obtain $\chi = 11$. The global gravity and that simulated from the topography agree over the range of 300–3000 km

determined spectra and structure functions are shown in Figs. 7a,b. This is substantially smaller than the observed topographic variations, and is not likely to be an important effect. However a similar calculation for the mantle gives a much larger effect because of the much greater mantle thickness, in this case we estimate the maximum buoyancy force to be the equivalent of $\approx 1.5\,\mathrm{km}$ of rock (Figs. 7c,d). Since extremes may be several times larger (especially since due to the likely multifractal nature of the density, we expect long or fat-tailed probability distributions), this may imply a direct role for mantle convection in orogenesis. Indeed, particularly large fluctuations – perhaps several times this value – might explain volcanic "hot spots".

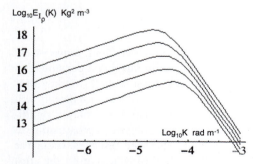

Fig. 7a Spectrum of total column integrated density for $z_c = 10, 20, 40, 80, 160\,\mathrm{km}$ (bottom to top) with $\rho_s = 250\,\mathrm{Kgm^{-3}}$, $l_s = 215\,\mathrm{km}$

Anisotropic Scaling Models of Rock Density and the Earth's Surface Gravity Field

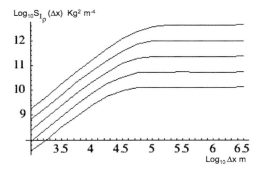

Fig. 7b The structure function corresponding to $z_c = 10, 20, 40, 80, 160$ km (bottom to top) with $\rho_s = 250$ Kgm^{-3}, $l_s = 215$ km. Using a mean lithosphere density of 3×10^3 Kgm^{-3}, the value $S_{I\rho} = 10^{11}$ Kg^2 m^{-4} corresponds to fluctuations of the order of 100 m of rock. The maximum at 160 km thick is $10^{12.4}$ i.e. about 500 m of rock

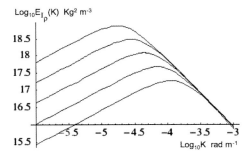

Fig. 7c Total vertically integrated fluctuations with $\rho_s = 30$ Kgm^{-3}, mantle thickness 3000 km, sphero – scale 10, 20, 40, 80, 160 km (bottom to top)

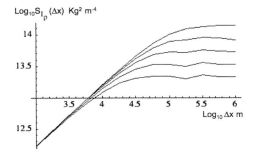

Fig. 7d The structure function of the vertical integral of the rock density with mantle thickness = 3000 km with sphero – scale = l_s = 10, 20, 40, 80, 160 km (bottom to top), with $\rho_s = 30$ Kgm^{-3}. If $\rho_0 = 4 \times 10^3$ Kgm^{-3}, then the equivalent thickness of the rock is $\Delta z = (S_{I\rho}(\Delta x))^{1/2}/\rho_0$ i.e. 1 km of rock corresponds to $S_{I\rho} \approx 10^{13.2}$

4 Conclusions

The recently published high resolution spectrum of the earth's geoid shows two breaks in the scaling at distances of roughly 3000 km and 100–200 km. The first clue to modeling the corresponding surface gravity field is to note that the contribution to the spectrum at horizontal wave number K falls off exponentially with the depth of the source. These breaks are therefore naturally associated with fundamental changes in the earth's internal structure: i.e. to the thicknesses of the crust and mantle. While this classical explanation is valid as far as it goes, it can do no more than explain the characteristic scales of the breaks. A second clue to modeling the geoid is to note the ubiquity of the horizontal and vertical scaling of geophysical fields including the properties of the rocks (e.g. density, magnetization, radioactivity, seismic velocity). This suggests the use of scaling models. For scales smaller than tens of kilometers in the crust, this approach has been adopted by several researchers (e.g. Maus and Dimri 1995,1996; Dimri and Vedanti, 2005) and would follow if over wide ranges, the nonlinear dynamical processes responsible for the variability had no characteristic scale; they were scale invariant or "scaling". The use of scaling models has the great advantage of automatically generating "red-noise" type scaling regimes similar to that observed in boreholes. In addition, scaling processes have long range correlations, huge nonclassical fluctuations, they display potentially realistic anomalies at all scales. The first models of this type were isotropic, "self-similar". However, as we showed in Sect. 2.3, the high variability (slow spectral fall-off, $\beta \approx 1$) in borehole rock spectra and the corresponding low variability (rapid spectral fall-off, $\beta_h \approx 5$) of the small scale (< 100 km) surface gravity are not compatible with self-similar (isotropic) models of rock density. A final flaw of the self-similar models is that they are not compatible with the observed horizontal/vertical stratification of the rocks.

Following the approach used in Lovejoy et al. (2001), Pecknold et al. (2001), Lovejoy and Schertzer (2007) we argue that while scaling is indeed a necessary ingredient in realistic models it must be anisotropic. Using a simple analytical model of the scaling spectrum, the first part of this paper (with various technical appendices) works out the consequences for second order statistics of the surface gravity field. In addition to the inner and outer breaks associated with any physical scaling regime, there is an additional source of scale breaking due to two incompatible symmetries: that of the isotropic gravitational Green's function (r^{-2} law for gravity) and the anisotropic rock density statistics. This introduces a scale break at the "sphero-scale" (l_s) where the density anomalies are roughly isotropic ("roundish").

We apply such anisotropic scaling models to both crust and mantle density fields. First, using (limited) borehole density data (from horizontal and vertical boreholes), combined with continental (Bouguer) gravity survey spectra, we estimate the fundamental exponents for the crust as $H_z \approx 3$, $s \approx 5.3$ and the sphero scale at $l_s \approx 250$ km with the corresponding density fluctuations ≈ 215 Kg m^{-3}. Since $H_z > 1$ at scales smaller than l_s, the rocks will be increasingly stratified. This model can thus readily explain both the horizontal and vertical density statistics and the surface gravity anomalies up to 100–300 km.

We then considered the contributions to surface gravity coming from the mantle (important at larger scales). For this purpose we developed a theoretical model of rock density variations in the mantle. This model was based primarily on dimensional analysis of the equations of convection at high Prandtl and Rayleigh numbers, and predicts $H_z = s = 3$ and $\rho_s = 30$ Kg m^{-3}, $l_s = 20$ km. The mantle contribution to the (mean) surface gravity spectrum predicted by this model was of the same order, but somewhat smaller than the observed surface gravity spectrum at about 1000 km scales.

The final source of surface gravity that we consider arises from the large crust-mantle density contrast and the topography. Using a simple model combined with the ETPO5 global topography data (at 5' of arc), as a surrogate for the crust thickness (corresponding to an assumption of Airy isostacy), we were able to quantitatively account for the surface gravity statistics over the range 300–3000 km. We did not explicitly discuss flexural rigidity since it presumably requires a scaling model which is outside our scope.

The overall combined density/gravity model advocated here (see Table 5 for a summary, see Lovejoy and Schertzer (2005) and the site www.physics.mcgill.ca/~gang/multifrac/index.htm corresponding multifractal simulations) thus involves separate anisotropic scaling regimes for the mantle and crust separated by a fractal density discontinuity and seems capable of explaining the surface gravity statistics from scale of at least meters out to about 3000 km (where core/mantle boundary and core contributions are important). Since the mantle contribution to surface gravity was found to be smaller than that due to the fractal crust-mantle boundary, the overall model involved four internal parameters (H_z, s, l_s, z_c), and one additional external parameter (the crust/topography thickness ratio χ, see Table 5). Since the model predicts the second order statistical behaviour of the density field in both horizontal and vertical directions, as well as the gravity spectrum, it is still fairly parsimonious. Although this study was deliberately confined to second order (spectral) statistics; the full scaling will likely show the density and the gravity to be multifractals (see Pecknold et al. 2001). They may enable us to make realistic multifractal models of the density and gravity.

Table 5 The various exponents and values used here. The mantle exponents are purely theoretical, P_s, $P_s l_s$ are from Table 4 whereas the crust values are purely empirical, mostly being obtained from Leary (1997) and Shiomi et al. (1997) density data with a single DNAG (Bouguer) gravity constraint

	H_z	s	l_s	ρ_s	$\rho_s l_s$
Mantle (from theory)	3	3	20 km	30 Kgm^{-3}	6×10^5 Kgm^{-2}
Crust (from Leary and Shiomi data)	3	5.3	215 km	250 Kgm^{-3}	5.4×10^7 Kgm^{-2}

Acknowledgments We especially thank J. C. Mareschal for numerous discussion and for support at various poionts in the project. We would also like to thank C. Jaupard, W. Peltier, M. Pilkington, J. Arkani-Hamed and J. Toedeschuk for helpful discussions. This research was performed only for scientific purposes, it was unfunded.

Appendix 1: Details of the Crust Density Formulae

1.1 The Basic Scaling of the Crust Density: Infinite Crust Thickness Results

For $H_z \neq s$, the horizontal spectrum is:

$$E_\rho(K) = 2\pi P_0 K \cdot 2 \int_{k_c}^{k_o} \frac{dk_z}{\left(\left(\frac{K}{k_s}\right)^{H_z} + \frac{k_z}{k_s}\right)^{s/H_z}}$$

$$= -4\pi P_0 K \frac{H_z}{s - H_z} k_s \left(\left(\frac{K}{k_s}\right)^{H_z} + \frac{k_z}{k_s}\right)^{-s/H_z + 1} \bigg|_{k_c}^{k_o} \quad (48)$$

where k_c, k_o are the inner and outer crust scales (there is a factor of 2 for the negative k_z). Taking $k_c = 0$, $k_o = \infty$, we have:

$$E_\rho(K) = C_c 2(2\pi)^4 \frac{H_z}{s - H_z} \rho_s^2 k_s^{-1} \left(\frac{K}{k_s}\right)^{1 + H_z - s} \quad ; s > H_z \quad (49)$$

i.e. $\beta_x = s - H_z - 1$. Similarly, for the vertical spectrum:

$$E_\rho(k_z) = 2\pi \int_0^\infty \frac{K dK}{\left(\left(\frac{K}{k_s}\right)^{H_z} + \left(\frac{k_z}{k_s}\right)\right)^{s/H_z}} \quad (50a)$$

$$E_\rho(k_z) = \frac{C_c (2\pi)^4 \Gamma\left(\frac{2}{H_z}\right) \Gamma\left(\frac{s-2}{H_z}\right)}{H_z \Gamma\left(\frac{s}{H_z}\right)} \rho_s^2 k_s^{-1} \left(\frac{k_z}{k_s}\right)^{(2-s)/H_z} \quad ; s > 2 \quad (50b)$$

i.e. $\beta_z = (s-2)/H_z$. We can calculate the (2-D) horizontal structure function using:

$$S(\underline{\Delta X}, 0) = \frac{2}{(2\pi)^2} \int_0^\infty (1 - J_0(K\Delta X)) E(K) dK \quad (51)$$

and the (1-D) vertical structure function with:

$$S(0, 0, \Delta z) = \frac{2}{\pi} \int_0^\infty (1 - \cos(k_z \Delta z)) E(k_z) dk_z \quad (52)$$

(there is an extra factor of 2 from the integration from $-\infty$ to 0). Using $l_s = \frac{2\pi}{k_s}$ we obtain:

$$S_\rho(\underline{\Delta X}, 0) = \rho_s^2 \left(\frac{\Delta X}{l_s}\right)^{s-H_z-2} ; \quad s > H_z + 2 \tag{53}$$

(i.e. $S_\rho(\underline{\Delta X}, 0) \propto \Delta X^{\beta_x - 1}$). If C_c is chosen to be equal to:

$$C_c = \frac{\pi^{H_z - s}(s - H_z)\Gamma\left(\frac{s-H_z}{2}\right)}{8H_z\left(-\Gamma\left(1 + \frac{H_z-s}{2}\right)\right)} \tag{54}$$

then $S_\rho(l_s, 0, 0) = \rho_s^2$ i.e. ρ_s^2 is the sphero-scale fluctuation variance. Since there is a low wave number divergence, this is the only natural choice of reference scale for defining ρ_s. Putting $s = 5.3$, $H_z = 3$, we obtain $C_c = 8.66 \times 10^{-4}$. With this choice for C_c in Eq. (54), we obtain the following for the vertical structure function:

$$S_\rho(0, 0, \Delta z) = B_{\rho z}\rho_s^2 \left(\frac{k_s \Delta z}{2\pi}\right)^{\left(\frac{s-2}{H_z}\right)-1}$$

$$= B_{\rho z}\rho_s^2 \left(\frac{\Delta z}{l_s}\right)^{\left(\frac{s-2}{H_z}\right)-1} ; \quad s - 2 > H_z \tag{55}$$

with the constant $B_{\rho z}$ given by:

$$B_{\rho z} = C_c(2\pi)^{2+\left(\frac{s-2}{H_z}\right)} \frac{\Gamma\left(\frac{2}{H_z}\right)\Gamma\left(\frac{s-2}{H_z}\right)\Gamma\left(\frac{s-2}{H_z}+1\right)\sin\left(\frac{\pi}{2}\left(\frac{s-2}{H_z}\right)\right)}{H_z\Gamma\left(\frac{s}{H_z}\right)} \tag{56}$$

(i.e. $S_\rho(0, 0, \Delta z) \propto \Delta z^{\beta_z - 1}$). Putting $s = 5.3$, $H_z = 3$, we obtain $B_{\rho z} = 0.124$. The fact that this dimensionless constant is not unity reflects the fact that the real space and Fourier space sphero-scales are not identical. Indeed, if we define the real space sphero-scale l_{rs} as the scale at which the vertical and horizontal structure functions are equal, $(S(l_{rs}, 0, 0) = S(0, l_{rs}, 0) = S(0, 0, l_{rs}))$ then we obtain:

$$\frac{l_{rs}}{l_s} = B_{\rho z}^{\frac{1}{\beta_x - \beta_z}} \tag{57}$$

which with the above values i.e. $\beta_x = 1.3$, $\beta_z = 1.1$ yields a ratio 2.96×10^{-5}. Using $l_s = 250$ km, this predicts $l_{rs} \approx 1$ m, but this value is so sensitive to the small difference $\beta_x - \beta_z$ that it should not be taken too seriously. In addition, this estimate does not take into account the finite crust thickness which will affect l_{rs} as determined by the structure functions as defined here (but will not affect l_s). However the smallness of the difference does indicate that direct (real space) measurements of l_{rs} (from rock outcrops for example) are likely to be highly variable.

1.2 The Effect of Finite Crust Thickness on Density and Gravity Statistics

The finite thickness of the lithosphere is important so that we must consider the case $k_c \neq 0$ (the results will be insensitive to the high wave number cut-off which we therefore put at ∞). For the horizontal density spectrum, from Eq. (48), we have:

$$E_\rho(K) = 4\pi P_0 \frac{H_z}{s-H_z} k_s K \left(\left(\frac{K}{k_s}\right)^{H_z} + \frac{k_c}{k_s} \right)^{-s/H_z+1} \quad ; \quad s > H_z \tag{58}$$

With respect to the infinite lithosphere behaviour, we see that there is a new $E_\rho(K) \approx K$ regime for $K < k_s(k_c/k_s)^{1/H_z}$.

For the corresponding horizontal structure function, there is no simple analytic expression, however, for the total variance:

$$R_\rho(0,0,0) = \langle \rho^2 \rangle = \int_0^\infty E_\rho(K) dK = \int_{k_c}^\infty E_\rho(k_z) dk_z$$

$$= 2C_c(2\pi)^4 \rho_s^2 \left(\frac{k_c}{k_s}\right)^{1+\frac{2-s}{H_z}} \frac{\Gamma\left(\frac{2}{H_z}\right)\Gamma\left(\frac{s-2}{H_z}-1\right)}{H_z \Gamma\left(\frac{s}{H_z}-1\right)} \tag{59}$$

This expression diverges as $k_c- > 0$ (explaining why we did not take ρ_s to be defined as the total variance). This determines the maximum of the density structure function.

The corresponding gravity spectrum is:

$$P_g(K) = (2\pi G)^2 2 \int_{k_i}^{k_o} P_\rho(K,k_z) \frac{dk_z}{K^2+k_z^2}$$

$$= 2(2\pi G)^2 \frac{P_{0c}}{K} (I_{s,H_z}(K,k_o) - I_{s,H_z}(K,k_i)) \tag{60}$$

where, for $s \neq H_z$:

$$I_{s,H_z}(K,k_z) = \left(\frac{H_z}{s-H_z}\right)$$
$$_2F_1 \text{Im} \left[\frac{\|(K,k_z)\|^{H_z-s}}{\|(K,-iK)\|^{H_z}} {}_2F_1\left(\left(1-\frac{s}{H_z}\right),1,\left(2-\frac{s}{H_z}\right),\frac{\|(K,k_z)\|^{H_z}}{\|(K,-iK)\|^{H_z}}\right) \right]$$

$$_2F_1(a,b,c,z) = \frac{\Gamma(c)}{\Gamma(b)\Gamma(c-b)} \int_0^1 t^{b-1}(1-t)^{c-b-1}(1-tz)^{-a} dt \tag{61}$$

(the K^{-1} is used in the definition of I so as to make the latter dimensionless) where $_2F_1$ is the hypergeometric function with integral representation indicated. For the case $k_o = \infty$, and large K, we obtain the simple result independent of H_z, as long as $H_z > 1$:

$$I_{s,H_z}(K,\infty) \approx \frac{\pi}{2}\left(\frac{K}{k_s}\right)^{-s}; \quad K \gg k_s$$

$$I_{s,H_z}(K,0) \approx O\left(\left(\frac{K}{k_s}\right)^{1-s-H_z}\log K\right); \quad K \gg k_s \tag{62}$$

Hence for any $s > 0$, $H_z > 1$ the low k_z contribution is negligible, so that:

$$E_g(K) = C_h\rho_s^2 k_s^{-3} G^2 \left(\frac{K}{k_s}\right)^{-s}; \quad K \gg k_s \tag{63}$$

with the dimensionless constant C_h, given by:

$$C_h = \frac{C_c}{2}(2\pi)^7 \tag{64}$$

Using $s = 5.3$, $H_z = 3$, we obtain $C_h = 167$.

Using this formula to estimate the structure function of the surface gravity and geoid, integrating in the horizontal from high frequencies down to k_s, we obtain:

$$S_g(\underline{\Delta X},0) \approx \frac{C_h}{8\pi^2}G^2\rho_s^2 k_s^{-2}\frac{(\Delta X k_s)^2}{s-3}; \quad \Delta X k_s < 1$$

$$S_{geoid}(\underline{\Delta X},0) \approx \frac{C_h}{8\pi^2}\frac{G^2}{g_z^2}\rho_s^2 k_s^{-2}\frac{(\Delta X k_s)^2}{s-1}; \quad \Delta X k_s < 1 \tag{65}$$

In the case of gravity, the structure function saturates at distances a little larger than $\Delta x = l_s$; for the geoid it changes to another power law.

Also, for the low frequency regime, we obtain (to leading order):

$$I_{s,H_z}(K,\infty) \approx \pi\frac{\text{Cos}\left(\frac{\pi}{2}\left(1-\frac{s}{H_z}\right)\right)}{\text{Cos}\left(\frac{\pi}{2}\left(1-\frac{2s}{H_z}\right)\right)}\left(\frac{K}{k_s}\right)^{-s/H_z}; \quad K \ll k_s; \tag{66}$$

whereas:

$$I_{s,H_z}(K,0) \approx -\frac{H_z}{s-H_z}\left(\frac{K}{k_s}\right)^{-s-1+H_z}; \quad K \ll k_s \tag{67}$$

hence the contribution from large k_z dominates for $s < H_z$, while for $s > H_z > 1$, the small k_z contribution dominates, overall, we obtain for the energy spectrum:

$$E_g(K) = C_l \rho_s^2 k_s^{-3} G^2 \left(\frac{K}{k_s}\right)^{-s/H_z}; \quad s < H_z$$

$$K << k_s \quad (68)$$

$$E_g(K) = C_l \rho_s^2 k_s^{-3} G^2 \left(\frac{K}{k_s}\right)^{-s-1+H_z}; \quad s > H_z$$

with the dimensionless constant C_l, given by:

$$C_l = C_c (2\pi)^7 \frac{\mathrm{Cos}\left(\frac{\pi}{2}\left(1 - \frac{s}{H_z}\right)\right)}{\mathrm{Cos}\left(\frac{\pi}{2}\left(1 - \frac{2s}{H_z}\right)\right)}; \quad s < H_z$$

$$C_l = 2C_c (2\pi)^6 \frac{H_z}{s - H_z}; \quad s > H_z \quad (69)$$

Using $s = 5.3$, $H_z = 3$, we obtain $C_l = 140$. Hence for the parameters $H_z > s$, we obtain:

$$\frac{E_\rho(K)}{E_g(K)} = \frac{H_z}{2\pi^3 (s - H_z) G^2} k_s^2 \left(\frac{K}{k_s}\right)^{1+H_z}; \quad K >> k_s$$

$$\frac{E_\rho(K)}{E_g(K)} = \frac{K^2}{(2\pi)^2 G^2}; \quad K << k_s \quad (70)$$

For the case $H_z = s = 3$, see below. This shows clearly that for $H_z > 1$, the stratification does not affect the low wave number regime, while on the contrary, it totally determines the high wave number behaviour.

In the case of the surface gravity, the finite cut-off at k_c leads to a third regime as outlined in Lovejoy et al. 2001 for the aeromagnetic case. This can be seen by defining the variable z:

$$z = \frac{\|(K, k_c)\|^{H_z}}{\|(K, -iK)\|^{H_z}} \quad (71)$$

we then have for $k_c > k_s$, have three regimes:

$$z \approx 1 \quad K >> k_s \left(\frac{k_c}{k_s}\right)^{1/H_z}$$

$$z \approx \frac{k_c}{k_c} \left(\frac{K}{k_s}\right)^{-H_z} \quad k_s \left(\frac{k_c}{k_s}\right)^{1/H_z} > K > k_s$$

$$z \approx \frac{ik_c}{k_c} \left(\frac{K}{k_s}\right)^{-1} \quad k_s < K \quad (72)$$

However using the estimate $k_c/k_s \approx 3$ ($k_s = 2\pi/250$ km, $k_c = 2\pi/80$ km) we find that the middle regime (which has $k^{-(s-1)}$ behaviour; see Lovejoy et al. (2001)) holds over a mere factor of $3^{1/3}$ and is thus is too limited in range to be noticeable.

Appendix 2: Details of the Mantle Density Statistics

2.1 The High Wavenumber Cut-Off

Contrary to the crust case with $s > H_z$, for the mantle, the convection model gives $s = H_z = 3$ which implies high wave number divergences. However, the sphero-scale plays the role of high wave number cut-off, with the result that many of the statistics are somewhat sensitive to the exact high wave number details. Given a model for the cut-off, a related problem is to find the most physically appropriate definition of the sphero-scale. In this appendix, we discuss both of these issues. The question of the cut-off will be illustrated by comparing two simple models.

1) *Wavenumber truncation*:

If we introduce a drastic cut-off in Fourier space at $k_z = k_s$, then we obtain the following horizontal density spectrum:

$$E_\rho(K) = 2\pi P_0 K \cdot 2 \int_0^{k_s} \frac{dk_z}{\left(\frac{K}{k_s}\right)^3 + \frac{k_z}{k_s}} = 2C_m(2\pi)^4 \rho_s^2 k_s^{-2} K \log\left(1 + \left(\frac{k_s}{K}\right)^3\right) \quad (73)$$

(the factor of 2 is from the integration over negative wave numbers) thus:

$$E_\rho(K) \approx 2C_m(2\pi)^4 \rho_s^2 k_s^{-2} K \log\left(\frac{k_s}{K}\right)^3 ; \quad K \ll k_s \quad (74)$$

the above is the (log corrected) $\beta_x = -1$ behaviour of the mantle convection theory. We can now calculate the total variance of the density fluctuations:

$$\langle \rho^2 \rangle = R(0,0,0) = \frac{1}{(2\pi)^2} \int_0^{k_s} E_\rho(K) dK = 2C_m \rho_s^2 (2\pi)^2 \sqrt{3} \frac{\pi}{6} \quad (75)$$

The drawback of this drastic Fourier space cut-off is that it is physically unrealistic and mathematically that it leads to a correlation function with artificial nonphysical oscillations about zero:

$$R_\rho(\Delta X, 0) = \frac{1}{(2\pi)^2} \int_0^\infty E_\rho(K) J_0(K\Delta X) dK$$

$$\approx \frac{2CK(2\pi)^2 \rho_s^2}{k_s \Delta X} J_1(k_s \Delta X); \quad \Delta X \gg k_s^{-1} \quad (76)$$

2) *Exponential cut-off*:

It is more physical to use an exponential cut-off which is less drastic and has the additional advantage of involving a more realistic correlation function. For this model, we take the modified spectrum:

$$P_\rho(K, k_z) = P_0 e^{-\|(K,k_z)\|^3} \|(K,k_z)\|^{-3} \qquad (77)$$

We thus obtain:

$$E_\rho(K) = 4\pi P_0 K k_s \Gamma\left(0, \left(\frac{K}{k_s}\right)^3\right) \qquad (78)$$

where $\Gamma(0,x) = \int_x^\infty t^{-1} e^{-t} dt$ is the incomplete Gamma function. For $K < k_s$, we have the following expansion:

$$E_\rho(K) \approx 4\pi P_0 K k_s \left(-\gamma_E - \log\left(\frac{K}{k_s}\right)^3 + \left(\frac{K}{k_s}\right)^3 + \ldots\right) \qquad (79)$$

Where $\gamma_E = 0.57\ldots$ is the Euler Gamma. Note that the leading behaviour for small K is $K \log K^3$ which is identical to the result for the truncated high frequency spectrum, but the $-2\pi P_0 K k_s \gamma_E$ is new. Also,

$$R_\rho(0,0,0) = \langle \rho^2 \rangle = \frac{1}{(2\pi)^2} \int_0^\infty E_\rho(K) dK = \frac{2 P_0 k_s^3 \Gamma\left(\frac{2}{3}\right)}{4\pi} \qquad (80)$$

We also have a more realistic (nonoscillating) correlation function:

$$R_\rho(\underline{\Delta X}, 0) = \frac{1}{(2\pi)^2} \int_0^\infty E_\rho(K) J_0(K \Delta X) dK \approx 2 H_z C_m \left(\frac{2\pi \rho_s}{k_s \Delta X}\right)^2$$

$$= 2 H_z C_m \left(\frac{\rho_s l_s}{\Delta X}\right)^2; \quad \Delta X \gg k_s^{-1} \qquad (81a)$$

(with $H_z = 3$). This exponential cut-off probably gives us the best estimate of C_m.

2.2 The Definition of ρ_s

Given the cut-off model, there are three obvious choices of definition of ρ_s:

$$\rho_{s1}^2 = R_\rho(0,0,0)$$

$$\rho_{s2}^2 = R_\rho(l_s, 0, 0) = R_\rho\left(\frac{2\pi}{k_s}\right)$$

$$\rho_{s3}^2 = \left(\frac{\Delta X}{l_s}\right)^2 R_\rho(\Delta X, 0, 0); \quad \Delta X \gg l_s \qquad (81b)$$

Depending on which we use, we obtain different constants C_m. $R_\rho(0,0,0)$ gives the physically significant total variance. The second and third definitions are not very different; they only differ because the power law behaviour of the correlation

Anisotropic Scaling Models of Rock Density and the Earth's Surface Gravity Field 189

Table 6 This table shows how two different definitions of ρ_s and different high wave number cut-offs affect the constant C_m. Since the exponential cut-off is probably more realistic, and physically, $\rho_s^2 =$ total variance is more significant, we use the value $C_m = 0.0187$

	$\rho_{s1}^2 = R_\rho(0,0,0)$	$\rho_{s3}^2 = \left(\frac{\Delta X}{l_s}\right) R_\rho(\Delta X)$	$\frac{\rho_{s2}}{\rho_{s1}} = \left(\frac{R_\rho(l_s,0,0)}{R_\rho(0,0,0)}\right)^{1/2}$
Exponential cut-off	$C_m = \frac{1}{(2\pi)^2 \Gamma(\frac{2}{3})} = 0.0187$	$C_m = 1/6 = 0.166$	2.98
truncation	$C_m = \frac{\sqrt{3}}{4\pi^3} = 0.0139$	$C_m = 0.25$	4.23

function is only asymptotically exact. The main choice is between either of these or the first, with the difference arising because of the non abrupt cutoff in the variance at the sphero-scale (a fourier space truncation will in fact have variance at scales $< l_s$ comparable to the exponential cut-off). We favour the first definition since it seems more physically relevant; in any case the differences are not large as Table 6 indicates.

Appendix 3: Column Buoyancy

The spectrum of the running vertical integral of the density responsible for the total column buoyancy force for a column thickness l_z is given by:

$$P_{Il_z,\rho}(K,k_z) = \left| \int_0^{l_z} e^{ik_z z} dz \right|^2 P_\rho(K,k_z) \tag{82}$$

where the first term is modulus squared of the indicator function of the Fourier transfrom of the integration interval. This yields:

$$P_{Il_z,\rho}(K,k_z) = \left(\frac{2\sin\left(\frac{k_z l_z}{2}\right)}{k_z}\right)^2 P_\rho(K,k_z) = \frac{4\sin^2\left(\frac{k_z l_z}{2}\right) P_0}{k_z^2 \left(\left(\frac{K}{k_s}\right)^{H_z} + \frac{k_z}{k_s}\right)^{s/H_z}} \tag{83}$$

To calculate the horizontal spectrum of the vertical integral, we integrate as usual over the entire crust (due to a low frequency divergence, a low frequency cut-off is indeed necessary). We use for this the simplest Fourier truncation model at wave number $k_c = 2\pi/l_c$, $l_c =$ crust thickness:

$$E_{Il_z,\rho}(K) = 4\pi K P_0 \int_{k_c}^{\infty} \frac{4\sin^2\left(\frac{k_z l_z}{2}\right) dk_z}{k_z^2 \left(\left(\frac{K}{k_s}\right)^{H_z} + \frac{k_z}{k_s}\right)^{s/H_z}} \tag{84}$$

If we integrate over the entire column, then $l_c = l_z$ and we obtain:

$$E_{I\rho}(K) = 4\pi K P_0 \int_{k_c}^{\infty} \frac{4\sin^2\left(\frac{k_z\pi}{k_c}\right) dk_z}{k_z^2 \left(\left(\frac{K}{k_s}\right)^{H_z} + \frac{k_z}{k_s}\right)^{s/H_z}} \tag{85}$$

The sine factor is mostly important for $k_z < k_c$, but the latter wave numbers are cut-off anyway, hence it makes a small change to the results. For many calculations we can therefore use the following approximation:

$$E_{I\rho}(K) \approx 4\pi K P_0 \int_{k_c}^{\infty} \frac{dk_z}{k_z^2 \left(\left(\frac{K}{k_s}\right)^{H_z} + \frac{k_z}{k_s}\right)^{s/H_z}} \tag{86}$$

Note that the large distance bound on the structure function is:

$$S_{I\rho}(\infty) = 2R_{I\rho}(0) = 2\langle \rho_I^2 \rangle = \frac{1}{\pi} \int_{k_c}^{\infty} E_\rho(k_z) \frac{dk_z}{k_z^2} \tag{87}$$

Using this approximation, we obtain the analytic result:

$$S_{I\rho}(\infty) = B_{I\rho}(2\pi)^2 \rho_s^2 k_s^{-2} \left(\frac{k_c}{k_s}\right)^{-\left(\frac{s-2}{H_z}\right)-1}$$

$$B_{I\rho} = C_c \frac{4\pi \Gamma\left(\frac{2}{H_z}\right)\Gamma\left(\frac{s-2}{H_z}\right)}{(s-2+H_z)\Gamma\left(\frac{s}{H_z}\right)} \tag{88}$$

If we include the sine factors, the corresponding expression is the same as the above but with corrections involving hypergeometric functions; numerically for $s = 5.3$, $H_z = 3$, the difference is a factor of 1.76 (see Fig. 7).

Putting the high frequency cut-off at ∞, we obtain:

$$E_{I\rho}(K) = 2C_c(2\pi)^4 \rho_s^2 k_s^{-3} \left(\frac{k_s}{k_c}\right)^{1+s/H_z} \left(\frac{K}{k_s}\right)_2$$

$$F_1\left(\frac{s}{H_z}, 1+\frac{s}{H_z}, 2+\frac{s}{H_z}, -\left(\frac{k_s}{k_c}\right)\left(\frac{K}{k_s}\right)^{H_z}\right) \quad s > -H_z \tag{89}$$

This has the following low and high wave number regimes:

$$E_{I\rho}(K) = 2C_c(2\pi)^4 \rho_s^2 k_s^{-3} \left(\frac{k_s}{k_c}\right)^{1+s/H_z} \left(\frac{K}{k_s}\right); \quad K \ll k_s \left(\frac{k_c}{k_s}\right)^{1/H_z}$$

$$E_{I\rho}(K) = 2C_c(2\pi)^4 \rho_s^2 k_s^{-3} \frac{H_z+s}{H_z} \left(\frac{k_s}{k_c}\right)\left(\frac{K}{k_s}\right)^{1-s}; \quad K \gg k_s \left(\frac{k_c}{k_s}\right)^{1/H_z} \tag{90}$$

Appendix 4: Estimating the Contribution from the Crust/Mantle Interface, Topography

If we now consider the case $|(\underline{R}-\underline{R}')| > h(R')$, we have the following approximation to the Green's function in Eq. 2:

$$g_z(\underline{R}|\underline{R}') \approx \frac{G\rho_c(\underline{R}')h(\underline{R}')^2}{2|(\underline{R}-\underline{R}')|^3}\left(1-\frac{3}{4}\left(\frac{h(\underline{R}')}{|(\underline{R}-\underline{R}')|}\right)^2+\ldots\right);$$
$$h(\underline{R}') \ll |(\underline{R}-\underline{R}')| \tag{91}$$

where $g_z(\underline{R}|\underline{R}')$ indicates the gravity at the surface location $\underline{R}$ due to a column at $\underline{R}'$. We now use the same approximation on a column in the mantle (density ρ_m), assumed to lie between depths h and $h_m \gg h$. Using the same approximation, and summing the contribution from the crust and mantle, we obtain:

$$g_z(\underline{R}|\underline{R}') \approx -\frac{G}{2|(\underline{R}-\underline{R}')|^3}(h(\underline{R}')^2\Delta\rho_m + h_m^2\rho_m); \quad h(\underline{R}') \ll |(\underline{R}-\underline{R}')| \tag{92}$$

The fluctuations in the column to column average mean column density ($h_m^2\rho_m$) can be neglected compared to the term $h(\underline{R}')^2\Delta\rho_m$ due to the contrast of the means ($\Delta\rho_m \approx 400$ Kgm^{-3}). This can be seen by estimating the statistics of the column integrated density fluctuations which for the crust yields $\rho_I \approx h_m\rho_m \approx 13$ Kgm^{-3} which is much smaller than $\Delta\rho_m$ (the analogous calculation for the mantle yields a column averaged variation of only 2–3 Kgm^{-3}).

Neglecting the $h_m^2\rho_m$ term we obtain:

$$g_z(\underline{R}) \approx -\frac{G\Delta\rho_m}{2}\int_{|\underline{R}-\underline{R}'|>H}\frac{h^2(\underline{R}')d^2\underline{R}'}{|\underline{R}-\underline{R}'|^3} \tag{93}$$

The range of integration must be such as to respect the thin crust approximation ($|(\underline{R}-\underline{R}')| > h(R') > H$; H is the typical thickness; it should be of the order of the largest h values encountered). For these scales the above power law convolution is a fractional differentiation of order 1 (integration order −1), so that in Fourier terms, we have the following relation between 2-D transforms:

$$\tilde{g}_z(\underline{K}) \approx -\frac{G\Delta\rho_m}{2}\tilde{h}^2(\underline{K})K; \quad K < 1/H \tag{94}$$

taking the complex conjugate equation and multiplying the two and ensemble averaging, we finally obtain:

$$E_g(K) \approx \frac{G^2\Delta\rho_m^2}{4}E_{h^2}(K)K^2; \quad K < 1/H \tag{95}$$

Where $E_{h^2}(K)$ is the horizontal spectrum of the square of the thickness. We can estimate the latter by considering that h is a constant thickness h_0 plus a fluctuating part h':

$$h(\underline{R}) = h_0 + h'(\underline{R}) \tag{96}$$

So that:

$$E_{h^2}(K) = E_{h_0^2}(K) + 4h_0^2 E_{h'}(K) + E_{h'^2}(K) \tag{97}$$

The term $E_{h_0^2}(K)$ is proportional to $\delta(K)$, and if h_0 is larger than the typical fluctuation, $E_{h'^2}(K) < E_h(K)$ so that:

$$E_{h^2}(K) \approx 4h_0^2 E_{h'}(K) \tag{98}$$

To test the consequences for the gravity spectrum, we can use the topography as surrogate for h'. If we assume that the "roots" of the topography are χ times larger, then we have:

$$h'(\underline{R}) = \chi h_t(\underline{R}) \tag{99}$$

where h_t is the topography. Overall, we obtain:

$$E_g(K) \approx G^2 \Delta\rho_m^2 h_0^2 \chi^2 E_{h_t}(K) K^2; \quad K < 1/H \tag{100}$$

i.e. in this range, $E_{h_t}(K) \propto E_{geoid}(K)$ (see Eq. 31), (see Fig. 6).

References

Blakely RJ (1995) Potential theory in gravity and magnetic applications. Cambridge University Press, Cambridge, pp 441
Bowin C (2000) Mass anomaly structure of the earth. Rev Geophys 38: 355–387
Bullard EC, Cooper RIB (1948) The determination of the mass necessary to produce a given gravitational field. Proc R Soc Lond A 194: 332–347
Busse FH (1989) In: Peltier WR (eds) Fundamentals of thermal convection, in mantle convection, plate tectonics and global dynamics—. Gordon and Breach publisher, New York, pp 23–95
Dimri VP, Vedanti N (2005) Scaling evidences of thermal properties in earth's crust and its implications. In: Dimri VP (ed) Fractal behaviour of the earth system, Springer, Heidelberg
Fedi M et al (2005) Regularity analysis applied to well log data. In: Dimri VP (ed) Fractal behaviour of the earth system, Springer, Heidelberg
Gagnon JS, Lovejoy S, Schertzer D (2003) Multifractal surfaces and topography. Europhys Lett 62: 801–807
Gagnon JS, Lovejoy S, Schertzer D (2006) Multifractal earth topography. Nonlinear Proc Geophys 13: 541–570
Kaula WM (1963) Elastic models of the mantle corresponding to varaitions in the external garvity field. J Geophys Res 68: 4967–4978
Jarvis GT, Pelletier WR (1989) In: Pelletier WR (ed) Convection models and geophysical observations, in Mantle convection: plate tectoonics and global dynamics, Gordon and Breach, New York, pp 479–592
Leary P (1997) Rock as a critical-point system and the inherent implausibility of realiable earthquake prediction. Geophys J Int 131: 451–466

Leary PC (2003) Fractures and physical heterogeneity in crustal rock. In Goff JA Hollinger K (ed) Heterogeneity in the crust and upper mantle. Kluwer Academic, New York, pp. 155–186

Lemoine FG, Kenyon SC, Factor JK, Trimmer RG, Pavlis NK, Chinn DS, Cox CM, Klosko SM, Luthcke SB, Torrence MH, Wang YM, Williamson RG, Pavlis EC, Rapp RH, Olson TR (1998) The Development of the joint NASA GSFC and NIMA geopotential Model EGM96, NASA Goddard Space Flight Center, Greenbelt, Maryland, 20771 USA, July

Lovejoy S, Pecknold S, Schertzer D (2001) Stratified multifractal magnetization and surface geomagnetic fields, part 1: spectral analysis and modelling. Geophys J Int 145: 112–126

Lovejoy S, Schertzer D (1998) Stochastic chaos and multifractal geophysics. In: Guindani FM, Salavadevi G (eds), Chaos, Fractals and models 96. Italian University Press, pp 38–52

Lovejoy S, Schertzer D (2007) Scaling and multifractal fields in the solid earth and topography. Nonlinear Proc Geophys 14: 465–502

Lovejoy S, Schertzer D, Gagnon JS (2005) Multifractal simulations of the Earth's surface and interior: anisotropic singularities and morphology. In: Cheng GB-CQ (ed) GIS and Spatial Analysis, Proceedings of the. International Association for Mathematical Geology, pp 37–54

Marsan D, Bean CJ (1999) Multiscaling nature of sonic elocities and lithography in the upper crystalline crust: evidence from the KTB main borehole. Geophys Res Lett 26: 275–278

Maus S, Dimri V (1995) Potential field power spectrum inversion for scaling geology. J Geophys Res 100: 12605–12616

Maus S, Dimri V (1996) Depth estimation from the scaling power spectrum of potential fields. Geophys J Int 124: 113–120

Maus S (1999) Variogram analysis of magnetic and gravity data. Geophysics 64: 776–784

Naidu P (1968) Spectrum of the potential field due to randomly distributed sources. Geophysics 33: 337–345

Pecknold S, Lovejoy S, Schertzer D (2001) Stratified multifractal magnetization and surface geomagnetic fields, part 2: multifractal analysis and simulation. Geophys Int J 145: 127–144

Pilkington M, Todoeschuck J (1993) Fractal magnetization of continental crust, Geophys Res Lett 20: 627–630

Pilkington M, Todoeschuck J (1995) Scaling nature of crustal susceptibilities. Geophys Res Lett 22: 779–782

Poirier JP (1991) Introduction to the physics of the earth's interior. Cambridge University Press, Cambridge, p 264

Schertzer D, Lovejoy S (1985a) The dimension and intermittency of atmospheric dynamics. In: Launder B (ed) Turbulent Shear Flow 4, Springer-Verlag, Berlin, pp 7–33

Schertzer D, Lovejoy S (1985b) Generalised scale invariance in turbulent phenomena. Physico-Chemical Hydrodynamics J. 6: 623–635

Schertzer D, Lovejoy S (1991) Nonlinear geodynamical variability: multiple singularities, universality and observables. In: Schertert D, Lovejoy S (eds) Scaling, fractals and non-linear variability in geophysics. Kluwer, Dordrecht, pp 41–82

Shiomi K, Sato H, Ohtake M (1997) Broad-band power-law spectra of well-log data in Japan. Geophys J Int 130: 57–64

Spector A, Grant FS (1970) Statistical models for interpreting aeromagnetic data. Geophysics 35: 293–302

Tchiguirinskaia I (2002) Scale invariance and stratification: the unified multifractal model of hydraulic conductivity. Fractals 10(3): 329–334

Vennig-Meinesz FA (1951) A remarkable feature of the Earth-s topography. Proc K Ned Akad Wet B Phys Sci 54: 212–228

Non-linear Theory and Power-Law Models for Information Integration and Mineral Resources Quantitative Assessments

Qiuming Cheng

Reprinted from *Mathematical Geosciences* DOI: 10.1007/s11004-008-9172-6, when citing this article please use the DOI number.

Abstract Singular physical or chemical processes may result in anomalous amounts of energy release or mass accumulation that, generally, are confined to narrow intervals in space or time. Singularity is a property of different types of non-linear natural processes including cloud formation, rainfall, hurricanes, flooding, landslides, earthquakes, wildfires and mineralization. The end products of these non-linear processes can be modeled as fractals or multifractals. Hydrothermal processes in the Earth's crust can result in ore deposits characterized by high concentrations of metals with fractal or multifractal properties. Here we show that the non-linear properties of the end products of singular mineralization processes can be applied for prediction of undiscovered mineral deposits and for quantitative mineral resource assessment, whether for mineral exploration or for regional, national and global planning for mineral resource utilization. In addition to the general theory and framework for the non-linear mineral resources assessment, this paper focuses on several power-law models proposed for characterizing non-linear properties of mineralization and for geoinformation extraction and integration. The theories, methods and computer system discussed in this paper were validated using a case study dealing with hydrothermal Au mineral potential in southern Nova Scotia, Canada.

Keywords Non-linear theory · singularity · mineralization · information integration · mineral potential mapping · GIS

1 Introduction

Prediction of undiscovered mineral deposits and quantitative assessments of mineral resources in an area of various scales ranging from a mineral district, to a region, to a nation, and to the globe often require delineation of target areas for undiscovered

---

Qiuming Cheng
State Key Lab of Geological Processes and Mineral Resources, China University of Geosciences, Wuhan & Beijing, China; Department of Earth and Space Science and Engineering, York University, Toronto, M3J 1P3, Canada, e-mail: qiuming@yorku.ca

mineral deposits, delineation possible areas of new mineral deposits, estimation of probable sizes of undiscovered mineral deposits, and the total mineral resource of various mineral deposit types (Cheng 1989). Identifying the target areas favorable for undiscovered mineral deposits is essential for not only estimating the total mineral resource potential in the study area but also for exploration for new mineral deposits. Previous studies have been devoted to development of new techniques and models facilitating the process of prediction of undiscovered mineral deposits (e.g. Bonham-Carter 1994). These studies have led to various approaches such as the weights of evidence, logistic regression, and artificial neural network for mapping mineral potential and identifying areas favorable for undiscovered mineral deposits (Bonham-Carter 1994, Singer 1993, Harris 1984, Zhao 1998). These methods integrate multiple geoscience layers of information (i.e. evidence) associated with the occurrence of mineral deposits. The layers of spatial evidence are often derived from multiple data sources, at multiple scales and in various formats. Integrating these types of geoinformation can reduce the spatial extent of target areas and update information about the sought-after mineral deposits. Depending on the integration methods used, the geoinformation can be expressed either as posterior probability of a unit area containing a mineral deposit as used in logistic models (Bonham-Carter 1994), as the favorability of a unit area for containing a specific mineral deposit as used in characteristic analysis method (McCammon et al. 1983), as a fuzzy membership value of a unit area containing mineral deposits as used in the fuzzy logic method (An et al. 1991), as a fuzzy probability as used in the fuzzy weights of evidence method (Cheng and Agterberg 1999), or finally as a simple additive score used in an index overlay model available in most GIS systems (Bonham-Carter 1994). Taking the posterior probability used in logistic models as an example, the more relevant layers of information are integrated, the smaller the extent of the target areas and the higher the posterior probability that a small area will contain one or more mineral deposits (Cheng 2004a).

Understanding the processes of information integration and the spatial-frequency distributions of the target areas is essential for development and application of mathematical models for mineral resource quantitative assessments. The current paper will address these issues and investigate new solutions from a non-linear point of view. It was shown that the significant information accumulation in the processes of integrating multiple layers of information is similar to the progressive concentration metal during multiplicative cascade processes. These integration processes can generate singularities following power-law frequency distribution between posterior probability or favorability and the size of target areas (Cheng 2007a). These types of power-law distributions may be useful not only for characterization of target areas as geological anomalies but also for delineation of target areas. Other types of non-linear models previously developed for characterizing singularity of mineralization can also be used in mineral resources assessment. For example, number – size model for describing mineral deposits (Agterberg 1995, Cheng and Agterberg 1996, Cheng 1997), density-distance model for modeling mineral deposit distribution (Agterberg et al. 1994, Cheng 1997, Singer 2008), local singularity model for mapping geochemical anomalies (Cheng 1999, 2007a, Cheng and Agterberg 2008), and the S-A model for separation of anomalies from background on the basis of generalized self-similarity (Cheng et al. 2001, Cheng 2004b), just to name a few.

The main proposition supporting the non-linear theory and application of power-law models is that mineralization can result from singular processes and mineral deposits can be regarded as the products of singular processes that may be characterized by power-law models (fractal models/mulitifractal models) (Cheng 2007a). The formation of mineral deposits is often associated with anomalous geological environments or geoanomalies (Zhao 1998). In this paper the applications of non-linear theories and models in the field of mineral deposits and mineral resources assessment will be reviewed followed by detailed discussions of new non-linear models. A case study will be used for mineral potential mapping for hydrothermal gold mineral deposits in southern Nova Scotia, Canada.

2 Singular Mineralization Processes and Singularity

Singular geo-processes including physical, chemical and biological processes may result in anomalous amounts of energy release or mass accumulation and concentration of matter that are, generally, confined to narrow intervals in space or time (Cheng 2007a). Singularity is a property of different types of non-linear natural processes including cloud formation (Schertzer and Lovejoy 1987), rainfall (Veneziano 2002), hurricanes (Sornette 2004), flooding (Malamud et al. 1996, Cheng 2008a,b), landslides (Malamud et al. 2004), forest fires (Malamud et al. 1996) and earthquakes (Turcotte 1997, Cheng et al. 1994a). The end products of these non-linear processes have in common that they can be modeled as fractals or multifractals. Hydrothermal processes are a special type of singular processes that occur in the Earth's crust, and can result in ore deposits characterized by high concentrations of metals with fractal or multifractal properties (Cheng et al. 1994b, Cheng and Agterberg 1996, Agterberg 1995, Mandelbrot 1989, Cheng 2007b). Non-linear processes which are widespread in nature are often characterized by positively skewed frequency distributions with Pareto upper-value tails. Total amount of ore and metals in hydrothermal ore deposits often have Pareto tails (Turcotte 1997, Cheng et al. 1994b, Agterberg 1995, Cheng 1997). Hydrothermal mineral deposits also can exhibit non-linear characteristics in ore-elements and associated toxic element concentration values in rock and related surface media such as water, soil, stream sediment, till, humus and vegetation (Cheng 2007a, Agterberg 2007a, Cheng and Agterberg 2008).

To characterize mineralization and its products (mineral deposits or geo-anomalies) from singular processes point of view is of general interest for not only for the study of mineral deposits, but also for mineral exploration and mineral resource assessment. This paper explores non-linear modeling techniques suitable for characterizing singular mineralization and for predicting locations of mineral deposits based on spatial data characterizing the geological features on the earth's surface or subsurface obtained by geological survey or other earth observation techniques such as geological, geophysical, geochemical and remote sensing techniques. The patterns or fields observed over a mineral district or metallogenic zone can be processed and analyzed using non-linear methods for identifying and characterizing local singularities associated with mineralization and locations of mineral deposits (Cheng 2007a, Cheng and Agterberg 2008).

To show how these models work some mathematical notations used in the singularity context need to be introduced. A measure μ representing either the amount of metal in a small area of linear measuring size ε (for 2-D problem) or the amount of energy released in a small time interval ε satisfies $<\mu(\varepsilon)> \propto <\varepsilon^\alpha>$, where $\propto$ stands for "proportional to" when scale ε approaches zero, and α is the singularity index also known as the Hölder exponent (Mandelbrot 1989). Usually this power-law exists statistically as represented in expectation form $<>$. According to the distribution of α the entire mapped area or time domain can be classified into subsets or fractals each with different singularity and accordingly different fractal dimensions (Cheng 1997). This is the reason that the field of μ has been described by the term multifractality. For simplicity the subsequent discussion of this paper will use 2-D field as an example introducing the principle the singularity. Similar discussions can be applied to temporal multifractals.

In a 2-D spatial problem, the small area with linear size ε, $A(\varepsilon)$, can be used as the area to define the measure $\mu[A(\varepsilon)]$. For example, A could be a square of size ε or any other shape with linear size ε, where $\varepsilon \propto \sqrt{A}$. With this notation the power-law relation between μ (or density ρ) and A can be expressed as

$$<\mu[A(\varepsilon)]> \propto <A^{\alpha/2}> \tag{1}$$

$$<\rho[A(\varepsilon)]> \propto <A^{\alpha/2-1}>. \tag{2}$$

The distribution of singularity α in the mapped area can be described by the fractal dimension spectrum function $f(\alpha)$ which implies that for a conservative field the majority of the area has values of α that are close to 2, whereas the areas with values $\alpha > 2$ or $\alpha < 2$ are more irregular or unusual (Cheng 1999). The statistical power-law distribution can be approximated numerically by sampling areas with different boxes of variable size. For example, in some locations the mean density values averaged for pixels at different resolutions might be independent of pixel size. In other situations the mean value might depend on pixel size. The former case indicates nonsingular background areas with singularity index α close to 2, whereas the latter corresponds to singular components where the value $\alpha > 2$ or < 2.

This singularity property has been frequently observed in exploratory geochemical and geophysical fields (Cheng et al. 1994b, Cheng 1997, 1999, 2000, Xie et al. 2007) and flood events (Cheng 2008a,b). Numerically, μ defined for a set of small areas, (i.e. squares of variable sizes (ε km on a side)) can be used to approximate the relations shown in (1) and (2) above. In this case the power-law relationship between measure μ and cell size ε can be expressed as $\mu(\varepsilon) = c \cdot \varepsilon^\alpha$ where c is a constant, and α is the estimate of singularity; the values of α and c can be estimated by measuring the slope of the straight line in a log-log plot of μ against ε. More complex windows can be used to quantify the anisotropic power-law. Keep in mind, if element concentration values are considered as realizations of a stationary random variable with a constant population mean, then $\alpha \approx 2$ represents non-singularity. "Singular" locations where $\alpha < 2$ may indicate enrichment of the element concentration and those with $\alpha > 2$ may indicate element concentration depletion. Further discussion on the existence and property of singularity can be found in Cheng (2007b, 2008c).

The index α can be estimated in several additional ways besides the above simple method using regular windows. For example, a set of contours can be used as A($\mathcal{E}$) or wavelet functions can be used to estimate the values of α and c (Cheng 2007c). The index, α is the exponent of the power-law relationship which can be expressed as:

$$\alpha = \frac{d\mu[A]}{dA} \frac{2}{\rho[A]} \quad (3)$$

or

$$\mu^{(n)}(A) = \frac{d^n \mu[A]}{d^n A} = cA^{\alpha/2-n} = cA^{-\Delta^n \alpha} \quad (4)$$

where $\mu^{(n)}(A)$ stands for the n-th derivative of $\mu(A)$, and the exponent, $-\Delta^n \alpha = \alpha/2 - n$, is the difference between the Holder exponent (singularity index, $\alpha/2$) and the orders of derivative n. This quantity characterizes the degree of singularity of the measure μ. From (3 to 4) we can see that the singularity index is related to a high-pass filter transformation of μ. The relation (4) indicates that only if n = $\alpha/2$ the derivative has a finite definition. If there is an integer n that satisfies $\Delta^n \alpha \leq 0$ and $\Delta^{n+1} \alpha > 0$, then we can term this α-order singularity which gives a zero value of nth order derivative, but an infinite value of the n + 1 order derivative. This indicates that if the α-value is not an integer then the power-law (1) always displays α-degree of singularity. For a problem with a 2D support, the α-value is usually around 2 and for those locations where the α-value is close to 2 the behavior of the distribution shows non-singularity or linearity. Therefore, the singularity index can be used to characterize the distribution of singularity. The constant c in the power-law relation stands for a type of density of μ measured in α-dimensional space and has a unit such as g/cm$^\alpha$. If the value of α is not integer then c stands for a fractal density. Singular locations correspond to areas with a fractal density of element concentration, whereas background areas have a normal element concentration density.

3 Power-law and Generalized Self-Similarity

It can be seen that the relations (1) and (2) involve only one scaling parameter $\mathcal{E}$, which usually corresponds to isotropic scaling or a power-law. More complex scaling transformations can be applied to these relations to take into account an anisotropic scaling property. For example, windows with area A can be defined by a self-affine transformation which involves two different scaling rates in horizontal and vertical directions. These types of power-law relations characterize self-affinity (stratification). More complex transformations can be applied such as GSI (generalized scale invariance transformation) (Schertzer and Lovejoy 1987) and irregular contours (Cheng 2005). These types of power-law relations characterize generalized self-similarity which shows diversity in the space domain and self-similarity in a special domain such as the frequency domain (Cheng 2004b) and Eigen domain (Cheng 2005). Identification of generalized self-similarity can be useful for separating anomalies according to distinctive generalized self-similarity observed in other domains. Several power-law models including C-A, S-A and N-λ have been developed in space, frequency and Eigen domains for separating anomalies from background for mineral resources assessments (Cheng et al. 1994b, Cheng et al. 2001, Cheng 2004b, 2005, Li and Cheng 2004). The S-A method will be applied to the case study in a following section.

4 Power-law and Fractal Dimension Spectrum

One of the unique properties of singular processes is the resulting power-law distribution of its end products. For example, the frequency distribution of the singularity, number-size distribution of mineral deposits, grade-tonnage of mineral deposits, and posterior probability of a unit area containing mineral deposits may follow power-law distributions (Cheng 2003). Recognition of these types of power-law distributions will be essential for mineral resources quantitative assessment. Several power-law models will be applied to the case study in this paper.

5 Multiplicative Cascade Processes and Multifractal Distributions

The theories and concepts of multiplicative cascade processes play a fundamental role in quantifying turbulent intermittency and other non-linear processes (Schertzer and Lovejoy 1985, Schertzer et al. 1997). A relatively simple 2-dimensional multiplicative cascade model is the model of de Wijs (de Wijs 1951, Agterberg 2001, 2007a). Other modified models are also available, for example, a cascade model with functional redistribution rate (Agterberg 2007b) and a cascade model with variable partition processes (Cheng 2005). The de Wijs cascade model involves partition of area into two sub-areas of constant shape and size. The concentration value (ρ) of a chemical element in the block then can be written as $(1+d) \cdot \rho$ for one half and $(1-d) \cdot \rho$ for the other half so that total mass is preserved. The coefficient of dispersion d is independent of block size. For the first cell at the beginning of the process, ρ can be set equal to unity. The index of dispersion (d) is independent of cell-size. In 2-D space, two successive subdivisions into quarters result in 4 and 16 cells with concentration values. The maximum element concentration value after k subdivisions is $(1+d)^k$, and the minimum value is $(1-d)^k$; k is even in 2-dimensional applications in order to preserve mass. In a random cascade, larger and smaller values are assigned to cells using a discrete random variable. The frequency distribution of the element concentrations at any stage of this process is called "logbinomial" because logarithmically transformed concentration values satisfy a binomial distribution. The logbinomial converges to a lognormal distribution although its upper and lower value tails remain weaker than those of the lognormal (Agterberg 2007a). Because of its property of self-similarity, the model of de Wijs was recognized to be a multifractal by Mandelbrot (1989) who adopted this approach for applications to the Earth's crust. The multifractal patterns generated by this cascade process have many local maxima and minima with singularity expressed as follows

$$\alpha = 2 - \frac{\xi \ln(1+d) + (1-\xi)\ln(1-d)}{\ln 2} = \xi a_{\max} + (1-\xi)a_{\min} \quad (5a)$$

$$f(\alpha) = -2\frac{\xi \ln \xi + (1-\xi)\ln(1-\xi)}{\ln 2} \quad (5b)$$

where ξ is a value with $0 \leq \xi \leq 1$. The maximum and minimum values of α from (5a) are $\alpha_{\max} = \alpha(0) = 2 - \log_2(1-d) = \log_2[4/(1-d)]$ and $\alpha_{\min} = \alpha(1) = \log_2[4/(1+d)]$. Form (5b) shows the fractal dimension spectrum $f(\alpha)$ which characterizes the distribution of singularity. It can be seen that the range of singularity α is related to the choice of d, $\Delta\alpha_{\max} = \alpha_{\max} - \alpha_{\min} = \log_2[(1+d)/(1-d)]$. As the value d approaches 0, the smaller the singularity value range. If d = 0 then $\Delta\alpha_{\max} = 0$. From the fractal dimension point of view, the areas with maximum and minimum singularity have dimension $f(\alpha(0)) = f(\alpha(1)) = 0$ and the areas with $\alpha(1/2) = 2 - \log_2[(1-d^2)] = 1/2(\alpha_{\max} + \alpha_{\min})$ have dimension $f(\alpha(1/2)) = 2$.

6 Multiplicative Cascade Process for Information Integration

6.1 Multiplicative Cascade Process

The weights of evidence method is a spatial decision support model integrating map layers of information for prediction of spatial events (often but not limited to point events) (Bonham-Carter et al. 1988). Successive overlay of evidential layers progressively partitions the study area into smaller sub-areas with updated posterior probability of containing points per unit area. In this paper it will be shown that the process of integrating layers of information using the weights of evidence method is similar to a non-linear multiplicative cascade process introduced in section 5. For convenience, we will use binary evidential layers as an example to illustrate this relationship. Each layer of a binary pattern when combined can divide the study area into two sub-classes: favorable and unfavorable areas for predicting mineral deposits. If we take the prior probability, P(D), as the initial measure of density of mineral deposits in the entire study area, then subsequent additions of an evidential layer of binary pattern (A or $\tilde{A}$) will result in one sub-area with increased posterior probability, $P(D|A) > P(D)$, and the other sub-area with decreased posterior probability, $P(D|\tilde{A}) < P(D)$. Suppose we have several evidence layers (A, B and C) represented as map layers that are positively correlated to the distribution of mineral deposits of the given type D, then the objective of the weights of evidence method is to integrate the information carried by each evidential layer to update the posterior probabilities of a small area containing mineral deposits, P(D|ABC) (Bonham-Carter et al. 1988, Agterberg 1989a,b, Agterberg and Bonham-Carter 1990, Bonham-Carter 1994, Cheng et al. 1996, Cheng and Agterberg 1999). Overlaying each evidence layer reduces the study area into successively small areas with different combinations of evidence (i.e. unique condition map). The posterior probabilities of these small areas having point D per unit area can be calculated according to the combination of evidence.

In order to apply the principle of multiplicative cascade processes to the process of information integration, a number of notations and operators must be introduced. If we assume a multiplicative cascade process applied to a unit area (in 2-D situation) with a series of partitions that split the areas at one step into smaller sub-areas at the next step is defined as

$$A_n = \beta A_{n-1} \qquad (6)$$

where A_n and A_{n-1} represent the two maps showing sub-areas at two steps and $\beta < 1$ are the coefficients ensuring that the sub-areas divided at the n-th generation is smaller than those in the (n-i)-th generation. The sizes of sub-areas on these two maps with respect to the total area can be denoted as their probabilities $P(A_n)$ and $P(A_{n-1})$. At each successive step, the unit mass quantity is redistributed into the smaller sub-areas on A_n and their portions are denoted as $\mu(A_n)$ and $\mu(A_{n-1})$, respectively. The relationship between the mass quantities in each of the sub-areas on these two maps can be associated and expressed as

$$\mu(A_n) = \lambda \mu(A_{n-1}) \qquad (7)$$

where $\mu(A_n)$ and $\mu(A_{n-1})$ represent the mass quantities defined in sub-areas on maps A_n and A_{n-1}, respectively, and $\lambda < 1$ are the coefficients. For simple binary partitions, one can assume each step of partition splits the areas on map A_{n-1} into two different cases corresponding to the presence and the absence of a binary pattern. Combining relations (6) and (7) gives

$$\mu(A_n) = (A_n/A_{n-1})^{\frac{\log \lambda}{\log \beta}} \mu(A_{n-1}) = (A_n/A_{n-1})^\alpha \mu(A_{n-1}) \qquad (8)$$

where α is the singularity index in the context of multifractals. The value of α indicates the changing rate of measure (λ) versus the changing rate of the area (β) applied to each sub-area on map A_n. If the changing rates of these two values are the same then $\alpha = 1$ and, otherwise, if the measure μ increases faster than that of the area when the area decreases, then $\alpha < 1$, and if the measure μ increases slower than that of area when the area decreases then $\alpha > 1$. In the case where $\alpha < 1$, this indicates that the density of mass quantity in the sub-area tends to increase. In terms of a mineralization process, those areas where mass quantity increases are delineated as areas favorable for mineral exploration. In the other case where $\alpha > 1$, this indicates that the density measure in the reduced area tends to decrease which implies element depletion during mineralization processes (i.e. unfavorable area for mineralization).

In order to use the principle of multiplicative cascade processes for integrating multiple layers of geoinformation, binary maps will be used again as an example. We assume each evidence layer to be integrated is a binary map with two types of patterns labeled as A and $\tilde{A}$, respectively. Each layer will further divide the areas on the map A_{n-1} into two groups with favorable patterns indicated as presence and absence, $A_n = AA_{n-1}$ or $A_n = \tilde{A}A_{n-1}$

$$P(A_n) = \begin{cases} P(A|A_{n-1})\,P(A_{n-1}) = \beta_1 P(A_{n-1}) \\ P(\tilde{A}|A_{n-1})\,P(A_{n-1}) = \beta_2 P(A_{n-1}) \end{cases} \tag{9}$$

where $\beta_1 = P(A|A_{n-1})$ and $\beta_2 = P(\tilde{A}|A_{n-1})$, representing the two coefficients of the transformation corresponding to the presence and absence of the pattern A, respectively.

If we define the mass quantity in relation (2) as the probability of mineral deposits on patterns A_n, $P(D|A_n)$ then

$$P(D|A_n) = \lambda P(D|A_{n-1}) \tag{10}$$

where λ has two values corresponding to the two cases of β.

Combining (10) and (9) gives

$$P(D|A_n) = P(D|A_{n-1}) \begin{cases} P(A|A_{n-1})^{a(A)-1} \\ P(\tilde{A}|A_{n-1})^{a(\tilde{A})-1} \end{cases} \tag{11}$$

where $\alpha = \log\lambda/\log\beta$. Form (11) gives the general form associating probabilities of D on A_n and A_{n-1}, respectively. We assume the layers follow the dominant ratio assumption (Cheng 2008d),

$$\frac{P(A_n|D)}{P(A_n|\tilde{D})} = \frac{P(A|D)P(A_{n-1}|D)}{P(A|\tilde{D})P(A_{n-1}|\tilde{D})} \tag{12}$$

which is weaker than the conditional independence assumption made in the ordinary weights of evidence method

$$P(A_n|D) = P(A|D)P(A_{n-1}|D), \quad P(A_n|\tilde{D}) = P(A|\tilde{D})P(A_{n-1}|\tilde{D}) \tag{13}$$

The Eq. (11) can be further simplified. For example, assume that A_1, A_2 and A_3 consist of three binary patterns each with two classes labeled as A, B, and C for presence and $\tilde{A}$, $\tilde{B}$ and $\tilde{C}$ for absence, we have the following relations (here only show relations for A but similar results are applicable to other patterns B and C):

$$P(A|D) = \lambda(A)P(A), \ P(\tilde{A}|D) = \lambda(\tilde{A})P(\tilde{A})$$
$$P(A|\tilde{D}) = \tilde{\lambda}(A)P(A), \ P(\tilde{A}|\tilde{D}) = \tilde{\lambda}(\tilde{A})P(\tilde{A}) \quad (14)$$

and

$$P(A) = \beta(A), \quad P(\tilde{A}) = \beta(\tilde{A}) \quad (15)$$

Therefore, the singularity indexes of these patterns are

$$a(A) = \frac{\log P(A|D)}{\log P(A)}, \quad a(\tilde{A}) = \frac{\log P(\tilde{A}|D)}{\log P(\tilde{A})}$$
$$\tilde{a}(A) = \frac{\log P(A|\tilde{D})}{\log P(A)}, \quad \tilde{a}(\tilde{A}) = \frac{\log P(\tilde{A}|\tilde{D})}{\log P(\tilde{A})} \quad (16)$$

By applying the odds transformation to the posterior probability (11) we can obtain

$$O(D|A^*B^*C^*) = \frac{P(D|A^*B^*C^*)}{P(\tilde{D}|A^*B^*C^*)}$$
$$= O(D)P(A^*)^{a(A^*)-\tilde{a}(A^*)}P(B^*)^{a(B^*)-\tilde{a}(B^*)}P(C^*)^{a(C^*)-\tilde{a}(C^*)} \quad (17)$$

where * stands for either presence (A, B, C) or absence ($\tilde{A}, \tilde{B}, \tilde{C}$). Further applying log-transformation to both sides of (17) gives

$$Logit(D|A^*B^*C^*) = \log it(D) + [a(A^*) - \tilde{a}(A^*)]\log P(A^*)$$
$$+ [a(B^*) - \tilde{a}(B^*)]\log P(B^*) + [a(C^*) - \tilde{a}(C^*)]\log P(C^*) \quad (18)$$

Applying the inverse odds transformation to form (17) gives the posterior probability as follows

$$P(D|A^*B^*C^*) = \frac{P(D)P(A^*B^*C^*|D)}{P(D)P(A^*B^*C^*|D) + P(\tilde{D})P(A^*B^*C^*|\tilde{D})}$$
$$= \frac{P(D)P(A^*)^{a(A^*)-1}P(B^*)^{a(B^*)-1}P(C^*)^{a(C^*)-1}}{P(D)P(A^*)^{a(A^*)-1}P(B^*)^{a(B^*)-1}P(C^*)^{a(C^*)-1} + P(\tilde{D})P(A^*)^{a(A^*)-1}P(B^*)^{a(B^*)-1}P(C^*)^{a(C^*)-1}} \quad (19)$$

Equation (19) gives a new model associating conditional probability $P(D|A^*B^*C^*)$ on patterns and the singularity index (weights α) of each individual layer.

6.2 Properties of the Weights and Implementation of the Model

We will examine the properties of the weights α in order to understand the model and to interpret the results. As explained previously, the weights index in (16) is

Non-linear Theory and Power-Law Models   205

similar to the singularity index defined in the multiplicative cascade multifractal models. It has the following properties:

(1) α has positive value varying around 1,
(2) $\alpha = 1$ iff D and A are independent, $P(A|D) = P(A)$
(3) $\alpha < 1$ iff D and A are positively correlated and $P(A|D) > P(A)$,
(4) $\alpha > 1$ iff D and A are negatively correlated and $P(A|D) < P(A)$.

The above properties indicate that α can be viewed as a statistical index measuring the correlation of pattern A and the deposits D. If these two patterns are independent then $\alpha = 1$. In this case $P(DA) = P(D)P(A)$ or $P(D|A) = P(D)$ implying the condition A does not provide any information to reduce the uncertainty of D. If $\alpha < 1$, then $P(D|A) > P(D)$ implying that given pattern A the conditional probability of D is higher than the prior probability of $P(D)$, implying that patterns A and D are positively correlated. On the contrary, if $\alpha > 1$ then $P(D|A) < P(D)$ implying that the conditional probability of D given pattern A is lower than the prior probability of $P(D)$. It shows pattern A and D are negatively correlated. Similarly the uncertainty of the weights and the posterior probability can be estimated by means of variances related to these statistics.

In order to calculate the posterior probability of mineral deposits given various combinations of layers, several situations can be considered:

(a) If we assume these binary maps A, B, and C are completely independent from each other as is usually assumed in multiplicative cascade processes, then the conditional posterior probability $P(D|A^*B^*C^*)$ can be simply calculated by

$$P(D|A^*B^*C^*) = P(D)P(A^*)^{a(A^*)-1}P(B^*)^{a(B^*)-1}P(C^*)^{a(C^*)-1} \quad (20)$$

(b) In Eq. (19) if we further assume $P(\tilde{D}) \approx 1$ and $P(A|\tilde{D}) - P(A) \approx 0$, then the conditional posterior probability can be calculated by the following simplified form

$$P(D|A,B,C,E)$$
$$\approx \frac{P(D)P(A)^{a(A)}P(B)^{a(B)}P(C)^{a(C)}P(E)^{a(E)}}{P(\tilde{D})P(A)P(B)P(C)P(E) + P(D)P(A)^{a(A)}P(B)^{a(B)}P(C)^{a(C)}P(E)^{a(E)}} \quad (21)$$

This form no longer involves $\tilde{a}$. This approximation is possible because in most situations the measuring unit chosen is so small that the number of units occupied by mineral deposits is much smaller than the number of units without mineral deposits.

In a special case, if all evidential layers are conditionally independent and their singularity values are close to a constant, then the posterior probability calculated using the Eq. (19) may follow a power-law distribution when several layers are integrated, $P(D|ABC) = P(D)P(ABC)^{\alpha-1}$, where α is related to the weighted average values of singularity values of all the evidential layers.

$$a = \frac{a(A)\log P(A) + a(B)\log P(B) + a(C)\log P(C)}{\log P(ABC)} \quad (22)$$

In the following section, the relationship between the singularity and the weights defined in the weights of evidence model will be discussed.

6.3 Singularity and Weights of Evidence Method

Comparing the singularity form defined in (16) and the ordinary weights defined in weights of evidence method (Bonham-Carter 1994)

$$W_A^+ = Log\frac{P(A|D)}{P(A|\tilde{D})}, \quad W_A^- = Log\frac{P(\tilde{A}|D)}{P(\tilde{A}|\tilde{D})} \tag{23}$$

gives

$$W_A^+ = \log P(A)[a(A) - \tilde{a}(A)], \quad W_A^- = \log P(\tilde{A})[a(\tilde{A}) - \tilde{a}(\tilde{A})] \tag{24}$$

This shows that the singularity is related to the weights of evidence.

7 Case Study – Application of Power-Law Models in Mineral Resources Assessment

7.1 Study Area

The case study deals with prediction of gold deposits and for mineral resources assessment in the southwestern Nova Scotia, Canada. The geology of the study area is illustrated in Fig. 1(A). The study area (≈ 7780 km^2) is mainly underlain by Cambro-Ordovician greenschist to amphibolites grade metamorphosed sedimentary rocks and Devonian granitoid rocks. The South Mountain Batholith (SMB) is a complex of multi-phase granites covering nearly one-third of the entire study area (Reynolds et al. 1987, MacDonald et al. 1992). A number of Au, U, W, and Sn deposits have been found in the area. About 20 Au mineral occurrences are found in the sedimentary rocks shown as dots in Fig. 1(A). The data available for the study include lake-sediment geochemical data, airborne magnetic data, ground based gravity data, airborne gamma ray-spectrometer data and a geological map. The geochemical lake-sediment data include 671 samples with concentration values of Cu, Pb, Zn, Ag, F, Li, Nb, Sn, Zr, Ti, Au, Sb, As, Th and W (Rogers et al. 1987). The sampling density was about 1 sample per 5 km^2. More information about the study area and dataset is referred to Xu and Cheng 2001).

7.2 Data Processing and Information Extraction

Data pre-processing is needed for most of the data layers in order to create patterns (i.e. binary patterns) directly associated with the location of mineral deposits. The

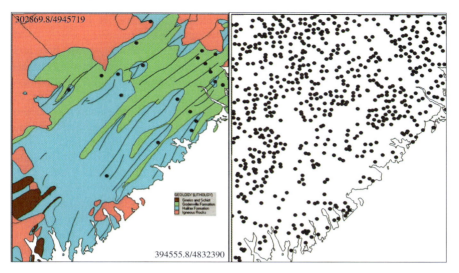

Fig. 1 Map on the left shows simplified geology and distribution of known mineral deposits in southern Nova Scotia, Canada (after Chatterjee, 1983). The solid black lines represent the anticline axes, and black dots for Au mineral deposits. Map on the right represents the locations of lake-sediment samples (N = 671) in the area. Data provided by Mineral Resources Division, Department of Mines and Energy, Nova Scotia, Canada.

following examples show some recently developed methods for data processing and information extraction.

7.2.1 Singularity Analysis and Anomaly Identification

The values of As, Cu, Pb, and Zn from the 671 lake sediment samples (shown in Fig. 1(B)) have been mapped both by the ordinary kriging and by the multifractal data interpolation method (Cheng 1999). Results can be found in Cheng (2006). The distribution of α-values was created using a moving window method (square window) with a maximum size $r_{max} = 15$ km (as half-side of the square) or 30 km (as the size of the square window). It was shown that patterns with $α < 2$ are mainly distributed either in the south of SMB as linear patterns with NW-SE orientation or aggregated around the contacts of SMB, especially in those places where faults or transition zones between different granitoid phases exist. Some of the clusters with low α-values show strong spatial correlation with the locations of Au deposits (more detailed discussions are referred to Cheng 2006).

7.2.2 Application of S-A Method for Separation of Multiple Element Anomalies from Background

In order to use multiple element data to delineate anomalous areas for prediction of location of undiscovered hydrothermal Au mineral deposits, a principal component analysis was applied to the 16 elements including Cu, Pb, Zn, Ag, F, Li, Nb, Sn, Zr, Ti, Au, Sb, As, Th, and W. Four components were created based on correlation coefficients. The results of loadings and scores on the first and third components are shown in Figs. 2– 3, respectively. From the loadings of elements on components one can see that the first component (Fig. 2(A)) is associated with most of the elements and the dominant elements include Cu, Pb, Zn, F, Li, Ru, Zr, Ti, As, and Th (Fig. 2(A)). The areas with positive scores as shown in Fig. 2(B) correspond to the regional anomaly of the mineralization and the areas favorable for mineralization. The loadings of elements on the third component (Fig. 3(A)) show that this component is mainly dominated by elements Sn, Au, As, and W, which is a mineralization-associated component. The areas with positive scores as shown in Fig. 3(B) correspond to the areas favorable for mineralization. Although these two components are associated with mineralization the patterns may reflect the overlapped effects of mineralization and background processes as well as random errors. Considering that the area of these anomalies is so large, further analyses may be required to reduce the anomalous areas with more well defined anomalies directly related to mineralization and locations of mineral deposits. Therefore, the score maps were further analyzed using the S-A method so that the maps were further decomposed into components. First, the score maps were converted into the frequency domain by means of Fourier transformation. Two components: power spectrum density and phases, were created for each score map by Fourier transformation. The power spectrum density can be used to define filters as explained previously and the implementation is also shown below. The results calculated by Fourier transformation from the two score maps are shown in Figs. 4(A) and 5(A), respectively. These two figures plot the data of S and $A(\geq S)$ on a log-log scale. In Fig. 4(A) the pattern shows a general linear trend and two straight-line segments were fitted to the data by means of a Least Square (LS) method. This gives two ranges of power energy spectrum S that maintain distinct scaling properties of the S-A relation. Two ranges of S were identified with a cutoff value $S_0 = 391$. The slopes and intercepts of the two straight-lines are -1.788 and -1.701 (slopes) and 14.635 and 14.143 (intercepts), respectively. The standard errors related to these two linear fits are 0.028 and 0.0023, respectively. The cutoff value $S = 391$ was used to define two filters: one consists of wave numbers with $S \leq 391$ as the anomaly filter and the wave number with $S > 391$ as background filter. Applying these two filters obtained from the S-A plot to the Fourier transformed functions and then converting them back to the spatial domain generated the two decomposed maps. The result obtained with the anomaly filter is shown in Fig. 4(B). According to the properties of Fourier transformation, the map obtained from the anomaly filter defined by $S < 391$ mainly contains the high frequency signal of the original scores of multiple elements on the first principal component. This may include local anomalous values of the first component as well as some random noise related to the data interpolation. These local anomalies depict a linear trend in the northwest-southeast direction and show a general spatial association with locations of most of the discovered mineral deposits. Comparing the filtered anomaly (Fig. 4(B)) with the score map (Fig. 2(B)), the anomalies on

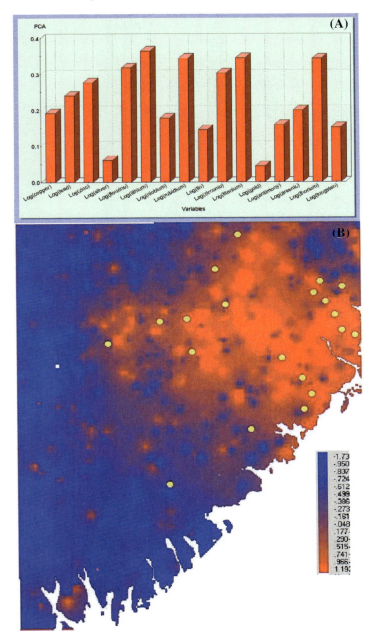

Fig. 2 Results obtained using Principal Component Analysis (PCA) on 16 elements. (**A**) Shows the loadings of elements on the first component and (**B**) shows the distribution of scores on the first component. Dots are mineral deposits for reference

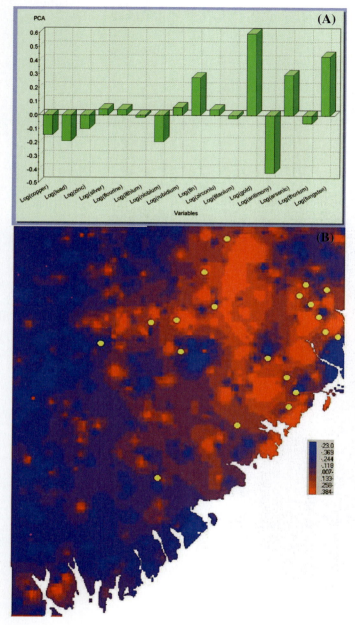

Fig. 3 Results obtained using Principal Component Analysis (PCA) on 16 elements. (**A**) Shows the loadings of elements on the third component and (**B**) shows the distribution of scores on the third component. Dots are mineral deposits for reference

Non-linear Theory and Power-Law Models

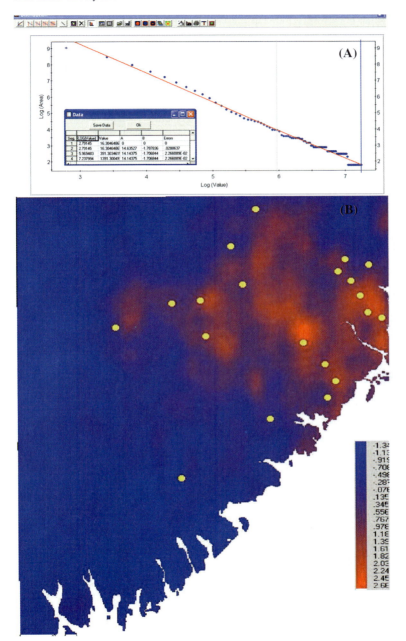

Fig. 4 Results obtained using S-A filtering method on the loadings of elements on the first component: (**A**) S-A plot; (**B**) Results obtained using high-pass filter

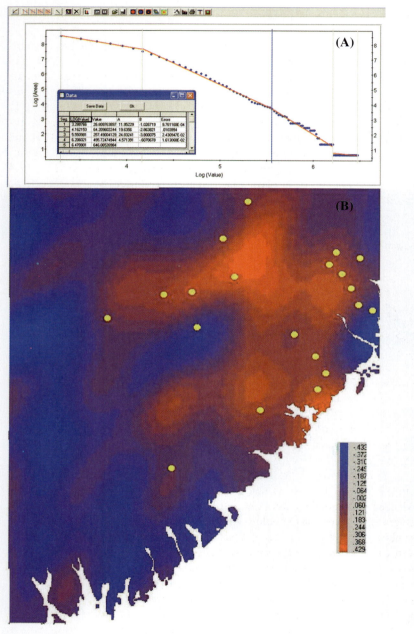

Fig. 5 Results obtained using S-A filtering method on the loadings of elements on the first component: (**A**) S-A plot; (**B**) Results obtained using low-pass filter

Non-linear Theory and Power-Law Models

the filtered map show clear patterns with an orientational distribution which might be related to the ore-controlling faults developed in the area.

Similar processes were applied to the score map related to the third principal component Fig. 3(B). The relationships between S and A($>$ S) are shown in Fig. 5(A) which were fitted with three straight-line segments by means of Least Square (LS) method. These three straight-line segments yield two main cutoff values $S_0 = 64.2$, and $S_1 = 257.5$. The slopes and intercepts of the three straight-lines are -1.03, -2.86, and -3.80 (slopes) and 11.95, 19.63 and 24.83 (intercepts), respectively. The standard errors related to these three LS fits are 0.0001, 0.0104 and 0.0024, respectively. After some experiments the second cutoff value 257.5 was chosen as the threshold forming two filters one with power spectrum density $S > 257.5$ and the other with $S < 257.5$. Applying these two filters allowed the score map to be decomposed into two components. The result obtained with the filter with $S > 257.5$ is shown in Fig. 5(B). The patterns shown in Fig. 5(B) mainly correspond to the dominant patterns of scores on the third principal component with the influence of high frequency signals reduced. The anomalies outline the areas surrounding the contacts between the Goldenville and Halifax Formations and contain most of the discovered mineral deposits.

So far we have extracted two patterns from multiple element concentration values using integrated principal components analysis and the S-A anomaly decomposition method. These two maps will be further used for mapping areas for undiscovered mineral deposits.

7.2.3 Optimum Distance Determination and Construction of Evidence Layers

In order to define binary patterns to be used for information integration, it often necessary to define an optimum threshold to classify maps. There are several methods for determination of optimum distance or cutoff values. One of these methods is to use the buffer function available in most GIS in conjunction with contrast statistics (C) or student statistics value (t) provided in the weights of evidence method (Bonham-Carter 1994). This method was used here for determining: (1) the optimum distance between location of mineral deposits and anticline axes interpreted from integrated information of geology and airborne magnetic data; (2) the optimum distance between location of mineral deposits and the contacts between Goldenville and Halifax formation; and (3) optimum cutoff values separating geochemical patterns on the bases of the decomposed geochemical anomaly maps. The results obtained are shown in Fig. 6(A–D). It was determined that 2.5 km and 4 km are the optimum distances between location of mineral deposits and anticline axes and the contacts between Goldenville and Halifax formation, respectively. The cutoff values were determined for separating the decomposed geochemical anomaly maps in Figs. 4(B) and 5(B) into binary evidence layers as shown in Fig. 6(C,D), respectively. The four binary maps obtained and shown in Fig. 6(A–D) will be overplayed to form a posterior probability map in the next section.

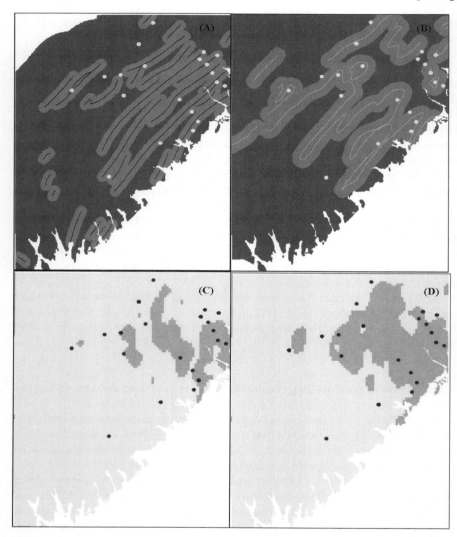

Fig. 6 Binary maps obtained on the basis of optimum distances and optimum cutoff values determined by the contrast provided in the weights of evidence method: (**A**) from anticline axes; (**B**) from contacts of Goldenville and Halifax Formations; (**C**) based on Fig. 4B; and (**D**) based on Fig. 5B

7.3 Map Integration and Information Updating

In order to delineate the areas for prediction of mineral deposits, we need to integrate all these four layers of binary patterns that were proven to be statistically associated with location of mineral deposits.

7.3.1 Cascade Processes

It was discussed previously that each information layer (binary pattern) to be combined divides the study area into two subclasses with areas reduced: favorable and unfavorable areas for predicting mineral deposits. If we define a square of size 1×1 km^2 as the measuring unit then the total study area occupies about 7779.9 units. The decimal value is due to incomplete squares occupied partially by the study area. There are in total 20 units occupied by mineral deposits from which the prior probability of a randomly chosen square from the area containing mineral deposits can be estimated as $P(D) = 20/7779.9 = 0.00258$. Similarly, one can calculate the numbers of units occupied by binary patterns and the units occupied by both binary patterns and mineral deposits. The Table 1 gives the results obtained from the four binary maps: optimum distance (2.5 km) from anticline axes, optimum distance (4 km) from the contacts between Goldenville and Halifax formations, and two geochemical anomaly maps based on factor 1 and factor 3, respectively. The second and third columns in Table 1 show the areas of the favorable patterns of each binary maps (A, B, C, E) and the numbers of mineral deposits occurring in each of the patterns (# points on pattern). ¿From these two columns one can calculate the probabilities of patterns of the fours binary maps, for example, $P(A) = 3065/7779.9 = 0.394$ and $P(\tilde{A}) = (7779.9 - 3065)/7779.9 = 0.606$, representing the probabilities of pattern within 2.5 km and beyond 2.5 km from anticline axes as defined in the first binary map. One can also calculate the probabilities of a unit area on binary patterns containing mineral deposits, for example, $P(A|D) = 15.4/20 = 0.773$ and $P(\tilde{A}|D) = 4.6/20 = 0.227$, representing the conditional probabilities of mineral deposits on and off pattern A, respectively. From these probabilities and conditional probabilities one can calculate the singularity indexes for each binary patterns, for example, $\alpha(A) = \log P(A|D)/\log P(A) = \log(0.773)/\log(0.394) = 0.276$ and $\alpha(\tilde{A}) = \log P(\tilde{A}|D)/\log P(\tilde{A}) = \log(0.227)/\log(0.605) = 2.960$. Similarly the statistics for other binary patterns are shown in Table 1.

¿From the results in the Table 1 we can see that these four binary maps all give strong singularity indexes $\alpha(A) < 1$ and $\alpha(\tilde{A}) > 1$, implying that the patterns A, B, C, and E are positively associated with the occurrence of mineral deposits whereas $\tilde{A}$, $\tilde{B}$, $\tilde{C}$ and $\tilde{E}$ are negatively associated with the occurrence of mineral deposits. This means that the areas covered by A, B, C, and E are favorable for occurrence of mineral deposits whereas the areas covered by $\tilde{A}$, $\tilde{B}$, $\tilde{C}$ and $\tilde{E}$

Table 1 The statistics obtained from each of the four layers of binary maps

	Area	# point	P(A)	P(Ã)	P(A|D)	P(Ã|D)	α(A)	α(Ã)
Anticline (A)	3065	15.5	0.394	0.606	0.773	0.227	0.276	2.960
Contact (B)	4418	19.1	0.568	0.432	0.953	0.046	0.084	3.657
Factor 1 (C)	3095	17.8	0.398	0.602	0.891	0.109	0.125	4.370
Factor 3 (E)	1508	10.5	0.194	0.806	0.526	0.474	0.391	3.465

are not favorable for occurrence of mineral deposits. This can be confirmed by calculating the conditional probability of mineral deposits on each of binary patterns of these maps from the data in Table 1, for example, $P(D|A) = 15.5/3065 = 0.0050$, about 2 times higher than the prior probability $P(D) = 0.00258$, and $P(D|\tilde{A}) = (20 - 15.5)/(7779.9 - 3065) = 0.000963$, about 3 times smaller than the prior probability. Integrating an additional layer of binary pattern (B or $\tilde{B}$) on A and $\tilde{A}$ further partition the area into four sub-areas labeled as AB, $A\tilde{B}$, $\tilde{A}B$, $\tilde{A}\tilde{B}$, respectively. Integration of these two layers of binary maps is similar to two generalizations of multiplicative cascade processes. The consequence of any generalization of cascading processes causes two sub-areas with posterior probabilities updated: one area with posterior probability increased and other areas with posterior probability decreased, $P(D|AB) > P(D|A)$ and $P(D|\tilde{A}B) > P(D|\tilde{A})$, and other two sub-areas with probability decreased $P(D|A\tilde{B}) < P(D|A)$ and $P(D|\tilde{A}\tilde{B}) < P(D|\tilde{A})$. These cascade processes involve irregular partition of area, meaning each partition does not guarantee the same proportions of sub-areas. The other difference between this process and the ordinary multiplicative cascade process is that the measure redistribution rate does not remain constant, in other words, the posterior probability changing rate differs from layer to layer. Therefore, this process can be considered as a more generalized form of a multiplicative cascade process. For comparison purposes, the results obtained by weights of evidence method were also calculated and shown in Table 2. The results show that the all four binary layers are significantly correlated with the location of mineral deposits.

Special consideration of this generalized multiplicative cascade process is that the partitions in different generations might be not independent. This is because the binary maps are generally not independent. Therefore the consequence of integrating these binary layers would not be the same as an ordinary multiplicative cascade process which usually involves independent consecutive partition processes. In order to calculate the posterior probability of mineral deposits given various combinations of layers, several situations can be considered: (a) If we assume that these binary maps are completely independent from each other, as is usually assumed in multiplicative cascade processes, then the conditional posterior probability $P(D|ABCE)$ can be simply calculated according to (20), $P(D|ABCE) = 0.0504$; (b) If we assume the binary maps follow the dominant ratio assumption then the posterior probability can be calculated according to (19), $P(D|ABCE) = 0.0497$; and (c) If we assume the maps meet the dominant ratio assumption and takes the approx-

Table 2 The statistics obtained using weights of evidence method

	Area	# point	W^+	$s(W^-)$	W^-	$s(W^-)$	C	s(C)	t-value
Anticline (A)	3065	15.5	0.67	0.25	−0.97	0.47	1.64	0.53	3.10
Contact (B)	4418	19.1	0.57	0.23	−2.23	1.00	2.81	1.03	2.73
Factor 1 (C)	3095	17.8	0.81	0.24	−1.68	0.67	2.49	0.71	3.52
Factor 3 (E)	1508	10.5	1.00	0.31	−0.53	0.32	1.53	0.45	3.41

Non-linear Theory and Power-Law Models

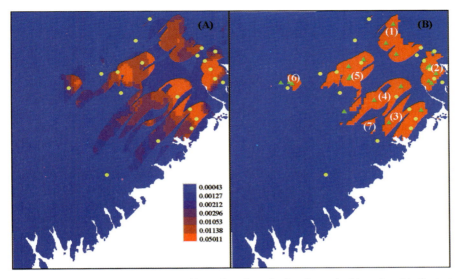

Fig. 7 (**A**) Posterior probability map with results adjusted using the method introduced in the text. (**B**) Areas delineated and labeled for further mineral exploration. Green triangles represent the location of intersections of anticline axes and contacts of Goldenville and Halifax Formations

imation $P(\tilde{D}) = 0.9974 \approx 1$ and the difference between conditional probabilities $P(.|\tilde{D})$ and $P(.)$ of all four binary maps $P(.|\tilde{D}) - P(.) \approx -0.001 \approx 0$ and the values of $\tilde{a}$ is close to 1, $\tilde{a}(.) \approx 1.0033$, then the conditional posterior probability can be calculated according to (22), $P(D|ABCE) = 0.0490$.

In the above discussion, one can see the difference between the posterior probabilities with or without the assumption is only 0.0007, which is much smaller than the prior probability 0.00258. This level of difference might be acceptable for this application. Using model (19), four layers of binary maps were combined and the posterior probability map was created and shown in Fig. 7(A). The highest posterior probability calculated is 0.0497 which is close to the highest value calculated in relation (19). The areas with high posterior probability values contain most of the known mineral deposits and can be considered as target areas for further mineral exploration. In order to make sense of these calculations, in the next section we will discuss some issues related to the results.

7.3.2 Conditional Independence and Adjustment of Posterior Probability

As we can see the posterior probability map in Fig. 7(A) was constructed according to the relation (19) under an assumption of binary layers meet the dominant ratio assumption (12). Due to the intrinsic association of these binary layers with respect to the occurrence of mineral deposits, the dominant ratio condition (12) or conditional independency (13) sometimes may not be met. Therefore, the values of the posterior probability may be biased. The assumption of conditional independency or dominant ratio assumption often needs to be validated if a large number of layers are used in the integration. Several statistical tests have been developed for testing the conditional independence hypothesis, for example, omnibus test (Bonham-Carter 1994) and CI test (Agterberg and Cheng 2002). Other approaches to overcome this problem include making combination of layers so that the reduced number of layers may meet the conditional independence or dominant ratio assumption (Agterberg 1989b, Cheng et al. 1994c). The other method is to adjust the posterior probability results so that it can overcome or reduce the effect due to the violation of this assumption (Kemp et al. 2001). It was noticed that the overlay processes involved in the integration of multiple layers may significantly affect the posterior probability, but be less significant on the general pattern of the posterior probability. On the other hand, overlay of layers can reduce the size of areas with updated information, as long as all layers are physically meaningful and they are associated with the location of mineral deposits. Therefore, integrating these layers will provide more information about the mineral deposits, unless these layers are not completely dependent. Otherwise if some layers are completely dependent on others, they should be removed since their contributions are already contained in those of the other layers. The patterns defined by the combination of layers are always useful even if the layers are not completely dependent. The actual posterior probability given each combination of layers has to be estimated. The models introduced in (19) to (21) only provide some simple ways to calculate the posterior probability under an assumption of independence. These models will inevitably involve some bias in approximating the real posterior probability. The question is whether the biased result is acceptable for the application.

Solutions to solving the problem need two considerations: on the one hand, if the estimation has some strongly biased values, then the results must be revised. On the other hand, if the layers are really close to conditionally independence, particularly if only a small number of mineral deposits are known and more deposits are expected, then the bias may be acceptable. Since it is impossible to fully validate the assumption, overlaying these layers assuming independence may provide extra information to show not only the posterior probabilities of areas with discovered mineral deposits used as training dataset, but also some new areas with elevated potential for new discovery. Probably the best advice is to understand the meaning of the layers and their contributions from a geological point of view. In Arc-SDM program an option is provided that rescales the posterior probability linearly to satisfy the omnibus test (Kemp et al. 2001). If there is a good training dataset, then the following direct estimation method can be used to adjust the results so that unbiased posterior probability values are ensured.

If we assume that the dataset includes a representative sample of mineral deposits, then the calculated posterior probability can be considered as new layer associated with occurrence of mineral deposits. Just like geochemical maps, the biased posterior probability values can be tested for association with the mineral deposits. If this posterior probability map is classified into discrete classes, then the area of

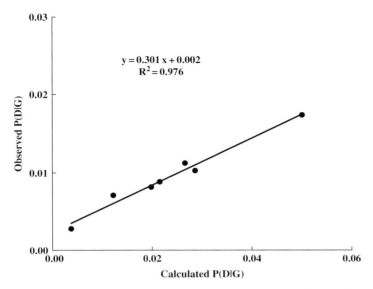

Fig. 8 Relationship between posterior probabilities calculated from weights of evidence model with assumption of weak conditional independency and from the actual training dataset, respectively. G stands for area classified on the basis of posterior probability map

each class and the number of mineral deposits in each class can be measured, and a revised posterior probability for each class can be estimated. For example, the pattern was reclassified into classes with a posterior probability interval of 0.000433. Seven classes were classified with posterior probability above the thresholds set as multiples of 0.000433. Then we calculate the adjusted posterior probability of deposits in each class using the number of actual deposits in each class and the area of the class. Figure 8 shows a general linear trend between these two sets of values. From the linear relation we can calculate the adjusted posterior probability from the unadjusted posterior probability values for each class. It can be seen that although the relation between these two sets of values is linear, the values obtained in Fig. 7(A) are generally larger than those obtained using the actual training dataset.

7.4 Non-linear Properties of Posterior Probability

7.4.1 Power-law Model for Frequency Distribution of Posterior Probability

In order to test whether the distributions of posterior probability show a singularity property, we plot the posterior probability against the area of each class. Figure 9(A,B) show the results obtained using original posterior probability and

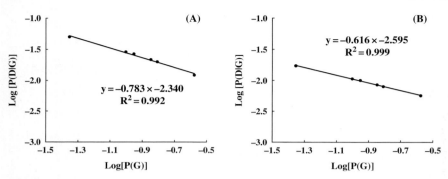

Fig. 9 Relationship between posterior probabilities of mineral deposits fell in the class G and the probability of G. Log-transform is 10-based. (**A**) The original posterior probability calculated using weights of evidence method and (**B**) the adjusted posterior probability

adjusted posterior probability against the area of each class. In order to reduce the effect of the classification of the classes, accumulative areas and corresponding posterior probabilities were converted and plotted in Fig. 9(A,B). The horizontal axes represents the log–transformation of area of class with posterior probability above a threshold divided by the total study area, which gives the estimation of P[G]. The vertical axes represent the log-transformation of posterior probability logP[D|G]. Both results obtained from original posterior probability and adjusted posterior probability show power-law relations between the accumulative area and the cutoff value of posterior probability. The relations obtained for the original posterior probability is $P[D|G] = 0.00457 P[G]^{-0.783}$. This result gives the estimated value, $\alpha(G) = 0.217$, which is close to the average value of $\alpha(A)$ to $\alpha(E)$ (0.219). The relations obtained from the adjusted posterior probability is $P[D|G] = 0.00254[G]^{-0.616}$. This result gives the estimated value, $\alpha(G) = 0.384$, which is greater than the average value of $\alpha(A)$ to $\alpha(E)$ (0.219). This discrepancy between the estimated singularity from the original posterior probability and that from the adjusted posterior probability may provide evidence indicating the four input layers are not completely conditionally independent.

7.4.2 Power-law Model for Predicting Undiscovered Mineral Deposits

For predicting undiscovered deposits and estimate the number of undiscovered mineral deposits, the unbiased posterior probability map should be used. The posterior probability map with the adjusted values can be used as unbiased estimator. From these results, we can establish the relationship between posterior probability and accumulative number of mineral deposits as seen in Fig. 10. Figure 10 gives two curves based on the posterior probability with and without the adjustment. Although these two results both yield power-law relations between the accumulative number of mineral deposits and the posterior probability, the model based

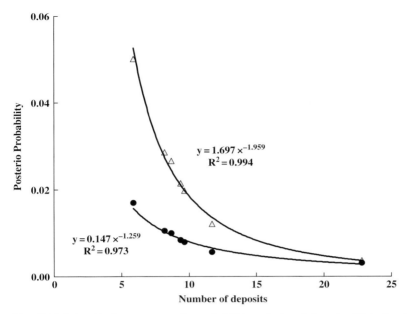

Fig. 10 Relationship between posterior probabilities of mineral deposits fall in the class G and the number of mineral deposits fell in each class (G). Dots represent the results obtained from the adjusted posterior probability and the triangles for the results obtained from the original posterior probability of each class G

on adjusted posterior probability generally gives fewer mineral deposits than that the model based on the original posterior probability values, at the given probability level. The model associating posterior probability and number of mineral deposits can be used in predicting the number of undiscovered mineral deposits in an area.

Based on the adjusted posterior probability, a classification identifies target areas as labeled in Fig. 7(B). Some areas are with known mineral deposits, but others have not yet discovered deposits, and these areas should be considered as favorable for further exploration. This model can also be used in geologically similar surrounding areas for mapping target areas.

The number of mineral deposits can be estimated for each area on the basis of posterior probability and the relation between the posterior probability and the number of mineral deposits. The predicted numbers of undiscovered mineral deposits can be further used for total mineral resource estimation using grade-tonnage model. The estimation of total mineral resource can then be conducted using Monte-Carlo simulation according to number of mineral deposits (N) and grade – tonnage (G-T). In this situation, these three parameters (N, G and T) are random variables, each following its own probability distribution. It has been shown that both grade and tonnage usually follow a power-law frequency distribution

(Turcotte 2002, Cheng 2003). Power-law distributions should be used for total mineral resource simulation.

8 Conclusions and Discussions

Singularity theory and power-law models have been demonstrated as suitable for characterizing hydrothermal mineralization and for mapping mineral potential. Singularity can be considered as a usual property of mineralization which involves enhancement and depletion of ore and associated elements in the Earth's crust, as well as in other relevant secondary media such as tills, soils, lake and stream sediments, humus and vegetation. Mapping such singularity is an effective way to delineate areas favorable for occurrence of mineral deposits and for estimating mineral resources.

The local singularity analysis provides an effective new tool for mapping locations and quantification of anomalies, especially for weak anomalies related to mineralization. Generalized self-similarity is useful for forming fractal filtering models for separating mixing anomalies caused by different geological processes especially between mineralization and other regional geological processes. The fractal spectrum provides necessary power-law distribution models for representing spatial and frequency distributions of mineralization in which the causative consequences are anomalies and mineral deposits.

To support decision making and to map areas for prediction purposes, multiple layers of information from diverse sources often need to be combined. Treating the process of information integration as a general form of a multiplicative cascade process makes it possible to interpret the information updating process from a nonlinear process viewpoint. The singularity index and weights of evidence are shown to be related indexes for measuring the spatial association between binary patterns and location of mineral deposits. The former involves ratio of log-transformations of probabilities and the value is independent of the type of log-transformation applied, whereas the latter is a log-transformation of ratio of conditional probabilities therefore the values of the weights are dependent on the log-transformation. Whereas the weights defined in the weights of evidence method characterize the statistical spatial association of the binary patterns and the location of mineral deposits, the singularity index explicitly characterizes the property of information enrichment or depletion due to the application of the binary patterns.

The layers involved in information integration often violate the conditional dependence or weak conditional independence (dominant ratio) assumption required by weights of evidence method for data integration. The solution proposed here is to adjust the results created by using multiple layers, even if dependency is present between layers. With this adjustment, the weights of evidence can be used to integrate multiple layers that can be dependent as long as these layers are associated with the occurrence of mineral deposits, and are geologically meaningful. Comparing posterior probabilities before and after adjustment, allows the conditional independence assumption to be tested.

Acknowledgments Thanks Drs. Graeme F. Bonham-Carter, Jeff Harris and one anonymous reviewer for their critical review of the paper and constructive comments. The research was financially supported by a Distinguished Young Researcher Grant (40525009)i±, a Strategic Research Grant (40638041) awarded by the Natural Science Foundation of China, a High-Tech Research and Development Grant (2006AA06Z115) by the Ministry of Science and Technology and a project on "New Methods for National Mineral Potential Quantitative Prediction" (121201063390110) awarded by the Ministry of Land and Resources of China.

References

Agterberg FP (1995) Multifractal modeling of the sizes and grades of giant and supergiant deposits. Int Geol Rev 37: 1–8

Agterberg FP (1989a) Computer programs for exploration. Science 245: 76–81

Agterberg FP (1989b) Systematic approach to dealing with uncertainty of geoscience information in mineral exploration. In: Weiss A (ed) Application of computers and operations in the mineral industry. Proceedings of the 21st APCOM symposium (Las Vegas, Nevada) Colorado Society of Mining Engineers, Littleton, Las Vegas, Nevada, pp 165–178

Agterberg FP (2001) Multifractal simulation of geochemical map patterns. In: Merriam DF, and Davis JC (eds) Geologic modeling and simulation: Sedimentary Systems. Kluwer, New York, pp 327–346

Agterberg FP (2007a) New applications of the model of de Wijs in regional geochemistry. Math Geol 39: 1–26

Agterberg FP (2007b) Mixtures of multiplicative cascade models in geochemistry. Nonlinear Process Geophy 14: 201–209

Agterberg FP, Bonham-Carter GF (1990) Deriving weights of evidence from geoscience contour maps for prediction of discrete events. Proceedings of the 22nd APCOM symposium (Berlin, Germany), Technical University of Berlin, vol 2, pp 381–396

Agterberg FP, Cheng Q (2002) Conditional independence test for weights of evidence modeling. Nat Resour Res 11: 249–255

Agterberg FP, Brown A, Cheng Q, Good D (1994) Multifractal modeling of fractures in Lac De Bonnet batholith, Manitoba. In: Proceedings of IAMG '94, Mont-Tremblant, Quebec, October, 1994, vol 1, pp 3–8

An P, Moon WM, Rencz A (1991) Application of fuzzy set theory for integration of geological, geophysical, and remote sensing data. Can J Explor Geophys 27: 1–11

Bonham-Carter GF (1994) Geographic information systems for geoscientists: modelling with GIS. Pergamon, Oxford, p 398

Bonham-Carter GF, Agterberg FP, Wright DF (1988) Integration of geological data sets for gold exploration in Nova Scotia. Photo. Eng. Remote Sensing 54:1585–1592

Cheng Q (1989) A quantitative method for evaluating mineral resource of multivariate populations. In: Wang S, Fan J, Cheng Q (eds) J Changchun Univ Earth Sci, Special Issue (in Chinese with English abstract) 19: 50–56

Cheng Q (1997) Discrete multifractals. Math Geol 29: 245–266

Cheng Q (1999) Multifractality and spatial statistics. Comp Geosci 25: 949–961

Cheng Q (2000) GeoData Analysis System (GeoDAS) for mineral exploration: user's guide and exercise manual. Material for the training workshop on GeoDAS held at York University, November 1–3, 2000, p 204, www.gisworld.org/geodat

Cheng Q (2003) Fractal and multifractal modeling of hydrothermal mineral deposit spectrum: application to gold deposits in the Abitibi Area, Can J China Univ of Geosci 14: 199–206

Cheng Q (2004b) A new model for quantifying anisotropic scale invariance and for decomposition of mixing patterns. Math Geol 36: 345–360

Cheng Q (2004a) Weights of evidence modeling of flowing wells in the Greater Toronto Area, Canada. Nat Resour Res 13: 77–86
Cheng Q (2005) Multifractal distribution of eigenvalues and eigenvectors from 2D multiplicative cascade multifractal fields. Math Geol 37: 915–927
Cheng Q (2006) GIS-based multifractal anomaly analysis for prediction of mineralization and mineral deposits. In: Harris J (ed) GIS For the Earth sciences. Geological Association of Canada, Tri-Co Group, Ottawa: 285–296
Cheng Q (2007a) Mapping singularities with stream sediment geochemical data for prediction of undiscovered mineral deposits in Gejiu, Yunnan Province, China. Ore Geol Reviews 32: 314–324
Cheng Q (2007b) Multifractal imaging filtering and decomposition methods in space, Fourier frequency and Eigen domains. Nonlinear Process Geophys 14: 293–303
Cheng Q (2007c) Singularity of mineralization process and power-law models for mineral resources quantitative assessments J China Univ of Geosci 18(special issue): 245–247
Cheng Q (2008a) A new combined model for prediction of river flow. J Hydrol 352: 157–167
Cheng Q (2008b) Local singularity analysis of river peak flow. Non-Linear Process Geophys (in press)
Cheng Q (2008c) Modeling local scaling properties for multi-scale mapping. Vadose Zone J 7(2): 525–532
Cheng Q (2008d) Comparison between Tau model and weights of evidence model. Submitted to Math Geosci
Cheng Q, Agterberg FP (1996) Multifractal modeling and spatial statistics. Math Geol 28: 1–16
Cheng Q, Agterberg FP (1999) Fuzzy weights of evidence method and its application in mineral potential mapping. Nat Resour Res 8: 27–35
Cheng Q, Agterberg FP (2008) Singularity analysis of ore-mineral and toxic trace elements in stream sediments. Comp Geosci (in press)
Cheng Q, Agterberg FP, Ballantyne SB (1994b) The separation of geochemical anomalies from background by fractal methods. J Geochem Explor 51: 109–130
Cheng Q, Agterberg FP, Bonham-Carter GF (1996) Fractal pattern integration method for mineral potential mapping. Nonrenew Res 5: 117–130
Cheng Q, Xu Y, Grunsky EC (2001) Integrated spatial and spectrum analysis for geochemical anomaly separation. Nat Resour Res 9: 43–51
Cheng Q, Bonham Carter GF, Agterberg FP, Wright DF (1994a) Fractal modeling in the geosciences and implementation with GIS. In: Proceedings of the 6th Canadian Conference on GIS, Ottawa, June 6–10, vol 1, pp 565–577
Cheng Q, Agterberg FB, Bonham-Carter GF, Sun J (1994c) Artificial intelligence model for integrating spatial patterns for mineral potential estimation with incomplete information. In: Proceedings of 6th Canadian Conference on GIS, Ottawa, June 6–10, 1994, vol 1, pp 261–274
Chatterjee AK (1983) Metallogenic map of Nova Scotia, ver. 1, scale 1: 500,000, Department of Mines and Energy, Nova Scotia, Canada
de Wijs HJ (1951) Statistics of ore distribution, part I. Geologie Mijnbouw 13: 365–375
Harris DP (1984) Mineral resources appraisal – mineral endowment, resources, and potential supply: concepts, methods, and cases. Oxford University Press, New York, p 455
Kemp LD, Bonham-Carter GF, Raines GL, Looney CG (2001) Arc-SDM: Arcview extension for spatial data modelling using weights of evidence, logistic regression, fuzzy logic and neural network analysis. http://www.ige.unicamp.br/sdm/
Li Q, Cheng Q (2004) Fractal singular value decomposition and anomaly reconstruction. Earth Sci (in Chinese with English abstract) 29: 109–118
MacDonald MA, Horne R, Corey MC, Ham L (1992) An overview of recent bedrock mapping and follow-up petrological studies of the South Mountain Batholith, southwestern Nova Scotia. Atlandtic Geol 2: 7–28
Malamud BD, Turcotte DL, Barton CC (1996) The 1993 Mississippi river flood: a one hundred or a one thousand year event? Environ Eng Geosci II: 479–486

Malamud BD, Turcotte DL, Guzzetti F, Reichenbach P (2004) Landslide inventories and their statistical properties. Earth Surf Process Landforms 29: 687–711

Mandelbrot BB (1989) Multifractal measures, especially for the geophysicist. Pure Appl Geophys 131: 5–42

McCammon RB, Botbol JM, Sinding-Larsen R, Bowen RW (1983) Characteristic Analysis – 1981: final program and a possible discovery. Math Geol 15: 59–83

Reynolds PH, Elias P, Muecke GK, Grist AM (1987) Thermal history of the southwestern Meguma zone, Nova Scotia, from an 40Ar/39Ar and fission track dating study of intrusive rocks. Can J Earth Sci 24: 1952–1965

Rogers PJ, Mills RF, Lombard PA (1987) Regional geochemical study in Nona Scotia. In: Bates JL, MacDonald DR (eds) Mines and mineral branch, report of activities 1986, 87–1: 147–154

Schertzer D, Lovejoy S (1985) The dimension and intermittency of atmospheric dynamics – multifractal cascade dynamics and turbulent intermittency. In: Launder B (ed) Turbulent Shear Flow. Springer-Verlag, New York, pp 7–33

Schertzer D, Lovejoy S (1987) Physical modeling and analysis of rain and clouds by anisotropic scaling of multiplicative processes. J Geophy Res 92: 9693–9714

Schertzer D, Lovejoy S, Schmitt F, Chigirinskaya Y, Marsan D (1997) Multifractal cascade dynamics and turbulent intermittency. Fractals 5: 427–471

Singer DA (1993) Basic concepts in three-part quantitative assessments of undiscovered mineral resources. Nonrenew Res 2: 69–81

Singer DA (2008) Mineral deposit densities for estimating mineral resources. Math Geosci 40: 33–46

Sornette D (2004) Critical phenomena in natural sciences: chaos, fractals, selforganization and disorder, 2nd edn. Springer, New York

Turcotte DL (1997) Fractals and chaos in geology and geophysics. 2nd edn. Cambridge University Press, Cambridge

Turcotte DL (2002) Fractals in petrology. Lithos 65: 261–271

Veneziano D (2002) Multifractality of rainfall and scaling of intensity-duration-frequency curves. Water Resour Res 38: 1–12

Xie S, Cheng Q, Chen G, Chen Z, Bao Z (2007) Application of local singularity in prospecting potential oil/gas targets. Nonlinear Process Geophy 14: 285–292

Xu Y, Cheng Q (2001) A multifractal filter technique for geochemical data analysis from Nova Scotia, Can J Geochem Explor, Analys Environ 1: 1–12

Zhao P (1998) Geoanomaly and mineral prediction: modern mineral resource assessment theory and method, Geological Publishing House, Beijing, p 300 (in Chinese)

Mineral Potential Modelling for the Greater Nahanni Ecosystem Using GIS Based Analytical Methods

J.R. Harris, D. Lemkow, C. Jefferson, D. Wright and H. Falck

Reprinted from *Natural Resources Research* DOI: 10.1007/s11053-008-9069-6, when citing this article please use the DOI number.

Abstract Mineral potential within the Greater Nahanni Ecosystem was modelled in a Geographic Information System (GIS) for four different deposit types: 1. SEDEX (stratiform shale-hosted sedimentary exhalative Zn-Pb-Ag), 2. "Carbonate-Fault" (carbonate-hosted zinc-lead-silver associated with major faults), 3. "Intrusion-Related" (includes skarn, rare metals and gemstones), and 4. Carlin-Type gold as lode and/or derived placer deposits. This mineral potential modelling study integrates data collected during the Nahanni Mineral and Energy Resource Assessment (MERA) undertaken from 2003 to 2007. The results have contributed to the process of determining the geographic boundaries of the proposed expansion of the Nahanni National Park Reserve.

Four mineral potential maps were produced (one for each deposit type) using a knowledge-driven approach. A weighting scheme based on integrated mineral deposit and regional geological knowledge was derived for the various evidence maps for each deposit model using expert opinion. The four potential maps were then combined into a final potential map using a maximum (MAX) operator. Plots showing the efficiency of the models (mineral potential maps) for predicting the known occurrences of the four deposit types show that partial data sets provide reasonable predictions of the remaining known mineral prospects, occurrences and deposits. Hydrocarbon potential from Nahanni MERA 1 was added to the final potential map

---

J.R. Harris
Geological Survey of Canada, 615 Booth St., Ottawa, ON K1A 0E9, Canada,
e-mail: harris@nrcan.gc.ca

D. Lemkow
Geological Survey of Canada, 615 Booth St., Ottawa, ON K1A 0E9, Canada,
e-mail: lemkow@nrcan.gc.ca

C. Jefferson
Geological Survey of Canada,615 Booth St., Ottawa, ON K1A 0E9, Canada,
e-mail: jefferson@nrcan.gc.ca

D. Wright
Geological Survey of Canada, 615 Booth St., Ottawa, ON K1A 0E9, Canada,
e-mail: dwright@nrcan.gc.ca

H. Falck
Northwest Territories Geoscience Office, 4601-B 52nd Ave, Yellowknife, NT Canada X1A 2R3

to ensure that both mineral and energy potential data were incorporated into the park configuration modelling.

Keywords GIS · spatial modelling · mineral potential · Nahanni · MERA

1 Introduction

This paper documents the application of a GIS (Geographic Information System)-based on a knowledge-driven approach to model mineral potential within the Greater Nahanni Ecosystem using one single and three composite deposit models: 1. SEDEX (stratiform shale-hosted sedimentary-exhalative Zn-Pb-Ag), 2. "Carbonate-Fault" (carbonate-hosted base-metals associated with major faults), 3. "Intrusion-Related" (includes skarn, rare metals and gemstones), and 4. "Carlin-Type" gold as lode and/or derived placer. The composite models are not formally recognized deposit types in the literature, but are informal amalgamations of several deposit types used specifically for this resource assessment, because their component types have so many attributes in common. They are capitalized throughout for ease of recognition by the reader, and defined in the DEPOSIT MODELS section later.

This work was carried out as part of the formal Mineral and Energy Resource Assessment (MERA) process (Government of Canada, 1995). An initial MERA (referred to herein as MERA 1) was conducted for two smaller regions adjacent to the existing Nahanni National Park Reserve between 1985 and 1988 (Jefferson and Spirito, 2003). MERA 2, conducted from 2003–2007 covers an expanded region including the entire catchment for the Nahanni River (Fig. 1).

The GIS analysis was undertaken with the purpose of producing mineral potential maps as systematically and quantitatively as possible from the suite of minerals-related studies completed under MERA 1 and MERA 2. The relative potential is estimated for the study area only, and there is no intent in this paper to imply absolute values of potential for undiscovered mineral resources. The results generated from the GIS mineral deposit modelling were transferred to Parks Canada for use in their GIS modelling related to defining park boundary options.

1.1 Mineral Potential Modelling Methods

Numerous modelling methods for producing mineral potential maps using a GIS have been developed over the past twenty years. These methods can be divided into two basic categories: data-driven and knowledge-driven techniques, for which reviews can be found in Bonham-Carter (1994) and Wright and Bonham-Carter (1996). A knowledge-driven approach was used in this study because only one producing mine (Canada Tungsten) is present in the study area. This is not a sufficient training set to enable a data-driven approach. The knowledge-driven technique used in

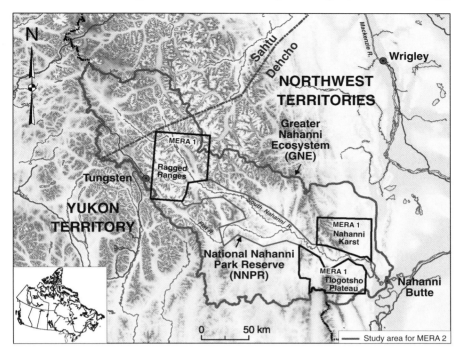

Fig. 1 Index and regional geographic context of Nahanni MERA 2 study area. MERA 1 study areas were: Nahanni Karst, Ragged Ranges, and Tlogotsho Plateau

this study consists of a simple weighting system in which each type of evidence for each mineral potential model was assigned a relative weight (reflecting importance) ranging in value from 1 to 10. This is based on expert opinion of the third and fifth authors of this paper and their exploration knowledge of the area.

Data-driven methods make use of the location of mineral deposits with respect to individual evidence maps to derive the weights for each evidence map (Bonham-Carter, 1994, Chap. 9). In this study, the location of mineral occurrences of a given deposit type were divided into four classes (defined resources, advanced prospect, showing and anomaly), assigned a weight and incorporated as evidence maps in the modelling process. Thus information on mineral occurrences was incorporated in the modelling process albeit using a different strategy than that used for data-driven approaches.

Hydrocarbon resource potential was not specifically assessed in the Nahanni MERA 2 but has been assessed by Osadetz et al. (2003). Three areas (see later figure) with significant hydrocarbon potential within the Greater Nahanni Ecosystem were identified in the southeast portion of the study area: Twisted Mountain Anticline, Mattson Anticline and the Etanda Dome. A hydrocarbon potential map was constructed comprising these three areas and converted to a binary map in which all three areas were given a high weight based on exploration knowledge (Osadetz, pers comm., 2007).

Particular weight was given to evidence derived from the studies supported by Nahanni MERA 2 that focused on significant local mineral occurrences and are reported in Barnes et al. (2007), Charbonneau (2007), Cousens (2007), Paradis (2007), Rasmussen et al. (2007), and Yuvan et al. (2007).

The assessment process used in this study employs GIS quantitative spatial analysis techniques to produce semi-quantitative relative mineral potential maps. This differs from all previous MERA studies which employed a completely subjective and much less systematic "light table" approach whereby analogue maps were overlaid and judged with respect to mineral potential. A powerful asset of the new GIS approach is to provide direct input to the GIS-based park planning system that does a cost-benefit analysis of biotic, physiographic and ecosystem attributes of importance to modern park sustainability to develop boundary options for public consultation purposes.

1.2 Study Area

The area being assessed for this study is located in the southwestern corner of the Northwest Territories (Fig. 1). It includes the entire length and catchment area of the South Nahanni River and a small extension to the northeast that includes the Ram Plateau. This assessment area is known as the Greater Nahanni Ecosystem (GNE). It is approximately 400 km long and 150 km across, from latitude 63°00' to 60°45'N and from longitude 123°50' to 129°45'W. The total area of the GNE is approximately 39,000 km^2. In this assessment the existing Nahanni National Park Reserve was excluded from the area of data collection, therefore the additive modelling process automatically produces an artificially depressed mineral potential within its boundaries. Data from the areas assessed in the MERA 1 study collected between 1985 and 1988 (Jefferson and Spirito, 2003) were integrated into this GIS study to provide seamless mineral potential maps throughout the balance of the GNE.

1.2.1 Geological Setting

In paleogeographic terms, this region was situated on the western margin of the North American Craton from more than 800 million years ago until the Middle Devonian, about 350 million years ago. Along this margin, the sedimentary rocks of the western two-thirds of this region were deposited in the shaly, deep-marine Selwyn Basin, whereas the eastern third of the strata are shallow marine carbonate shelf facies of the MacDonald Platform. After the Middle Devonian, conglomerate and black shale were derived from colliding terranes to the west and silty shale was derived from distal orogens to the east. These two different shale successions were roughly coeval and their interface in the central part of the study area is poorly known. The western part of the study area was uplifted due to continued mountain building in the west, and a mixture of foreland shale, limestone, and coal-bearing sandstone were deposited in the southern and eastern areas.

The next geological phase of this region was dominated by structural and igneous activity. The Mackenzie Mountains were produced by compressional folding and thrusting of the above sedimentary units during the Jurassic (~170 Ma) to Cretaceous (~100 Ma). Major granitoid batholiths (Selwyn Plutonic Suite) intruded these rocks during a Cretaceous collision of an island arc with western North America. The majority of the plutons are quartzofeldspathic and have accessory minerals within a narrow range of compositions that permit the plutons to be classified as biotite-muscovite ('two-mica'), biotite-hornblende or biotite (Gordey and Anderson, 1993; Rasmussen et al., 2007). Right-lateral strike-slip motion along major faults began in the Tertiary (~ 40 Ma) and continues today.

The present landscape and exposed rock units (Fig. 2) are a result of recent weathering and erosion by water and ice during and after tectonic uplift. The eastern section of the study area is a dissected plateau exposing broadly folded karst (vertical dissolution of carbonate rocks by rain water produced sinkholes and caves) on the north side of the Nahanni River and coal-bearing sandstone and shale with minor limestone on the south side. Continental glaciers carried till and mud east from the Canadian Shield into the eastern part of the study area. Mountain glaciers carved the U-shaped valleys and cirques into the Selwyn Ranges (Duk-Rodkin et al., 2007).

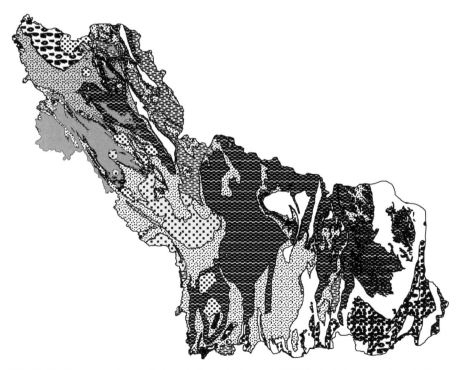

Fig. 2 Geology used for spatial modelling of Nahanni MERA 2 study area, after Jefferson et al. (2003), Okulitch (2005a,b) and Rasmussen et al. (2007). Units are combined in different ways (favourite rock units) for different mineral deposit models as listed in Tables 1–4

LEGEND

CRETACEOUS
SELWYN GRANITIC SUPER SUITE

Tay River, and Tombstone suites; grandodiorite to quartz monzonite (LKgd, LKmzg, CEKgd, Kgd, Kqm, Kdi)

Tungsten suite: biotite + muscovite + garnet + tourmaline grandiorite to quartz monzonite (CEKgd, Lkgd, Kqm)

Platform to intracratonic basin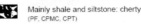

TRIASSIC to CRETACEOUS

Shale to sandstone
(KSu, KSk, KL, KSc, KGr, TrJL)

CARBONIFEROUS to PERMIAN

Mainly shale and siltstone: cherty
(PF, CPMC, CPT)

EARLY CARBONIFEROUS

Coal-bearing sandstone and shale (Mattson Fm) overlying dolomitic limestone, with minor shale and siltstone (CM, CFG, CF, CPr, CYC, CY)

Transitional to deep basinal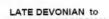

LATE DEVONIAN to EARLY MISSISSIPPIAN

Siliceous shale, minor siltstone, sandstone and conglomerate (Earn Group and Portrait Lake formation) (DCE-p, DCE, DE-PL)

MIDDLE DEVONIAN to CARBONIFEROUS

Predominantly shale: (Besa River and Horn River, and Fort Simpson formations)
(DCBR, mDHR, uDs, DHlc, uDFS, uDpc, uDspc)

MIDDLE DEVONIAN to CRETACEOUS

 Diagenetic shelf facies, dolostone and hydrothermal dolomite (Manetoe, Presqu'ile, and Grizzly Bear) (Pp, PM, PGB)

MIDDLE DEVONIAN

 Calcareous, bioclastic shale, minor limestone and dolostone (mDFH, mDF)

 Limestone, minor dolostone and calcareous shale (MDN, mDL, lmDGB, lmDA, lmDAL, lmDNa, MDHN, MDFHN, MDHe)

CAMBRIAN to EARLY DEVONIAN

Calcareous, graptolitic shale, minor quartz arenite and thin basal conglomerate (Road River Fm) (SDD-cd, SRR, SDSa, SDRR, O-DRR, ODRR-h, COR, ICGL, ICBR)

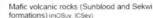 Dolostone with minor limestone and quartz arenite (Whittaker and Broken Skull formations)
(IDSo, IDD-CM, IDD-V, SD-RR, OSW, OH, mOSu, COBS, SDD, O-DWD, C-SBW, COBSu, uCH, mCA, mCR, ICSe, C-Sh, lmDC)

 Mafic volcanic rocks (Sunbloood and Sekwi formations) (mOSuv, ICSev)

NEOPROTEROZOIC
WINDERMERE SUPERGROUP

 Shale, siltstone and sandstone
(ECV, ECH, ECH-N, ECH-Y)

 Dolostone, minor limestone; minor shale
(ER, EGt, Esh, nPK)

Fig. 2 (continued)

More details on the geology of the study area are provided by Jefferson et al. (2003), Okulitch (2005a, in review) and Falck (2007).

1.3 Data Collection

A critical component of the Nahanni MERA was to compile existing data in addition to data collected during the MERA 2 project into a digital, georeferenced GIS database. A comprehensive digital database covering the entire study area is essential to conduct the mineral and energy resource assessment and mineral potential modelling using GIS-based analytical methods. In total the current GIS database contains over 200 layers of digital geoscience data including bedrock geology, surficial geology, geochemistry, geophysics, mineral deposits, and topographic data (Lemkow et al., 2007).

A selection of the bedrock geology compiled by Okulitch (2005b; 2005a, in review) at a scale of 1:1,000 000 is included in the digital GIS database. The bedrock geology map (Fig. 2) is a simplified classification of the major bedrock lithological groupings based on Okulitch (2005a). This fundamental framework locates key geological attributes associated with the four deposit models that in turn embrace the most likely deposit types to be found in the study area. These lithologic attributes include primary sedimentary rock types, products of diagenetic alteration (e.g. Manetoe facies; Morrow et al., 1990), structure (faults and folds) and intrusive rock types (granitic plutons). All of these lithologic attributes provide indirect evidence for sources, transport, and deposition of elements that may have formed mineral deposits.

A total of 2463 stream sediment samples and 2068 stream water samples were collected (Fig. 3a) by various surveys within the Greater Nahanni Ecosystem and geochemically analyzed to help model mineral potential. Over half these sam-

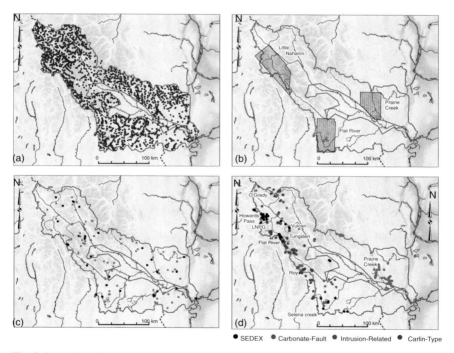

Fig. 3 Locations of data collected for Nahanni MERA 1 and 2. (**a**) Locations of stream sediment and water samples collected by Spirito and Jefferson (2003) in MERA 1 areas and by McCurdy et al. (2007) for MERA 2 under National Geochemical Reconnaissance (NGR) program. (**b**) Locations of geophysical surveys as reported by Charbonneau (2007). (**c**) Locations of spring water sample sites, from Hamilton et al. (2003) for MERA 1 areas, and from Caron et al. (2007). Anomalous springs are shown by black dots. (**d**) Locations of mineral occurrences from NORMIN.db. Deposit models are explained in text

ples (1374 stream sediment and 1378 stream water samples) were collected by McCurdy et al. (2007) as part of MERA 2 during the 2004 and 2005 field seasons and combined with additional geochemical data for stream sediments and waters from previous surveys: 396 sediment samples from the first Nahanni MERA (Jefferson and Spirito, 2003); 693 sediment samples and 683 water samples from previous National Geochemical Reconnaissance surveys (Friske et al., 2001; Day et al., 2005). All of these data are included in the GIS database and were spatially analysed herein.

Detailed public domain airborne geophysical data were not available for the study area prior to this MERA study. Therefore, as part of the Nahanni MERA 2 project, three airborne surveys were flown at 135 m nominal terrain clearance along lines spaced at 500 m for a total of 16,000 line kilometres over three areas shown in Fig. 3b. These areas were considered most likely to generate conflicts between mineral potential and other competing land uses. The geophysical data provide additional evidence to help build the mineral deposit knowledge base, refine the mineral potential modelling, and may help resolve future planning issues. For each area, ten themes of data were generated including total radioactivity, potassium (K%), equivalent uranium (eU, ppm), equivalent thorium (eTh, ppm), eU/eTh, eU/K, eTh/K, ternary radioactive plots (K, eU, eTh), residual magnetic total field (nT), and calculated magnetic vertical gradient (nT/m). Key aspects of these surveys are discussed and illustrated by Charbonneau (2007). The full primary data sets have been published (Carson et al., 2007a,b,c) and are included in the GIS database (Lemkow et al., 2007).

Spring water geochemistry can be an effective tool for investigating regional compositional trends and for indirectly detecting buried mineral deposits that were transected by sub-terranean flow paths. Spring water geochemistry surveys reported by Hamilton et al. (2003) and Caron et al. (2007) contributed to the Nahanni MERA 1 and MERA 2 respectively. A total of 95 samples (Fig. 3c) collected from 78 springs for those studies were compiled and integrated for this GIS analysis.

Mineral deposit data (Fig. 3d) were compiled and rationalized for the Nahanni MERA 1 over the entire watershed (Jefferson and Spirito, 2003). This data, a significant contribution to the NORMIN database (Department of Indian Affairs and Northern Development, 2006), was augmented by mineral deposit studies conducted as part of the Nahanni MERA 2 (Rasmussen et al., 2006; Paradis, 2007; Barnes et al., 2007; Cousens, 2007; Yuvan et al., 2007, Rasmussen et al., 2007). All known mineral showings in the study area as determined from the NORMIN database and the current MERA study were classified into the four deposit models outlined in the following section.

2 Deposit Models

Four different deposit models were designed to incorporate the most important mineral deposit types for this region (Fig. 3d), using the minimum number of GIS operations. The SEDEX (stratiform sediment-hosted exhalative Zn-Pb-Ag) model is a single deposit type that is well documented and pre-eminent in the region. The three other models are composites of two or more deposit types with spatial and genetic affinities, Carbonate-Fault Type, Intrusion-Related Type, and Carlin-Type gold (Carlin or other types of intrusion-related lode gold and/or derived placer deposits). Each of these models is reviewed as follows.

2.1 SEDEX (Stratiform Sediment-hosted Exhalative Zn-Pb-Ag)

The SEDEX mineral potential map created in this study is based on a well known set of attributes summarized by Carne and Cathro (1982), Lydon (1995), Goodfellow (2007), and Cousens (2007). These include favourable host lithology (black, graphitic, siliceous shale with rapid lateral facies changes including conglomerate tongues), structural features (known faults, and faults inferred from facies changes), as well as direct and indirect pathfinder elements that are detected by stream sediment, stream water and spring water geochemistry surveys, and the locations of SEDEX-type mineral occurrences (Table 1).

Table 1 The **SEDEX** mineral potential map created in this study is based on a well known set of attributes summarized by Carne (1982), Lydon (1995), Goodfellow (2007), and Cousens (2007)

Favorable rock units	Weight
1. Transitional to deep basinal strata (Cambrian to Early Devonian + Middle Devonian to Early Mississippian are established producers)	10
2. Basinal Windermere strata (theoretical producing rock package)	3
Structural Features	
3. Cambrian and Ordovician volcanic units (like fault, represents deep crustal structure favouring hydrothermal fluid flow)	10
4. Volcanic: 10 km buffer *outside units*	3
5. Growth faults (north-northwesterly faults along Broken Skull River and parallels, parallels to the carbonate-shale facies change (#5), and parallels to the northeasterly Leith Ridge and Fort Norman structures are interpreted as Laramide expressions of deep crustal weaknesses by steeply dipping breaks discordant to regional fold and thrust trends): 3 km buffer	2
6. Carbonate-shale facies change (Ordovician-Devonian platform edge from Morrow, represents depositional structure favourable for SEDEX): 3 km buffer	2
7. All other faults (mainly southwest-dipping thrusts): 2 km buffer	1
Geochemistry; anomalous direct and indirect pathfinder elements	
8. Stream sediment, direct: Ag, Cu, Pb, Zn. (each element map was normalized (to provide equal influence) and then all elements were summed and the mean value was calculated – this map was divided into a 5 class map reflecting high to low concentration – class thresholds based on natural breaks in the data histogram)	**10, 7, 2, 1, 0** by class

Table 1 (continued)

Favorable rock units	Weight
9. Stream sediment, indirect: (As, Ba, Bi, Cd, Cr, F, Hg, Mn, Mo, Ni, P, Sb, V) (each element map was normalized (to provide equal influence) and then all elements were summed and the mean value was calculated – this map was divided into a 5 class map reflecting high to low concentration – class thresholds based on natural breaks in the data histogram)	**6, 4, 2, 1, 0** by class
10. Stream water, direct: Ag, Cu, Pb, Zn. (each element map was normalized (to provide equal influence) and then all elements were summed and the mean value was calculated – this map was divided into a 5 class map reflecting high to low concentration – class thresholds based on natural breaks in the data histogram)	**10, 7, 2, 1, 0** by class
11. Stream water, indirect: As, Ba, Cd, Cr, F, Mn, Mo, Ni, P, pH, Sb, V (each element map was normalized (to provide equal influence) and then all elements were summed and the mean value was calculated – this map was divided into a 5 class map reflecting high to low concentration – class thresholds based on natural breaks in the data histogram)	**6, 4, 2, 1, 0** by class
12. Spring-water sample sites: all SEDEX from Table 5.12 of Hamilton et al., 2003 and selected springs from Table 6 of Caron et al. (2007) anomalous in one or more of 14 pathfinder elements: direct: Ag, Cu, Pb, Zn; and indirect: As, Ba, Cd, Cr, Mn, Mo, Ni, pH, Sb, V. buffer = 3 km.	**2**
13. Ph – sampled from water chemistry – data were interpolated and map divided into 8 classes reflecting acid to basic conditions	**10, 9, 8, 7, 4, 3, 2, 0**
14. Mineral occurrences of the SEDEX type were assigned to one of 4 categories: defined resources (weight 10), advanced prospect (weight 7), showing (weight 4), anomaly (weight 2). Buffer = 3 km.	**10, 7, 4, 2**

TOTAL of 17 evidence maps used for producing the SEDEX potential map.

SEDEX deposits are most likely to occur within the early Cambrian to Middle Devonian Selwyn Basin that comprises the deep-water euxinic (oxygen-poor) portions of the western North American passive margin sedimentary sequence. Two main districts at Macmillan Pass and Howard's Pass are defined by linear belts of such occurrences near the margins of the Selwyn Basin. The only past-producing SEDEX deposits in the Selwyn Basin are those of the Anvil District hosted by basal Cambrian shay strata near Whitehorse. The deposits in the Howard's Pass district contain some of the largest base metal accumulations in the world.

2.2 Carbonate-Fault

Favourable rock units, structural features, geochemistry and the location of Carbonate-Fault-associate mineral occurrences (Table 2) were used to produce a mineral potential map.

The composite Carbonate-Fault deposit model accounts for the spatial association of zinc-lead-silver deposits of at least three types: (1) vein quartz-carbonate,

Table 2 "CARBONATE – FAULT" (3 spatially associated types of zinc-lead-silver massive sulphide: stratabound, Mississippi Valley, and vein quartz-carbonate of the Prairie Creek camp as summarized by Paradis, 2007)

Favourable rock units	*Weight*
1. Diagenetic Facies (vuggy recrystallized dolostone of the Devonian to Cretaceous Manetoe (PM), Presqu'ile (Pp), and Grizzly Bear (PGB) formations)	**10**
2. Middle Devonian (all strata: limestone, dolostone and shale)	**5**
3. Cambrian to Early Devonian Platformal (limestone and dolostone)	**5**
4. Cambrian to Early Devonian Transitional to Deep Basinal (shale)	**2**
5. Buffer around Diagenetic Facies: 3 km buffer only in #3 to 5	**5**
Structural Features	
6. Growth faults (as for SEDEX): 3 km buffer	**2**
7. Carbonate-shale facies change (as for SEDEX): 3 km buffer	**2**
8. All other faults (as for SEDEX): 2 km buffer	**1**
Geochemistry; anomalous direct and indirect pathfinder elements	
9. Stream sediment, direct: Ag, Cu, Pb, Zn (each element map was normalized (to provide equal influence) and then all elements were summed and the mean value was calculated – this map was divided into a 5 class map reflecting high to low concentration – class thresholds based on natural breaks in the data histogram)	**10, 7, 2, 1, 0** by class
10. Stream sediment, indirect: Cd, F, Hg, Sb (each element map was normalized (to provide equal influence) and then all elements were summed and the mean value was calculated – this map was divided into a 5 class map reflecting high to low concentration – class thresholds based on natural breaks in the data histogram)	**6, 4, 2, 1, 0** by class
11. Stream water, direct: Ag, Cu, Pb, Zn. (each element map was normalized (to provide equal influence) and then all elements were summed and the mean value was calculated – this map was divided into a 5 class map reflecting high to low concentration – class thresholds based on natural breaks in the data histogram)	**10, 7, 2, 1, 0** by class
12. Stream water, indirect: Cd, F, Sb. (each element map was normalized (to provide equal influence) and then all elements were summed and the mean value was calculated – this map was divided into a 5 class map reflecting high to low concentration – class thresholds based on natural breaks in the data histogram)	**6, 4, 2, 1, 0** by class
13. Spring-water sample sites: Prairie Creek mine water from Table 5.12 of Hamilton et al., 2003 and selected anomalous springs from Table 6 of Caron et al., 2007) anomalous in one or more of 6 pathfinder elements: direct: Ag, Cu, Pb, Zn and indirect: Cd and Sb. Buffer = 3 km.	**2**
14. Mineral occurrences of the CARBONATE-FAULT type were assigned to one of 4 categories: defined resources (weight 10), showing (weight 4); buffer = 3 km.	**10, 4**

TOTAL of 15 evidence maps used for producing the CARBONATE – FAULT potential map.

(2) SEDEX and, (3) Mississippi Valley. Fluids trapped during sedimentation combined with fluids derived from deeply buried basement rocks may have leached metals from basement rocks as they flowed along more permeable zones, driven by pressure due to increased burial near basin centres, by differential tectonic uplift or subsidence, or by thermal convection. Where the fluids encountered fault zones in competent rocks, the permeability was locally enhanced, thereby focusing the fluids. As these fluids passed through other layers including different aquifers, reaction with the wall rock or mixing with fluids from other aquifers may have altered the parameters of the fluid sufficiently for metals to precipitate. The Manetoe digenetic alteration facies (Morrow et al., 1990) is an example of the alteration of carbonate layers due to basinal fluids. Mississippi Valley-type base-metal deposits (MVT) such as the classic past-producing Pine Point district are typically associated with this style of alteration. Faults especially reactivated vertical growth-faults, locally facilitated fluid transportation and mixing at Pine Point, as well as in variants of MVT, such as the Irish-type of carbonate-hosted Zn-Pb-Ag deposits. Faults may also have allowed great fluctuations in fluid pressure regimes, allowing the formation of breccias, stockworks and veins.

2.3 *Intrusion-Related (Includes a Variety of Skarn, Rare Metal and/or Gemstone Types)*

Favourable host lithology, distance to plutons of various types, structural features, favourable pathfinder elements in stream sediments, stream waters, and springs, and the locations of Intrusion-Related mineral occurrences (Table 3) were used to produce an Intrusion-Related potential map.

Intrusions provided both primary metaliferous fluids and heat for metal redistribution and reconcentration. In addition to mineral concentrations which are direct products of the magma (rare element-bearing pegmatites and gemstones) the introduction of magmatic and meteoric fluids driven by heat energy from the intrusion and their interaction with reactive host rocks can produce many different styles of mineral deposits. These include the extensive skarn family, mantoe, chimney, epithermal, endoskarn, porphyry and high temperature replacement deposits. A wide range of metal assemblages and rare metals is associated with the Intrusion-Related family of deposits, including one or more of W, Cu, Zn, Pb, Ag, Au, As, Sb, Li, Ta, Cs, and Sn (refer to Barnes et al., 2007). For this Nahanni MERA 2 GIS analysis, gold (Au) is treated separately as described in the next section.

In order to form intrusion-associated metal concentrations a number of additional factors are important, to optimize the effects of heat energy and chemical reactions with host rocks. Fluid pathways are essential to channel fluids through and away from metal sources and focus them in zones where the deposition of metals can build up economic concentrations. Because the plutons intruded deformed sedimentary

Table 3 INTRUSION (Pluton) – RELATED: skarn tungsten or zinc-lead-silver, rare metals and/or gemstones (excludes Carlin Gold), modelled after Dawson et al. (1992), Barnes et al. (2007), Rasmussen et al. (2007), and Yuvan et al. (2007)

Favourable rock units	Weight
1. High-to very high potential granitic bodies (include hornfels): 1, 5 and 20 km buffers; weights 10, 5 and 2 respectively (inside and outside of margins only)	**10, 5, 2**
2. Moderate potential granitic bodies (include hornfels): 1, 5 and 10 km buffers; weights 5, 2 and 1 respectively (inside and outside of margins only)	**5, 2, 1**
3. Low potential granitic bodies: 1 and 5 km buffers, weights 3 and 1 respectively	**3, 1**
4. Shaly limestone units. Sedimentary units within a 20 km buffer of all granite bodies are weighted 1–10 for SKARN potential based on the presence of impure limestone units, with the #10 rank going to heterogeneous units (Vampire and Sekwi). Sedimentary unit values are added to the buffer values of the respective plutons	
Structural Features	
5. Fault density map based on line density algorithm (assigned weights based on low, medium and high fault density)	**5, 3, 1**
Geochemistry	
6. Stream sediment, direct: Cu, Pb, W, Zn (each element map was normalized (to provide equal influence) and then all elements were summed and the mean value was calculated – this map was divided into a 5 class map reflecting high to low concentration – class thresholds based on natural breaks in the data histogram)	**10, 7, 2, 1, 0** by class
7. Stream sediment, indirect: Ag, As, Au, Bi, Sb, Sn, Te. (each element map was normalized (to provide equal influence) and then all elements were summed and the mean value was calculated – this map was divided into a 5 class map reflecting high to low concentration – class thresholds based on natural breaks in the data histogram)	**6, 4, 2, 1, 0** by class
8. Stream water, direct: Cu, Pb, W, Zn. (each element map was normalized (to provide equal influence) and then all elements were summed and the mean value was calculated – this map was divided into a 5 class map reflecting high to low concentration – class thresholds based on natural breaks in the data histogram)	**10, 7, 2, 1, 0** by class
9. Stream water, indirect: Ag, As, Au, Bi, Sb, Sn, Te (each element map was normalized (to provide equal influence) and then all elements were summed and the mean value was calculated – this map was divided into a 5 class map reflecting high to low concentration – class thresholds based on natural breaks in the data histogram)	**6, 4, 2, 1, 0** by class
10. Spring-water sample sites: three "tungsten skarn" from Table 5.12 of Hamilton et al., 2003 and selected springs from Table 6 of Caron et al. (2007) anomalous in one or more of 6 direct pathfinder elements: W, Cu, Pb, Zn; and 8 indirect: Ag, As, Au, Bi, F, Sb, Sn, Te. Buffer = 3 km.	**2**
11. Mineral occurrences of the INTRUSION type (after Rasmussen et al., 2007 and NORMIN, 2007) were assigned to one of 4 categories: defined resources (weight 10), advanced prospect (weight 5), showing (weight 2), anomaly (weight 1). Buffer = 3 km.	**10, 5, 2, 1**

TOTAL of 14 evidence maps used for producing the INTRUSION potential map.

rocks, such pathways were generally along pre-existing structures (faults, fold axes and permeable strata) that may also have constrained the upward ascent of magma to form the intrusions.

Another critical aspect of these deposits is that the focusing parameters under which metals are carried in solution must have changed in order to form an ore deposit rather than disperse the metals distally from the source and dilute them. Mechanisms to focus precipitation include chemical reaction with the wall rock (such as changes in pH (acidity), Eh (oxidation), trace element contents, cooling of the fluid and/or depressurization). For example, a hot, acidic, high-pressure fluid passing through a carbonate unit will be cooled and neutralized. This explains why limestone, dolomite and calcareous shale layers are common hosts to skarn. Skarn deposits tend to be best developed in the first relatively pure limestone layer in a mixed stratigraphic assemblage that the hydrothermal fluids encounter on their ascent from the related intrusion.

Depressurization of mineralizing fluid related to an active fault system can also be important; thus the presence of faults as a transport mechanism and a concentration mechanism is also an important factor for localizing intrusion-related deposits. Ponding due to physical "traps" may also concentrate metals as the movement of fluids is impeded giving them time to cool.

Some deposits are located within a pluton, caused by ingress of fluid from country rock, concentrating alteration within the upper portion of the pluton (termed endoskarn). Endoskarn may be recognised by the presence of metamorphosed remnants of wall rock preserved in the pluton as roof pendants. Such alteration introduces more potassium into the pluton resulting in crystallization of potassic feldspar, muscovite and/or sericite clay, and registered by high K/U, and K/Th ratios. Gamma ray spectrometer data is useful for detecting potassium enrichment, and thereby targeting alteration associated with such deposits. Gamma ray data were acquired in three strategic areas located in Fig. 3b and interpreted as reported by Charbonneau (2007). The gamma ray data (K channel) were used as signature of potassic alteration.

Skarn deposits in the Canadian Cordillera are abundant, diverse and economically significant. In the study region skarn occurrences are localised where the Middle Cretaceous Selwyn Plutonic Suite discordantly intrudes the lowest and/or thickest limestone unit of an upper Proterozoic to lower Paleozoic shelf-carbonate pelite sequence. The broad thermal aureole at the contact between a Cretaceous quartz monzonite stock and a Lower Cambrian limestone unit is the setting of the world-class Cantung Mine. A belt of W, Cu (Zn, Mo) skarn showings to deposits follows an arcuate trend of generally small mid Cretaceous granitoid plutons from the southwestern Northwest Territories to the Dublin Gulch District of the Yukon Territory. This trend contains some of the largest and highest grade resources of skarn tungsten in the world, mainly in the Cantung Mine and in the Mac Tung deposit at Macmillan Pass. Thus proximity to a pluton, especially where in contact with an appropriate carbonate unit, is an important vector for Intrusion-Related deposits.

2.4 Carlin-Type (Intrusion- or Fault-Associated Lode Gold and/or Related Placer Deposits)

Favourable rock units, structural features, favourable stream sediment, stream water and spring water geochemistry, potassium anomalies detected by airborne gamma ray spectrometer data and the locations of intrusion-associated placer and lode gold mineral occurrences (Table 4) were used to produce a Carlin-Type mineral potential map.

Table 4 CARLIN &/or PLACER: Intrusion-related lode gold and associated placer deposits (after Cline et al., 2005; Emsbo et al., 2006)

Favourable rock units	*Weight*
1. High-to very high potential granitic bodies (include hornfels): 1, 5 and 20 km buffers; weights 10, 5 and 2 respectively (inside and outside of margins only)	**10, 5, 2**
2. Moderate potential granitic bodies (include hornfels): 1, 5 and 10 km buffers; weights 5, 2 and 1 respectively (inside and outside of margins only)	**5, 2, 1**
3. Low potential granitic bodies: 1 and 5 km buffers, weights 3 and 1 respectively.	**3, 1**
4. Reactive limestone units. Carlin uses same sedimentary units and additive weighting as for SKARN	
Structural Features (Hydrothermal flow corridors)	
5. Growth faults as for SEDEX, within 20 km of granite bodies, treat as vertical: 2 km buffer each side	1
6. All other faults, within 20 km of granite bodies 2 km buffer down dip to SW	1
7. Fault intersections, within 20 km of granite bodies, treat as vertical: 2 km buffer	2
8. Anticlines: 2 km buffer	1
9. Volcanic units regardless of age: 5 km buffer	2
Geochemistry; anomalous direct and indirect pathfinder elements	
10. Stream sediment, direct: Au. (interpolated map – divided into 5 classes based on high to low concentration)	**10, 8, 2, 1, 0**
11. Stream sediment, key indirect: As, Sb, Hg, Te. (each element map was normalized (to provide equal influence) and then all elements were summed and the mean value was calculated – this map was divided into a 5 class map reflecting high to low concentration – class thresholds based on natural breaks in the data histogram)	**6, 4, 2, 1, 0** by class
12. Stream sediment, possible indirect: Ag, Ba, Cs, Cu, Fe, Mo, Pb, Se, Si, Te, Tl, W, Zn. (each element map was normalized (to provide equal influence) and then all elements were summed and the mean value was calculated – this map was divided into a 5 class map reflecting high to low concentration – class thresholds based on natural breaks in the data histogram)	**4, 3, 2, 1, 0** by class

Table 4 (continued)

Favourable rock units	Weight
13. Stream water, key indirect: As, Sb, Hg, Te. (each element map was normalized (to provide equal influence) and then all elements were summed and the mean value was calculated – this map was divided into a 5 class map reflecting high to low concentration – class thresholds based on natural breaks in the data histogram)	**6, 4, 2, 1, 0** by class
14. Stream water, possible indirect: Ag, Ba, Cs, Cu, Fe, Mo, Pb, Se, Si, Te, Tl, W, Zn (each element map was normalized (to provide equal influence) and then all elements were summed and the mean value was calculated – this map was divided into a 5 class map reflecting high to low concentration – class thresholds based on natural breaks in the data histogram)	**4, 3, 2, 1, 0** by class
15. Springs with CARLIN signature anomalous in Ag, Sb, As, Au, Ba, Hg, Ti, Tl. : Buffer 3 km	**2**
16. Placer gold occurrences only where they fit Rasmussen's local derivation criteria: around Selwyn Plutonic Suite and within buffer of springs with Carlin signature. Buffer = 5 km	**5**
17. Lode mineral occurrences of the Carlin and other granite-related types (after Rasmussen, (2007); Cline et al., 2005; and NORMIN) were assigned to one of 2 categories: showing (weight 2), or anomaly (weight 1); 3 km buffer	**2, 1**
18. Airborne gamma ray spectrometry should detect potassic alteration zones that may be expressed as illite, dickite or potassium feldspar. (binary map of anomalous potassium concentration	**1**
19. Ph – sampled from water chemistry – data were interpolated and map divided into 8 classes reflecting acid to basic conditions	**10, 9, 8, 7, 4, 3, 2, 0**

TOTAL of 20 evidence maps used for producing the CARLIN potential map.

Explanatory notes for Tables 1–4 inclusive:

- Anomalous cut-offs were determined by separate statistical analysis of frequency distribution of results for samples in each of the rock units used for each deposit model.
- The "Buffer" for each sample is calculated by interpolation within a data-determined zone of influence.
- Each sample location is assigned to a rock unit based on straightforward spatial intersection. Although most individual stream catchment areas and spring water flow paths intersect multiple rock types, their relative proportions were NOT taken into account in determining the background rock influence on each sample. In a few cases, the location is just inside a rock unit that did not contribute significantly to the sample medium.
- Stream water pH data were included in stream sediment geochemistry as well as stream water geochemistry for spatial modelling (see discussion in text under SEDEX).

Carlin-Type gold is a special, possibly Intrusion-Related deposit model that for this study comprises at least two basic genetic hypotheses. Recent research has determined many of the associated empirical factors involved in these highly economic and commonly very large gold deposits (Poulsen, 1996; Cline et al., 2005; Emsbo et al., 2006). The characteristics of these gold deposits range from stratiform disseminations, breccia zones, sinters, and vein stockworks to bonanza-style vein systems. Richards (1989) and Rowan (1989) recognized some of the geochemical attributes of such deposits in the Selena Creek area. Work under MERA 2 reported by McCurdy et al. (2007) and Charbonneau (2007) focused on this area to further test the premise that As, Sb, Hg, Ti $+/-$ Ba bearing placer-gold occurrences here were derived from one or more Carlin-Type lode-gold deposits.

Depending on the deposit being examined, the genesis of Carlin-Type gold has been ascribed to a number of different mechanisms. The varied appearances and settings of the different deposits have resulted in a range of hypotheses describing their possible origins, and these are summarized here by two generalized hypotheses. The first hypothesis emphasizes the common proximity of the deposits to granitoid intrusions (plutons) which are considered to have been the thermal engines that drove fluid flow and were themselves the source for much of the fluid that carried the gold and associated trace metals.

The second hypothesis for Carlin deposits emphasizes the regional linear trends of the gold deposits that geologists have divided into regional "trends" (i.e. The Carlin trend, the Battle Mountain trend, etc.). According to this hypothesis the plutons were not responsible for the heat, but were simply accessories. Instead, the extensional tectonic regime resulted in high heat flow that mobilized metal-bearing fluids to form the deposits. The fluids themselves were derived from meteoric water that penetrated the crust along deep-seated fracture systems created by the same extensional tectonism.

The metal contents of Carlin deposits, regardless of their hypothetical origins, tend to be distinctive with As, Sb, Hg, Ti and Ba accompanying Au and Ag. The suite of pathfinder elements is controlled by the geochemistry of the transporting fluids, especially at the cool temperatures at which these deposits tend to form.

The host rocks are primarily impure carbonate strata, typically argillaceous limestone and dolostone that were effective at channeling fluids while at the same time providing the reactive elements ($CaCO_3$) to alter the pH of the transporting fluids resulting in the precipitation of sulphide minerals.

Active fault systems are also thought to play a role in the formation of Carlin-Type deposits by localizing and concentrating deposits along fault traces with a reduction in ore thickness away from the fault zones. The faults act as fluid conduits transporting and concentrating fluids from large regions into focused flow zones. Intersections of conjugate fault systems where permeability is the greatest are highly prospective zones.

Deformation structures such as anticlines may also play a significant role in the formation of Carlin-Type deposits. Although these structures may not have been active during the mineralization events, folding and modification of the sedimentary sequence in many cases resulted in an increase in permeability, especially in the dilational, anticlinal axis zones.

In the search for the Carlin-Type of deposit, the presence of other mineral deposits is often used as a vector, including linear strings of seemingly unrelated metal concentrations. Long-lived, deep-seated crustal sutures along which continents may have amalgamated are commonly the locus of a variety of metal-bearing fluids derived from intrusive and extrusive magma, regional metamorphic events and from the surface. As such deeply buried structures are rarely exposed and difficult to identify using geophysical techniques, they are often identified by the linear pattern formed by the deposits themselves.

More direct indicators of Carlin-Type deposits are fine-grained placer gold deposits with pathfinder elements that represent weathered and locally transported lode gold deposits. Placer-gold occurrences in many places have provided the first evidence of the presence of Carlin and other types of lode gold deposits.

The westernmost Nahanni River region has been identified as one of the most prospective in Canada for hosting Carlin-Type deposits (Rowan, 1989; Richards, 1989; Poulsen, 1996; Rasmussen et al., 2007). All of the elements identified as critical to the formation of a fine-grained disseminated sediment-hosted gold deposit (Carlin-Type) discussed above have been identified in the study area, except for the regional extensional tectonism.

3 Applied Methodology

3.1 Overview: Production of Mineral Potential Maps

Figure 4 provides a summary of the modelling process used to produce the various potential maps. For each deposit model, vectors (indicators) to mineralization were selected based on the exploration criteria discussed previously, in concert with exploration knowledge of the area. Evidence maps (thematic data layers) were created based on these vectors and each map was weighted (Tables 1–4). The weights for each evidence map were subjective and are based on expert opinion (of the 3rd and 5th authors) in conjunction with regional exploration knowledge derived from literature searches and discussions with geologists from industry and provincial and territorial governments. It should be noted that weighted mineral occurrences for each style of mineralization were also used as evidence maps in the modelling process.

One primary potential map was produced for each of the four deposit models and one other for hydrocarbon potential This was accomplished by combining the weighted evidence maps for each model, using a simple additive technique (weighted index-overlay) followed by normalization to a scale of 1–50 to ensure each of the four deposit maps had equal weight in the final, combined model. The hydrocarbon map was also normalized so that each of the three high potential areas had a weight of 50 reflecting their high potential for hydrocarbons (K. Osadetz, pers comm., 2007). These four individual potential maps (one for each model) were divided into 5 classes based on natural breaks in the histograms with 5 being the highest mineral potential. The 4 multi-class maps and the fifth hydrocarbon potential map (already converted into 1 class of very high potential) were then combined into one final mineral potential map using a MAX operator as outlined in Fig. 4. The rationale behind the MAX operator is that high potential for any one deposit model is sufficient on its own. If the additive function were to be used at this stage, this process would subdue a high potential rating that is based on only one deposit model compared to a similar high potential rating that is derived for two or more

Mineral Potential Modelling for the Greater Nahanni Ecosystem Using GIS

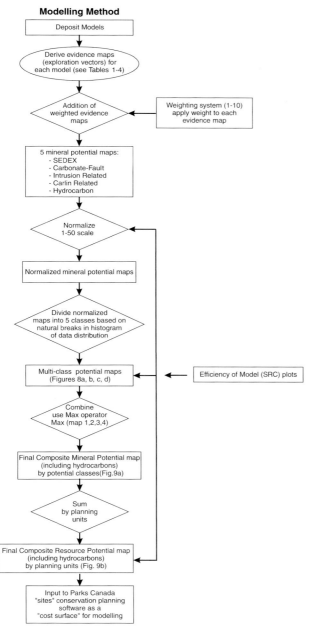

Fig. 4 Flow chart for GIS modelling operations reported in this paper

deposit models.The final step was to partition the geologically based mineral potential map into drainage-defined planning units developed by Parks Canada for park planning purposes.

Plots showing the efficiency of the model (i.e. how well the potential map predicts the known occurrences; termed an SRC plot) for the final additive potential map was generated for each of the four deposit models to assess how well the map predicts the individual occurrences as organized by deposit type. These graphs were produced by plotting cumulative area of each potential map (from highest to lowest potential) against the number of deposits predicted (Chung and Fabbri, 2003 and Harris et al., 2006a,b).

Space does not permit showing the individual evidence maps for all four models. However to demonstrate the methodology, a complete set of evidence maps is shown in Fig. 5 for the SEDEX model only.

3.2 SEDEX Model

A total of 17 weighted evidence maps (Table 1; Fig. 5) were used to create an additive SEDEX mineral potential map (see Final Model section). The individual evidence maps are described below.

3.2.1 Favourable Rock Packages

Favourable rock packages included Cambrian to mid-Devonian transitional to deep basinal strata (Fig. 5a) which are established SEDEX hosts, as well as basinal Neoproterozoic Windermere Supergroup strata (shaly sandstone) (Fig. 5b). These strata may have been the source of metals to form the overlying SEDEX deposits as well as having potential to host SEDEX deposits in their own right. The former rock package is highly weighted (see Table 2) whereas the latter is assigned a lower weight due to its speculative genetic ties to the SEDEX model.

The early Silurian portion of the generally black, calcareous, graptolitic, Cambrian to Silurian shale in the Road River Group hosts the Howard's Pass deposits. These collectively represent one of the few giant to super giant SEDEX-type camps of the world. Of similar age but different in lithology, the carbonaceous cherty black shale of the Whittaker Formation also hosts similar stratabound zinc and lead sulphide minerals at Prairie Creek (one of the three types that constitute the fault-associated composite model). The total extent of the SEDEX portion of Prairie Creek is not known, but its presence supports the concept of a broader mineralizing event in the late Silurian. A second series of SEDEX-type mineral deposits, characterized by the Tom and Jason deposits in the Macmillan Pass area, is hosted by late Devonian to Mississippian siliceous black shale and conglomerate of the Earn Group. All of the Cambrian to Mississippian basinal strata together constitute the highly weighted first lithological evidence map.

Mineral Potential Modelling for the Greater Nahanni Ecosystem Using GIS

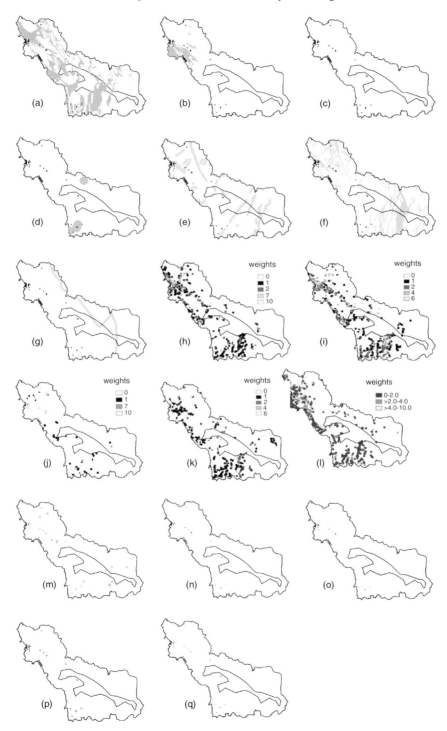

Fig. 5 (continued)

East of the Earn Group tract, the contemporaneous grey silty shale units and the Besa River and Horn formations are not known to host mineral deposits. Neither formations display facies changes such as siliceous or highly carbonaceous zones, or conglomerate. Therefore these units are characterized as an unfavourable rock package and given no weight in this analysis.

3.2.2 Structural Features

Cambrian and Ordovician volcanic units (Fig. 5c), as well as proximity to these units (Fig. 5d), are modelled in a similar fashion to faults, in that they represent deep crustal zones of weakness that may have channeled hydrothermal fluid flow. Growth faults (Fig. 5e) are interpreted as Laramide expressions of northeast-trending deep crustal offsets, the Leith Ridge and Fort Norman structures (Aitken and Pugh, 1984). These are interpreted as zones of weakness that caused subtle lateral facies changes and could have focused the transport of hydrothermal fluids during sedimentation. They are now interpreted as reactivated in terms of steeply dipping cross faults that are discordant to northwesterly regional fold and thrust trends.

All other faults (excluding growth faults) (Fig. 5f) are mainly southwest dipping thrusts, some of which may have been primary depositional extensional faults that could have channeled mineralizing fluids but were reversed during the compressional Laramide deformation that formed the Selwyn and Mackenzie mountains.

A major carbonate-shale facies change (Fig. 5g) along the Ordovician to Devonian platform edge is interpreted as yet another favourable location for the development of structures that could have transported hydrothermal fluids.

3.2.3 Stream Sediment and Stream Water Geochemistry

Elements considered to reflect both direct and indirect evidence for SEDEX deposits (Table 1) were extracted from the stream sediment and water geochemical data sets (Fig. 3a). Variograms were constructed for each pathfinder element in order to determine a "zone of influence" around each sample point (see Harris

Fig. 5 Evidence maps used for modelling SEDEX deposit type, (**a**) transitional to deep basinal strata, wt = 10; (**b**) basinal Windermere strata, wt = 3; (**c**) Cambrian to Ordovician volcanic units, wt = 10; (**d**) 3 km buffer around volcanic units, wt = 5; (**e**) 3 km buffer around growth faults, wt = 2; (**f**) 2 km buffer around faults, wt = 1; (**g**) carbonate-shelf facies change, 3 km buffer, wt = 1; (**h**) stream sediment geochemistry, direct indicators; (**i**) stream sediment geochemistry, indirect (pathfinder) indicators; (**j**) stream water geochemistry, director indicators; (**k**) stream water geochemistry, indirect (pathfinder) indicators; (**l**) pH, water geochemistry; (**m**) 3 km buffer around anomalous springs, wt = 2; (**n**) 3 km buffer around SEDEX occurrences, defined resources, wt = 10; (**o**) 3 km buffer around SEDEX occurrences, wt = 7; advanced prospects, wt = 5; (**p**) 3 km buffer around SEDEX occurrences, showings, wt = 2; and (**q**) 3 km buffer around SEDEX occurrences, anomalies, wt = 1

et al., 2001). The zone of influence (typically 3 km) was used as a search radius for an inverse-distance-weighted (IDW) algorithm used to interpolate the data over selected favourable rock units. An advantage of this approach which utilizes a fixed radius is that areas characterized by a low density of sample points are not interpolated, thus reducing the introduction of artifacts and thereby increasing the certainty in the interpolation process. The interpolated map of each element was then normalized from either ppm or ppb to byte data (0–255) so that in the additive processes of combining the data all of the included elements would be equally weighted preventing any one element to dominate based on a wider data distribution range. A map showing mean normalized concentration was then produced by summing the normalized interpolated element maps and dividing by the number of elements. For example, the evidence map of direct SEDEX pathfinder elements obtained from stream sediment data was created by using the following formula:

$$(nAg + nCu + nPb + nCu)/4$$

where n = normalized value

This map was then broken into 5 classes based on natural breaks in the histograms of the normalized data. The classes were assigned subjective weights of 10, 7, 2, 1, and 0 for the highest to lowest mean concentrations respectively. The entire data processing method (applied to the geochemical data for all deposit models) is summarized in Fig. 6.

This same process was repeated for the indirect SEDEX pathfinder elements: As, Ba, Cd, Cr, F, Mn, Mo, Ni, Sb, and U. However the lower weights assigned to the 5 class normalized mean map (6, 4, 2, 1, 0) reflect the lower predictive power of the indirect pathfinder elements.

The same process was applied to the water geochemical data, excluding those pathfinder elements that were not determined (see Table 1). Fig. 5h,i,j,k show the final 5 class normalized mean maps for the direct and indirect elements from the stream sediment and water geochemical data used to produce the SEDEX mineral potential map

3.2.4 pH (Acidity of Stream Water)

The pH is lowered by the oxidation of sulphidic minerals under aqueous conditions to produce sulphuric acid. The pH in turn influences the transport of metals in water and their adsorption onto particulate matter in stream sediments. Low pH (acidic) water has enhanced capacity to transport most metals and may thereby enhance their geochemical expression. In neutral to alkaline environments the chemical mobility of many metals, including Zn, Ag, and Pb, and to a lesser extent As and Cd, is severely restricted (Plant and Raiswell, 1994). Conversely, metals such as U, Mo, and V are relatively mobile. As the pH of stream waters exceeds 7.0 in much of the survey area because of the dominance of limestone, the geochemical dispersion of many elements is reduced in those areas. It is expected that pH will tend to be buffered consistently within rock units except for anomalous sites.

Fig. 6 GIS modelling process to determine a "zone of influence" around sample points for interpolation of geochemical data between sites of stream sediment, stream water and spring water samples

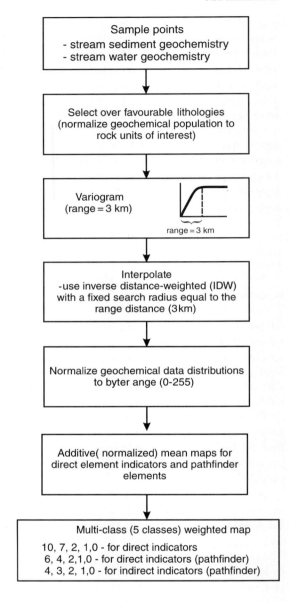

Using the above knowledge, an evidence map of pH was produced from the water geochemical data using the following method. The point data were interpolated using IDW with a 3 km fixed search radius as determined from a variogram (Fig. 6). This interpolated map was then divided into 8 equal classes based on quartiles with the following weights (10, 9, 8, 7, 4, 3, 2, 0) representing extremely acidic to basic pH conditions, respectively (Fig. 5l). The more acidic conditions were given a higher weight to reflect more favourable environments assuming that the cause of the acidity was sulphide minerals such as sphalerite (zinc + / − silver), galena (lead + / − silver) and tetrahedrite (copper + silver), recognizing that pyrite and carbonaceous shale that contain no base metals also reduce pH.

3.2.5 Spring Water Geochemistry

Spring water sample sites were subjectively classified as favourable for SEDEX, based on anomalous pathfinder elements (all SEDEX anomalous springs from Table 5.12 of Hamilton et al., (2003) and selected anomalous springs characterized as SEDEX from Table 6 of Caron et al. (2007)). The following pathfinder elements were considered as direct evidence: Ag, Cu, Pb, Zn; and as indirect evidence: As, Ba, Cd, Cr, Mn, Mo, Ni, pH, Sb, V. Those sites considered as anomalous for one or more elements indicating SEDEX were buffered to 3 km and assigned a weight of two (Fig. 5m).

3.2.6 Mineral Occurrences

Mineral occurrences of the SEDEX-type were assigned to one of four categories: defined resources (weight 10), advanced prospect (weight 7), showing (weight 4), and anomaly (weight 2). The weighted points were then buffered to 3 km representing a "zone of influence" (Fig. 5n,o,p,q). A distance of 3 km was chosen based on observations with respect to mineralization and alteration made in the field.

3.3 Carbonate-Fault Model

A total of 15 weighted evidence maps (Table 2) were used to create an additive Carbonate-Fault mineral potential map. This map indicates relative potential for the presence of undiscovered zinc-lead-silver massive sulphide deposits that fit this composite model as described above in the deposit model overview section. The evidence maps are as follows.

3.3.1 Favourable Rock Units

Favourable host rocks include dolostone of the Devonian to Cretaceous Manetoe, Presqu'ile and Grizzly Bear formations (diagenetic facies) (weight 10), middle Devonian strata (weight 5), Cambrian to early Devonian carbonate platform strata (limestone, dolostone) (weight 5), Cambrian to early Devonian calcareous shale (weight 2) and proximity to the diagenetic facies listed above (3 km buffer, weight 5).

The Manetoe and Presqu'ile facies are diagenetic and/or hydrothermal alteration products of carbonate lithofacies (limestone and dolostone) that constitute the Mackenzie Platform (Morrow et al., 1990). This style of alteration is associated with reef-related dolomite diagenesis and with the passage of hydrothermal fluids, an alteration style that is present in and around significant carbonate-hosted base-metal deposits in western Canada (i.e. Pine Point, Robb Lake, and near Prairie Creek (Paradis, 2007)).

3.3.2 Structural Features

Growth faults (3 km buffer, weight 2), all other faults (2 km buffer, weight 1) and the trend of the facies change between the Devonian carbonate and shale assemblages, as per the SEDEX model (2 km buffer, weight 1) were used as evidence maps for the presence of undiscovered Carbonate-Fault-related deposits. As for the SEDEX deposit model, these features were likely to have focused the flow of hydrothermal fluids.

3.3.3 Stream Sediment and Stream Water Geochemistry

Four evidence maps were created based on direct and indirect elements for Carbonate-Fault- related mineralization from the stream sediment and stream water geochemical data. The processing and weighting strategy was the same as described for the SEDEX model, however, different suites of pathfinder elements were used as listed in Table 2.

3.3.4 Spring Water Geochemistry

Spring water sample sites were subjectively classified as favourable for Carbonate-Fault deposits based on anomalous pathfinder elements (Prairie Creek mine water from Table 5.12 of Hamilton et al., (2003) and selected anomalous springs from Table 6 of Caron et al., (2007)). The following pathfinder elements were considered: direct (Ag, Cu, Pb, Zn) and indirect (Cd and Sb). Those considered anomalous for one or more of these elements were assigned a weight of 2 with a 3 km buffer.

3.3.5 Mineral Occurrences

Mineral occurrences classified in the Carbonate-Fault model were assigned to one of two categories: defined resources (weight 10) and showing (weight 4). These classified points were buffered to 3 km representing a potential "zone of influence" reflecting the same distance rationale as discussed above for the SEDEX model.

3.4 Intrusion-Related Model

A total of 14 weighted evidence maps (Table 3) were used to create an additive Intrusion-Related mineral potential map. These are described below.

3.4.1 Favourable Rock Units

Each intrusion (pluton) was assigned to one of three groups based on qualitative geological assessment of their potential (high, medium, or low) to host or be spatially associated with undiscovered Intrusion-Related deposits, following the classification of Rasmussen et al. (2007). Additional geological knowledge provided by Barnes et al. (2007) and Yuvan et al. (2007) was also considered. For modelling purposes, any mapped hornfels associated with the margin of a pluton was included as part of that pluton, based on the assumption that hornfels (contact metamorphosed sedimentary rock) was caused by a shallowly underlying portion of a pluton below the hornfels. The plutons were buffered at 3 different distance intervals and within each pluton itself reflecting zones of different alteration intensity as observed from various field studies and mapping initiatives. The proximal zone (1 km buffer outside and inside each pluton representing a zone of maximum alteration) was assigned the highest weight.

Shaly limestone units situated within 20 km of any pluton were also weighted for their potential to host undiscovered skarn deposits (Fig. 7). Units thought to have the highest potential for skarn deposits are defined by their content of impure limestone. The highest rank of 10 was assigned to the heterogeneous limestone-bearing units (Vampire and Sekwi formations) that include the immediate host of the Cantung Mine.

3.4.2 Structural Features

A fault density map was produced from a vector file of faults using a line density algorithm in the GIS. The basic premise is that faults acted as conduits for mineralizing fluids either from the plutons themselves or remobilized from the host rocks as a result of pluton emplacement. This algorithm produces a density map based on the number of faults within a 5 by 5 km neighbourhood. This map was divided into 3 classes based on natural breaks in the; high, medium and low density; and assigned weights of 5, 3 and 1, respectively.

3.4.3 Stream Sediment and Stream Water Geochemistry

Two multi-class additive and normalized maps were produced showing direct and indirect pathfinder elements from the stream sediment data. Two maps were similarly

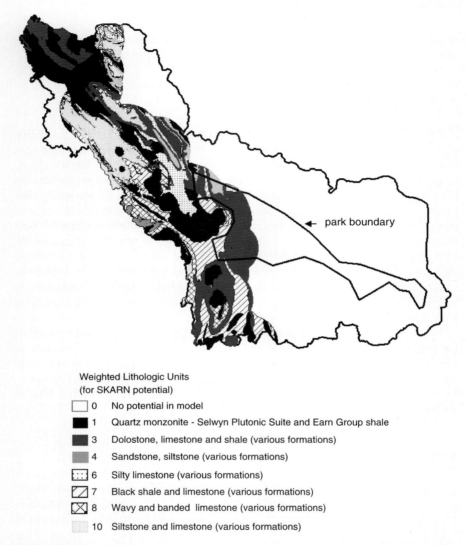

Fig. 7 Rated geology within areas of influence of intrusions in Selwyn Plutonic Suite, used for modelling Intrusion-Related and Carlin-Type gold deposit potentials. Dark shades are assigned to low potential, light shades to high

produced from the water geochemical data. The same geochemical processing method described for the SEDEX model (Fig. 6), using a different suite elements for direct and indirect evidence (see Table 3), was adopted for the Intrusion-Related model.

3.4.4 Spring Water Geochemistry

Spring-water sample sites were subjectively classified as favourable for Intrusion-Related mineralization based on anomalous pathfinder elements (three examples labeled "tungsten skarn" from Table 5.12 of Hamilton et al., (2003) and selected anomalous springs from Table 6 of Caron et al., (2007)). Direct pathfinder elements include W, Cu, Pb, Zn; and indirect pathfinder elements include Ag, As, Au, Bi, F, Sb, Sn, and Te. As for the SEDEX model a 3 km buffer (zone of influence) around each anomalous sample point was produced and a weight of 2 was assigned for this evidence map.

3.4.5 Mineral Occurrences

Mineral occurrences of the Intrusion-Related model (after NORMIN, Barnes et al., 2007; Rasmussen et al., 2007, and Yuvan et al., 2007) were assigned to one of four categories: defined resources (weight 10), advanced prospect (weight 5), showing (weight 2), and anomaly (weight 1). As for the SEDEX model, each weighted occurrence was buffered to a distance of 3 km representing a "zone of influence", again using the distance rational discussed for the SEDEX model.

3.5 Carlin Model

A total of 20 weighted evidence maps (Table 4) were used to create an additive mineral potential map for undiscovered Carlin-Type lode gold and/or related placer deposits. The evidence maps are described below.

3.5.1 Favourable Rock Units

Many researchers feel that Carlin-Type mineral deposits share a strong association with late Cretaceous plutons (e.g. Poulsen, 1996; Rasmussen et al., 2006, 2007). In the Nevada Carlin-type locality, carbonate units are essential as host rocks. Thus, the same three pluton groups and variegated limestone units, with the same buffering and weighting strategies as the Intrusion-Related model, were applied to the Carlin-Type model.

3.5.2 Structural Features

With the initiation of fluid systems, transportation corridors were also needed to focus fluid flow. Additional dilational zones such as anticline axes may have enhanced permeability and therefore are also an important structural feature to model. Thus, as with the Intrusion-Related model, growth faults within 20 km of a granite body were buffered to 2 km reflecting a zone of influence and assigned a weight of 1. All other faults were buffered to a distance of 2 km but only on the down-dip side (i.e. toward the southwest) and weighted as 1. Fault intersections were buffered to 2 km with a weight of 2 and within 20 km of a granite body. Anticlinal axes were buffered to 2 km with a weight of 1. A 5 km buffer (weight 2) was assigned to all volcanic units and the units themselves were weighted 2. Regardless of the age of extrusion, volcanic units are modelled as indicating deep-seated basement structures. All of the above structural evidence maps highlight features that may have served as conduits for mineralizing fluids. Rational for buffer distances were based on field observations and exploration knowledge.

3.5.3 Stream Sediment and Stream Water Geochemistry

Three evidence maps were created from the stream sediment data; one based on direct pathfinder elements, one on indirect pathfinder elements and one on possible pathfinder elements (see Table 4). The same processing and weighting strategy described for the SEDEX model was also used for the Carlin model. The only exception was the addition of a possible pathfinder map whose 5 normalized classes are assigned weights of 4, 3, 2, 1 and 0. This map was lesser ranked due its lower predictive power. The only stream sediment element used for direct evidence was Au.

The same procedure was repeated for the stream water geochemical data using a modified suite of elements (Table 4). Only two evidence maps, indirect pathfinders and possible pathfinders, were produced from the water geochemical dataset, because no direct evidence (Au) was analyzed.

3.5.4 pH

The same strategy and weights used to create pH evidence maps for the SEDEX and Intrusion-Related models were applied to the Carlin modelling.

3.5.5 Spring Water Geochemistry

Springs with a "Carlin signature" in anomalous pathfinder elements (Ag, Sb, As, Au, Ba, Hg, Ti, Tl) were buffered to a distance of 3 km and assigned a weight of 2.

3.5.6 Placer Gold Occurrences

Placer gold occurrences are historically the first direct indicator in the discovery history of many Carlin-Type deposits and these were selected from the mineral deposits database (NORMIN). Given the importance of intrusions to the composite

Mineral Potential Modelling for the Greater Nahanni Ecosystem Using GIS

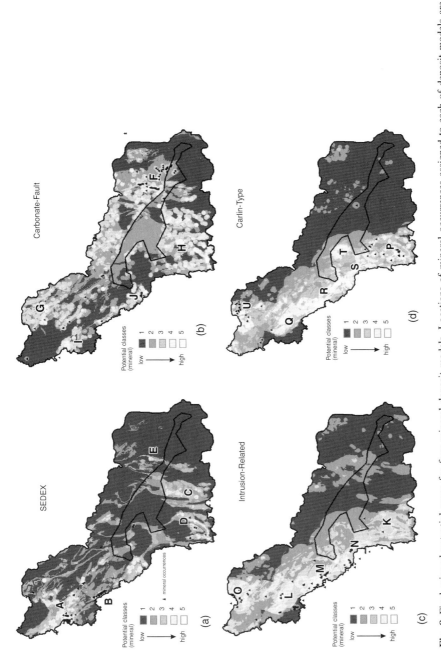

Fig. 8 Final mineral potential maps for four mineral deposit models. Locations of mineral occurrences assigned to each of deposit models are shown for reference. Simplified 5-class maps (derived from normalized, additive potential maps which are not shown) used to create final potential map are shown in (**a**) SEDEX, (**b**) Carbonate- Fault, (**c**) Intrusion (Pluton) related, (**d**) Carlin-Type. Areas of highest potential are indicated by UNIQUE CAPITAL LETTERS and explained in text

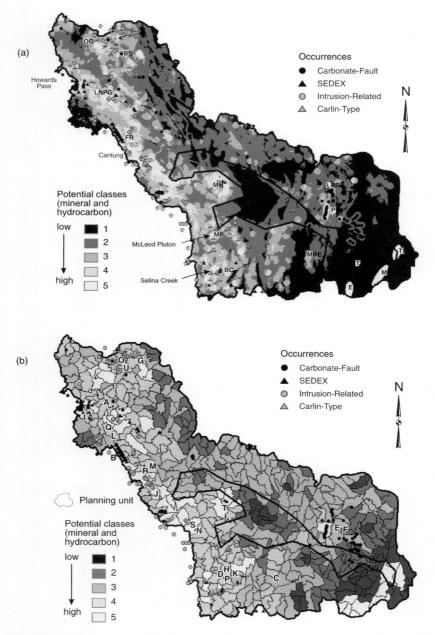

Fig. 9 Composite mineral potential maps. Mineral occurrences by type are shown for reference. (**a**) Final mineral potential map generated by combining maximum values of individual 5 class potential maps for each deposit model as well as potential map for Hyrdocarbons, (**b**) Resource potential apportioned to park planning units using MAX operator. See text for explanation

Carlin model used in this analysis, placer gold occurrences were included in the model only where they fit local derivation criteria within the buffer of a pluton or a spring with plutonic characteristics. Thus not all of the eastern Nahanni gold occurrences were included. A 5 km buffer around each placer gold occurrence with a weight of 5 was employed to create the evidence map.

3.5.7 Lode Gold Occurrences

Lode gold occurrences that are compatible with the Carlin model (after Rasmussen et al., 2007; Cline et al., 2005; and NORMIN) were assigned to one of two categories; showing (weight 2), or anomaly (weight 1) and each group of points was buffered to 3 km creating two evidence maps.

3.6 Composite Model

The four additive-normalized potential maps (not shown) and associated five-class potential maps derived from the additive normalized maps are shown in Fig. 8a,b,c,d (see Fig. 4 and previous discussion for method) The five-class maps (Fig. 8a,b,c,d) and the single class hydrocarbon potential map of Osadetz et al. (2003) were used to create the composite potential map using the following equation:
MAX (Carlin, Fault, Intrusion, SEDEX, Hydrocarbon)

– where the input maps represent the 5-class potential maps and the Hydrocarbon map.

This composite potential map (Fig. 9a) was used as the basis for selecting conservation areas as it formed the "cost layer" that was used in the site selection process undertaken by Parks Canada. Figure 9b shows a map of Parks Canada's planning units in which the maximum value (based on Fig. 9a) for each planning unit was calculated by using a zonal analysis in the GIS.

The rational for using a MAX operator was to ensure that a planning unit was labeled high potential (thus a high cost) even if only one of the four potential maps showed the area to be of high potential. Thus the planning units shown in light grey on Fig. 9b represent areas where one or more of the five potential maps indicate high mineral or energy resource potential. This strategy provides a comprehensive estimate of mineral potential.

4 Results

4.1 Validation

Plots showing how well each mineral potential map predicts the known occurrences (known as efficiency of classification or SRC plots) were generated (Fig. 10a–d) for

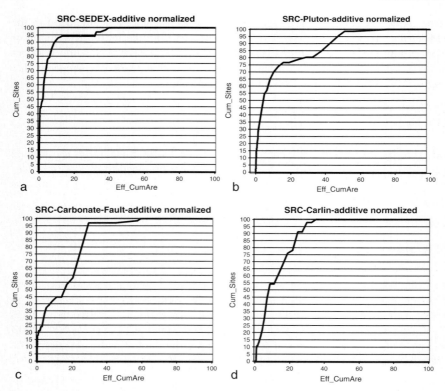

Fig. 10 Efficiency of classification (model) graphs (SRC) for each of four deposit models in this GIS study, showing how well potential maps predict individual occurrences classified by deposit type. (**a**) Sedex additive – normalized (**b**) Plutonic additive – normalized (**c**) Carbonate-Fault – additive – normalized (**d**) Carlin additive – normalized. These plots show number of occurrences predicted as a function of area, ranked from high to low potential, on each of additive – normalized mineral potential maps for each deposit type. *Cum_Sites* represents number of occurrences (% of total) along Y axis and *Eff_CumAre* represents cumulative area of each potential map along X axis

the purpose of validating the models. These plots show the number of occurrences predicted as a function of area, ranked from high to low potential, on each of the additive – normalized mineral potential maps for each deposit type. Considering the top 5% of the highest potential areas, the SEDEX model is the best predictor with approximately 80% of the SEDEX-style mineral occurrences being predicted. The Intrusion-Related (Pluton) model is next best, predicting about 60%, the Carlin-Type model slightly weaker at 40%, and the Carbonate-Fault model is the least predictive, with only 30% success. Considering the top 1% of the high potential area, the SEDEX model is again the best, predicting close to 50% of the SEDEX occurrences. The Intrusion-Related and Carbonate-Fault models predict 20% of its style of mineral occurrences whereas the Carlin-Type model that predicts only 11%. Each of the individual models is at best only moderately predictive, although all four models predict 100% of the associated mineral occurrences in approximately 30%

of the high potential area. The consistently poorer performance of the Carlin-Type potential map suggests that the deposit model is not particularly characteristic of these types of mineral occurrences at least within the study area.

4.2 Geological Interpretation

This section relates the areas of highest mineral potential as established by GIS analysis and modelling, in qualitative terms, to mineral deposit models and the geological and geochemical evidence supporting them. In a number of places there is overlap between the evidence for different deposit models and their quantitative GIS model results. Geographic and mineral potential areas of interest referred to below are represented in Fig. 8.

4.2.1 SEDEX zinc-lead-silver

Five areas of interest on the final SEDEX potential map (Fig. 8a) are discussed next.

Area A in the northwest section of the study area contains some of the Howard's Pass deposits. The high potential area containing these deposits highlights the northwest part of a southeast-northwest-trending linear zone within the study area, informally known as the "zinc corridor", extending from the northwest end of the existing park reserve and the Flat River Valley to the Howard's Pass deposits. A separate sub-parallel strand of this corridor contains a high potential zone (area B) northwest of the Cantung deposit.

Area C represents a broad area of relatively high potential in the Meilleur River Embayment of the Selwyn Basin, hosted by the Road River Group, but partially masked by the younger Besa River Formation. This area is generally under-explored due to its remoteness but was previously highlighted by the Nahanni MERA 1, along the western edge of the Tlogotsho Plateau. The Meilleur River Embayment had earlier been identified as an important eastern extension of the Selwyn Basin by Morrow and Cook (1987). Despite the lack of exploration interest, this MERA 2 study has quantitatively reinforced evidence that the Meilleur River Embayment has relatively high potential for the presence of undiscovered SEDEX deposits based on the additive favourability of stream sediment, stream water and spring water geochemistry, northeasterly growth faults associated with the Leith Ridge basement structure, and favourable rock units throughout the area.

Area D in the vicinity of the Selena Creek deposit represents an area of relatively high potential, very much like that of Area C and underlain by the Road River Group within the Selwyn Basin proper. This area is characterized by rolling vegetated hills with little exposure except where dissected by deep canyons. Some zinc showings have been identified by the limited exploration efforts in this area. The regional northeasterly trend of high mineral potential from the Prairie Creek to Meilleur River areas continues toward the southwest into this area, parallel to the deep-seated Leith Ridge structure.

Area E represents the northeastern extension of the Leith Ridge trend established by areas C and D, into the MacDonald Platform. This area is enhanced in potential by the growth fault attribute contributed by the carbonate-shale facies change that trends northerly along the Prairie Creek sub-embayment of the Meilleur River Embayment. Stream sediment, stream water and spring water geochemical data further support this designation.

4.2.2 Carbonate-Fault

Figure 8b shows mineral potential maps for the Carbonate-Fault zinc-lead-silver deposit model. A large area around the Prairie Creek deposit (Area F) containing many fault-related prospects has been classified as high potential based on this model. This rating is supported by favourable stream sediment chemistry, a high density of mapped faults and the presence of diagenetically altered limestone and dolostone units (Manetoe facies).

The Broken Skull area (Area G) is in the vicinity of the Broken Skull Fault, which is interpreted as a reactivated growth fault due to lithological changes across it. A number of anomalous hot springs in this area, with geochemistry suggestive of mantle fluids, support the inference that this fault is long-lived and deep-seated. This area is also characterized by anomalous stream sediment geochemistry.

Area H in the vicinity of the Selena Creek deposit is classified as higher in potential due to anomalous sediment geochemistry, favourable rock units, and the presence of a number of north-trending faults. Some SEDEX prospects have been identified in this area as well as the high potential for Intrusion-Related mineralization may reflect an overlap in the genetic deposit models. The plutons in this vicinity may have acted as heat sources for re-mobilizing mineralized fluids along major faults.

Areas I (Howard's Pass) and J (southeast of Cantung) are located along the western strand of the "zinc corridor", delineated here primarily by anomalous stream sediment geochemistry, favourable basinal shale packages and a high density of faults.

A southeast-northwest trending divide is apparent between the high potential zones F and G in strata of the Mackenzie Platform, and areas H, I and J within the Selwyn Basin, that may reflect an overlap of Intrusion-Related models with SEDEX parameters. There are some examples of Carbonate-Fault occurrences in the southwestern area, but most of these show evidence of higher temperatures, favouring classification as Intrusion-Related instead.

4.2.3 Intrusion-Related

Figure 8c shows the Intrusion-Related mineral potential map. Obvious high potential areas are restricted to the western portion of the region in proximity to the plutons. Area K around and to the north of Selena Creek is modelled as high in potential by virtue of its proximity to the "McLeod and Big Charlie plutons", intersecting carbonate units, anomalous geochemistry (particularly stream sediment), anomalies in spring waters, and anomalous geophysical signatures. This area also contains a number of skarn occurrences.

Area L-M-N is a broad zone of high potential related directly to the Tay River plutonic suite. Its favourability is supported in part by anomalous stream sediment geochemistry, as well as carbonate units within the pluton buffers.

Area O is centered near O'Grady Lakes/Moose Ponds, covering the O'Grady batholith of the Tombstone plutonic suite and a number of high-ranking Intrusion-Related mineral prospects. In addition to anomalous stream sediment geochemistry, the pluton-buffered areas also contain favourable carbonate units and several Intrusion-Related anomalies represented by pathfinder elements in spring waters.

4.2.4 Carlin-Type

Figure 8d shows the Carlin-type Au potential map. An expansive area of high potential occurs in the vicinity of the Selena Creek placer deposit (Area P). This area has been classified as high potential based on a series of placer gold showings, favourable pathfinder elements (As, Sb, Hg, Ti and Ba accompanying Au and Ag), lithology, proximity to plutons and position at the southern end of a regional trend of existing gold prospects and deposits extending to Alaska.

The linear "zinc corridor" (Area Q-R-S) which contains both the Howard's Pass and the Cantung regions is divided and partly overlapped by the high potential area for Carlin gold-type mineralization based on the proximity to plutons, a high density of anticlines, and along a trend of numerous showings to deposits including one active mine. This area illustrates an overlap in parameters with the Intrusion-related model.

The high potential rating of Area T is a result of the intersection of the Mount Hamilton pluton buffer with favourable carbonate lithology, anomalous geochemical data from stream sediment, stream water and springs, as well as minor placer gold occurrences.

Area U, around and to the east of the O'Grady batholith, is similar to Area T with the added influence of favourable structural parameters such as faults parallel to the Broken Skull River.

5 Discussion and Summary

Figure 9a,b present the composite mineral (and hydrocarbon) potential based on all of the models discussed above. Twelve per cent of the park expansion study area is characterized by moderate to high non-renewable resource potential, with restricted areas of very high potential. The high potential ratings are consistent with previously published, aerially or topically restricted resource assessments by Jefferson and Spirito (2003) and Lariviere et al. (2006). Although still not all-encompassing, this GIS analysis achieves more comprehensive and objective results than the previous assessments by:

- incorporating broader and more advanced sampling and analytical techniques for the field and laboratory geochemistry,
- enhancing regional geological and geophysical knowledge,
- inclusion of quantitative spatial modelling in a GIS that has applied state-of-the-art integration methods to combine a knowledge-driven weighted dataset of deposit models, with the geochemical and geological data.

However, the mineral potential maps are constrained by:

- the present knowledge base
- the use of only four enhanced composite deposit models that together with the hydrocarbon map represent all non-renewable resource types known or thought to be important in the Nahanni Park area
- a limited number of evidence maps used to produce each map
- a subjective knowledge-driven weighting system for evidence layers.

This entire resource assessment process is knowledge-driven and the maps generated represent "our best estimate" of relative mineral potential given the limitations discussed above. However during our experiments to generate data-driven weightings we found that the knowledge-driven estimates are reasonable although adjustment of the weights and inclusion of other evidence may improve the performance of each map as the predictive power of the maps were moderate at best (Fig. 10). These maps would likely change in various areas with increased exploration knowledge, the availability of more (uniformly distributed) geoscience data for additional evidence maps, and further data-driven modelling to refine the weighting systems.

There is much evidence to indicate that where there is sufficient number of mineral occurrences, data-driven methods typically out-perform knowledge-driven techniques (Harris et al., 2001). Subjectivity and uncertainty were introduced in the initial definition of at least two of the deposit models: Intrusion-Related and Carbonate-Fault, each of which is a composite of at least three individual deposit types. It is likely that if each individual deposit type was modelled separately, the results may have been more discriminating. Subjectivity was required to select the various evidence maps for each of the models, as well as the applied weighting system. This added uncertainty to the modelling process which in turn may have resulted in less predictive maps. Furthermore the assignment of each mineral occurrence to one of the four deposit models introduced subjectivity.

In this paper the mineral occurrences were used in the modelling process to help establish potential for each deposit type although in a different manner than used in typical data-driven modelling processes where the weighting system is defined by the number of occurrences falling on the favourable areas of each evidence map. If more occurrences fall on the favourable areas than would be expected to occur by chance then the evidence map would receive a high weight in the modelling process. In this paper as stated previously a knowledge-based approach was used to define the weights. However, the mineral deposit data divided into classes of importance and weighted accordingly were used directly as evidence maps in the modelling process. Recent work by Harris (in press) indicates that this approach in which the deposits are used as evidence maps performs as well as traditional data-driven approaches in which the location of the deposits are used to drive the modelling process.

Several areas within the study area were identified as having very high mineral potential. For SEDEX zinc-lead-silver type deposits, five areas were identified: (A) Howards Pass at the northwest end of the "zinc corridor"; (B) another strand of this corridor northwest of Cantung; (C) the Meilleur River Embayment; (D) Selena Creek; and (E) east of Prairie Creek. For carbonate fault zinc-lead-silver deposits, five areas are highlighted: (F) Prairie Creek; (G) Broken Skull Fault; (H) Selena Creek; (I) Howard's Pass; and (J) southeast of Cantung. For intrusion-related deposits, three areas stand out: (K) Selena Creek; (L, M, N) a broad zone intruded by the Tay River plutonic suite including Cantung; and the (O) O'Grady batholith. Finally, for the Carlin-type lode and/or placer gold deposits, six areas have high ratings: (P) Selena Creek; (Q, R, S) a broad linear zone of intrusions dividing and overlapping the "zinc corridor'; (T) surrounding the Mount Hamilton pluton; and (U) in and around the O'Grady batholith. High natural gas potential was assessed under MERA 1 p in the southeast corner of the study area, over the untested Etanda Dome and Twisted Mountain Anticline, as well as over the partially explored Mattson Anticline.

The use of GIS analysis procedures has greatly facilitated the process of producing mineral potential maps and transferring the results to Parks Canada as part of the MERA process. This method is an improvement over the traditional "light table" approach of overlaying maps and making visual assessments of mineral potential that have been historically used for MERA. The strictly GIS-based cost-benefit analysis utilizing these results partitioned into park planning units should not however replace round-table discussion and deliberation. Economic and strategic analysis will help the stakeholders understand the relative values of the various park attributes and cost factors, such as non-renewable resources with the aid of these GIS results.

Acknowledgments This paper builds on all of the contributions made by the Nahanni MERA 2 participants, as acknowledged by the citations of their 2007 papers in this volume. Parks Canada provided substantial funding of this study through the MERA Process, with interdepartmental guidance from Indian and Northern Affairs Canada, Government of the Northwest Territories, Natural Resources Canada and Environment Canada. Program leadership within Natural Resources Canada was provided by Margo Burgess, Manager of the Legislated Environmental and Resource Assessment Program of Earth Sciences Sector. Interaction with David Murray and Phil Wilson of Parks Canada facilitated GIS partitioning of results into the planning units. Critical reviews by Mark Mihalasky, Vesa Nykanen, Eric Grunsky, Tracy Lynds and Barham Daneshfar are much appreciated and have led to a much clearer presentation of this work. Editing by Elizabeth Ambrose and drafting assistance by Sue Davis are gratefully acknowledged. This is Geological Survey of Canada (GSC) contribution # 20070576.

References

Aitken JD, Pugh DC (1984) The fort Norman and Leith Ridge structures: major, buried, Precambrian features underlying Franklin Mountains and Great Bear and Mackenzie plains. Bull Can Petroleum Geolo 32: 139–146

Barnes EM, Groat LA, Falck H (2007) A review of the late Cretaceous Little Nahanni Pegmatite Group and associated rare-element mineralization in the Selwyn Basin area, Northwest Territories. In: Wright DF, Lemkow D, Harris JR (eds) Mineral and energy resource assessment of the Greater Nahanni Ecosystem under consideration for the expansion of the Nahanni National Park Reserve, Northwest Territories. Geological Survey of Canada, Open File 5344, pp 191–202

Bonham-Carter GF (1994) Geographic information systems for geoscientists: modelling with GIS. Pergamon, Oxford, p 398

Carne RC, Cathro RJ (1982) Sedimentary exhalative (SEDEX) zinc-lead-silver deposits, northern Canadian Cordillera. Can Inst Min Metall Bull 75: 66–78

Caron M-E, Grasby SE, Ryan MC (2007) Spring geochemistry: a tool for exploration in the South Nahanni River basin of the Mackenzie Mountains, Northwest Territories. In: Falck H, Wright DF, Harris J (eds) Mineral and energy resource potential of the proposed expansion to the Nahanni National Park Reserve, North Cordillera, Northwest Territories,; Geological Survey of Canada, Open File 5344, pp 31–73

Carson JM, Dumont R, Potvin J, Buckle J, Shives RBK, Harvey B (2007a) Geophysical Series – NTS 95F – Virginia Falls, Northwest Territories; Geological Survey of Canada, Open File 5154, 1 CD-ROM

Carson JM, Dumont R, Potvin J, Buckle J, Shives RBK, Harvey B (2007b) Geophysical Series – NTS 95D and 95E – Flat River, Northwest Territories; Geological Survey of Canada, Open File 5160, 1 CD-ROM

Carson JM, Dumont R, Potvin J, Buckle J, Shives RBK, Harvey B (2007c) Geophysical Series – NTS 105H and 105I – Little Nahanni River, Northwest Territories; Geological Survey of Canada, Open File 5164, 1 CD-ROM

Charbonneau BW (2007) Evaluation of airborne radiometric and magnetic data in the vicinity of the Nahanni National Park Reserve, Northwest Territories, Canada. In: Wright DF, Lemkow D, Harris JR (eds) Mineral and energy resource assessment of the Greater Nahanni Ecosystem under consideration for the expansion of the Nahanni National Park Reserve, Northwest Territories. Geological Survey of Canada, Open File 5344, pp 99–124

Chung CF, Fabbri A (2003) Validation of spatial prediction models for landslide hazards mapping; Natural Hazards 30: 451–472

Cline J, Hofstra A, Muntean J, Tosdal RM, and Hickey KA (2005) Carlin-type gold deposits in Nevada: critical geologic characteristics and viable models. In: Hedenquist W, Thompson JFH, Goldfarb RJ, Richards JP (eds) Economic geology 100th anniversary volume. Society of Economic Geologist, pp 451–484

Cousens B (2007) Radiogenic isotope studies of Pb-Zn mineralization in the Howards Pass area, Selwyn Basin. In: Wright DF, Lemkow D, Harris JR (eds) Mineral and energy resource assessment of the Greater Nahanni Ecosystem under consideration for the expansion of the Nahanni National Park Reserve, Northwest Territories. Geological Survey of Canada, Open File 5344, pp 279–292

Dawson KM, Panteleyev A, Sutherland Brown A, Woodsworth GJ (1992) Regional metallogeny, Chapter 19. In: Gabrielse H, Yorath CJ (eds) Geology of the cordilleran orogen in Canada. Geological Survey of Canada, Geology of Canada No. 4, pp 707–768 (also Geological Society of America, The Geology of North America, v. G2)

Day SJA, Lariviere JM, Friske PWB, Gochnauer KM, MacFarlane KE, McCurdy MW, McNeil RJ (2005) National Geochemical Reconnaissance (NGR): Regional stream sediment and water geochemical data, Macmillan Pass–Sekwi Mountains, Northwest Territories; Geological Survey of Canada, Open File 4949, 1 CD-ROM

Department of Indian Affairs and Northern Development (2006) NORMIN.DB, The Northern Minerals Database; Northwest, Territories Geoscience Office, Yellowknife, Northwest Territories

Duk-Rodkin A, Huntley D, Smith R (2007) Quaternary geology and glacial limits of the Nahanni National Park Reserve and adjacent areas, Northwest Territories, Canada. In: Wright DF, Lemkow D, Harris JR (eds) Mineral and energy resource assessment of the Greater Nahanni Ecosystem under consideration for the expansion of the Nahanni National Park Reserve, Northwest Territories. Geological Survey of Canada, Open File 5344, pp 125–130

Emsbo P, Groves DI, Hofstra AH, Bierlein FP (2006) The giant Carlin gold province: a protracted interplay of orogenic, basinal, and hydrothermal processes above a lithospheric boundary; Mineralium Deposita 41: 517–525

Falck H (2007) Geological overview, Appenidx 1. In: Wright DF, Lemkow D, Harris JR (eds) Mineral and energy resource assessment of the Greater Nahanni Ecosystem under consideration for the expansion of the Nahanni National Park Reserve, Northwest Territories. Geological Survey of Canada, Open File 5344, pp 327–365

Friske PWB, McCurdy MW, Day SJA (2001) Regional stream sediment and water geochemical data, eastern Yukon and western Northwest Territories; Geological Survey of Canada, Open File 4016, 1 CD-ROM

Goodfellow WD (ed) (2007) Mineral deposits of Canada, Geological Association of Canada, MDD Special Paper 5, pp 1068

Gordey S, Anderson RG (1993) Evolution of the Northern Cordilleran Miogeocline, Nahanni Map Area (105I), Yukon and Northwest Territories; Geological Survey of Canada Memoir 428, p 214

Government of Canada (1995) Terms of Reference, Mineral and Energy Resource Assessment (MERA) of Proposed National Parks in Northern Canada; commissioned and approved by the Senior MERA committee, Governments of Canada, Yukon and Northwest Territories, Ottawa; website http://www.nrcan.gc.ca/ms/pdf/merae.pdf

Hamilton SM, Michel FA, Jefferson CW, Power-Fardy D (2003) Spring water geochemistry and hydrogeology for mineral resource assessment of the South Nahanni River Region. In: Jefferson CW, Spirito W (eds) Non-renewable mineral and energy resource potential of the proposed extensions to Nahanni National Park Reserve, Northern Cordillera, Northwest Territories. Geological Survey of Canada, Open File 1686, pp 5–1 to 5–40

Harris JR, Sanborn-Barrie M (2006a) Mineral potential mapping: examples from the Red Lake Greenstone Belt, Northwest Ontario. In: Harris JR (ed) GIS for the Earth sciences. Geological Association of Canada, Special Volume 44, pp 1–21

Harris JR, Sanborn-Barrie M, Panagapko D, Skulski T, Parker JR (2006b) Gold prospectivity maps of the Red Lake greenstone belt: application of GIS technology. Can J Earth Sci 43: 865–893

Harris JR, Wilkinson L, Heather K, Fumerton S, Bernier M, Dahn R (2001) Gold potential of the Swayze Greenstone Belt, Ontario: application of GIS technology. Nat Resour Res 10: 91–124

Jefferson CW, Spirito W (eds) (2003) Non-renewable mineral and energy resource potential of the proposed extensions to Nahanni National Park Reserve, Northern Cordillera, Northwest Territories. Geological Survey of Canada, Open File 1686, p 253

Jefferson CW, Spirito WA, Hamilton SM (2003) Geological setting, Chapter 3. In: Jefferson CW, Spirito W (eds) Non-renewable mineral and energy resource potential of the proposed extensions to Nahanni National Park Reserve, Northern Cordillera, Northwest Territories. Geological Survey of Canada, Open File 1686, pp 3–1 to 3–21

Lariviere JM, Eddy BG, Udell A, Slack T (2006) Mackenzie valley mineral potential map; Northwest Territories Geoscience Office, Open File 2006-03, scale 1:1,500,000

Lemkow D, Harris JR, Slack T (comp.) (2007) Digital geoscience database: a contribution to the mineral and energy resource assessment of the greater Nahanni ecosystem, Northwest Territories, Appendix 4. In: Wright DF (eds) Mineral and energy resource assessment of the

Greater Nahanni Ecosystem under consideration for the expansion of the Nahanni National Park Reserve, Northwest Territories

Lydon J (1995) Sedimentary exhalative sulphides (SEDEX). In: Eckstrand OR, Sinclair WD, Thorpe RI (eds) Geology of Canadian Mineral Deposit Types No. 8. Geological Survey of Canada, pp 130–152 (also Geological Society of America, The Geology of North America, v. P-1)

McCurdy MW, McNeil RJ, Friske PWB, Day SJB, Wilson RS (2007) Stream sediment geochemistry in the proposed extension to the Nahanni Park Reserve. In: Wright DF, Lemkow D, Harris J (eds) Mineral and energy resource potential of the proposed expansion to the Nahanni National Park Reserve, North Cordillera, Northwest Territories. Geological Survey of Canada, Open File 5344, pp 75–98

Morrow DW, Cook DG (1987) The Prairie Creek Embayment and Lower Paleozoic Strata of the Southern Mackenzie Mountains. Geol Surv Can Memoir 412: 195

Morrow DW, Cumming GL, Aulstead KL (1990) The Gas-Bearing Devonian Manetoe Facies, Yukon and Northwest Territories. Geol Surv Can Bull 400: 87

Okulitch A (2005a) Geology of the Redstone River Area, National Earth Science Series, Geological Atlas; National Resources Canada, Map NP-9/10-G, scale 1:1 million

Okulitch A (2005b) Geology of the Ross River Area, National Earth Science Series, Geological Atlas; National Resources Canada, scale 1:1 million

Osadetz KG, Chen Z, Morrow DW (2003) Petroleum resource assessment of the Tlogotsho Plateau, Nahanni Karst, and adjacent areas under consideration for expansion of Nahanni National Park Reserve, Northern Cordillera and District of Mackenzie. In: Jefferson CW, Spirito W (eds) Non-renewable mineral and energy resource potential of the proposed extensions to Nahanni National Park Reserve, Northern Cordillera, Northwest Territories. Geological Survey of Canada, Open File 1686, pp 8–1 to 8–36

Paradis S (2007) Isotope geochemistry of the Prairie Creek carbonate-hosted zinc-lead-silver deposit, southern Mackenzie Mountains, Northwest Territories. In: Wright DF, Lemkow D, Harris JR (eds) Mineral and energy resource assessment of the Greater Nahanni Ecosystem under consideration for the expansion of the Nahanni National Park Reserve, Northwest Territories. Geological Survey of Canada, Open File 5344, pp 131–176

Poulsen KH (1996) Carlin-type gold deposits and their potential occurrence in the Canadian Cordillera. in Cordillera and Pacific Margin; Geological Survey of Canada, Current Research 1996-A, pp 1–9

Plant JA, Raiswell RW (1994) Modifications to the geochemical signatures of ore deposits and their associated rocks in different surface environments. In: Hale M, Plant JA (eds) Handbook of exploration geochemistry, vol 6: Drainage Geochemistry. Elsevier, pp 73–109

Rasmussen KL, Mortensen JK, Falck H (2006) Morphological and compositional analysis of placer gold in the South Nahanni River drainage, Northwest Territories; Yukon Exploration and Geology 2006, pp 237–250

Rasmussen KL, Mortensen JK, Falck H, Ullrich TD (2007) The potential for intrusion-related mineralization within the South Nahanni River MERA area, Selwyn and Mackenzie Mountains, Northwest Territories. In: Wright DF, Lemkow D, Harris JR (eds) Mineral and energy resource assessment of the Greater Nahanni Ecosystem Under consideration for the expansion of the Nahanni National Park Reserve, Northwest Territories. Geological Survey of Canada, Open File 5344, pp 203–278

Richards BG (1989) Report on the Selena Creek placer property, Nahanni Mining Division, Northwest Territories, Latitude: 60°55'N; Longitude: 126°40'W; unpub. report for Sirius Resource Corporation and Verdstone Gold Corporation; Dynamin Engineering Limited, Vancouver, p 19

Rowan L (1989) Similarities of geological environment and potential for Carlin-style mineralization at Selena Creek; INAC Assessment Files, November 27, 1989, unpublished memorandum

Spirito WA, Jefferson CW (2003) Regional surficial geochemistry. In: Jefferson CW, Spirito W (eds) Non-renewable mineral and energy resource potential of the proposed extensions to Nahanni National Park Reserve, Northern Cordillera, Northwest Territories. Geological Survey of Canada, Open File 1686, pp 4–1 to 4–79

Wright DF, Bonham-Carter GF (1996) VHMS favourability mapping with GIS-based integration models, Chisel Lake-Anderson Lake area. In: Bonham-Carter GF, Galley AG, Hall GEM (eds) EXTECH I: A multidisciplinary approach to massive Sulphide Research in the Rusty Lake-Snow Lake Greenstone Belts, Manitoba. Geol Surv Can Bull 426: pp 339–376

Yuvan J, Shelton K, Falck H (2007) Geochemical investigations of the high-grade quartz- in Mineral and energy resource assessment of the Greater Nahanni Ecosystem under consideration for the expansion of the Nahanni National Park Reserve, Northwest Territories. Wright DF, Lemkow D, Harris JR (eds) Geological Survey of Canada, Open File 5344, pp 177–190

Map Scale Effects on Estimating the Number of Undiscovered Mineral Deposits

Donald A. Singer and W. David Menzie

Reprinted from *Natural Resources Research* DOI: 10.1007/s11053-008-9068-7, when citing this article please use the DOI number.

Abstract Estimates of numbers of undiscovered mineral deposits, fundamental to assessing mineral resources, are affected by map scale. Where consistently defined deposits of a particular type are estimated, spatial and frequency distributions of deposits are linked in that some frequency distributions can be generated by processes randomly in space whereas others are generated by processes suggesting clustering in space. Possible spatial distributions of mineral deposits and their related frequency distributions are affected by map scale and associated inclusions of non-permissive or covered geological settings. More generalized map scales are more likely to cause inclusion of geologic settings that are not really permissive for the deposit type, or that include unreported cover over permissive areas, resulting in the appearance of deposit clustering. Thus overly generalized map scales can cause deposits to appear clustered. We propose a model that captures the effects of map scale and the related inclusion of non-permissive geologic settings on numbers of deposits estimates, the zero-inflated Poisson distribution. Effects of map scale as represented by the zero-inflated Poisson distribution suggest that the appearance of deposit clustering should diminish as mapping becomes more detailed because the number of inflated zeros would decrease with more detailed maps. Based on observed worldwide relationships between map scale and areas permissive for deposit types, mapping at a scale with twice the detail should cut permissive area size of a porphyry copper tract to 29 percent and a volcanic-hosted massive sulfide tract to 50 percent of their original sizes. Thus some direct benefits of mapping an area at a more detailed scale are indicated by significant reductions in areas permissive for deposit types, increased deposit density and, as a consequence, reduced uncertainty in the estimate of number of undiscovered deposits. Exploration enterprises benefit from reduced areas requiring detailed and expensive exploration, and land-use planners benefit from reduced areas of concern.

Donald A. Singer
U.S. Geological Survey, 345 Middlefield Road, Menlo Park, California 94025, USA
e-mail: singer@usgs.gov

W. David Menzie
U.S. Geological Survey, 12201 Sunrise Valley Dr., Reston, Virginia 20192, USA
e-mail: dmenzie@uogo.gov

Keywords Mineral resource assessments · mineral deposit density · zero-inflated Poisson · mapping benefits

1 Introduction

Undiscovered mineral resource estimation can be captured by three components: the number of deposits, their grades and tonnages, and their locations (Singer, 1993). For many deposit types, geologic controls on locations and grades and tonnages can be represented by proper use of mineral deposit models. Estimating numbers of undiscovered mineral deposits could be based on deposit density models, or expert judgment, or both. Stochastic processes and related spatial distributions of deposits offer other statistical guidelines for increasing specification of these estimates. Numbers of deposits estimates are linked to grade and tonnage models and to delineation of permissive tracts and are therefore affected by the scale of maps used for delineation.

Permissive areas are delineated where geology allows the existence of deposits of one or more specified types and are based on geologic criteria derived from deposit models that are themselves based on studies of known deposits within and, more frequently, outside the study area. A geologic map is the primary local source of information identifying tracts that are permissive for different deposit types and for delineating the tracts.

When assessing undiscovered mineral deposits by type, the base map selected is typically affected by there being only a small number of geologic map scales of the area available, publication scale of the assessment, and time and space limits on the assessors. An effect of these limits is the often-occurring situation where the map scale selected is not ideal, because the delineated permissive tract containing the geologic units that could host the deposits may also include unreported units that could not contain the deposits. The delineated tract may also contain unreported geologic units that cover the geologic units of interest or unreported parts that are too deeply buried to be permissive. In these situations, areas of delineated tracts are larger than necessary due to inflation by unaccounted for non-permissive areas, or by covered areas that are poorly explored. In assessments, estimates of the number of undiscovered deposits should be adjusted to account for the added parts of the tracts that may not contain deposits.

These effects could be reflected in apparent or real spatial distributions of deposits in tracts, and also in the frequency distributions of known deposits and estimates of the number of undiscovered deposits. Spatial and frequency distributions of deposits are linked, in that some frequency distributions can be generated by processes randomly in space, whereas others are generated by processes that suggest clustering in space. Here we examine possible spatial distributions of mineral deposits, their related frequency distributions, and consider the effects of map scale on these distributions and on reductions of areas that could contain deposits. Finally, a model that captures the effects of scale on numbers of deposits estimates is provided.

2 Spatial Distributions of Deposits

When asked, most economic geologists will say that the locations of mineral deposits of any given type are clustered. On a page-sized map of North America, the locations of kuroko massive sulfide deposits or of porphyry copper deposits appear to be clustered. Of course there are large parts of North America that geologically could not contain one or the other of these kinds of deposit. Perhaps what we perceive as clustering of mineral deposits is really clustering of geologic settings. In a more detailed map that only includes geologic settings that are permissive for the deposit type, the deposit locations might be interpreted by some geologists as clustered and by others as not clustered. Even within permissive settings, mineral deposits may appear to be clustered due to uneven exploration caused by covering materials over possible deposits. Clustering of mineral deposits can have a strong effect on estimates of the number of undiscovered deposits. Clustering implies that in an assessment where a deposit is known, the probability that there will be additional deposit(s) is increased by the knowledge of deposits and, of course, the opposite is true when a deposit in not known. Even where the assessed tract is sparsely explored, clustering of deposits means more uncertainty needs to be expressed in the estimate of number of deposits. The appearance of clustering can reasonably be assumed to be affected by map scale.

In order to consider properly whether mineral deposits are clustered in space, it is necessary to define unambiguously what we mean by a deposit and what are the sampling conditions. A mineral deposit is defined as a mineral occurrence of sufficient size and grade that might, under the most favorable circumstances, be considered economic (Cox et al., 1986). When thinking about clustering of mineral deposits at a continental scale, the deposits can be considered as points. However, all mineral deposits have some two-dimensional size and, on a detailed-scale map, the deposits can no longer be considered as points in space.

An important consideration is the question of what the sampling unit should be to define a deposit. Mineral deposit data are available to varying degrees for districts, deposits, mines, adits, and shafts. Perception of clustering can be affected by which of these is studied. For example, it is not uncommon to see multiple prospects or mines located on the same deposit or in the same mineralized system being used as evidence of clustering of mineral deposits. For porphyry copper deposits the following operational rule has consistently been applied in our studies to determine which orebodies should be combined. All mineralized rock or alteration separated by less than two kilometers of unmineralized or unaltered rock is combined into one deposit. Thus, if the alteration zones of two deposits are within two kilometers of each other, the deposits are combined. For kuroko and Cyprus massive sulfide deposits, the rule used is that each deposit must be separated from other deposits by at least 500 m of barren rock in order to be counted as a separate deposit, otherwise it is combined with its neighbor. In resource assessments, or studies of clustering, such an operational spatial rule is necessary to avoid any bias caused by legal boundaries or sparse information. Making the issue more interesting, projected surface areas of deposits of each type have highly skewed frequency distributions such as the lognormal. Consequentially, deposits cannot be placed into cells of the same size such that one and only one deposit is possible. However, it is possible to use cells that are large enough to hold the largest deposit and multiple deposits in other cells in order to count frequencies of deposits to test against known probability distributions.

The other part of defining what we mean by deposits and sampling conditions is clarifying where we would test for the spatial distribution of deposits. If we limit the analysis to a specific deposit type, it is illogical to examine spatial distributions of deposits where they could not possibly exist. This means that geologic settings where a deposit type could not exist should not be considered. More difficult is deciding how to deal with permissive geologic settings that are covered. Because most permissive areas that are covered are poorly explored, numbers of deposits that are reported from covered areas are typically biased downward and therefore are not representative of the densities (ie., deposits/area) or spatial distributions of deposits. In studies of clustering of deposits or of deposit densities, part of permissive areas that are covered should, in most cases, be excluded.

Given these definitions of deposits and sampling conditions, what is known of the spatial distributions of mineral deposits? Unfortunately, little is known. Studies on the spatial distribution of mineral deposits that have been conducted to date have either contained more than one deposit type, did not confine the sample to permissive geologic settings, or did not exclude covered areas (Agterberg, 1977, 1984; Bliss and Menzie, 1993). An important reason why studies that follow the definitions of deposits and proper sampling conditions presented here have not been conducted is the rareness of situations where the conditions are met and yet a large number of deposits are known. When the rules are applied, the polygons containing deposits are typically irregular in shape with boundaries where significant "edge effects" must be addressed. After adjustments for edge effects (Ripley, 1984), few polygons are likely to remain that contain a large enough number of deposits for statistical analysis. However, we can learn some things about clustering and scale effects from consideration of the frequency distributions of deposits in space and from some simulations.

3 Possible Frequency Distributions of Deposits in Space

Traditionally, frequency distributions of mineral deposits have been modeled as Poisson or negative binominal distributions. Discussions about the applicability of these and some other discrete distributions to the spatial distribution of mineral deposits are provided by Agterberg (1984) and Harris (1984). The Poisson and negative binominal distributions are discussed here as models that provide insights into observed mineral deposit distributions, and as a foundation for a possible alternative distribution presented later.

The Poisson distribution is frequently used to characterize counts in spatial distributions of plants, animals, bacteria and, has been suggested for counts of mineral deposits. The Poisson distribution has several advantages over some alternative discrete distributions, such as requiring estimation of only one parameter and simple calculation of the probability of individual numbers of deposits. Also the Poisson distribution is easy to work with mathematically. Where a Poisson distribution is appropriate, the probability of a particular number of deposits (x) can be estimated by:

$$P[x, \lambda] = \lambda^x \exp\{-\lambda\}/x! \qquad (1)$$

where λ is the expected number of deposits per unit area. The standard deviation is $\sqrt{\lambda}$.

Poisson distributions, which result from stochastic processes that distribute events randomly in space, have the property that the mean is equal to the variance, and each outcome is independent from sample to sample, that is, the probability that any cell contains a deposit is independent of whether an adjacent cell contains a deposit. Studies of deposit distributions suggest this assumption of independence may not be appropriate; that is, deposits seem to be clustered in space (e.g. Agterberg, 1977). Two stochastic processes can generate clustered spatial patterns; one process distributes clusters in the sample space; the second distributes events in clusters. For example, if clusters of deposits are generated by a Poisson process and individual deposits within a cluster by a log series, the individual deposits will fit a negative binomial distribution.

A negative binomial distribution (Johnson et al., 1993) is appropriate where the presence or absence of a deposit affects the probability of a deposit nearby—that is, when p (probability of deposit) varies systematically, or when some form of clustering exists. Where a negative binomial distribution is appropriate, the probability of a particular number of deposits (x) can be estimated by:

$$Nb[x, p, k] = \{(x+k-1)!/(k-1)!x!\}\, p^k\, (1-p)^x \qquad (2)$$

where $p = $ a probability, and $k > 0$, a parameter. The expected number of deposits, (λ), is $k\,(1-p)/p$, and the standard deviation is $\sqrt{(k\,(1-p)/p^2)}$.

A clustering distribution can be generated when the mean of a random process varies systematically across the space, as Griffiths and Ondrick (1970) demonstrated with simulations in which they introduced cover masking part of the area containing events (e.g. deposits) generated by a Poisson process. The resulting distribution of events in cells led to rejection of a Poisson and acceptance of a negative binomial distribution even though the underlying distribution was Poisson.

Rejection of the Poisson distribution due to cover in the experiments by Griffiths and Ondrick points to a possible explanation of some observed results of indirect measures of spatial distributions of deposits such as deposit densities.

4 Deposit Densities and Effects of Scale

Mineral deposit densities provide models useful to guide estimates of the number of undiscovered deposits in the same way that tonnage and grade frequencies are models of sizes and qualities of undiscovered deposits (Singer, 1993; Singer et al., 2001).

Estimates of deposit density are based on frequencies of deposits per unit of permissive area in well-explored control areas around the world. Deposit density estimates are made for a particular deposit type in a tract determined to be permissive for that type based on mapped geology and that have an associated grade and tonnage model. Internally consistent descriptive, grade and tonnage, and deposit density models are required in order to prevent biased estimates of undiscovered resources. For example, the same spatial rules that apply to deposits used in grade and tonnage models must be applied to the deposits used to develop deposit density models. The number of deposits per unit area can be illustrated in histograms that show variation of frequencies by deposit type.

Little information is available concerning the variability of deposit densities within deposit types. In one study (Singer, 1994), the area (in logarithms base 10) of permissive ultramafic rock was found to be a useful predictor of the number (in logarithms base 10) of podiform chromite deposits. The variability of deposit densities within porphyry copper type deposits, along with questions about effects of map or assessment scale on estimates of densities, were addressed and a density model was developed (Singer et al., 2005; Singer and Menzie, 2005). Recently, a study of deposit density of volcanic-hosted massive sulfide deposits was completed (Mosier et al., 2007). One recent study considers deposit densities across deposit types (Singer, 2008)

Densities of podiform chromite, porphyry copper, and volcanic-hosted massive sulfide deposits were estimated per 100,000 km^2 of exposed permissive rock and mapped cover was excluded from the control area estimates. Each deposit density histogram is skewed to high densities of deposits (Fig. 1) and, in each situation, the variance is much larger than the mean indicating more variance than is expected in a Poisson distribution, which suggests clustering of deposits. Due to exclusion of permissive areas containing zero deposits and the variable sizes of the permissive tracts, proper testing of the frequency distributions to determine if the Poisson distribution is consistent with the observed distributions was not attempted.

Deposit density is inversely related to the exposed area of permissive rock for each of the studied deposit types (Fig. 2). Density of deposits decreases with increasing size of delineated tract for each of the types. When tract size for porphyry copper deposits increases by 100 percent (doubles), one would expect the number of deposits to increase by 100 percent, but it instead it increases by 33 percent. Similar decreases in density are shown for podiform chromite (Singer, 1994) and volcanic-hosted massive sulfide deposit types (Mosier et al., 2007).

As a general rule, geologic maps published at a detailed scale allow the delineation of smaller tracts. To some extent the reverse is also true, that is, smaller tracts suggest that a more detailed base geologic map has been used. For both porphyry copper (Singer et al., 2005) and volcanic-hosted massive sulfide deposits (Mosier et al., 2007) there is a significant positive relationship between map scale and permissive area (Fig. 3) and a negative relationship between map scale and deposit density (Fig. 4). Based on the relationships represented in Fig. 3, mapping at half the scale (say 1:1,000,000 to 1:500,000) would cut the permissive size of a porphyry copper tract to 29 percent and a volcanic-hosted massive sulfide tract to 50 percent

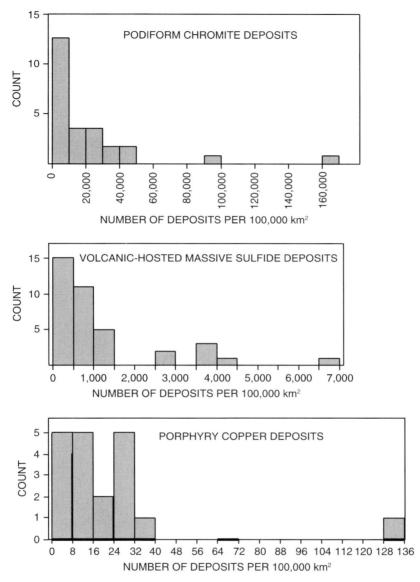

Fig. 1 Histograms of the number of deposits per 100,000 km² of permissive rock. Top, podiform chromite deposits, middle, volcanic-hosted massive sulfide deposits, bottom, porphyry copper deposits

of their original sizes. Density of podiform chromite deposits (Singer, 1994) was determined from maps that were all at the same published scale.

So where are the missing deposits in the larger tracts? No obvious genetic reasons are known to explain why densities of deposits should decrease as tracts become larger—genetically one would expect that the density would not change with

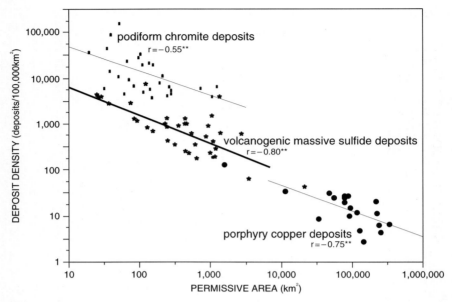

Fig. 2 Plot of the density of deposits per 100,000 km^2 versus the permissive area in km^2 along with the regression lines. Top, podiform chromite deposits, middle, volcanic-hosted massive sulfide (VMS) deposits, bottom, porphyry copper deposits

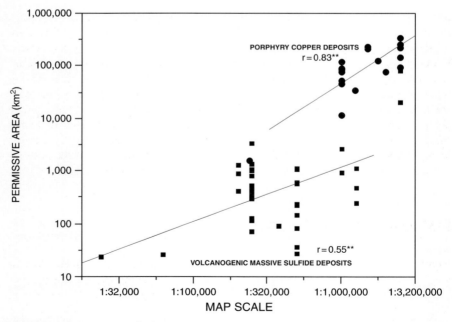

Fig. 3 Permissive area in km^2 versus map scale along with the regression lines for porphyry copper and volcanic-hosted massive sulfide deposits

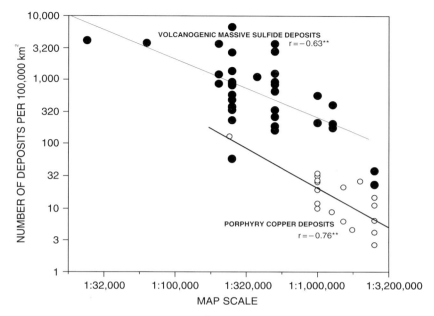

Fig. 4 Number of deposits per 100,000 km² versus map scale along with the regression lines for volcanic-hosted massive sulfide and porphyry copper deposits

tract size. Two possibilities exist to account for the discrepancies; either larger permissive tracts have more area of geologic settings that are not really permissive for the deposit type, or proportionally they have more covered area than smaller tracts. For both possibilities the added settings are not recorded in the maps used to delineate the larger tracts. If larger tracts had a larger proportion of cover than smaller tracts, we might expect to see some relationship between reported percent cover and map scale, yet no significant relationship is observed in the only study where percent cover in permissive tracts was recorded—porphyry copper deposit density (Singer et al., 2005). Thus, the more likely explanation for lower density of deposits in larger tracts is that larger permissive tracts include more area of geologic settings that are not really permissive for the deposit type and are not indicated on the map.

5 Modeling Map Scale Effects

Number of mineral deposits conditioned on size of permissive tracts seems to provide an excellent means of estimating the number of undiscovered mineral deposits by type. It does not directly address the issue of deposit clustering or map scale effects on densities, however. The results of the experiments of Griffiths and Ondrick (1970) and the previous explanation that larger tracts probably contain more geologic settings that are not permissive suggest a possible model that could connect map scale and apparent clustering. Suppose that there is a tract mapped in great detail that contains only permissive rock and that it is divided into cells where the probability that any cell contains a deposit is independent of whether an adjacent cell contains a deposit. If this same tract is mapped at a scale such that some geologic settings that are not permissive are not distinguished from those that are permissive, the effect would be the same as adding cells that contain no deposits thereby inflating the number of cells with zero deposits. This would generate a zero-inflated Poisson distribution (Johnson et al., 1993; Puig and Jordi, 2006) with:

$$P[x = 0] = w + (1-w)\exp\{-\lambda\},$$
$$P[x] = (1-w)\lambda^x \exp\{-\lambda\}/x!, \quad x \geq 1 \qquad (3)$$

where, w is the proportion of cells where the zero probability is inflated and λ is the expected number of deposits per cell without the inflation of zeros. With the inflation of zeros the expected number of deposits is:

$$\mu = (1-w)\lambda. \qquad (4)$$

The expected number of deposits decreases in proportion to the proportion of cells where the zero probability is inflated. The relative variation, represented by the coefficient of variation, increases as the zero probability is inflated. Examples of several proportions of zero inflation of a Poisson distribution are shown in Fig. 5.

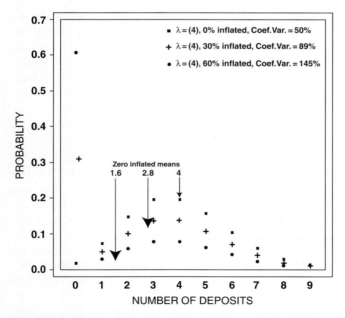

Fig. 5 Probability versus number of deposits for zero-inflated Poisson distributions with the same mean (λ) of 4 and zero percent inflations, 30 percent inflated zeros, and 60 percent inflated zeros, all with coefficients of variation

To see how this might work, suppose that a control tract based on a 1:1,000,000 scale map is delineated as permissive for a deposit type and the proportion mapped as covered is subtracted from the area. If 60 percent of the permissive area is actually not permissive because of the necessary lumping of geologic units at that scale, then the control tract density would be deflated by the additional area that is really not permissive. If this tract was remapped at 1:500,000 scale we would expect that the total permissive area would decrease due to the recognition and exclusion of some portion of the non permissive geologic settings that was identified with the more detailed mapping. Perhaps the proportion of area that is unavoidably included as permissive would drop to 40 percent. Because the total number of deposits remains the same and permissive area decreases, deposit density increases. This pattern would continue with diminishing effects with each more detailed map. Thus with control areas from across the world mapped at different scales, we would expect to see inverse relationships between area permissive and deposit density and between deposit density and map scale. A discrete lognormal distribution of deposits across control areas and statistics that suggest clustering of deposits would be consistent with this pattern also. We have not found a properly sampled set of locations of typed mineral deposits to test the fit of a zero-inflated Poisson distribution.

6 Conclusions

Estimates of number of mineral deposits are directly related to the size of permissive tract. Deposit density (deposits/area) decreases as the size of permissive tract increases for each of the three deposit types studied. In the control areas used to estimate densities, mapped cover and non-permissive rock units were excluded from calculations. Permissive area is positively related to map scale for both volcanic-hosted massive sulfide and porphyry copper deposit types. Control areas for podiform chromite deposits were nominally mapped at the same scale.

Deposit density decreases as the map scale increases for both volcanic-hosted massive sulfide and porphyry copper deposit types. More generalized map scales mean that areas of geologic units that could contain a particular deposit type appear to be larger than the same areas on detailed maps, because of the inclusion of non permissive units, but are not differentiated on the more generalized geologic map. Not all of the effects on density are necessarily due to map scale. More detailed geological mapping associated with a more recent vintage map has been shown to be a better predictor of density in some situations (Bernknopf et al., 2007). Typically, newer vintage and more detailed maps go hand-in-hand however. Nevertheless, some maps are of little value in delineating permissive areas at any scale due to concepts and approaches used in mapping where geologic settings are not clearly indicated. Although it is possible that more generalized maps contain proportionally more cover over permissive rocks, the only study testing this relationship (Singer et al., 2005) suggests that it is not necessarily correct. If the scale-effect area added in an assessment contains only geologic units that are not permissive, the density of deposits estimated should decrease because no additional area that is really permissive has been added, but the variability about the estimate should be larger by the addition of area containing zero deposits.

The zero-inflated Poisson distribution can capture the effects of unreported non-permissive geologic units and perhaps unreported cover. Apparent effects of map scale as determined by logic and as represented by the zero-inflated Poisson distribution suggest that the appearance of deposit clustering should diminish as mapping becomes more detailed. With more detailed mapping, the expected number of deposits per unit area should increase because of the removal of non-permissive area and the relative variability would decrease. It is possible that the mapping scale needed to remove all hidden non-permissive geologic units would be so detailed that only one deposit of average size could be present. This deposit would be limited by the deposit type's operational spatial rule. If so, clustering of deposits is likely to be observed under all realistic sampling situations even if the underlying distribution is not clustered. Thus a zero-inflated Poisson, negative binominal, or some other distribution that represents clustering will probably be appropriate for mineral deposits under most realistic situations.

Based on observed relationships between map scale and areas permissive for deposit types, mapping at half the scale (say 1:1,000,000 to 1:500,000) would cut the permissive size of a porphyry copper tract to 29 percent and a volcanic-hosted massive sulfide tract to 50 percent of their original sizes. Thus some direct benefits of mapping an area at a more detailed scale are indicated by the significant reduction in areas permissive for deposit types and the reduced uncertainty in the estimate of number of undiscovered deposits. The benefits for exploration enterprises are in reduced areas requiring detailed and expensive exploration, and for land-use planning, the benefits are in reduced areas of concern.

References

Agterberg FP (1977) Frequency distributions and spatial variability of geological variables. In: Ramani RV (ed) Application of computer methods in the mineral industry. Society of Mining Engineers of AIME, New York, pp 287–298

Agterberg FP (1984) Discrete probability distributions for mineral deposits in cells. In: Proceedings, 27th international geological congress, vol 20. VNU Science Press, Utrecht, pp 205–225

Bernknopf R, Wein A, St-Onge M, Lucas S (2007) Analysis of improved government geological map information for mineral exploration: incorporating efficiency, productivity, effectiveness and risk considerations. Geological survey of Canada bulletin 593. US Geological Survey Professional Paper 1721, p 45

Bliss JD, Menzie, WD (1993) Spatial mineral-deposit models and the prediction of undiscovered mineral deposits. In: Kirkham RV, Sinclair WD, Thorpe RI, Duke JM (eds) Mineral deposit modeling: Geological Association Canada Special Paper 40, pp 693–706

Cox DP, Barton PR, Singer DA (1986) Introduction. In: Cox DP, Singer DA, (eds) Mineral deposit models: US Geological Survey Bulletin 1693, pp 1–10

Griffiths JC, Ondrick CW (1970) Structure by sampling in the geosciences. In: Patil GP (ed) Random counts in physical science, geoscience, and business. The Penn State Statistics Series, The Pennsylvania State University Press, University Park and London, pp 31–55

Harris DP (1984) Mineral resources appraisal—mineral endowment, resources, and potential supply—concepts, methods, and cases. Oxford University Press, London, p 445

Johnson NL, Kotz S, Kemp AW (1993) Univariate discrete distributions, 2nd edn. John Wiley & Sons, New York, p 565

Mosier DL, Singer DA, Berger VI (2007) Volcanogenic massive sulfide deposit density. US Geological Survey Scientific Investigations Report 2007–5082, p 21 (http://pubs.usgs.gov/sir/2007/5082/)

Puig P, Valero J (2006) Count data distributions—some characterizations with applications. J Am Stat Assoc 101 (473): 332–340

Ripley BD (1984) Spatial statistics—developments 1980–3. Int Stat Rev 52 (2): 141–150

Singer DA (1993) Basic concepts in three-part quantitative assessments of undiscovered mineral resources. Nonrenew Resour 2 (2): 69–81

Singer DA (1994) Conditional estimates of the number of podiform chromite deposits. Nonrenew Resour 3 (3): 200–204

Singer DA (2008) Mineral deposit densities for estimating mineral resources. Math Geosci 40 (1): 33–46

Singer DA, Berger VI, Menzie WD, Berger BB (2005) Porphyry copper density. Econ Geol 100 (3): 491–514

Singer DA, Menzie WD (2005) Statistical guides to estimating the number of undiscovered mineral deposits: an example with porphyry copper deposits. In: Cheng Q, Bonham-Carter G (eds) Proceedings of IAMG—the annual conference of the international association for mathematical geology, Geomatics Research Laboratory, York University, Toronto, Canada, pp 1028–1033

Singer DA, Menzie WD, Sutphin D, Mosier DL, Bliss JD (2001) Mineral deposit density—an update. In: Schulz KJ (ed) Contributions to global mineral resource assessment research. US Geological Survey Professional Paper 1640–A, A1–A13. (Also available online at http://pubs.usgs.gov/prof/p1640a/)

Are Fractal Dimensions of the Spatial Distribution of Mineral Deposits Meaningful?

Gary L. Raines

Reprinted from *Natural Resources Research* DOI: 10.1007/s11053-008-9067-8, when citing this article please use the DOI number.

Abstract It has been proposed that the spatial distribution of mineral deposits is bifractal. An implication of this property is that the number of deposits in a permissive area is a function of the shape of the area. This is because the fractal density functions of deposits are dependent on the distance from known deposits. A long thin permissive area with most of the deposits in one end, such as the Alaskan porphyry permissive area, has a major portion of the area far from known deposits and consequently a low density of deposits associated with most of the permissive area. On the other hand a more equi-dimensioned permissive area, such as the Arizona porphyry permissive area, has a more uniform density of deposits. Another implication of the fractal distribution is that the Poisson assumption typically used for estimating deposit numbers is invalid. Based on data sets of mineral deposits classified by type as inputs, the distributions of many different deposit types are found to have characteristically two fractal dimensions over separate non-overlapping spatial scales in the range of 5–1,000 km. In particular, one typically observes a local dimension at spatial scales less than 30–60 km, and a regional dimension at larger spatial scales. The deposit type, geologic setting, and sample size influence the fractal dimensions. The consequence of the geologic setting can be diminished by using deposits classified by type. The crossover point between the two fractal domains is proportional to the median size of the deposit type. A plot of the crossover points for porphyry copper deposits from different geologic domains against median deposit sizes defines linear relationships and identifies regions that are significantly under explored. Plots of the fractal dimension can also be used to define density functions from which the number of undiscovered deposits can be estimated. This density function is only dependent on the distribution of deposits and is independent of the definition of the permissive area. Density functions for porphyry copper deposits appear to be significantly different for regions in the Andes, Mexico, United States, and western Canada. Consequently, depending on which regional density function is used, quite different estimates of numbers of undiscovered deposits can be obtained. These fractal properties suggest that geologic studies based on mapping at

---

Gary L. Raines
U.S.Geological Survey (retired), c/o Mackay School of Earth Sciences, UNR, MS 176, Reno, NV 89557, e-mail: garyraines@earthlink.net

scales of 1:24,000–1:100,000 may not recognize processes that are important in the formation of mineral deposits at scales larger than the crossover points at 30–60 km.

Keywords Fractal dimension · spatial analysis · mineral deposits · deposit density

1 Introduction

Carlson (1991) presented an analysis of the spatial distribution of Great Basin precious metal deposits that had two fractal dimensions with a crossover point occurring at approximately 30 km. Based on these fractal properties, Carlson raised interesting questions about the formation of mineral deposits and the assessment of mineral endowment. Carlson's work has, however, been ignored by many economic geologists or dismissed as a measure of fractal dimensions of the Great Basin, at best, or dismissed as an artifact of sample size. Others (for example Agterberg et al., 1993; Cheng and Agterberg, 1995; McCaffrey and Johnston, 1996; Blenkinsop and Sanderson, 1999) have investigated fractal dimensions of deposits and obtained interesting results similar to or supporting those of Carlson (1991). Sanderson et al. (1994), Johnston and McCaffrey (1996), Roberts et al. (1998 and 1999) report fractal dimensions for veins, suggesting that additional different fractal dimensions occur at larger scales than addressed here. Herzfeld (1993) presents an overview of some further fractal literature.

The purpose of this paper is simply to test the proposals made by Carlson (1991) concerning the fractal properties of the distributions of mineral deposits using mineral deposits classified by deposit type. A large number of mineral deposits in Nevada (Raines et al., 1996), the United States (Long et al., 1998) and compilations of significant porphyry deposits of the world (Mutschler et al., 1999; Singer and others, 2005) have been classified by deposit types. So Carlson's fractal analysis can be repeated now with diverse deposit types, sample sizes, and geographic locations in order to evaluate the criticisms. The new analysis shows that whereas the criticisms are at least partially valid due to the mixed types of gold deposits used by Carlson; his conclusion that the spatial distribution of mineral deposits is fractal is valid and useful insights about the degree of exploration and numbers of undiscovered deposits can be obtained from knowledge of these fractal dimensions. There are many implications and questions raised by these conclusions. The implications are not addressed fully here. This paper is intended simply to stimulate interest in the consideration of the fractal distributions of mineral deposits.

For the study reported, the box counting method of Mandelbrot (1985) was used to assess the fractal properties of the spatial distributions of mineral deposits. The necessary measurements as discussed in Carlson (1991) in a modern GIS are simply a matter of converting the mineral-deposit points to rasters using varying cell size, and then counting the number of cells. Using this GIS approach to box counting avoids the edge effects associated with traditional mathematical approach to box counting. Then using a graphical curve fitting approach available in Excel allows for simple and accurate definition of the bifractal dimensions. Using this Excel graphical curve-fitting method demonstrates that Agterberg et al. (1993, Table 1) also have bifractal results similar to those reported here. The appendix discusses some special considerations for box counting and creating the fractal plots. Examples of the bifractal plots are shown in Fig. 1. To assess whether Carlson was measuring a property of the geology or mineral deposits or simply an artifact, fractal dimensions were calculated for areas in Nevada using several random sets of points, map boundaries from the 1:2,500,000 US geologic map (King and Beikman, 1974) and the 1:500,000 Nevada geologic map (Stewart and Carlson, 1978), and the faults from the Nevada geologic map. These random and geologic fractal measurements are compared to those of mineral deposits in order to demonstrate that useful information is indeed in the fractal dimensions of different types of mineral deposits. The fractal dimensions measured are summarized in Fig. 2.

2 Nevada Deposits

Using mineral deposits classified by deposit type from Raines et al. (1996), all of the deposit types considered have two fractal dimensions for their spatial distribution in the range from 5 to 1,000 km. What, however, was Carlson measuring when he combined all precious metal deposits for the Great Basin? First, note in Fig. 2, that the fractal dimensions of the group precious metal deposits for Nevada have almost the same values as those obtained by Carlson for the Great Basin. Therefore, the measurements are repeatable. An understanding of these measurements can be obtained by comparison with the fractal measurements for the geologic boundaries for Nevada from the King and Beikman (1974) US geology and Stewart and Carlson (1978) Nevada-geology boundaries and faults. These geologic features have fractal dimensions near 1 for local scale and 2 for regional scale. This indicates these features are approximately normal Euclidean lines and points at these scale ranges. Similarly, when random points are located in the areas of bedrock or in the area of Nevada, fractal dimensions of less than 0.1 (local) and near 1.7 (regional) are obtained. Therefore, these random points are behaving approximately like Euclidean points but not quite like lines. These measurements suggest that indeed by combining deposit types by commodity a spatial distribution that confounds the bedrock geology with the spatial distribution of the mix-type deposits of Nevada is a primary influence in the measured fractal dimensions. A similar result is found for metallic commodities. Therefore, Carlson's measurement seems to be strongly confounded with the spatial properties of geology of Nevada and precious metal deposits, resulting in a fractal dimension intermediate between that of the geology and deposit type.

A different result is obtained if the spatial distribution is measured for the deposits grouped by deposit type. This seems reasonable because the assumption of deposit types is that each deposit type is formed by a distinct process. Thus, if the fractal dimensions are influenced by the processes forming the spatial pattern, the

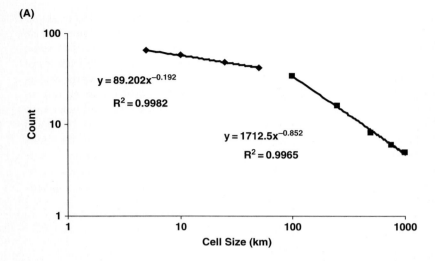

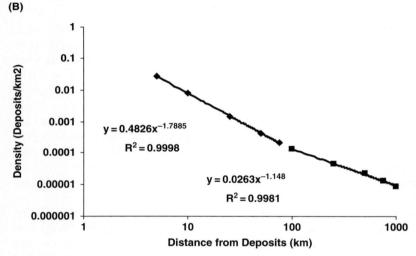

Fig. 1 Log-Log plots of box-counting results for western Canada significant porphyries. In the regression equations for Cell size versus Count, the two portion of the plot define a crossover point at 88 km. For scales less than and greater than 88 km the fractal dimensions are 0.19, referred to as the local dimension, and 0.85, referred to as the regional dimension. In the Density versus Distance plot, the fractal dimensions are two minus the exponent and typically are slightly different from the other plot

measurement should be made on deposits of one type. The fractal dimensions of the distributions of different deposits types summarized in Fig. 2 show that each deposit type has fractal dimensions that are different and different from the geology or a random set of points.

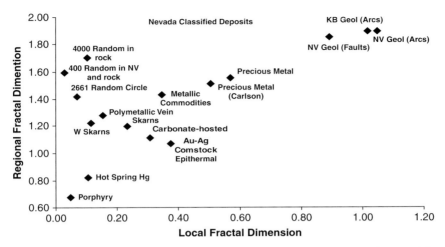

Fig. 2 Fractal dimensions plot for Nevada deposits, geology, and random-sample tests. A fractal dimension of zero is a Euclidean point, one is a Euclidean line, and two is a Euclidean plane. Therefore, the geology lines are fractal with properties of lines and planes. The random points show an example for 400 and 4,000 random points in the Nevada and in geology bedrock. For 400 random samples, the same dimensions were found whether restricted to rock outcrop or statewide. Therefore, the random points behave almost like points and lines. The precious metal calculation almost duplicates Carlson's measurement for the Great Basin. Metallic commodities plot at a slightly different position. Mineral deposits by type plot at distinctly different positions

It has been suggested that the fractal dimensions are some artifact of sample size. Figure 3 summarizes the fractal dimensions versus sample size for the Nevada deposits, the Nevada random points, and the significant deposits. Because different deposit types have different fractal dimensions, there is a spread in the fractal dimension axis in Fig. 3. Also different deposit types form at different crustal densities and thus the processes of formation create differing numbers of sites. Thus, the deposit types with larger sample sizes have slightly higher fractal dimensions because these deposit types are more space-filling and consequently have higher dimensions (Barnsley, 1988). This space-filling aspect is demonstrated by the random points and gives a sense of how this changes fractal dimensions. It seems that although there is a small consequence of sample size, which is a consequence of the deposit type, the spread of fractal dimensions observed cannot be accounted for by sample size alone.

The random experiment where a smaller number of points are randomly located suggests that the local fractal dimension of less than 0.05 is a cantor dust as discussed by Carlson (1991). However, when there are as many as 4,000 points controlled by line-like rock outcrop, the fractal dimensions of the outcrop influence the fractal dimensions. Thus, the local fractal dimension of porphyries might be interpreted to indicate that at scales less than about 60 km, the spatial process is a Cantor dust. There are only regional processes acting at distances greater than 60 km that can be observed in these data.

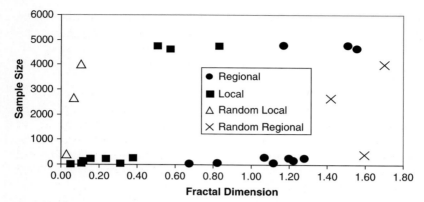

Fig. 3 Fractal dimensions and sample size for Nevada measurements. This plot addresses the question of correlation of fractal dimensions and sample size. Two clusters with no correlation between sample size and fractal dimensions. Fractal dimension is independent of sample size; however, sample size is dependent on deposit type

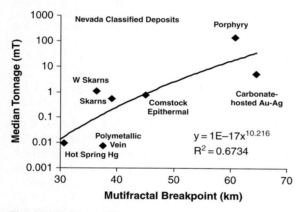

Fig. 4 Correlation of Nevada classified deposits fractal-dimension crossover point with median tonnage of the deposit types. These median values are from Singer, Berger, and Moring (2005) and refer to the global median, not the median for Nevada, which it is assumed would increase the correlation

A consequence of having two fractal dimensions is that a crossover point occurs between the two fractal domains. In a plot of the crossover point versus the log of the median deposit size for the various deposits, the relationship in Fig. 4 is obtained. The coefficient of explanation is relatively large for this relationship, supporting the conclusion that the crossover point is related to deposit size. If the crossover point has a physical meaning, then it is reasonable to assume the fractal dimensions, which define the crossover point, have significance.

Finally, consider the distribution of deposit types in Fig. 2. The skarns and polymetallic veins are clustered around 0.20 (local) and 1.25 (regional), gold deposits are clustered about 0.35 (local) and 1.15 (regional), and porphyry and hot spring mercury are together near 0.1 (local) and 0.7 (regional). The deposit types are not randomly distributed in this fractal dimension space. This suggests process similarities between groups of deposit types. It is thought provoking to consider hot spring mercury deposits grouping with porphyry deposits.

Thus, these data support Carlson's (1991) conclusions. The spatial distributions of mineral deposits are fractal. In order to minimize the confounding with geologic setting, the fractal dimensions of mineral deposits should be measured by deposit type. The fractal dimensions are also influenced by the sample size because the fractal dimensions are a measure of space filling, but the sample size is influenced by the processes that create the deposits. For example, large deposit types such as porphyries occur less frequently than polymetallic veins simply because of the size of the processes involved. Therefore, the fractal dimensions reflect this size difference. Similarly the scale at which the fractal dimensions change is related to the size of the deposits, and the larger the deposit then the larger the process to produce that deposit. Therefore, the fractal dimensions carry information about the deposits and their associated creation processes.

3 Porphyry Deposits of the Americas

The previous analysis suggests it is acceptable to consider the fractal dimensions of the spatial distributions of mineral deposits, and three questions are addressed here. How do these numbers change for different geologic environments? How do fractal dimensions change for different samples of mineral deposits? Can useful insights be obtained from these differences? Considering significant deposits from the United States as compiled by Long et al. (1998), giant copper camps (Mutschler et al., 1999), and global porphyry copper deposits (Singer and others (2005), Fig. 5 summarizes the fractal dimensions of significant deposits (Long et al., 1998) and giant copper camps (Mutschler et al., 1999). If the crossover points are plotted against median size, (Fig. 6) a linear relationship is observed for the giant copper camps of western Canada, the Basin and Range, and Andean cordillera. Mexico and Alaska are outliers from this linear trend.

The fractal dimensions of significant and giant copper camps (Fig. 5) show some similarities and differences to the deposits in Nevada (Fig. 2). Generally, the local fractal dimensions are small suggesting point or local processes control the spatial patterns at local scales. The giant copper camps, however, have smaller fractal dimensions than the significant deposits. The decrease in fractal dimensions can be explained because the sample densities are smaller than those of Nevada. The smaller samples are less space-filling because the authors that created these datasets were more selective in what is included, that is the dimensions are different because different criteria were used for what deposits to include. It is interesting to note that the Alaskan followed by Mexican porphyries have the smallest fractal dimensions. In Fig. 6 considering only the giant copper camps, Alaska and Mexico appear to

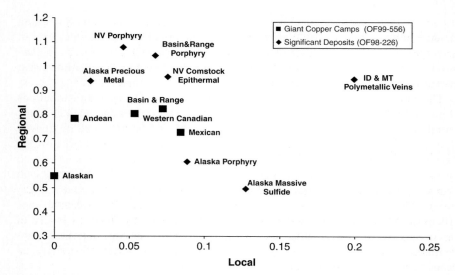

Fig. 5 Fractal-dimension plot for significant deposits for the United States and giant porphyries from Alaska, western Canada, the Basin and Range, Mexico, and Andean cordillera. It seems that giant deposits have smaller fractal dimensions than significant deposits. Alaskan giant porphyries are the only ones having a local fractal dimension of zero

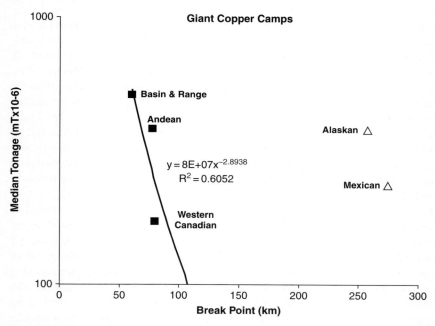

Fig. 6 Crossover points versus median size of porphyries for important areas in the Americas. These are from the giant copper camps (Mutschler, 1999). The moderately good regression line is for western Canada, the Basin and Range, and Andean giant porphyries. Based this correlation, Alaska and Mexico are considered outliers

be outliers to the linear trend defined by Basin and Range, Andean, and western Canada giant copper camps. Other groupings of these points do not define a linear trend with such a large R^2. This might be a consequence of degree of exploration, development style, or the geology. Have the smaller deposits not been explored for in Alaska and Mexico or is there something fundamentally different about Alaskan and Mexican copper camps?

3.1 Number of Undiscovered Deposits

Estimates of undiscovered deposits alone does not consider all of the factors required to decide if an area is worthy of further exploration. Besides the political issues, the size (grade and tonnage) of deposits must be considered. Thus, for example the size of Alaskan deposits might be smaller than Chilean deposits. Consequently, even though there may be many undiscovered deposits in an area, if those deposits are all small the exploration might better focus where fewer, but larger, deposits occur. For examples of these considerations, see Singer (1994). In the analysis presented here, only one aspect of the decision to explore is considered, that is how many are there to be found.

If Alaska and Mexico are underexplored, then the fractal density function (referred to as PDF in Carlson, 1991; Fig. 1 for an example) can be used to make an estimate of the number of undiscovered deposits as suggested by Carlson (1991). Because the fractal dimensions are clearly dependent on the sample of deposits analyzed, a very robust dataset is probably best. The global porphyry deposits as defined by Singer, Berger, and Moring (2005) has the strictest definitions of porphyry deposits and benefited from the previous summaries; so it was used to address these density considerations. The fractal dimensions of these data are shown in Fig. 7. The fractal dimensions and crossover point versus deposit size are again different from previous examples. In this example, the crossover point versus deposit size seems to define several linear trends. Other groupings did not form trends with large R^2. There is a good trend for Mexico, Great Basin, and western Canada. Arizona and Peru suggest another parallel trend, but there are only two points; so this "trend" is suspect. If Arizona and Peru are added to the first trend, the fit is insignificant. Based on this analysis, Chile and Alaska are outliers.

The PDFs are shown in Fig. 8 and the regression equations are summarized in Table 1. Additionally in Table 1, a regression equation based on all the data except Alaska and the Great Basin as an estimate of a typical PDF is included. The method used to calculate these PDFs implies that a distance of 1 km should have a density of one because a known deposit will be contained in that 1 km radius. Alaska and the Great Basin have projected densities greater than 1 at distance of 1 km. The other curves are in an envelope defined by Chile and western Canada with Chile having the highest density and western Canada the lowest. Only the local portions of the Alaskan and Great Basin density curves are outside this Chile-western Canada

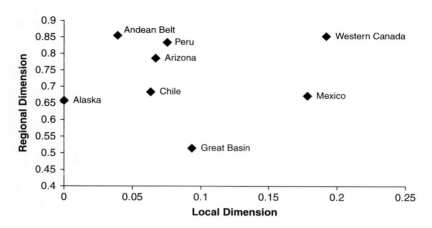

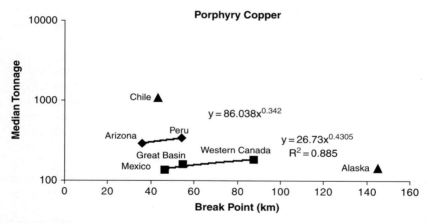

Fig. 7 Porphyry copper deposits of the Americas. In the fractal dimensions plot (**A**) the Andean and Arizona deposit seem to form a cluster. The others, western Canada, Mexico, Great Basin, and Alaska seem to be outside this cluster. Similarly, in the tonnage versus crossover point plot, The Arizona and Chilean and western Canada, Great Basin, and Mexico appear to form separate clusters based on the sizes of the deposits. Alaska and Chile are outliners to these clusters. It seems reasonable to assume Alaska is an outlier due to under exploration. Chile is a class by itself

envelope. For these reasons, Alaska and the Great Basin were excluded from the regression equation.

Based on a selection of these PDFs, estimates have been made of the number of undiscovered deposits in Arizona, Alaska, and the Great Basin. These calculations were selected because permissive areas for porphyry deposits were provided by the USGS (USGS National Mineral Assessment Team, 1998). To demonstrate the sensitivity to the different PDFs but not to advocate which is the best estimate,

Fractal Dimensions of the Spatial Distribution

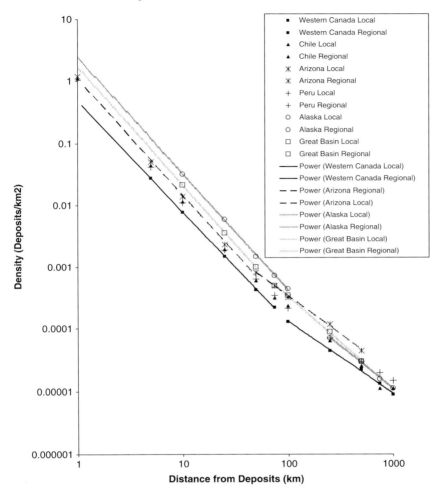

Fig. 8 Comparison of PDF for different areas. Western Canada and Chile might define one envelope of PDF. Alaska and the Great Basin might be considered outliers at the local scale because Alaska and Great Basin have PDF that project significantly greater than a density of 1 at a distance of 1 km. Arizona is slightly larger (1.16) than a density of 1 at a distance of 1 km. All the others have computed densities of slightly less than one at a distance of 1 km

estimates of the number of undiscovered porphyry deposits are shown in Table 2. Besides the small differences in PDFs, the shape of the permissive area, as suggested by Carlson (1991) influences the different estimates. This shape-related difference is because the fractal PDF is a function of distance of a cell from a known deposit. Based on tests with the Alaskan data, removing one known deposit did not make a significant change in the estimated number.

Table 1 Summary of probability density functions for various areas in the Americas. In all cases the R^2 of the PDF was greater than 0.99. The crossover point is the distance at which the density changes from the Local PDF to the Regional PDF. The median tonnage is from Singer, Berger, and Moring 2005) and used in the plots above. The units of tonnage are millions of metric tons. Regression is done using Arizona, Chile, Peru, and western Canada

Area	Local PDF	Crossover point (km)	Regional PDF	Medan Tonnage
Alaska	y = 2.4012x^ 1.873	234	y = 0.1286x^ − 1.3532	145
Great Basin	y = 1.70851x^ − 1.9064	56	y = 0.3173x^ − 1.4864	159
Mexico	y = 4.2379x^ − 1.8218	47.1	y = 0.6393x^ − 1.3292	134
Arizona	y = 1.1696x^ − 1.9328	33.1	y = 0.1033x^ − 1.2388	292.5
Peru	y = 0.7684x^ − 1.8315	61.7	y = 0.0587x^ − 1.2076	337.5
Chile	y = 1.0323x^ − 1.9301	48	y = 0.123x^ − 1.3689	1090
Canada	y = 0.4826x^ − 1.7885	94	y = 0.0263x^ − 1.148	181
Andean Belt	y = 0.3279x^ − 1.926	59.4	y = 0.0157x^ − 1.1822	493
Regression	Y = 0.9486x^ − 1.9261 ($R^2 = 0.99$)	Average = 59.2 Median = 54.8	y = 0.1087x^ 1.332 ($R^2 = 0.9492$)	

Table 2 Calculated estimates of number of undiscovered deposits based on PDF from a selection of areas. In all cases, the permissive area is as defined by U.S.Geological Survey National Mineral Assessment Team (1998). All numbers are arbitrarily rounded to the nearest integer. The fractal estimates using the Arizona and Chilean PDF also seem too large. In the case of Arizona, this might be because of extension after formation of the deposits

Area	Fractal estimate			
	Western Canada PDF	Arizona PDF	Chilean PDF	Regression PDF
Arizona	24	92	87	67
Alaska	11	36	16	27
Great Basin	10	44	30	52

Further, the fractal properties of these distributions implies that the probability that any cell contains a deposit is not independent of whether an adjacent cell contains a deposit: that is, the Poisson basis of the Singer and Menzie (2005) density function is disputed by the fractal measurements.

4 Conclusions

Based on this analysis, the fractal dimensions obtained are a function of the geologic setting, the type of deposit and the deposit sample size.

The fractal dimensions of the spatial distributions of mineral deposits based on deposit types are meaningful numbers. However, these numbers are definitely influenced by how well the deposits are classified. One typically observes two fractal or bifractal domains in the range 5–1,000 km, a local domain associated with spatial scales less than 30–60 km, and a larger domain associated with spatial scales up to 1,000 km. Therefore, Carlson (1991) with deposits classified by type was fundamentally correct.

Because the sample size influences the results of the computation, the findings provide information on the degree of exploration for a given deposit type and size. The influence of sample size is still included, but sample size is a consequence of degree of exploration and the size of the deposit type (larger deposit types occur less frequently). The crossover points between these two fractal domains show a linear relationship with the size of deposits. The plot of crossover points versus the size of the deposits seems to be linearly correlated. Outliers to these linear relationships seem to be related to the degree of exploration. It follows that such fractal analysis may be a useful tool for the identification of underexplored areas.

If the spatial distributions of mineral deposits are statistically self-similar, including undiscovered deposits, then the Poisson assumption of independence of adjacent cells is not valid. Therefore, the fractal approach provides an alternative estimate of the number of undiscovered deposits. This means that the shape and the area of a permissive tract are significant factors in a model of undiscovered deposits rather than just the area of the tract alone. The shape and area at the time of formation of the deposits might provide a more consistent density, not the current shape and area, such as that of the Basin and Range, which has been extended subsequent to formation of the porphyry deposits. There is, of course, the problem of which density function to use, especially for a poorly explored area.

Crossover points in the range of 30–60 km (Figs. 6 and 7) raise questions about the scale of studies in geology. Do we really understand processes that occur at scales greater than 30–60 km based on mapping at 12 m (map scale of 1:24,000)? What might such a process be?

There are at least two approaches, resolution and extent, to transforming the crossover scale at 30–60 km to map scales. If this crossover point is considered the minimum resolution of the measurements, then the implied map scale of measurement is 1:60,000,000–1:120,000,000. If 30–60 km is considered the minimum extent of the measurement and assuming Nyquist sampling is appropriate, then map scales near 1:100,000 define the map scale of the crossover point. Modern government geologic field mapping, which is typically at map scales of 1:24,000 or 1:100,000, may not be recording processes of the regional fractal domain greater than 30 to 60 km because maps appropriate to such regional scales are typically made by generalization of larger map-scale information. Thus, we focus on processes at the largest map scales (typically near 1:24,000) of the local fractal domain but make no primary observations in the regional fractal domain. What processes might have not been observed, and what measurements might be appropriate to address such regional processes?

Acknowledgments I wish to acknowledge Neal Fordyce of Placer Dome who felt that Carl Carlson's papers raised interesting questions and asked if fractal dimensions might provide insights on where to explore. Steve Ludington, Mike Zientek, Vic Mossotti, and Don Singer have provided useful discussions and suggestions during the course of the research for this paper. Vic Mossotti, Frits Agterberg, Quiming Cheng, and P. Gumeil provided helpful reviews of this paper. I would also like to acknowledge the work of Berger, Bookstom, Ludington, Moring, Mutschler, Long, DeYoung, and Singer who compiled the databases of mineral deposits that made this analysis possible.

Appendix: Calculation Methods

The calculations reported in this paper utilized ArcGIS for the box counting and distance to nearest known deposit. The calculations of fractal dimensions, and associated graphic, and estimations of numbers of undiscovered deposits utilized Excel. While most of the calculations are straight forward as discussed in Carlson (1991), some of the processing for the calculations and a few of the calculations are clarified here. In addition, a graphical approach using Excel to define the fractal dimensions is described here that solves the edge-effects problems with more classical fitting approaches. The following description outlines the ArcGIS processing steps and the Excel equations.

Box Counting and Crossover Point

Box counting of points is simple in ArcGIS. It is simply a matter of converting the points as a shapefile to a raster using different appropriate cell sizes to determine the count of cells. ArcGIS converts point shapefiles to a raster by creating cells that contains at least one point. Cells that contain no points are treated as "No Data" area and no cell appears for these locations in the raster attribute table. For fractal dimensions of geographic points, only a few measurements at intervals such as 1, 10, 25, 50, 75, 100,000, 250,000, 500,000, 750,000, and 1,000,000 km will provide a convenient sample for plotting. It was found that if narrower intervals of cell sizes were used the curves had a stair-step because there was often no difference in the count between, for example 10 and 15 km. These stair-steps simply add noise to the measurement. The density is calculated from the cell size, count, and the total number of points used as shown in (1). Cell size is the length of the sides of a square box, which is the cell that contains at least one point. Cell size is thought of as the scale, meaning extent, of the measurement.

$$Density = \frac{T}{Count^*(Cell\ Size)^2} \qquad (1)$$

where T = total number of known deposits.

Then a log–log plot of cell size versus count made in Excel with a trend line using a power function and providing the coefficient of explanation, R^2, will provide the fractal dimensions and coefficient of explanation as in Fig. 1A. The fractal dimensions are simply the exponent in the fitted power function (Carlson, 1991). Generally, these plots have two straight-line segments that indicate the fractal domains. In order to obtain a good trend line, it is necessary to decide what the minimum cell size is and where the transition occurs from one fractal domain to the next. With real data, it is possible to make cell sizes so small or large that no difference is observed. These measurements should be excluded. The trend lines can be found interactively by selecting a cell-size interval that maximizes the R^2. Typically, an R^2 value close to 0.99 can be obtained. If the interval includes cell sizes transitional to the other fractal domain, that is points near the crossover, the R^2 will decrease rapidly. Therefore, the best estimate of the fractal dimension is obtained where box counts near the crossover point are excluded. Thus, an iterative selection process that maximizes the R^2 does the curve fitting. This graphical approach to curve fitting avoids the edge problems encountered in traditional approaches, and with a bit of experience is fast.

As discussed in the paper, the crossover point appears to be roughly correlated with the size of the deposits. The exact crossover point between fractal domains can then be computed in Excel by finding the point at which the two power functions defined in the graph are equal. A similar definition of the fractal dimensions, which was referred to by Carlson (1991) as the probability density function, PDF, can be obtained by plotting the cell size versus the density. Cell size in this situation is interpreted to mean distance to the nearest known deposit. In the PDF, the fractal dimensions are 2 minus the exponent (Carlson, 1991). In both of these plots, two fractal domains are defined that are referred to here as local and regional scales. "Local" refers to the smaller cell sizes, so to the smaller extent. Each fractal dimension is thought to reflect processes functioning in the scale range for which that fractal dimension applies.

Integration for Number of Deposits

To apply the probability density function, PDF, to make an estimate of the number of deposits it is necessary to define the spatial area where the deposit can occur. In the discussion below, the permissive tracks as defined by the USGS national assessment (U.S Geological Survey National Mineral Assessment Team, 1998) are used. Then using the known deposit points from Singer, Berger, and Moring (2005), the distance to the nearest known deposit is computed with the Distance function in ArcGIS with a mask defined by the permissive areas. The distance in the PDF is interpreted as the distance to the nearest known mineral deposit, so it is necessary to integrate the appropriate PDF over space as measured by distance to the nearest known deposit. This integration of distance leads to a dependency on the shape of the permissive area.

¿From the Distance function in ArcGIS, a table with distance in kilometers and counts of cells at that distance can be generated and used in Excel for the integration. The integration equations that would be integrated from Fig. 1B are shown in (Equation 2).

For these calculations, a cell size of 100 m was used. Thus, a multiplier of 0.01 was used to

$$N = 0.4826^*(\text{Distance})^{-1.7885*}(0.01^*(\text{Count}))$$
/10 for distance less than break point
$$N = 0.0263^*(Dis\tan ce)^{-1.148*}(0.01^*(Count))$$
/10 for distance greater than break point \hfill (2)

where N = number of deposits,

$$\text{total number of deposits} = \sum_{Range\ of\ Distance} N,$$
number of undiscovered deposits = total number of deposits
−number of known deposits

convert the count of 100 × 100 m cells to area in square kilometers. For reasons that apparently have to do with how ArcGIS generates the distance to the nearest known deposits as a rectangular array of cells, a division by 10 of the area is required. There is an outstanding question about the influence of the cell size on these measurements. A cell size of 100 m was used here based on previous experience with the types of data used for the national assessments. A few tests gave slightly different estimates of total number of deposits where 1 km cells were used.

References

Agterberg FP, Cheng Q, Wright DF (1993) Fractal modeling of mineral deposits. In: Proceeding 24th APCOM symposium, vol 1. Canadian Institute of Mining, Metallurgy, and Petroleum Engineers, Montréal, Canada, pp 43–53
Barnsley MF (1988) Fractals everywhere. Academic Press, Inc., Boston, p 394
Blenkinsop TG, Sanderson DJ (1999) Are gold deposits in the crust fractals? A study of gold mines in the Zimbabwe craton. In: McCaffrey KJW, Lonergan L, Wilkinson JJ (eds) Fractures, fluid flow and mineralization: geological society. Special Publication 155, London, pp 141–151
Carlson CA (1991) Spatial distributions of ore deposits. Geology 19: 111–114
Cheng Q, Agterberg FP (1995) Multifractal modeling and spatial point processes. Math Geol 27 (7): 831–845
Herzfeld UC (1993) Fractals in geosciences—challenges and concerns. In: Davis JC, Herzfeld UC (eds) Computers in geology—25 years of progress. Oxford University Press, New York, pp 217–230
Johnston JD, McCaffrey KJW (1996) Fractal properties of vein systems and the variation of scaling relationships with mechanism. J Struct Geol 18: 349–358
King PB, Beikman HM (1974) Geologic map of United States: U.S. Geological Survey, 3 sheets, scale 1:2,500,000
Long KR, DeYoung J, Ludington SD (1998) Database of significant deposits of gold, silver, copper, lead, and zinc in the United States: U.S.Geological Survey Open-File Report 98-206, Online version 1.0 (http://geopubs.wr.usgs.gov/open-file/of98-206/)

Mandelbrot BB (1985) Self-affine fractals and fractal dimension. Phys Scr 32: 257–260
McCaffrey KJW, Johnston JD (1996) Fractal analysis of mineralized vein deposit: Curraghinalt gold deposit, County Tyrone. Mineral Deposita 31: 52–58
Mutschler FE, Ludington S, Bookstrom AA (1999) Giant porphyry-related metal camps of the world—a database: http://geopubs.wr.usgs.gov/open-file/of99-556/ and database
Raines GL, Sawatzky DL, Connors KA (1996) Great Basin geoscience data base: U.S.Geological Survey Digital Data Series 41
Roberts S, Sanderson DJ, Gumiel P (1998) Fractal analysis of Sn-W mineralization from Central Iberia-insights into the role of fracture connectivity in the formation of an ore deposit. Econ Geol 93 (3): 360–365
Roberts S, Sanderson DJ, Gumiel P (1999) Fractal analysis and percolation properties of veins. In: McCaffrey KJW, Lonergan L, Wilkinson JJ (eds) Fractures, fluid flow and mineralization: geological society. Special Publication 155, London, pp 7–16
Sanderson DJ, Roberts S, Gumiel P (1994) A fractal relationship between vein thickness and gold grade in drill-core from La Codosera, Spain. Econ Geol 89: 68–173
Singer DA (1994) Conditional estimates of the number of podiform chromite deposits. Nonrenew Resour 2 (2): 69–81
Singer DA, Berger VI, Moring BC (2005) Porphyry copper deposits of the world: database, map, and grade and tonnage models: U.S.Geological Survey Open File Report 2005–1060, p 9 and database, http://pubs.usgs.gov/of/2005/1060
Singer DA, Berger VI, Menzie WD, Berger BR (2005) Porphyry copper deposit density. Econ Geol 100: 491–514
Singer DA, Menzie WD (2005) Statistical guides to estimating the number of undiscovered mineral deposits: an example with porphyry copper deposits. In: Proceeding of IAMG'05, GIS Spatial Analysis, vol 2, pp 1028–1029
Stewart JH, Carlson JE (1978) Geologic map of Nevada: U.S. Geological Survey, 2 sheets, scale 1: 500,000
U.S.Geological Survey National Mineral Assessment Team (1998) Assessment of undiscovered deposits of gold, silver, copper, lead, and zinc in the United States: U.S. Geological Survey Circular 1178, p 29 and database

African Neoproterozoic Mineral Deposits and Pan African Metallogenesis

Christien Thiart and Maarten de Wit

Reprinted from *Natural Resources Research* DOI:10.1007/s11053-008-9063-z, when citing this article please use the DOI number.

Abstract Amalgamation of a number of continental fragments during the Late Neoproterozoic resulted in a united Gondwana continent. The time period in question, at the end of the Precambrian, spans about 250 million years between ~800 and 550 Ma. Geological activity focused along orogenic belts in Africa during that time period is generally referred to as 'Pan African'. We identify three age-related classes of tectonic terranes within these orogenic belts, differentiated on the basis of the formation-age of their crust: juvenile (e.g. mantle derived at or near the time of the orogenesis, ~0.5–0.8 Ga), Paleoproterozoic (~1.8–2.5 Ga), Archean (>2.5 Ga). We combine African mineral deposits data of these terranes on a new Neoproterozoic tectonic map of Africa. The spatial correlation between geological terranes in the belts and mineral occurrences are determined in order to define the metallogenic character of each terrane, which we refer to as their 'metallogenic fingerprint'. We use these fingerprints to evaluate the effectiveness of mobilisation ('recycling') of mineral deposits within old crustal fragments during Pan African orogenesis. This analysis involves normalization factors derived from the average metallogenic fingerprints of pristine older crust (e.g. Palaeoproterozoic shields and Archean cratons not affected by Pan African orogenesis) and of juvenile Pan African crust (e.g. the Nubian Shield). We find that mineral deposit patterns are distinctly different in older crust that has been remobilized in the Pan African belts compared to those in juvenile crust of Neoproterozoic age, and that the concentration of deposits in remobilised older crust is in all cases significantly depleted relative to that in their pristine age-equivalents. Lower crustal sections (granulite domains) within the Pan African belts are also strongly depleted in mineral deposits relative to the upper crustal sections of juvenile Neoproterozoic terranes. A depletion factor for all terranes in Pan African orogens is derived with which to evaluate the role of mineral

Christien Thiart
AEON – Africa Earth Observatory Network, and Departments of Statistical Sciences University of Cape Town, Private Bag, Rhodes Gift, Rondebosch 7701, Cape Town, South Africa,
e-mail: christien.thiart@uct.ac.za

Maarten de Wit
AEON – Africa Earth Observatory Network, and Departments of Geological Sciences, University of Cape Town, Private Bag, Rhodes Gift, Rondebosch 7701, Cape Town, South Africa,
e-mail: maarten.dewit@uct.ac.za

deposit recycling during orogenesis. We conclude that recycling of old mineral deposits in younger orogenic belts contributes, on average, to secular decrease of the total mineral endowment of continental crust. This could be of value when formulating exploration strategies.

Keywords Pan-African · mineral deposits · spatial coefficients · metallogenesis

1 Introduction

The formation of Gondwana occurred through the assembly of different crustal terranes (blocks) fragmented from an earlier supercontinent Rodina (e.g. de Wit et al., 1988, 2008; Unrug, 1997). Throughout a period of ~250 million years, between about 800–550 Ma, the Gondwana sector that later emerged as 'Africa' was widely affected by this period of amalgamation along four major orogenic systems, each in turn comprising a number of orogenic belts (Fig. 1; de Wit et al., 2008). During these Pan African (PA) accretion and collision events, some old continental blocks (e.g. Proterozoic shields and Archean cratons, or parts of these) that were caught up within the PA orogenic zones, became tectonically modified and metamorphosed in their form and structure, and often significantly rejuvenated through new, PA-age, magmatism. Large domains of these orogenic belts were also formed almost entirely from juvenile (relatively new) material derived directly from the mantle at that time, including major continental and oceanic arcs, oceanic plateaus, ophiolites of various tectonic settings, and other related igneous and metamorphic rocks (e.g. Stern, 1994). In addition, contemporaneous sediments derived from various sources were reworked in the PA orogenic belts. A significant number of juvenile Neoproterozoic mineral deposits formed throughout these PA belts, but existing deposits, inherited from the older continental terranes, were also incorporated. The fate of the latter is of interest to the general topic of crustal recycling: can we recognise signatures of older rejuvenated deposits, and if so, to what degree are such older deposits upgraded, diluted or even completely removed during subsequent orogenic processes, in our study example, Pan African events? This question is of fundamental importance for resource evaluation of different crustal segments of continental crust within (PA) orogenic belts, for understanding metallogenesis, and for estimating total mineral endowment of the crust in general (de Wit et al., 1999; Kesler and Wilkinson, 2008; Mabidi et al., 2007; Veizer et al., 1989).

In the past our analyses have focussed mostly on pristine crustal domains, those unmodified by later tectonic events. Here we build on our previous studies and investigate how we might evaluate mineral potential of old crust reworked during (PA) thermo-tectonism, and explore how we might differentiate this from mineral potential of juvenile crustal terranes within PA belts. Because PA orogenesis affected crust on a continental scale, it offers a unique opportunity to evaluate variations in large scale metallogenic processes in orogenic belts.

(A) # African Mineral Deposits and Pan African Metallogenesis

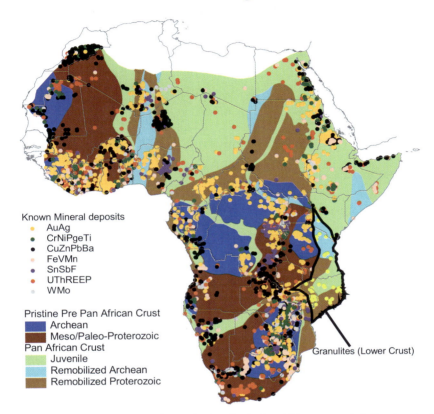

Fig. 1 (A) Mineral deposit map of tectonic domains in Pan African orogenic belts, showing locations of deposits used in this study. Data from mineral deposit GIS database housed at AEON (Thiart and de Wit 2000, 2006) http://www.uct.ac.za/depts/cigces/gondmin.htm). Tectonic subdivisions are modified from new Pan African Tectonic map of de Wit et al. (2008). (B). Schematic representation of the four major Neoproterozoic (Pan African) orogenic belts of Africa. Archean cratons (not shown, but see 1A) are embedded in stable shield areas

The mineral database on which we base our analysis contains ~8000 African deposits of which ~30% (2420) occur within the Pan African orogenic belts (Fig. 1). First, we characterise the metallogenic fingerprints of three different Pan African domains within these orogenic belts using selected elements and mineral deposit groups (c.f. de Wit and Thiart, 2005; Thiart and de Wit, 2006). The three Pan African tectonic domains we identify are: (1) Juvenile crust generated within a period of about 250 million years, at or near the time of the Pan African

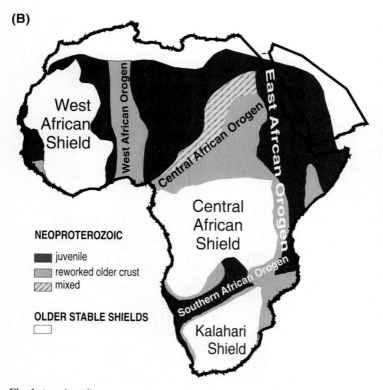

Fig. 1 (continued)

orogenesis (~0.5–0.8 Ga); (2) Crust inherited dominantly from Paleoproterozoic shields (~1.8–2.5 Ga); and (3) Crust inherited from old (Archean) cratonic fragments (>2.5 Ga). The geological and isotopic basis for these subdivisions are discussed in detail elsewhere (de Wit et al., 2008, and references therein), and are depicted in Fig. 1. Second, we compare the metallogenic fingerprints of the PA terranes to areas unaffected by PA orogenesis.

Metallogenic regions often retain some of their character over the time span of 3000 million years or more (de Wit et al., 1999, and references therein). In addition, in Africa, we have also quantified a distinct secular change in the degree of mineralisation during formation of its juvenile crust between the Mesoarchean (3500 Ma) and Pan African times (Mabidi et al., 2007). In principle then, we can evaluate if older terranes remobilised within younger orogenic zones retain their inherited mineral potential. We test this here by examining if older terranes within PA belts are relatively enriched or depleted in mineral deposits during PA orogenesis. We also test if there is any difference in mineralisation in regional lower crustal sections (granulite terranes) relative to upper crustal sections in the PA belts.

2 Data and Methods

The geological and mineral deposit data used for this study is incorporated in a GIS relational database housed at AEON, as described in detail elsewhere (Thiart and de Wit, 2000, 2006). This database classifies mineral deposits in five size groups, ranging from occurrences to very large deposits (world class). This is the dataset we explore in this study. Distinction is made in our database between mineral deposit types, and ~5% of the deposits contain information on grade and tonnage (production and reserve data), but this information is not exploited here. Rather, we restrict ourselves in this study to spatial distribution of deposits ('mineral hotspots'), in the crust, and do not quantify the total mineral endowment of this crust. The latter will be explored in future work.

The distribution of mineral deposits of the different types of PA terranes, as well as across older shields and cratons is shown in Fig. 1. Twenty one elements were selected for our analyses, and these were divided into seven element groups according to their geochemical affinities (e.g. lithophile, chalcophile, siderophile), as well as their relative abundance in the database. PA-age deposits of each of the seven element groups in the three age-differentiated terranes within the PA belts, and the total number of all these PA deposits (2420) are given in Table 1. This table also lists the number of deposits in the Archean cratons and Proterozoic shields not affected by the PA orogenesis, and those deposits within a distinct granulite facies region exposed in the PA mobile belt of East Africa assumed to represent exhumed lower PA crust (Fig. 1).

Previously we defined metallogenic fingerprints of fragments of continental crust through their spatial association with a combination of element groups, and applied this at three scales (cratons, continents, and super-continents; de Wit and Thiart, 2005). The measure of spatial association (normalized to area) we termed the spatial coefficient (Mihalasky and Bonham-Carter, 2001), r_{ij}. The spatial coefficient (r_{ij}) is based on the weight of evidence approach (Bonham-Carter, 1994), where

$$r_{ij} = \frac{N(T_i \cap D_j)/N(D_j)}{A(T_i)/A(T_\bullet)}, \tag{1}$$

in which $A(T_i)$ is the area of the ith terrane (T_i), and $A(T_\bullet)$ is the total area of all the terranes in the study ($A(T_\bullet) = \sum A(T_i)$). $N(D_j)$ is the total number of deposits in the jth element group (D_j). $N(T_i \cap D_j)$ represents the number of deposits of group D_j in terrane T_i.

The spatial coefficient [(1)] represents the proportion of deposits (say gold – j) of all the jth deposits (in geological domains) that occur in the specified geological terrane (i) per unit area of all similar geological terranes. The values of the spatial coefficient range from 0 to infinity; it is equal to 1 if there is no spatial association between a terrane and an element group (e.g. if the proportion of jth mineral is the same as the proportion of area occupied by the ith terrane). For values of $r_{ij} > 1$ (expected number of deposits is greater than by chance), there is a positive association between mineral j and geological terrane i; $r_{ij} < 1$ (fewer deposits expected than

Table 1 Number of mineral deposits with selected elements on Pan African rocks

	AuAg	CrNiPgeTi	CuZnPbBa	SnSbF	FeVMn	WMo	UThREE	Total	Area in km^2
Pan African age									
Remobilized Archean	72	7	15	37	9	12	36	188	1372574
Remobilized Proterozoic	247	19	127	35	26	6	46	506	4542788
Pan African Juvenile	693	119	462	75	95	31	191	1666	8912554
Pan African Lower crust	15	4	9	3	4		25	60	995698
Subtotal	1027	149	613	150	134	49	298	2420	15823615
Other age									
Archean	1718	433	426	197	398	331	177	3680	5141783
Meso/Paleo-Proterozoic	606	76	407	139	81	122	324	1755	7223013
Total	3351	658	1446	486	613	502	799	7855	28188411

by chance) indicates a negative association. Because all negative associations are compressed in the range from 0 to 1, and all positive associations fall in the range of 1 to infinity, we use the natural log of r_{ij} to eliminate this skewness. $\ln(r_{ij})$ is now a symmetric value around 0: positive associations are greater than 0, and negative associations are less than 0. We plot these values with their approximated standard error of $\ln(r_{ij})$ (Bonham-Carter et al., 1989).

3 Results

The natural log of the spatial coefficients between the element group and each geological terrane ($\ln(r_{ij})$) are given in Table 2, and plotted in Fig. 2 to illustrate the effects of Pan African remobilisation on the mineral endowment of various different types of terranes. Clearly, on average, Archean terranes within the Pan African belts are depleted relative to Archean cratons unaffected by PA tectonism (Fig. 2C). There is, for example, a strong positive spatial association for five of the element groups in the unaffected Archean cratons compared to the strong negative association for the Archean terranes remobilized in the PA belts (suggesting a much lower chance of finding such mineral deposits in these latter terranes). If we ignore the two strongly lithophile element groups (e.g. tin and uranium), the two Archean terranes plotted in Fig. 2C are sub-parallel to each other but separated by roughly $2\,(\ln r_{ij})$ units. This implies a significant depletion of deposits in Archean crust remobilised in PA belts. The smaller degree of relative depletion in the strong lithophile element groups (commonly concentrated in granites) might be intuitively expected because PA granites derived through melting of older crust occur frequently throughout the PA belts (e.g. Dada, 2008).

The metallogenic fingerprint for the unaffected Paleoproterozoic terranes oscillates around zero (no obvious spatial association) (Fig. 2B). By contrast that of the remobilized Proterozoic terranes is strongly negative (Fig. 2B). This negative association illustrates that there is a lower chance of finding mineral deposits in these remobilized terranes. If we ignore the value for tungsten (W) because of the large standard error, the two Proterozoic terranes plotted in Fig. 2B are sub-parallel to each other but are separated by roughly $1\,(\ln r_{ij})$ unit. Again, this implies a significant depletion of deposits in Proterozoic crust during PA remobilisation. To explore such depletion during remobilization within PA orogens in 3-D, we analysed a large section of granulite grade rocks of the southern part of the Neoproterozoic orogen in East Africa (ca 1 million km^2) from Tanzania to Mozambique. This region represents an exhumed section of mid-lower crust in one of the major PA orogens – i.e. the East African Orogen, de Wit et al., 2001, 2008) that is more depleted in mineral deposits than all other crustal terranes analysed in this study (Figs. 1A and 2A). Since this part of the belt comprises an ensemble of both Neoproterozic and ancient crustal domains (remobilised Archean and Proterozoic terranes), the small number of mineral deposits here support models that suggest granulite facies metamorphism and associated dehydration processes in lower crustal domains play a significant role

Table 2 Natural log of the spatial coefficient (r_{ij}) between element groups and Pan African rocks. Values below and in parentheses is standard error

	AuAg	CrNiPgeTi	CuZnPbBa	SnSbF	FeVMn	WMo	UThREE	Total
Pan African Age								
Remobilized Archean	−0.8181	−1.5211	−1.5463	0.4469	−1.1989	−0.7115	−0.0776	−0.7102
	(0.1179)	(0.3780)	(0.2582)	(0.1644)	(0.3333)	(0.2887)	(0.1667)	(0.0729)
Remobilized Proterozoic	−0.7823	−1.7194	−0.6070	−0.8055	−1.3349	−2.6015	−1.0293	−0.9170
	(0.0636)	(0.2294)	(0.0887)	(0.1690)	(0.1961)	(0.4082)	(0.1474)	(0.0445)
Pan African Juvenile	−0.4245	−0.5586	0.0105	−0.7173	−0.7130	−1.6332	−0.2796	−0.3993
	(0.0380)	(0.0917)	(0.0465)	(0.1155)	(0.1026)	(0.1796)	(0.0724)	(0.0245)
Pan African Lower crust	−2.0657	−1.7597	−1.7361	−1.7444	−1.6888		−0.1213	−1.5313
	(0.2582)	(0.5000)	(0.3333)	(0.5774)	(0.5000)		(0.2000)	(0.1291)
Other age								
Archean	1.0334	1.2830	0.4794	0.7985	1.2696	1.2850	0.1943	0.9433
	(0.0241)	(0.0481)	(0.0485)	(0.0712)	(0.0501)	(0.0550)	(0.0752)	(0.0165)
Meso/Paleo-Proterozoic	−0.3485	−0.7968	0.0939	0.1099	−0.6623	−0.0529	0.4590	−0.1370
	(0.0406)	(0.1147)	(0.0496)	(0.0848)	(0.1111)	(0.0905)	(0.0556)	(0.0239)

African Neoproterozoic Mineral Deposits

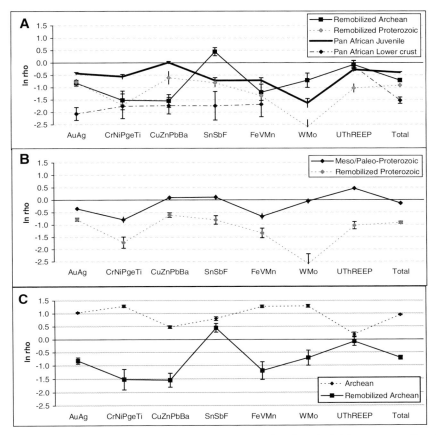

Fig. 2 Metallogenic fingerprints of Pan African tectonic domain and of older areas not affected by Pan African orogenesis. (**A**) All terranes in Pan Africa orogenic belts, (**B**) comparison between unaffected Proterozoic shields and the PA remobilized Proterozoic terranes, (**C**) comparison between unaltered Archean craton and remobilized Archean terranes. Error bars shown (unless smaller than symbols) represent one standard deviation. Note that values for older terranes are more stable (standard error smallest) than that of remobilized terranes. Also note that standard error is high for lower crustal section, because there are few deposits in this depleted section

in the depletion processes of ore forming elements in the lower crust during orogenesis (e.g. Cameron, 1988; Kerrich, 1989; Blenkinsop et al., 2004).

Within the PA orogenic belts, the total relative depletion in the older terranes and in lower continental crust appears, in general, greater than the total new mineral enrichment in the juvenile Pan African terranes (Fig. 2A). This implies that in the PA belts, depletion of mineral deposits in the older remobilized terranes is not balanced against enrichment (formation and preservation) of new mineral deposits in the juvenile terranes. Thus, in Africa, there is probably a reduction in the total number of mineral deposits in PA crust, relative to older crust, related to Neoproterozoic orogenesis. In turn this implies that recycling of African crustal mineral-riches back into the mantle is likely related to uplift and erosion of its former, and widespread, Pan African mountain systems (e.g. Burke et al., 2003). This is consistent with findings that the continental crust of the present Earth (with the exception of the strong lithophile element groups) is depleted in ore forming elements relative to that in the past (e.g. Mabidi et al., 2007).

4 Conclusion

Specific metallogenic regions often retain their metallogenic character over similar time spans even when subjected to later remobilisation (de Wit et al., 1999; Thiart and de Wit, 2006). In addition, distinct secular change in mineralisation during formation of juvenile crust over about 3000 million years, between the Mesoarchean (3500 Ma) and Pan African times (Mabidi et al., 2007), shows that the younger juvenile African terranes are less endowed with mineral deposits relative to older juvenile terranes. However, in this study we find that older terranes that were remobilized during PA orogenesis are depleted in the number of mineral deposits relative to equivalent-age terranes that were unaffected by PA orogenesis. Of these, lower crustal terranes affected by PA orogenesis are the most depleted crustal segments of all. Our new analyses of the existing mineral deposit data from PA terranes also indicates that during metallogenic processes in PA orogens, the number of mineral deposits in old remobilised terranes is depleted to a greater degree than the formation of new mineral deposits in juvenile sections of PA crust. This is likely related to erosion of upper crustal deposits within the PA orogenic belts. Therefore recycling of old mineral deposits in younger orogenic belts likely contributes, on average, to secular decrease of the total mineral endowment of continental crust.

In ending, we would like to re-emphasize some shortcomings in our analyses, such as a lack of distinction between deposit types. For example, in the base metal group (Cu, Zn, Pb, Ba) we do not distinguish MVT deposits from carbonate vein types or porphyry deposits. Differentiating between these (and/or their host rock types that may be susceptible to different recycling rates) may bear significantly on the results and our interpretations thereof. This will be tested in future work.

Acknowledgments We thank AngloGold Ashanti for stimulating this research. We are also grateful to Danny Wright and Neil Fordyce for their excellent reviews that helped improve our communication. We thank Graeme Bonham-Carter and Frits Agterberg for their encouragement and many interesting discussions about the statistics of mineral deposits. This is AEON contribution 41.

References

Blenkinsop TG, Kröner A, Chiwara V (2004) Single stage, late Archean exhumation of granulites in the Northern marginal Zone. Limpopo Belt, Zimbabwe and relevance to gold mineralization at Renco mine. South Afr J Geol 197: 337–395

Bonham-Carter GF (1994) Geographic information systems for geoscientists: modelling with GIS. Pergamon, Oxford, p 398

Bonham-Carter GF, Agterberg FP, Wright DF (1989) Weights of evidence modelling: a new approach to mapping mineral potential. In: Agterberg FP, Bonham-Carter GF (eds) Statistical applications in the Earth sciences. Geological Survey of Canada, Paper 89-9, pp 171–183

Burke K, MacGregor DS, Cameron NR (2003) Africa's petroleum systems: four tectonic 'Aces' in the past 600 million years. In: Arthur TJ, MacGregor DS, Cameron NR (eds) Petroleum geology of Africa: new themes and developing technologies. Geological Society, London, Special Publication 207, pp 21–60

Cameron EM (1988) Archean gold: relation to granulite formation and redox zoning in the crust. Geology 16: 109–112

Dada SS (2008) Proterozoic evolution of the Nigeria–Boborema province. In: Pankhurst RJ, Trouw RAJ, de Brito Neves BB, de Wit MJ (eds) West Gondwana: pre-Cenozoic correlations across the south Atlantic region. Geological Society, London, Special Publication 294, pp 121–136, doi: 10.1144/SP294.7

de Wit MJ, Bowring S, Ashwal LD Randrianasolo LG, Morel VPI, Rambeloson RA (2001) Age and tectonic evolution of Neoproterozoic ductile shear zones in southwestern Madagascar, with implications for Gondwana studies. Tectonics 20:1–45

de Wit MJ, Jeffery M, Bergh H, Nicolaysen L (1988) Geological map of sectors of Gondwana reconstructed to their disposition ~150 Ma, scale 1: 10,000,000. American Association Petroleum Geologists, Tulsa, Oklahoma

de Wit MJ, Stankiewicz J, Reeves C (2008) Restoring Pan-African-Brasiliano connections: more Gondwana control, less Trans-Atlantic corruption. In: Pankhurst RJ, Trouw RAJ, de Brito Neves BB, de Wit MJ (eds) West Gondwana: pre-Cenozoic correlations across the south Atlantic region. Geological Society, London, Special Publication 294, pp 399–412, doi: 10.1144/SP294.20

de Wit MJ, Thiart C (2005) Metallogenic fingerprints of Archean cratons. In: McDonald I, Boyce AJ, Butler IB, Herrington RJ, Polya DA (eds) Mineral deposits and Earth evolution. Geological Society, London, Special Publication 248, pp 59–70

de Wit MJ, Thiart C, Doucouré M (1999) Scent of a supercontinent: Gondwana's ores as chemical tracers, tin, tungsten and the Neoproterozoic Laurentia-Gondwana connection. J Afr Earth Sci 28: 35–51

Kerrich R (1989) Archean gold: relation to granulite formation or felsic intrusions? Geology 17: 1011–1015

Kesler S, Wilkinson, B (2008) Earth's copper resources estimated from tectonic diffusion of porphyry copper deposits. Geology 36: 255–258

Mabidi T, Thiart C, de Wit MJ (2007) Secular changes recorded in mineralization of African crust. J Afr Earth Sci 47: 88–94, doi:10.1016/j.jafrearsci.2006.12.002

Mihalasky MJ, Bonham-Carter GF (2001) Lithodiversity and itsspatial association with metallic mineral sites, Nevada great basin. Nat Resour Res 10: 209–226

Stern RJ (1994) Arc assembly and continental collision in the Neoproterozoic East African orogen: implications for the consolidation of Gondwana(land). Annual Rev Earth Planet Sci 22: 319–351

Thiart C, de Wit MJ (2000) Linking spatial statistics to GIS: exploring potential gold and tin models of Africa. South Afr J Geol 103: 215–230

Thiart C, de Wit MJ (2006) Fingerprinting the metal endowment of early continental crust to test for secular changes in global mineralization. In: Kesler SE, Ohmoto H (eds) Evolution of early Earth's atmosphere, hydrosphere and biosphere – constraints from ore deposits. Geol soc am memoir vol 198, pp 53–66, doi:10.1130/2006.1198(03)

Unrug R (1997) Rodinia to Gondwana. The geodynamic map of Gondwana supercontinent assembly. GSA Today 2: 2–6

Veizer J, Laznicka P, Jansen SL (1989) Mineralization through geological time: recycling perspective. Am J Sci 289: 484–524

On Blind Tests and Spatial Prediction Models

Andrea G. Fabbri and Chang-Jo Chung

Reprinted from *Natural Resources Research* DOI: 10.1007/s11004-008-9179-z, when citing this article please use the DOI number.

Abstract This contribution discusses the usage of blind tests, BT, to cross-validate and interpret the results of predictions by statistical models applied to spatial databases. Models such as Bayesian probability, empirical likelihood ratio, fuzzy sets, or neural networks, were and are being applied to identify areas likely to contain events such as undiscovered mineral resources, zones of high natural hazard, or sites with high potential environmental impact. Processing the information in a spatial database, the models establish the relationships between the distribution of known events and their contextual settings, described by both thematic and continuous data layers. The relationships are to locate situations where similar events are likely to occur. Maps of predicted relative resource potential or of relative hazard/impact levels are generated. They consist of relative values that need careful quantitative scrutiny to be interpreted for taking decisions on further action in exploration or on hazard/impact mitigation and avoidance. The only meaning of such relative values is their rank. Obviously, to assess the reliability of the predicted ranks, tests are indispensable. This is also a consequence of the impracticality of waiting for the future to reveal the goodness of our prediction. During the past decade only a few attempts have been made by some researchers to cross-validate the results of spatial predictions. Furthermore, assumptions and applications of cross-validations differ considerably in a number of recent case studies. A perspective for all such experiments is provided using two specific examples, one in mineral exploration and the other in landslide hazard, to answer the fundamental question: how good is my prediction?

Keywords Blind test · cross-validation · spatial prediction · favorability · mineral exploration · hazard

---

Andrea G. Fabbri
Università di Milano-Bicocca, Piazza della Scienza, 1, 20126 Milan, Italy and Vrije Universiteit, de Boelelaan 1087, 1081 HV Amsterdam, The Netherlands, e-mail: andrea.fabbri@unimib.it

Chang-Jo Chung
Geological Survey of Canada, 601 Booth Street, Ottawa, K1A 0E8 Canada,
e-mail: chung@NRCan.gc.ca

1 Introduction

Constructing a map is an exercise to capture the distribution of observed objects or measurements, in conveniently simplified terms, to comprehend and communicate their spatial significance. In the geosciences such objects are considered as indicators of processes of concern, either physical, or social or combined. During the past half a century mapped objects have become more factual, i.e., less interpretative, the capture increasingly less manual with the development of remote sensors, and their representation has turned from analog to digital. The latter is a fact that has encouraged the practice of overlaying different types of maps covering a same study area to derive specific themes with combined features of varying spatial continuity, connectivity and other desirable spatial properties.

Given the observed distribution of mineral occurrences of particular commodities, for instance, their typical (recognizable) spatial setting could be reflected as a "non-random" distribution over preferred combinations of mapping units partly revealing their environment of deposition. This was the starting point of many applications of statistical models to predict future discoveries in mineral exploration.

Some of the drawbacks of such experiments, however, have been that: (1) the established relationships are limited to the study area selected and its assumed time relevance, thus providing only relative characterization; (2) several quantitative models and associated assumptions can be used to express the spatial relationships in different ways; and (3) the quality of the prediction results is difficult or sometime impossible to assess.

Similar considerations can be made for the spatial prediction of natural hazards and of environmental impacts. There too, map data layers of thematic units and continuous values are overlaid and the resulting aggregated values are transformed and modeled to express the likelihood of hazard or of impact occurrence, so that study areas can indicate priority locations for detailed inspection in view of hazard prevention, avoidance or mitigation.

This contribution discusses how empirical measures of relative quality, termed cross-validation can be and should be obtained through blind tests, BT, of spatial predicted values from map overlays. Such measures require specific assumptions, scenarios and analytical strategies. For simplicity we will use the terms *event occurrence* to refer equally to resource discovery or hazard-impact, even if the former implies a process already occurred (deposition) while the latter a process to occur in the future.

2 Spatial Prediction Models and Associated Assumptions

Various statistical models can be used to establish spatial relationships between the distribution of point-like or patch-like events, and the mapping units in which they tend to occur. The latter can be categorical, such as lithologies or land uses, or can express continuous values, such as geophysical anomalies or terrain slope values.

The events are preferably instances of a specific type so that consistency of origin and context can be expected when relating them to the categorical units and continuous value maps.

Commonly used spatial prediction models are based on: (1) Bayesian Probability Theory's Joint Conditional Probability function and the Likelihood Ratio function, and its derived monotonic functions such as the Certainty Factor, and the Weights of Evidence; (2) Zadeh's Fuzzy Set membership function; and (3) Dempster-Shafer Evidential Theory's Belief Function. A mathematical unified framework for these models has been provided by Chung and Fabbri (1993), together with criteria to construct them and to estimate predicted values. One main assumption of the above spatial prediction models is that each map data layer provides "independent" evidence of favorable setting. A general term used for the models is Favorability Functions.

Additional assumptions that support the application of the models to predict further discoveries or future hazards, is a degree of similarity between the observed-constructed settings of the known event occurrences and those of the future ones. It is that "degree of similarity" that allows to extrapolation in space and possibly in time.

Another set of inherent assumptions is the causal relationships between the mapping units and the events. Such relationships are the result of expert's knowledge, i.e., of the opinion of scientists specialized on the commodities or the hazards. The experts are to provide guidance for the construction of the spatial databases and for their interpretation. Another general assumption is that the spatial database constructed for a study area sufficiently documents the above relationships so that the statistics obtained from it can be used to support the spatial prediction. Inherent to this assumption is a degree of uniformity of detail and consistency or "granularity" between the map data layers. Such layers in general consist of a mixture of categorical and continuous values.

3 Relative Indexes and Their Measures

The statistics from the spatial database are considered as partial evidence in favor or against the occurrence of events. The assumptions on the relevance of those statistics to represent the condition of future occurrences provide essentially a way to obtain a relative ranking of all units within a map and later of all points with aggregate values ranked from a set of overlaid map layers, based on a spatial prediction model. Using the model means that, given two separate points in a study area, one point will have a higher aggregate value than the other. The relative ranks are the only interpretable evidence obtainable from the model and the database. It is doubtful that the original scores have any direct meaning other that the relative ranking.

After constructing a "favorability function" as the spatial model, a relative potential level is estimated at every pixel by computing the score of the favorability function at every pixel in the study area. We will be using the term "potential" to refer to either "resource discovery" or to "hazard" to indicate the relative predicted index scores. These computed scores normally range from 0 to infinity. Because they express relative levels of potential, they can be replaced by ranks (or orders) instead of the actual scores. In a study area, suppose that there are *n* pixels. We expect to have *n* estimated scores, one at each pixel. These *n* values are sorted in decreasing order and replaced by their rank, ranging now from 1 to *n*. Dividing each rank by the number of pixels *n* standardizes the ranks. The resulting standardized ranks ranging from 1/*n* to 1 were termed as "predicted relative potential indices", or PRP indices; with the pixel having the highest PRP index being assigned the value 1, and the pixel having the lowest PRP index being assigned a value of 1/*n*. By plotting the PRP index at each pixel, a PRP map is generated. To illustrate the PRP index, let us consider a pixel with 0.95 as the index. It means that the pixels whose favorability function scores are greater than the score of the pixel with 0.95 as the PRP index cover 5% of the study area. We will later use the PRP indices to evaluate the prediction maps through "fitting-rate curves" and "prediction-rate curves" using cross-validation.

Simple ways to analyze rank statistics were discussed by Chung and Fabbri (2003) who have described several benefits of using such ranking procedure to generate the potential classes for a prediction map. For example, suppose that we wish to generate 100 equal-size prediction classes where each class covers 1% of the study area. Then such PRP indices obtained by the ranking procedure provide a useful tool. The 100 equal-size classes are generated in the following manner. Assign "Class 100" consisting of the pixels with the PRP indices larger than 0.99 and less than or equal to 1. The pixels in "Class 100" cover 1% of the study area. "Class 99" is assigned the pixels with the predicted potential indexes larger than 0.98 and less than or equal to 0.99. Similarly "Class 1" consists of the pixels with the PRP indices less than or equal to 0.01. For considering an appropriate number of potential classes, however, the meaningful number of classes depends on the quality of information available in the database and on the significance of the model used. To illustrate the relationship between computed favorability function values and corresponding PRP indices, a scatter plot can be used.

The first step to evaluate a prediction map is to compare the predicted potential indices of the pixels with the known occurrences of the events (note that these events were used to generate the prediction map) and such comparison generates the "fitting-rate curve" of the prediction map. Suppose that we have *m* known events. To produce the fitting rate curve of a prediction map, simply obtain *m* predicted potential indexes at *m* known events and then sort the *m* values in decreasing order; $(q_1, q_2, \ldots, q_m)$, where q_1 indicates the largest PRP index. We generate the following m pairs:

$$\{(1-q_1),\ 1/m\},\ \{(1-q_2)\ 2/m\},\ldots,\ \{(1-q_m), 1\}$$

The scatter plot of these *m* pairs constitutes the fitting-rate curve where the X-axis represents the portion of the study area assigned to a "potential" class and the Y-axis represents the proportion of the known events that have occurred within the assigned "potential" class. Such a fitting, however, only reflects how the classes discriminate between the settings identified using the distribution of the observed events, and does not necessarily reflect the distribution of future occurrences. For that, other techniques and assumptions are necessary, as we will see later on through the blind-test procedures.

4 How Good is the Predicted Relative Potential Index as a Predictor?

Potential indices as we have described, are to reflect not just the fitting to the prediction classes but the likelihood of future event occurrence, given the combined presence of the map unit data layers. Such a likelihood, however, is restricted so far to the distribution of the past events and the associated database of the study area. To study and interpret their effectiveness as predictors of future occurrences we have to assume a similarity of conditions between what has been observed in the past and what will occur in the future (e.g., new resource discoveries, new hazardous events, etc.). Saying that the past is the key to the future is only a starting point that means that we are willing to infer, given the observed trends, that within a given time interval and within a given study area, we expect as many events (or say twice as many, or some other larger or smaller number) as observed in the database. Alternatively but impractically, we could wait for a sufficiently long time and see how many events would occur with respect to our prediction.

Another more convenient empirical way to study the effectiveness of our initial prediction that used the distribution of all past events is to perform a cross-validation of the prediction results by partitioning the set of observed events into a prediction subset and a testing subset. With the former we can obtain a second prediction and the relative ranked equal area classes. With the latter we can verify how the testing subset of events is distributed across those new classes. A "good" prediction should show a strong clustering of testing events in the higher value classes. This second clustering, however, will be different from that of the fitting classes mentioned earlier. Nevertheless, it is a measure of its effectiveness.

The next section describes how to interpret the prediction results via blind tests.

5 What is a Blind Test and What is it Telling Us?

A BT is a fundamental way to cross-validate the results of spatial predictions empirically, short of waiting for events to occur. A BT is obtained, for instance, by pretending that part of the known events is unknown. It will be used to test the prediction results generated using the other part of the known events to establish the spatial relationships. The probability table estimated via BT depends entirely on how the partition is selected and the interpretation of the probability is again solely contingent and subject to the partition. The event partition can be obtained in various ways, depending on the quality and quantity of the event data available.

(i) *Only very few events are known that cannot be separated in different periods or sub-areas.* One event out of *m* is used to BT and all the *m-1* remaining ones are to generate a prediction to be cross-validated by the excluded event. Using the *m-1* remaining events, a prediction map based on the PRP indices is constructed. The PRP index is obtained at the pixel containing the excluded event. The operation is iterated *m* times, once for each of the *m* excluded events. This leads to generating *m* PRP indices showing how well each future event can be predicted, as the "next" event to occur, by all the other existing ones. To produce the "prediction-rate curve", simply sort the *m* indices in decreasing order; $(p_1, p_2, \ldots, p_m)$, where p_1 indicates the largest PRP index. We generate the following *m* pairs:

$$\{(1-p_1), 1/m\}, \{(1-p_2) 2/m\}, \ldots, \{(1-p_m), 1\}$$

The scatter plot of these *m* pairs constitutes the prediction rate curve where the X-axis represents the proportion of the study area assigned to a "potential" class and the Y-axis may be regarded as the representation of the proportion of the "future" events that have occurred within the assigned "potential" class. In contrast with the fitting-rate curve that only reflects how the classes discriminate between the settings identified using the distribution of the observed events, the prediction-rate curve reflects the distribution of future occurrences in the prediction map.

(ii) *Numerous events are known but cannot be separated in different periods or sub-areas.* A random half of the events is used to BT and the other random half is used to predict. The BT can be repeated inverting the role of the two random halves or it can be repeated several times with newly generated random halves, to obtain integrated statistics on the stability and reliability of the prediction results.

(iii) *Numerous events are known that can be separated in several temporal subgroups.* A BT is performed using the older set of events to predict and the younger set for testing. The statistics from the BTs provide true time prediction results. In such cases the quality of the prediction results should reflect the stability in time of the thematic map units subjected to transformation (e.g., climatic or human-induced) such as land use or land cover.

(iv) *Numerous events are known that can be separated in several spatial sub-groups.* The event distribution in some sub-areas is used to BT the results of a prediction obtained from an adjacent sub-area, in which the spatial relationships have been established. It means that the statistics on the relationships is obtained from one area and then is applied to another area. The BT is dependent on the similarity of conditions and events in the areas analyzed and compared. In some situations, for instance, the spatial data allow a combination of Strategies *(iii)* and *(iv)*.

(v) Other types of BTs can be performed. Changing the combination of thematic and continuous data layers or the quality-resolution, BT are obtained in experiments corresponding to one or more of the previous types of BT just described.

To produce the "prediction-rate curve" for *(ii), (iii), (iv) and (v)*, as described in *(i)*, we have to obtain PRP indices from the pixels that contain the observed events but that were not used in constructing the prediction map in BT. Supposing that we obtain k indices and sort them in decreasing order; $(p_1, p_2, \ldots, p_k)$, where p_1 indicates the largest predicted potential index. We generate the following k pairs:

$$\{(1-p_1), 1/k\}, \{(1-p_2) 2/k\}, \ldots, \{(1-p_k), 1\}$$

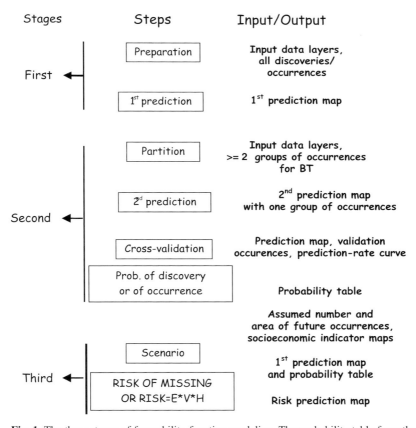

Fig. 1 The three stages of favorability function modeling. The probability table from the Second Stage, which depends entirely on how the partition for BT is made for validation, is the most critical piece of information to interpret the prediction results in the First Stage and to obtain the risk prediction map in the Third Stage, where E is element at risk, V is vulnerability and H is hazard. The term discovery is used interchangeably with the term occurrence to refer to exploration and to hazard/impact applications, respectively

The scatter plot of these k pairs constitutes the prediction rate-curve where the X-axis represents the proportion of the study area assigned to a "potential" class and the Y-axis may be regarded as the representation for the proportion of the "future" events, which occurred within the "potential" class. Performing BTs appears as a practical way of interpreting many aspects of prediction modeling: (1) quality of data layers (categorical and continuous), distribution of types of known events/discoveries, and expert's knowledge of the spatial database; (2) significance of the predicted relative PRP index maps; (3) effect of database partitioning in modeling; (4) comparisons of the results of different prediction models; and (5) assessment of scenarios for exploration or for risk analysis.

A general purpose strategy for favorability function predictive modeling is shown in Fig. 1 as an operational flowchart with three stages. The distribution of known discoveries or of the hazardous occurrences is used to establish their spatial relationship with the units of the input map data layers. The terms discoveries or occurrences are used interchangeably to refer to exploration or to hazard/impact applications. The interpretation of the probability table obtained depends entirely on how the partition for BT was made. To perform analyses according to the strategies listed earlier, iterations can be executed looping back one or more steps. In the next section examples of applications with and without BTs are discussed. Dedicated software based on cross-validation has been discussed by Fabbri et al. (2004).

6 Spatial Predictions with Event Partitions and Blind Tests

6.1 Some General Purpose Applications

Once a unified framework for favorability function models had been set up (Chung and Fabbri, 1993) and applications of various models were developed, it became evident that to interpret the results of predictions, either mineral potential maps or hazard maps, empirical tests were necessary to obtain scientific measures of success and decision values of the prediction results. BTs were used, for instance, in cross-validations for the following purposes:

- assessment of predictive power of landslide hazard (Chung et al., 1995); a first application of BT to interpret the "goodness" of spatial prediction results;
- comparisons of the performance of different prediction models, and their integration with expert's knowledge (Chung and Fabbri, 1998, 1999);
- estimation of probability of mineral discovery by an operational unit area for exploration (Chung et al., 2002a);
- separation of influential and non influential data layers in landslide hazard predictions (Chung et al., 2002b); it enabled to identify predictions of greater reliability due to the higher empirical support to characterize the settings of landslide occurrence;

- assessment of uncertainty in landslide hazard predictions (Chung, 2006); iterating many times the selection of random halves of the events, prediction-rates were obtained to express the level of uncertainty associated with the predicted classes;
- comparisons between spatial, temporal and spatial/temporal predictions (Chung and Fabbri, 2008);
- cost-benefit analysis of prediction-rate curves of landslide hazard (Chung and Fabbri, 2003); a ratio of effectiveness was applied to identify the most reliable parts of the prediction-rate curves;
- landslide risk assessment via probability of occurrence estimation (Chung and Fabbri, 2004; Chung et al., 2005a); the introduction of socioeconomic indicator maps led to the assessment of landslide risks to people, infrastructures and valuable land uses.

6.2 Two Examples of BT Strategies

To clarify in some detail the usefulness of BT, one recent application of spatial prediction modeling in mineral exploration, with only six known discoveries is discussed, followed by a second application to landslide hazard for which 92 known occurrences are used. The two BT strategies are different so are the results obtained and their significance.

A spatial database for diamond exploration in the Lac de Gras area of the Northwest Territories, in Canada, was used by Chung et al. (2005b) and Chung and Fabbri (2005) to obtain the prediction-rate curves shown in Fig. 2. The study area covers 34.6 km × 22.9 km (692 × 450 pixels of 50 m resolution) and contains six diamondiferous kimberlite ore bodies (Beartooth, Panda, Koala, Koala North, Fox and Misery). Additionally, fifteen kimberlites with only micro-diamonds were known. Radiometric and magnetic sensor maps interpolated from parallel flights, proximity maps to faults and dikes (as continuous data layers) and the presence of two indicator minerals, chromium-spinel and chromium-diopside, were used in the study. In addition, a bedrock lithology map (categorical data layer) was employed to characterize the spatial associations of the ore bodies and of the other kimberlites with micro-diamonds.

A fuzzy set prediction model based on the likelihood ratio function was instrumental to obtain and interpret the prediction maps following strategy (i) in Sect. 6. A first prediction map was obtained using the locations of all the six deposits. It was then interpreted with the prediction table estimated from the cross-validation procedures using six blind tests. Six more prediction maps were so obtained from the BT. Figure 2 shows parts of the prediction-rate curve in light gray with circles from the latter six prediction maps. For a comparison, two additional experiments in which different inputs were performed: (1) instead of seven data layers, only one data layer, the magnetic total field, with the six ore bodies was used in an additional BT, using the same strategy (i) in Sect. 6, to study the effects of input data

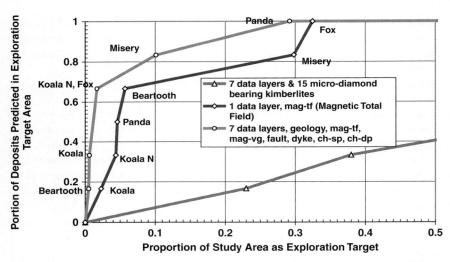

Fig. 2 Example of prediction-rate curves obtained from cross-validation in the Lac de Grass area, Northwest Territories, Canada. It was obtained using strategy (i) in Sect. 6. The cumulative plots allow to compute the probability of discovery of a new deposit location within the high potential area of the corresponding prediction maps, not shown here (Chung et al., 2005b). Description is in the text. Vertical gradient is vg, total field is tf, chromite is ch, spinel is sp, and diopside is dp

layers, and (2) the same seven data layers in the earlier BT, but using the 15 kimberlites with micro-diamonds instead of the six ore bodies, to test whether kimberlites with micro-diamonds can "predict" the locations of the six ore bodies. The cross-validation results were also plotted in Fig. 2.

As discussed in Chung et al. (2005b), even without seeing the 13 prediction maps generated, we can compare in Fig. 2 the prediction results. The comparison is made by considering the area proportion of the higher prediction classes containing the ore bodies, each predicted as "next" to occur by the other five, as the light gray with circles and the dark gray with diamonds prediction-rate curves. Obviously, the prediction of the six ore bodies by the locations of the fifteen kimberlites with micro-diamonds is poor! The BT shows that statistically the two types of kimberlites have different characterizations. It suggests that the locations of kimberlites with micro-diamonds do not provide any useful information to locate undiscovered ore bodies in this case study. In a second application, to hazard modeling, a greater number of known occurrences allowed a different strategy to be selected.

A spatial database for landslide hazard studies was constructed for the Fanhões-Trancão area, north of Lisbon in Portugal. The study area is $13.3\,km^2$. Detailed geologic-geomorphologic mapping at 1:2,000 identified 92 shallow translational slides. They were compiled and digitized into a 5m × 5m resolution spatial database consisting of digital images of 760 × 700 pixels. The causal factors (i.e., related to the occurrence of landslides) are: continuous data layers, i.e., elevation, slope angle, aspect angle obtained from the digital elevation model (DEM), and categorical ones, i.e., geology map, surficial deposit map and land-use map.

The 92 landslides in the study area consist of 43 pre-1980 landslides and of the remaining 49 post-1980 landslides. The region has been the focus of numerous geomorphologic analyses for hazard zonation by Zêzere et al. (2004).

A landslide hazard (potential) prediction map of the Fanhões-Trancão area, Portugal was obtained by Chung and Fabbri (2008), using the Fuzzy Set membership function of the Likelihood Ratio Function. The same function has been used in the other prediction experiments. Input data were the locations of the polygons of the 92 shallow translational landslides and the six geomorphologic and topographic map layers. In that application, the number of the 92 landslide polygons that fell into each of 200 hazard classes was counted. Each class corresponded to 0.5% of the study area. To be counted within a class, at least 50% of the pixels in a landslide polygon must be included in the class. The counts are weighted by the numbers of pixels in the polygons. Weighted counts of the landslide polygons form the "fitting-rate table" or curve that was plotted as the gray line with triangles in Fig. 3 with the horizontal axis representing the proportion of the study area predicted as hazardous, and the vertical axis showing the cumulative proportion of landslides falling within each class. A second experiment generated another prediction map using only 43 pre-1979 landslides and its fitting-rate curve is also shown in Fig. 3, falling below the previous fitting-rate curve based on all the 92 landslides. The third curve in the illustration is the prediction-rate curve from the latter experiment that provides a measure of "goodness" of the classes obtained in the two preceding predictions using the time partition of the landslide occurrences. Strategy (iii) of Sect. 6 was used in this experiment. Here the assumption was made that the 49 post-1980

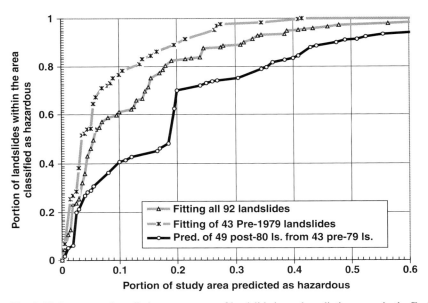

Fig. 3 Fitting-rate and prediction-rate curves of landslide hazard prediction maps in the Fanhões-Trancão area, Lisbon, Portugal (modified after Chung and Fabbri, 2008)

landslides are unknown and represent the future occurrences during a 25-year period (1980–2004). Additionally, we assumed that the prediction rate obtained represents the prediction power of the first prediction that used all the 92 occurrences for the next 25 years, i.e., the period 2005–2030.

The 10% of the study area with the highest predicted values (Fig. 3) corresponds to a prediction rate of 41% whereas the fitting rates are 61 and 77%, respectively. The latter two would overestimate the "goodness" of the prediction. They only indicate the "goodness" of fit between the landslides and the causal factors.

In another experiment, the study area was divided into two mutually exclusive sub-areas, the left region and the right region, as in strategy (iv) of Sect. 6. An experiment of this type would enable the similarity of geomorphologic settings or of climatic conditions to be tested. The left region contains 38 landslides (13 pre-1979 and 25 post-1980) and the right region includes 54 landslides (30 pre-1979 and 24 post-1980). Lower prediction-rate curves are compared (Fig. 4) to the prediction-rate curve from Fig. 3. There the previous prediction of the 49 post-1980 from the 43 pre-1979 landslides is compared with two spatial predictions in the right half using the landslides from the left half region and vice versa. Corresponding values for the 10% of highest predicted classes are 41% versus 37%. A mosaic of two prediction images is the result of this validated prediction image.

An extensive discussion of these and more experiments can be found in Chung and Fabbri (2008), who also combined strategies *(iii)* and *(iv)* that provided even lower prediction-rate values.

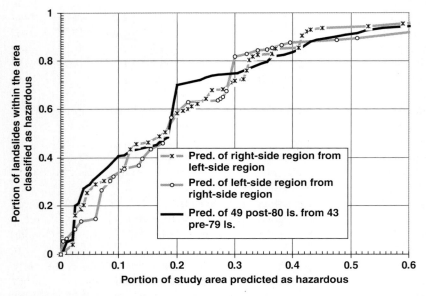

Fig. 4 Some prediction-rate curves of landslide hazard prediction maps in the Fanhões-Trancão area, Lisbon, Portugal (modified after Chung and Fabbri, 2008)

Clearly, all the above mentioned characteristics of "goodness" of the prediction images generated would be totally unknown without cross-validation via BT. Consequently, the BTs lead to considerations and introspections on the similarity of occurrences in time, of settings in time and in space, of comparability between adjacent study areas, between prediction models, and on how to use the prediction-rate values for estimating the probabilities of occurrences for each class or for each pixel. Far from trivial consequences follow the use of BT!

7 Considerations on Recent Spatial Predictions in the Geosciences

Having explored the information that must be extracted from spatial databases by BT of the prediction results, it should be instructive to consider a number of research papers in spatial modeling that would greatly benefit from BT and/or from more extensive applications of BT. Since cross-validations of spatial prediction results have been initially proposed (Chung et al., 1995; Chung and Fabbri, 1999) to interpret the predictions and for a variety of related applications, some of these listed in Sect. 6.1, relatively few examples of BT can be found either in mineral exploration or in natural hazard studies.

In the past 12 years or so interest in empirical validation or BT for prediction modeling in mineral exploration has varied from complete absence to considerable concern. However, there does not seem to be a consistent systematic or standardized approach to the application of cross-validation techniques. For instance, the evaluation of spatial modeling for epithermal gold deposit prediction by Raines (1999) rightly saw the prediction results as a "relative ordinal rank" but no BT was reported. The separation of favorability values into favorable, permissive, and non permissive was obtained by identifying breaks in the cumulative area ranks. That corresponds to using the fitting rates of the deposits used to predict and not to the prediction-rates from a cross-validation.

A different approach is the one by Singer and Kouda (1999) who compared several probabilistic models in the prediction of mineral deposits. They analyzed a test data set of 15 volcanic hosted massive sulfide deposits in a study area with 23 binary map data layers in the Province of Manitoba, Canada. Considered as wise by the authors was to perform independent validation tests by dividing the entire study area in two parts, one for predictive modeling and the other for validation. A random subset of 8 of the 15 deposits was selected together with a randomized half of the 6460 unique-condition polygons covering the study area and containing the 8 modeling deposits. The other half contained the remaining 7 deposits. Predictions were compared in terms of correctly classified polygons as deposit polygons or as barren polygons. Interestingly, they observed that very few deposits were correctly recognized in the independent tests whereas in the initial prediction modeling a high percentage had been recognized. Those authors made efforts to discuss in depth the pros and cons of the methods used, including the loss of information caused by binarizing all map data even when continuous. Nevertheless, also in that case, their analyses could be further expanded by applying strategy (i) of Sect. 5 (i.e., the take-one-out procedure) also used for the application described in Fig. 2.

An illustrative instance of a successful application is the one by Cheng (2004), who applied spatial modeling to predict the potential distribution of artesian aquifers in the Oak Ridge Moraine study area, near Toronto, Canada. As training points for modeling he used the spatial distribution of 353 wells with water level above the surface. Binary expressions of surficial geology map, distance from thick drift layers, distance from the Oak Ridge Moraine, and distance from steep slope zones were used as evidential data layers. Buffer zones with unequal intervals were generated to obtain binary units from distance maps. Purpose was the identification of combinations of conditions to reduce the prediction areas of having flowing wells by two thirds by generating a posterior probability map. BT of the results was not described, nevertheless the application would appear promising even if it cannot be certified how much so. It can be suggested that, the use of strategy (ii) of Sect. 5, and the repetition of the analysis, say 30–40 times, with new random half partitions of the training and validation points, would provide empirical means to interpret the "goodness" of the relative posterior probability value ranking obtained. In addition, a comparison of the 30–40 results would help to assess their robustness. The applications considered are just used here to exemplify the likely benefits of BT even in innovative and successful contributions, independently from the prediction models used.

Other more recent works dealt with problems such as the assessment of the quality of the prediction results (Porwal et al., 2003a, b), and the comparison of different predicting methods and models when analyzing the same data set (de Quadros et al., 2006; Brown et al., 2003; and Porwal et al., 2006a, b). Much of the emphasis in those works, however, was more on experimenting with new advanced techniques than on the interpretation of the significance of their application results. In addition, the strategies and specific assumptions of those cross-validations techniques were so different that it is not possible to evaluate or compare their usefulness in more general experiments or situations. For instance, lumping together fitting and prediction rates complicates the evaluation of the prediction quality. Thresholds to transform multi-value prediction maps into binary or tertiary maps are likely to weaken the cross-validation. In addition, some cross-validation experiments appeared limited to the training of classifiers and not directed to interpret the final prediction results.

Applications of spatial prediction models to natural hazard show a similar trend in the last few years. For instance in a special issue of Natural Hazards there are contributions without validation of prediction results (e.g., Corominas et al., 2003), one in which validation has been avoided, in favor of fitting curves, with the argument of unavailability of the time of occurrence of landslides in the database (van Westen et al., 2003), three studies in which validation was considered as integral part of the interpretation of hazard predictions (Santacana et al., 2003; Remondo et al., 2003a, b), and two more studies in which validation was used to explore and compare prediction powers or to eliminate misunderstandings on perceived obstacles to spatial predictive modeling (Chung and Fabbri, 2003; Fabbri et al., 2003).

Indicative of the degree of confusion now remaining about validation in spatial prediction modeling, is a recent collection of papers on spatial modeling in GIS. In it four contributions deal with prediction of hazards (landslides) or vulnerability (aquifer), and six with the prediction of natural resources (metals, aggregates and soils). All contributions claim to perform validations of modeling results; however, entirely different strategies are followed and assumptions made. Some approaches use fitting-rate curves (success rates) to identify "natural breaks" in them and obtain interpretable classes (Arthur et al., 2007; Masetti et al., 2007; Poli and Sterlacchini, 2007; Behnia, 2007; and Nelson et al., 2007). Generally weak comparisons are made of different prediction results by using either too few classes or too few occurrences to verify limited numbers of predictions (e.g., Nykanen and Ojala 2007; Coolbaugh et al., 2007; and Tissari et al., 2007). No effective validation of the prediction results appears in those contributions. Robinson and Larkins (2007) provide the only instance of prediction-rate curves in a diagram with proportions of sites correctly predicted (sensitivity) on the vertical axis and the cumulative area fraction (of study area) on the horizontal axis. Following a technique applied by Begueria (2006), they use a function of the area under the curve to establish the quality of the model prediction results. No further discussion is provided of the significance of such curve pattern in prediction modeling.

Applications that seem to lead to a more consistent approach of BT in mineral exploration are those by Chung (2003), Harris et al., 2003, Agterberg and Bonham-Carter (2005), and Skabar, (2005). Recent works on landslide hazard based on cross-validations are the ones by Zêzere et al. (2004) and by Lee et al. (2006). In natural hazard studies the approach by Chung (2006) and Chung and Fabbri (2003, 2004, and 2008) are targeting a more consistent way to use cross-validation techniques to estimate probabilities of occurrence of hazardous events.

8 Concluding Remarks

We have discussed how in spatial prediction modeling only relative ranks can be obtained using prediction models and their assumptions. We have dealt with the problem of assessing the "goodness" of the prediction results via a variety of empirical blind tests. A three-stage strategy for favorability function modeling has been proposed for which dedicated software is available that is soundly based on cross-validation. Examples of general purpose spatial prediction were listed, followed by two applications of BT that use prediction-rate curves to interpret the prediction results and proceed with the estimation of probabilities of occurrence. A number of recent applications were pointed out in which varying degrees and strategies of validation were attempted, while other ones seem to use ad hoc scenarios of limited effectiveness. Some additional applications appeared to potentially lead to a standardization of validation techniques.

A few recommendations can now be made for further research. In order to establish standards to interpret and compare the results of spatial predictions, three initiatives must be initiated in the geosciences: (1) identify one or two spatial databases to be distributed and analyzed by many researchers with different models to achieve agreement on how to construct BTs; then (2) organize a special meeting on the standardization of validation strategies; and (3) focus on representing and assessing by BT the uncertainties associated with the prediction results. The authors of this contribution are committed to the last initiative.

There is now a wealth of different prediction methods and many applications have been attempted, however, scientific progress at present is perhaps needed more in assessing the significance and stability of the predictions obtained than in devising additional ways to establish spatial relationships with sophisticated new prediction models whose effectiveness may not be easily evaluated.

Acknowledgments Thank to Prof. Carlos Roberto de Souza Filho, University of Campinas, Brazil, Dr. Graeme F. Bonham-Carter, Geological Survey of Canada, and two anonymous reviewers for helpful suggestions to improve the manuscript.

References

Agterberg FP, Bonham-Carter GF (2005) Measuring the performance of mineral potential maps. Nat Resour Res 14 (1): 1–17

Arthur JD, Wood HAR, Baker AF, Cichon JR, Raines GL (2007) Development and implementation of a Bayesian-based aquifer vulnerability assessment in Florida. Nat Resour Res 16 (2): 93–107

Begueria S (2006) Validation and evaluation of predictive models in hazard assessment and risk management. Nat Hazards 37: 315–329

Behnia P (2007) Application of Radial Basis Functional Link Networks to exploration for Proterozoic mineral deposits in central Iran. Nat Resour Res 16 (2): 147–158

Brown W, Groves D, Gedeon T (2003) Use of fuzzy membership input layers to combine subjective geological knowledge and empirical data in a neural network method for mineral-potential mapping. Nat Resour Res 12 (3): 183–200

Cheng Q (2004) Application of weights of evidence methods for assessment of flowing wells in the Greater Toronto Area, Canada. Nat Resour Res 13 (2): 77–86

Chung CF (2003) Use of airborne geophysical surveys for constructing mineral potential maps. Econ Geol Monogr 11: 879–891

Chung CF (2006) Using likelihood ratio functions for modeling the conditional probability of occurrence of future landslides for risk assessment. Comput Geosci 32 (8): 1052–1068

Chung CJ, Bonin D, Fannin RJ, Fabbri AG, Journeay M (2006) Uncertainty in predicting landslide hazard in the Coquitlam reservoir area, B.C., Canada. In: Proceedings of IAMG'06, Liege, Belgium, 3–8 Sept 2006, Session S09-07, p 4, CD

Chung CF, Fabbri AG (1993) Representation of geoscience data for information integration. J Nonrenew Res 2 (2): 122–139

Chung CF, Fabbri AG (1998) Three Bayesian prediction models for landslide hazard. In: Buccianti A (ed) Proceedings of international association for mathematical geology 1998 annual meeting (IAMG'98), Ischia, Italy, 3–7 October 1998, pp 204–211

Chung CF, Fabbri AG (1999) Probabilistic prediction models for landslide hazard mapping. Photogramm Eng Remote Sens 65 (12): 1389–1399

Chung CF, Fabbri AG (2003) Validation of spatial prediction models for landslide hazard mapping. In: Chacon J, Corominas J (eds) Special issue on landslides and GIS. Nat Hazards 30 (3): 451–472

Chung CF, Fabbri AG (2004) Systematic procedures of landslide hazard mapping for risk assessment using spatial prediction models. In: Glade T, Anderson MG, Crozier MJ (eds) Landslide hazard and risk. John Wiley & Sons, New York, pp 139–174

Chung CF, Fabbri AG (2005) On mineral potential maps and how to make them useful. In: Cheng Q, Bonham-Carter G (eds) GIS and spatial analysis, proceedings of IAMG '05, vol 1, Toronto, Canada, 21–26 August 2005, pp 533–538

Chung CF, Fabbri AG (2008) Predicting future landslides for risk analysis – spatial models and cross-validation of their results. Geomorphology 94 (3–4): 438–452

Chung CF, Fabbri AG, Chi KH (2002a) A strategy for sustainable development of nonrenewable resources using spatial prediction models. In: Fabbri AG, Gáal G, McCammon RB (eds) Geoenvironmental deposit models for resource exploitation and environmental security. Kluwer, Dordrecht, pp 101–118

Chung CF, Fabbri AG, Jang DH, Scholten HJ (2005a) Risk assessment using spatial prediction model for natural disaster preparedness. In: van Oosterom P, Zlatanova S, Fendel EM (eds) Geo-informastion for disaster management. Springer, Berlin, pp 619–640. Proceedings of Gi4DM, the first symposium on geo-information for disaster management, Delft, Netherlands, 21–23 March 2005, pp 619–640

Chung CF, Fabbri AG, van Westen CJ (1995) Multivariate regression analysis for landslide hazard zonation. In: Carrara A, Guzzetti F (eds) Geographical information systems in assessing natural hazards. Kluwer Academic Publishers, Dordrecht, pp 107–133

Chung CF, Harris JR, Keating P, Kjarsgaard B, Parsons S (2005b) Preliminary report on developing a mineral potential map for diamond exploration in Lac de Gras area, NWT. Unpublished manuscript

Chung CF, Kojima H, Fabbri AG (2002b) Stability analysis of prediction models for landslide hazard mapping. In: Allison RJ (ed) Applied geomorphology: theory and practice. John Wiley and Sons, Ltd., New York, pp 3–19

Coolbaugh MF, Raines GL, Zehner E (2007) Assessment of exploration bias in data-driven predictive models and the estimation of undiscovered resources. Nat Resour Res 16 (2): 199–207

Corominas J, Copons R, Vilaplana JM Altimir J, Amigó J (2003) Integrated landslide susceptibility analysis and hazard assessment in the principality of Andorra. In: Chacon J, Corominas J (eds) Special issue on landslides and GIS. Nat Hazards 30 (3): 421–435

de Quadros TFP, Koppe JC, Strieder AJ, Coste JFCL (2006) Mineral potential mapping: a comparison of weights-of-evidence and fuzzy methods. Nat Resour Res 15 (1): 49–65

Fabbri AG, Chung CF, Cendrero A, Remondo J (2003) Is prediction of future landslides possible with a GIS? In: Chacon J, Corominas J (eds) Special issue on landslides and GIS. Nat Hazards 30 (3): 487–499

Fabbri AG, Chung CF, Jang DH (2004) A software approach to spatial predictions of natural hazards and consequent risks. In: Brebbia CA (ed) Risk Analysis IV. WIT Press, Southampton, Boston, pp 289–305

Harris DV, Zurcher L, Stanley M, Marlowe J, Pan G (2003) A comparative analysis of favorability mapping by Weights of Evidence, Probabilistic Neural Networks, Discriminant Analysis and Logistic Regression. Nat Resour Res 12 (4): 241–255

Lee S, Ryu JH, Lee MJ, Won JS (2006) The application of artificial neural networks for landslide susceptibility mapping at Jangshung, Korea. Mat Geol 38 (2): 199–220

Masetti M, Poli S, Sterlacchini S (2007) The use of Weights-of-Evidence modeling technique to estimate the vulnerability of groundwater to nitrate contamination. Nat Resour Res 16 (2): 109–119

Nelson EP, Connors KA, Suárez CS (2007) GIS-based slope stability analysis, Chuquicamata open pit copper mine, Chile. Nat Resour Res 16 (2): 171–190

Nykänen V, Ojala VJ (2007) Spatial analysis techniques as successful mineral-potential mapping tools for orogenic gold deposits in the northern Fennoscandian Shield, Finland. Nat Resour Res 16 (2): 85–92

Poli S, Sterlacchini S (2007) Landslide representation strategies in susceptibility studies using Weights-of-Evidence modeling technique. Nat Resour Res 16 (2): 121–134

Porwal A, Carranza EJM, Hale M (2003a) Knowledge-driven and data-driven fuzzy models for predictive mineral potential mapping. Nat Resour Res 12 (1): 1–25

Porwal A, Carranza EJM, Hale M (2003b) Artificial neural networks for mineral-potential mapping: a case study from Aravalli province, western India. Nat Resour Res, 12 (3): 155–171

Porwal A, Carranza EJM, Hale M (2006a) Bayesian network classifiers for mineral potential mapping. Comput Geosci 32: 1–16

Porwal A, Carranza EJM, Hale M (2006b) A hybrid fuzzy weights-of-evidence model for mineral potential mapping. Nat Resour Res 15 (1): 1–14

Raines GL (1999) Evaluation of weights of evidence to predict epithermal-gold deposits in the great basin of the western United States. Nat Resour Res 8 (4): 257–276

Remondo J, González-Díez A, Díaz de Terán JR, Cendrero A (2003a) Quantitative landslide susceptibility models by means of spatial data analysis techniques; a case study in the lower Deva valley, Guipuzcoa (Spain). In: Chacon J, Corominas J (eds) Special issue on landslides and GIS. Nat Hazards 30 (3): 267–279

Remondo J, González-Díez A, Díaz de Terán JR, Cendrero A, Fabbri AG, Chung CF (2003b) Validation of landslide susceptibility maps; examples and applications from a case study in Northern Spain. In: Chacon J, Corominas J (eds) Special issue on landslides and GIS. Nat Hazards 30 (3): 437–449

Robinson GR Jr, Larkins PM (2007) Probabilistic prediction models for aggregate quarry siting. Nat Resour Res 16 (2): 135–146

Santacana N, Baeza C, Corominas J, de Paz A, Marturiá J (2003) A GIS-based multivariate statistical analysis for shallow landslide susceptibility mapping in La Pobla de Lillet Area (Eastern Pyrenees, Spain). In: Chacon J, Corominas J (eds) Special issue on landslides and GIS. Nat Hazards 30 (3): 281–295

Singer DA, Kouda R (1999) A comparison of weight-of-evidence method and probabilistic neural networks. Nat Resour Res 8 (4): 287–298

Skabar AA (2005) Mapping mineralization probabilities using multilayer perceptrons. Nat Resour Res 14 (2): 109–123

Tissari S, Nykänen V, Lerssi J, Kolehmainen M (2007) Classification of soil groups using Weights-of-Evidence method and RBFLN-neural nets. Nat Resour Res 16 (2): 159–169

van Westen CJ, Rengers N, Soeters R (2003) Use of geomorphological information in indirect landslide susceptibility assessment. In: Chacon J, Corominas J (eds) Special issue on landslides and GIS. Nat Hazards 30 (3): 399–419

Zêzere JL, Reis E, Garcia R, Oliveira S, Rodrigues ML, Vieira G, Ferreira AB (2004) Integration of spatial and temporal data for the definition of different landslide hazard scenarios in the area north of Lisbon (Portugal). Nat Hazards Earth Syst Sci 4: 133–146

Strip Transect Sampling to Estimate Object Abundance in Homogeneous and Non-Homogeneous Poisson Fields: A Simulation Study of the Effects of Changing Transect Width and Number

Timothy C. Coburn, Sean A. McKenna and Hirotaka Saito

Reprinted from *Mathematical Geosciences* DOI: 10.1007/s11004-008-9147-7, when citing this article please use the DOI number.

Abstract This paper investigates the use of strip transect sampling to estimate object abundance when the underlying spatial distribution is assumed to be Poisson. A design- rather than model-based approach to estimation is investigated through computer simulation, with both homogeneous and non-homogeneous fields representing individual realizations of spatial point processes being considered. Of particular interest are the effects of changing the number of transects and transect width (or alternatively, coverage percent or fraction) on the quality of the estimate. A specific application to the characterization of unexploded ordnance (UXO) in the subsurface at former military firing ranges is discussed. The results may be extended to the investigation of outcrop characteristics as well as subsurface geological features.

Keywords Strip transect sampling · unexploded ordnance · object abundance · spatial Poisson field · design-based computer simulation

1 Introduction

Physical, biological, and environmental phenomena often occur in nature as manifestations of spatial point processes, and the Poisson distribution is generally a reasonable model for their rates of occurrence. Many scientific problems involve the estimation of the rate of occurrence (intensity) or abundance of some population or species of interest through sampling studies. A common field-based survey approach

---

Timothy C. Coburn
Department of Management Science, Abilene Christian University, Box 29315, Abilene, TX 79699, USA, e-mail: coburnt@acu.edu

Sean A. McKenna
Geoscience Research and Applications Group, Sandia National Laboratories, Albuquerque, NM 87185, USA, e-mail: samcken@sandia.gov

Hirotaka Saito
Department of Ecoregion Science, Tokyo University of Agriculture and Technology, Fuchu, Tokyo 183-8509, Japan, e-mail: hiros@cc.tuat.ac.jp

is to record measurements along a series of transects, estimating the total number of objects in the entire field, or their rate of occurrence, from the resulting data.

While the number, size, and shape of the transects necessary to obtain high quality estimates of object abundance is always of concern, the transect sampling design is sometimes developed in a subjective way to provide "adequate coverage," to facilitate ease of data collection, or to identify some presumed spatial pattern, rather than to optimize statistical performance. Using a design- rather than model-based framework, this paper specifically addresses the effects of changing number of transects and transect width (or alternatively, coverage percent or fraction) under random sampling on the accuracy and precision of estimated object abundance.

Computer simulation is employed to conduct a sampling experiment in which different numbers and widths of transects are repeatedly selected from both homogeneous and non-homogeneous Poisson fields and used to estimate object abundance. Sample-based statistics are computed and compared for various combinations of sample size and transect width. The results of the study can be used to develop a strategy for optimizing accuracy and precision of the estimate assuming fixed cost and uniform detection in a specific operational setting.

2 Motivating Example

While transect sampling is most often associated with agricultural, biological, or ecological studies of species abundance, it has important applications in other scientific arenas, particularly in the earth, geological, and soil sciences, and in environmental disciplines. The results reported here are specifically motivated by the problem of characterizing unexploded ordnance (UXO) in the subsurface at former military firing ranges prior to returning them to the public domain.

In this situation, the presence or absence of objects is determined by traversing the range of interest along pre-determined transects with a device designed to measure the geophysical response of those objects at the surface or buried in the near subsurface, and then applying an appropriate threshold to the response. Such responses are generally due to ordnance-related fragments associated with previous target areas. The measurement device can be aerial- (e.g. helicopter mounted) or land-based, with a specified observational "footprint" or coverage width. At any given field location there may be multiple former target areas, and UXO and associated fragments are assumed to be concentrated at those sites but with varying and unknown intensities. Consequently, the occurrence of UXO (and fragments) superimposed on a more uniformly distributed background of non-UXO related clutter (metallic litter, old fencing, iron rich rocks, naturally occurring objects, etc.) is conceptualized as a non-homogeneous Poisson process. The situation might also be conceptualized in terms of other statistical distributions, such as the Negative Binomial. Since the intensity of UXO (and associated fragments) can exhibit spatial correlation, the ultimate objective is to precisely estimate its abundance, as well as the extent of the areas of elevated object intensity that might contain UXO, over the entire field of interest using geostatistical methods (Saito et al., 2005; Saito and McKenna, 2007). However, because knowledge of the subsurface is limited, the distributional characteristics of UXO can never be fully determined; and hence it is important to validate the empirical performance of any estimation procedures under a variety of known conditions before applying them to live situations.

Characterization of UXO sites is related to a number of other geoscience problems. Among these are the detection of live mines in former mine fields and estimation of the rate of fractures, fissures, or caves in geological formations. Jensen et al. (2007) consider similar issues relative to the characterization of petroleum reservoirs and Shieh et al. (2005) present an interactive sampling strategy for locating hot spot regions such as those that might be associated with the presence of heavy metals. Further, Watson and Barnes (1995) provide a broader discussion of criteria that can be used to propose sampling networks for the purpose of locating extremes in problems of this nature.

3 Prior Work on Transect Sampling

There are at least two forms of transect sampling (Buckland et al., 1993, 2001). *Line* transects consist of randomly positioned paths that are traversed by an observer who records sitings of population elements, or pods of elements, along either side of the path. In this case, the transect often has no specifically defined width and serves only as an operational mechanism to facilitate observation. The efficiency of the method (sometimes referred to as distance sampling) depends on the differential probability of detection at varying distances and orientations from the line. *Strip* or *belt* transects, on the other hand, are plot-like geographic entities having fixed length and width that are sampled at random from the collection of all such entities within a region of interest. In this case, the strips constitute the sampling units for purposes of observing, measuring, or recording the population elements contained within them. A complete census of the population elements within each strip is obtained; and hence, from a finite population sampling standpoint, strip transect sampling is quite different from line transect sampling. Both strip and line transects are often oriented parallel to one another, and perpendicular to a boundary of some regularly shaped (e.g. rectangular) region, although such an arrangement is by no means required.

There has been considerable work on the problem of estimating population intensity or object/species abundance using the line transect method. Burnham et al. (1980), Ripley (1981), and Upton and Fingleton (1985) provide excellent discussions of both theory and applications, and summarize the available literature up through the middle 1980s. The line transect method has continued to attract interest from a variety of disciplines, particularly the ecological and biological sciences, where a fairly robust amount of additional research has been undertaken (e.g. Quang and Becker 1996). Both Cowling (1998) and Hedley and Buckland (2004) provide insight into spatial modeling based on line transect sampling, while Barry and Welsh (2001) provide an elegant presentation of the theoretical properties of distance sampling in both the design- and model-based frameworks. Buckland et al. (1993) give additional historical details about the line transect method. Line transect sampling has also been widely used in geological studies of both outcrop and subsurface characteristics (e.g. Laslett, 1982); but in this context the transects are seldom selected or positioned in a random manner.

Less research has been devoted to estimation based on strip/belt transects, with Upton and Fingleton (1985) providing, perhaps, the most comprehensive coverage. One reason stems from the fact that strip/belt transects can be regarded as contiguous sets of equal-sized quadrats, and quadrat sampling has been studied in considerable detail. However, to achieve economic or operational efficiencies, some situations and applications necessitate the use of strip/belt transects in and of themselves without resorting to an artificial subdivision into quadrats. Estimating the abundance of ordnance- related clutter is one such application. Questions naturally arise as to how many and what size transects should be selected to achieve specified precision of the estimates. Despite the long history of strip transects, we know of no investigations that specifically address the effects of changing transect number and width, specifically in an environmental restoration context. Consequently, this idea is the subject of the present work.

3.1 Design- and Model-Based Considerations

We note at this juncture that the operational aspects of transect sampling in the context of UXO site characterization (i.e. optimization of the number and size of the transects) have more to do with sample design than with modeling procedures and assumptions; and yet we acknowledge that model-based approaches to spatial investigations (e.g. those derived from geostatistcal methods; Shieh et al., 2005; Watson and Barnes, 1995) have become somewhat more prevalent in recent years. The differentiation primarily stems from an understanding of how randomness is to be treated. Borchers et al. (2002) describe three kinds of randomness that can impact the estimation of object abundance: (1) randomness due to changes in location or features of the objects (state model), (2) randomness associated with the ability to detect or observe the objects (observation model or detection function), (3) and randomness attributable to the survey design.

In the case of UXO or other discrete subsurface geological features, a design-based formulation focused primarily on the sampling scheme seems most appropriate (although, as Barry and Welsh (2001) suggest, the associated randomness probably derives from a mixture of all three sources identified above). The rationale for this assertion follows logically from an assessment of the physical conditions, or environment, of observation. First, note that the locations of the subsurface objects or features exist in a state of immobility (i.e. object position is non-stochastic, in contrast to randomness in the locations of animals which can move about) and their physical characteristics presumably do not change during the time span of the survey. Second, while the objects may clearly adhere to a two- or three-dimensional spatial trend, such a trend is only observable through sampling, and then, perhaps, only partially so. Third, as previously indicated, the strip sampling approach assumes a census will be conducted within each strip, suggesting the probability of detection for each object is 1. This is not to imply that every object can be perfectly detected, although we assume perfect detection in the present context. For objects that exist in the subsurface, the detection function must, of necessity, depend on the reliability of the observation mechanism (e.g. a geophysical device or a mining operation), but an investigation of these types of functions is beyond the scope of the current study. Thompson and Ramsay (1987) provide an excellent discussion of detectability functions associated with the different field sampling approaches for observing spatial point processes. Finally, the design-based formulation requires fewer statistical assumptions which better fits the situation of having restricted knowledge about the subsurface.

De Gruitjer et al. (2006) provide a particularly lucid discussion of the difference between the design- and model-based approaches, underscoring the need to appropriately assign the basis of inference and account for randomness in the system of observation. Additional details are given in Särndal et al. (1992) and de Gruitjer and ter Braak (1990).

4 Estimation Theory

For the reasons discussed above, we propose a design- rather than model-based approach to estimate object abundance at sites potentially contaminated with UXO. We believe the approach is directly extendable to estimation of the abundance of other subsurface features. Assume a spatial point process can be represented as the distribution of points in a rectangular field (or range) of interest having X–Y dimensions W and L (Fig. 1A), respectively. The distribution is assumed to be Poisson with intensity parameter $\lambda = T/A$, where T is the total number of objects in the field and $A = WL$ is the area of the field. Let y be a location in two- or three-dimensional space. After Thompson and Ramsay (1987), the detectability function, $g(y)$, is the conditional probability that, given an object exists at location y in the field, it is detected.

Completely subdivide the field into a set of equal-sized non-overlapping strips of width w, representing the observation "footprint," such that $N = W/w$, an integer, is the total number of strips. As depicted in Fig. 1B, we assume the transects are straight, oriented north-south (top-bottom), and have equal length and width (i.e. the strips are equally dimensioned). We also assume the boundaries of the region are static and that there are no edge effects that impact the sampling and/or estimation.

Since N is finite for a finite sized region or domain, N decreases as w increases; i.e. selecting a value of w uniquely determines the value of N. Indeed, in a rectangular field of fixed size, N can be very small when w is large. If the true number of objects in the field is known and each object can be uniquely located and counted, then the number of objects per strip can be completely determined.

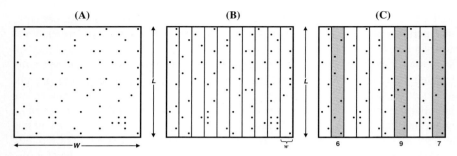

Fig. 1 (**A**) Example spatial point process in a rectangular field; (**B**) equal-sized transects superimposed on the field; (**C**) sampled transects (gray; with object count shown below)

Let n of the N available transects of width w be selected at random (i.e. n is the sample size) without replacement, and compute T_i, the total objects in the ith sampled transect (Fig. 1C). In the context of UXO site characterization, the presence of an object is associated with a geophysical anomaly or threshold exceedance. We make the strong assumptions that all such objects of interest along each transect can be uniquely located, correctly identified, and incontrovertibly counted so that $g(y)$ is defined to be the indicator function $g(y) = 1$ over the strip and 0 elsewhere. A census is effectively conducted within each sampled strip.

After Borchers et al. (2002) we designate the surface area of the rectangular region covered by the sampled strips as a ($a = n \cdot w \cdot L$), and refer to it as the "covered area." If the surface area of the entire region is A, then the coverage percent is $a/A \times 100$ (henceforth we refer to this quantity as "coverage fraction," a term that is more commonly used in the literature).

Under the foregoing conditions, the N strips comprise a finite population of sampling units of equal size and dimension, and the probability of selection is constant. We conceptualize each strip as a cluster in the sense of conventional cluster sampling. Here, the term cluster sampling refers to the physical size of the sampling unit (the strip) rather than a group of co-located objects (e.g. a geographical concentration of metallic fragments at a former ordnance range or a swarm of geological fractures). This formulation facilitates a subsequent extension to strips of differing lengths that can be treated as clusters of unequal size, or to aggregation/agglomeration of strips into larger units. Others (e.g. Scheaffer et al. 1996) have suggested this same approach.

Estimation theory based on cluster sampling is straightforward and well-known; but results from extensive sampling experiments for the present context have not been reported. After Lohr (1999), an unbiased estimate, $\hat{T}$, of the total objects, T, in the entire field is given by

$$\hat{T} = \frac{N}{n} \sum_{i=1}^{n} T_i. \tag{1}$$

Using this expression, an unbiased estimate, $\hat{\lambda}$, of the Poisson intensity λ over the entire region is given by

$$\hat{\lambda} = \frac{\hat{T}}{A}. \qquad (2)$$

Re-writing Equation (1) as $\hat{T} = \left(\sum_{i=1}^{n} T_i \right) / (n/N)$, designating n/N as the inclusion probability or sampling fraction, yields a form of the Horvitz-Thompson estimator (Horvitz and Thompson 1952). The variance of $\hat{T}$ is given by

$$\sigma_{\hat{T}}^2 = V(\hat{T}) = N^2 \left(1 - \frac{n}{N}\right) \frac{\sigma_T^2}{n}, \qquad (3)$$

and the variance of $\hat{\lambda}$ is given by

$$\sigma_{\hat{\lambda}}^2 = V(\hat{\lambda}) = \frac{1}{A^2} V(\hat{T}) = \frac{N^2}{A^2} \left(1 - \frac{n}{N}\right) \frac{\sigma_T^2}{n}, \qquad (4)$$

where

$$\sigma_T^2 = \frac{1}{N-1} \sum_{i=1}^{N} \left(T_i - \frac{T}{N} \right)^2, \qquad (5)$$

Since σ_T^2 is unknown, it must be estimated by the variance of the totals from the sampled transects given by

$$s_T^2 = \frac{1}{n-1} \sum_{i=1}^{n} \left(T_i - \frac{\hat{T}}{N} \right)^2, \qquad (6)$$

which is equivalent to the somewhat different expression given in Arvanitis and Portier (1997). Consequently, the estimated variance of the estimated total objects in the field based on a random sample of n transects of width w is given by

$$s_{\hat{T}}^2 = N^2 \left(1 - \frac{n}{N}\right) \frac{s_T^2}{n}. \qquad (7)$$

When the strips do not have the same dimensions or area, the estimation theory is not as straightforward and (1)–(7) must be modified, or an alternate approach must be used. The difference is analogous to the distinction between cluster sampling with and without equal-sized sampling units, with strips being selected using a probability- proportional-to-size (*pps*) approach in the latter case. It is well known that the more general form of the Horvitz-Thompson estimator plays an important role when sampling units are selected using *pps* (Borchers et al., 2002; Scheaffer et al., 1996; Lohr, 1999; and many others). In the context of estimating the abundance of objects at a potential UXO site, strip transects of the same length but differing widths may be undesirable, and even operationally impossible, if the "footprint" of the measuring device is fixed; whereas transects of the same width but differing lengths may be inevitable, and perhaps even desirable, particularly if the region of interest is irregularly shaped.

Other estimators and designs have been proposed when the sampling units are defined to be strip or belt transects. Stehman and Salzer (2000) compare two different ratio estimators for use in an ecological context when strips are selected using systematic sampling. Thompson (1992) also discusses applications of ratio estimation and systematic sampling, as well as systematic and strip adaptive cluster sampling which may be useful in the context of estimating object abundance in the subsurface.

5 Simulation Methodology and Results

As suggested earlier, before geostatistical methods can be reasonably used to estimate object abundance in a field with a spatially varying intensity when the exact distribution is *unknown,* information must be obtained about the statistical performance of the estimator under *known* conditions. Described here are the results of evaluating various scenarios using repeated sampling of transects from simulated spatial fields whose distributions are known.

5.1 Homogeneous Case

The first case involves the estimation of object abundance in a homogeneous Poisson field. A spatial point process within a 5000 m × 5000 m field was created through computer simulation where the intensity of the process, λ, was pre-determined. Three different fields were constructed to investigate the span of possible results— one each for $\lambda = 5 \times 10^{-4}$ objects/m^2, $\lambda = 5 \times 10^{-5}$ objects/m^2, and $\lambda = 5 \times 10^{-6}$ objects/m^2. These values of λ are thought to encompass the range of values that might be observed in actual situations. Figure 2A depicts the field constructed for $\lambda = 5 \times 10^{-5}$. Regular strip grids such as the one depicted in Fig. 1B were superimposed on each of these fields using a series of different transect widths ranging from 10 to 500 m.

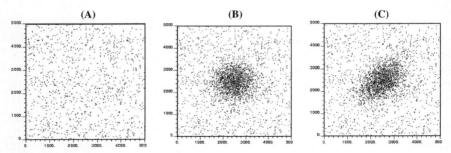

Fig. 2 Three simulated Poisson fields: (**A**) homogeneous field, $\lambda = 5 \times 10^{-5}$; (**B**) non-homogeneous field consisting of the homogeneous field with a single isotropic target superimposed; (**C**) non-homogeneous field consisting of the homogeneous field with a single anisotropic target superimposed

For each spatial field, samples of transects of each width were randomly selected without replacement. Sample size, n, was varied from 2 to 20 in increments of 1, and 180 sampling replications were performed for each sample size. The number of replicates is restricted to 180 because the number of unique combinations of transects at higher widths is limited (e.g. in Fig. 1B, if $W = 5000$ and $w = 250$, N must be 20; consequently, if $n = 18$ there are only 190 possible combinations from which to choose since, applying the Binomial coefficient, $\binom{20}{18} = 190$).

For each individual sample, $\hat{T}$ and the estimated variance of $\hat{T}$ were computed using (1) and (7), respectively. Each $\hat{T}$ was compared to the respective (ground truth) value of T and the difference, $\hat{T} - T$, and the percent difference, $\frac{\hat{T}-T}{T} \cdot 100$, were computed. For every w, $\hat{T}$ and $s_{\hat{T}}^2$ were then averaged for each set of 180 samples generated for each of $n = 2, 3, \ldots, 20$. Table 1 reports the average $\hat{T}$ for selected combinations of λ, n, and w, along with the average difference and the average percent difference relative to the respective value of T.

Table 1 indicates that, for most combinations of n and w, the total number of objects estimated using (1) is, on average, within 1% of the respective true total objects in the field, regardless of λ. As might be expected, for all three homogeneous Poisson fields ($\lambda = 5 \times 10^{-6}$, $\lambda = 5 \times 10^{-5}$, $\lambda = 5 \times 10^{-4}$) $\hat{T}$ is least accurate (in the sense of average percent difference from T, in absolute value) for the smallest samples of the narrowest transects. For example, when $\lambda = 5 \times 10^{-6}$, the average percent difference between T and $\hat{T}$ ranges from about 3 to 13% when w is 25 m or less and n is 8 or less. In general, for a specified value of λ, average percent difference between T and $\hat{T}$ improves as *both* n and w increase; and it also improves as λ increases. While improvement due to increasing n and w is at least partly attributable to the Central Limit Theorem, it also suggests that coverage fraction may be more important than the number of transects and their width.

Figure 3 (A–C) shows the average percent difference between T and $\hat{T}$ versus coverage fraction for the three fields. Three observations can be made: (1) there is more scatter in the values at low coverage fractions, (2) the average value dampens to zero as coverage fraction increases, and (3) the dampening occurs more rapidly as λ increases. The dampening to zero is a direct manifestation of the statistically unbiased nature of the estimator, $\hat{T}$; that is, $E(\hat{T} - T) = 0$.

The results shown in Table 1 provide a somewhat different perspective when viewed in terms of the raw difference, $\hat{T} - T$. In almost all cases, the average value of $\hat{T} - T$ deviates from zero, with the amount again decreasing as both n and w increase. Further, larger average differences are associated with higher values of λ. For example, when $\lambda = 5 \times 10^{-4}$ the average difference between T and $\hat{T}$ exceeds 20 objects for several combinations of n and w.

It is worth noting that, although unbiased, $\hat{T}$ cannot be expected to obtain the true value, T, in any specific sample, nor would the average of $\hat{T} - T$ in repeated sampling be expected to be zero unless all possible samples of the same size were considered. Nonetheless, in the specific context of remediating UXO sites where only a single sample of n transects of width w would be selected due to operational constraints, it is desirable for $\hat{T} - T$ to be as small as possible to insure against

Table 1 Estimated versus true total objects in a homogeneous Poisson field for selected combinations of intensity, λ, transect width, w, and numbers of transects, n. All values are based on 180 replications

n		$\lambda = 5 \times 10^{-6}$					$\lambda = 5 \times 10^{-5}$					$\lambda = 5 \times 10^{-4}$				
		10 m	25 m	50 m	100 m	200 m	250 m	10 m	25 m	50 m	100 m	200 m	250 m	10 m	25 m	50 m	100 m	200 m	250 m
2	Avg Est	120.8	108.3	120.0	126.3	117.8	123.1	1283.3	1283.3	1244.4	1236.7	1249.6	1247.4	12298.6	12385.0	12503.3	12480.4	12494.1	12497.0
	Avg Diff	−3.2	−15.7	−4.0	2.3	−6.2	−0.9	34.3	34.3	−4.6	−12.3	0.6	−1.6	−201.4	−115.0	3.3	−19.6	−5.9	−3.0
	Avg %Diff	−2.58	−12.66	−3.23	1.85	−5.00	−0.73	2.75	2.75	−0.37	−0.98	0.05	−0.13	−1.61	−0.92	0.03	−0.16	−0.05	−0.02
5	Avg Est	109.4	118.9	122.9	118.6	124.6	121.5	1277.8	1259.3	1257.4	1258.7	1243.6	1249.0	12462.2	12564.0	12487.1	12505.1	12493.1	12519.9
	Avg Diff	−14.6	−5.1	−1.1	−5.4	0.6	−2.5	28.8	10.3	8.4	9.7	−5.4	0.0	−37.8	64.0	−12.9	5.1	−6.9	19.9
	Avg %Diff	−11.77	−4.11	−0.89	−4.35	0.48	−2.02	2.31	0.82	0.67	0.78	−0.43	0.00	−0.30	0.51	−0.10	0.04	−0.06	0.16
8	Avg Est	128.5	128.1	125.6	123.5	124.3	124.5	1230.6	1238.9	1251.6	1252.8	1246.1	1248.9	12495.1	12524.0	12466.1	12498.5	12504.4	12495.8
	Avg Diff	4.5	4.1	1.6	−0.5	0.3	0.5	−18.4	−10.1	2.6	3.8	−2.9	−0.1	−4.9	24.0	−33.9	−1.5	4.4	−4.2
	Avg %Diff	3.63	3.31	1.29	−0.40	0.24	0.40	−1.47	−0.81	0.21	0.30	−0.23	−0.01	−0.04	0.19	−0.27	−0.01	0.04	−0.03
12	Avg Est	124.8	122.6	123.2	125.9	124.3	124.3	1251.2	1255.0	1248.4	1253.6	1249.1	1251.7	12559.0	12521.3	12479.2	12499.0	12498.8	12487.8
	Avg Diff	0.8	−1.4	−0.8	1.9	0.3	0.3	2.2	6.0	−0.6	4.6	0.1	2.7	59.0	21.3	−20.8	−1.0	−1.2	−12.2
	Avg %Diff	0.65	−1.13	−0.65	1.53	0.24	0.24	0.18	0.48	−0.05	0.37	0.01	0.22	0.47	0.17	−0.17	−0.01	−0.01	−0.10
15	Avg Est	127.4	127.5	128.7	124.4	124.3	123.6	1254.3	1243.3	1249.4	1253.5	1245.5	1244.0	12552.6	12527.2	12492.6	12505.0	12499.7	12500.5
	Avg Diff	3.4	3.5	4.7	0.4	0.3	−0.4	5.3	−5.7	0.4	4.5	−3.5	−5.0	52.6	27.2	−7.4	5.0	−0.3	0.5
	Avg %Diff	2.74	2.82	3.79	0.32	0.24	−0.32	0.42	−0.46	0.03	0.36	−0.28	−0.40	0.42	0.22	−0.06	0.04	0.00	0.00
18	Avg Est	129.8	125.1	123.7	125.0	124.3	123.7	1266.0	1238.3	1255.4	1250.3	1249.0	1249.0	12502.8	12505.9	12475.7	12509.7	12492.7	12500.0
	Avg Diff	5.8	1.1	−0.3	1.0	0.3	−0.3	17.0	−10.7	6.4	1.3	0.0	0.0	2.8	5.9	−24.3	9.7	−7.3	0.0
	Avg %Diff	4.68	0.89	−0.24	0.81	0.24	−0.24	1.36	−0.86	0.51	0.10	0.00	0.00	0.02	0.05	−0.19	0.08	−0.06	0.00

Notes: When $\lambda = 5 \times 10^{-6}$, T, the true number of objects in the field is 124; when $\lambda = 5 \times 10^{-5}$, T = 1249; and, when $\lambda = 5 \times 10^{-4}$, T = 12500. Est = Total objects in the field estimated from Eq. (1), Avg Est = $(1/180) \sum$Est, Diff = Est − T, Avg Diff = $(1/180) \sum$Diff, %Diff = [(Est − T)/T]*100, Avg %Diff = $(1/180) \sum$%Diff.

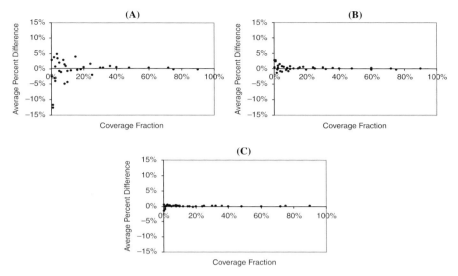

Fig. 3 Average percent difference in estimated and true total objects for three simulated homogeneous Poisson fields versus coverage fraction: (**A**) $\lambda = 5 \times 10^{-6}$, (**B**) $\lambda = 5 \times 10^{-5}$, and (**C**) $\lambda = 5 \times 10^{-4}$. Each point represents the average over 180 replications

(1) increased physical risk (T is larger than $\hat{T}$) or (2) unnecessary expenditure of remediation resources (T is smaller than $\hat{T}$). Figure 3 suggests that in fields with smaller λ the concern would be greater than in fields with higher λ, since a larger coverage fraction would be required to drive the average difference (or average percent difference) to 0.

Equation (7) provides a way to estimate the uncertainty, or precision, associated with the value of $\hat{T}$ obtained from any individual sample. Figure 4 (A–C) presents the average $s_{\hat{T}}^2$ versus coverage fraction for the three homogeneous fields. The results for selected w are shown. As in Table 1, these figures indicate that $\hat{T}$ is most imprecise for small samples of narrow transects, again suggesting that coverage fraction may be more important than n or w. In fact, as Fig. 4 illustrates, average $s_{\hat{T}}^2$ monotonically declines along the same curve as the coverage fraction increases, irrespective of w, dampening out as coverage fraction approaches about 10–15%.

Figure 4 (A and B) also illustrates the effect on precision of changing λ in homogeneous fields. As λ increases by an order of magnitude, average $s_{\hat{T}}^2$ increases accordingly for essentially the same field coverage.

5.2 Non-Homogeneous Case

The second case involves the estimation of object abundance in non-homogeneous Poisson fields. Similar to the homogeneous case described above, a spatial point

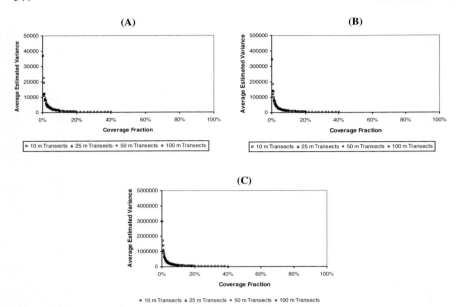

Fig. 4 Average estimated variance of estimated total objects ((7)) versus coverage fraction for each of the three homogeneous Poisson fields: (**A**) $\lambda = 5 \times 10^{-6}$, (**B**) $\lambda = 5 \times 10^{-5}$, and (**C**) $\lambda = 5 \times 10^{-4}$. Each point represents the average over 180 replications. Note the change in vertical scale

process within a 5000 m × 5000 m field was created through computer simulation where the intensity of the process was pre-set to $\lambda = 5 \times 10^{-5}$ (background). A second process was superimposed on the first one to create a non-homogeneous pattern characterized by a single central feature having increased intensity (Fig. 2B and C). In the UXO context, such features would represent target areas. The feature of increased intensity in Fig. 2B is isotropic, whereas the one in Fig. 2C is anisotropic with a major axis at 45°. Both non-homogeneous fields contain the same number of objects, but the comparison homogeneous field (Fig. 2A) has fewer objects. Regular strip grids such as the one depicted in Fig. 1B were superimposed on each of these fields using the identical transect widths identified above. Estimation proceeded in the same way outlined for the homogeneous case.

Table 2 reports the average $\hat{T}$ for selected w for each of three comparison fields: background (homogeneous case, $\lambda = 5 \times 10^{-5}$), background plus a single isotropic target, and background plus a single anisotropic target. Table 2 also reports the average difference in $\hat{T}$ and T, as well as the average percent difference, associated with each respective combination of n and w.

Over all combinations of n and w, average percent difference between $\hat{T}$ and T ranges between ±6% for both of the non-homogeneous fields; for the comparison homogeneous field the range is about ±3%. There is no obvious pattern in the values that can be specifically attributed to the difference in the spatial arrangements of the fields, and the overall effect of changing n and w is not as readily apparent as for

Table 2 Estimated versus true total objects in one homogeneous and two non-homogeneous Poisson fields for selected combinations of transect width, w, and number of transects, n. All values are based on 180 replications

n		Background Only, $\lambda = 5 \times 10^{-5}$				Background + 1 Isotropic Target				Background + 1 Anisotropic Target			
		10 m	25 m	50 m	100 m	10 m	25 m	50 m	100 m	10 m	25 m	50 m	100 m
2	Avg Est	1283.3	1283.3	1244.4	1236.7	3169.4	3110.0	3094.7	2915.4	2890.3	2936.1	2922.2	2970.7
	Avg Diff	34.3	34.3	−4.6	−12.3	178.4	119.0	103.7	−75.6	−100.7	−54.9	−68.8	−20.3
	Avg %Diff	2.75	2.75	−0.37	−0.98	5.96	3.98	3.47	−2.53	−3.37	−1.84	−2.30	−0.68
5	Avg Est	1277.8	1259.3	1257.4	1258.7	2911.7	2970.2	2985.6	2934.7	2869.4	2974.4	3086.3	2847.7
	Avg Diff	28.8	10.3	8.4	9.7	−79.3	−20.8	−5.4	−56.3	−121.6	−16.6	95.3	−143.3
	Avg %Diff	2.31	0.82	0.67	0.78	−2.65	−0.70	−0.18	−1.88	−4.07	−0.55	3.19	−4.79
8	Avg Est	1230.6	1238.9	1251.6	1252.8	2981.9	2903.1	2946.1	3046.0	3037.8	3021.3	2926.5	3022.2
	Avg Diff	−18.4	−10.1	2.6	3.8	−9.1	−87.9	−44.9	55.0	46.8	30.3	−64.5	31.2
	Avg %Diff	−1.47	−0.81	0.21	0.30	−0.30	−2.94	−1.50	1.84	1.56	1.01	−2.16	1.04
12	Avg Est	1251.2	1255.0	1248.4	1253.6	2974.1	3037.1	2945.5	3042.9	3075.7	2918.4	3019.7	3045.6
	Avg Diff	2.2	6.0	−0.6	4.6	−16.9	46.1	−45.5	51.9	84.7	−72.6	28.7	54.6
	Avg %Diff	0.18	0.48	−0.05	0.37	−0.57	1.54	−1.52	1.74	2.83	−2.43	0.96	1.83
15	Avg Est	1254.3	1243.3	1249.4	1253.5	3063.7	3061.6	2968.6	2971.7	3013.0	3014.5	3007.0	3041.0
	Avg Diff	5.3	−5.7	0.4	4.5	72.7	70.6	−22.4	−19.3	22.0	23.5	16.0	50.0
	Avg %Diff	0.42	−0.46	0.03	0.36	2.43	2.36	−0.75	−0.65	0.74	0.79	0.53	1.67
18	Avg Est	1266.0	1238.3	1255.4	1250.3	3014.5	3017.3	2943.0	2980.4	3005.6	3008.2	2967.2	3005.6
	Avg Diff	17.0	−10.7	6.4	1.3	23.5	26.3	−48.0	−10.6	14.6	17.2	−23.8	14.6
	Avg %Diff	1.36	−0.86	0.51	0.10	0.79	0.88	−1.60	−0.35	0.49	0.58	−0.80	0.49

Notes: For Background Only, T, the true number of objects in the field is 1249; for Background + 1 Isotropic Target, T = 2991; and, for Background + 1 Anisotropic Target, T = 2991.
Est = Total objects in the field estimated from Eq. (1), Avg Est = $(1/180)\sum$Est, Diff = Est − T, Avg Diff = $(1/180)\sum$Diff, %Diff = [(Est − T)/T]*100, Avg %Diff = $(1/180)\sum$%Diff. Values for Background Only are identical to those shown in Table 1.

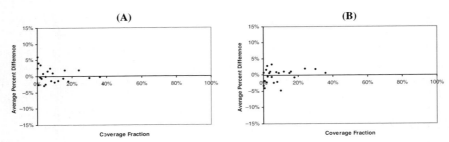

Fig. 5 Average percent difference in estimated and true total objects for two simulated non-homogeneous Poisson fields versus coverage fraction: (**A**) homogeneous field with $\lambda = 5 \times 10^{-5}$ plus a single isotropic target superimposed (Fig. 2B); (**B**) homogeneous field with $\lambda = 5 \times 10^{-5}$ plus a single anisotropic target superimposed (Fig. 2C). Each point represents the average over 180 replications

the comparison homogeneous field. On the other hand, average percent difference is somewhat higher, for all combinations of n and w, than the corresponding values for the homogeneous field, and the average raw differences are uniformly higher. This inability to obtain values of $\hat{T}$ closer to T, on average, reflects both the larger number of objects in the non-homogeneous fields than in the homogeneous field and the presence of the area of higher object intensity.

Figure 5 (A and B) shows the average percent difference between T and $\hat{T}$ versus coverage fraction. As in the case of the homogeneous field (Fig. 3B), the values dampen to zero as coverage fraction increases, but not as rapidly. As might be expected, this suggests that a higher coverage fraction may be needed for any specific sample in order for the estimator to get as close to T (say, within 1%) as in the homogeneous case. However, the values plotted in Fig. 5A and B are more scattered or inconsistent than those in Fig. 3B, which makes the assessment of the effect of increasing coverage fraction somewhat less clear cut.

Figure 6 (A and B) presents average $s_{\hat{T}}^2$ versus coverage fraction for the two non-homogeneous fields. In Fig. 4A–C, which present the corresponding graphs for the three homogeneous fields, all the values of average $s_{\hat{T}}^2$ associated with various w lie

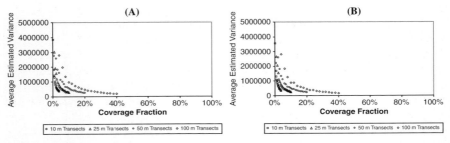

Fig. 6 Average estimated variance of estimated total objects ((7)) versus coverage fraction for two simulated non-homogeneous Poisson fields: (**A**) homogeneous field with $\lambda = 5 \times 10^{-5}$ plus a single isotropic target superimposed (Fig. 2B); (**B**) homogeneous field with $\lambda = 5 \times 10^{-5}$ plus a single anisotropic target superimposed (Fig. 2C). Each point represents the average over 180 replications

along a single, common, monotonically declining curve. This is not the case for the two non-homogeneous fields. Whereas w (or alternatively, n) has no apparent effect on the precision of $\hat{T}$ in the homogeneous fields, it definitely impacts the precision of $\hat{T}$ in the two non-homogeneous fields. As w increases, a higher coverage fraction (or a larger n) is required to achieve the same precision obtained with narrower transects. Further, the difference between these two particular non-homogeneous fields with regard to type of feature (isotropic or anisotropic) apparently has little effect on the precision of $\hat{T}$ since the graphs in Fig. 6A and B are nearly identical.

6 Discussion

As noted earlier, the objective of this work was to investigate the effects of changing w and n when strip transect sampling is used to estimate object abundance (specifically, geophysical anomalies associated with characterization of UXO sites) in spatial Poisson fields. Collectively, the results indicate that, along with other factors, the ability to precisely and accurately estimate total objects depends on the amount of the field covered by the sample rather than on a specific n and/or w. This has important implications for the design of strip transect surveys because it implies that the n and w can be selected to accommodate operational survey costs once an appropriate coverage fraction is determined. Conversely, for a fixed w such as may be defined by the footprint of the sampling instrument, the value of n necessary to achieve the appropriate coverage fraction can be determined.

In the case of homogeneous Poisson fields, achieving a reasonably good estimate of total objects (in terms of average percent difference between $\hat{T}$ and T) depends largely on λ, in conjunction with coverage fraction. For example, when $\lambda = 5 \times 10^{-4}$, sampling 10% of the field yields a maximum average percent difference of less than $\pm.5\%$ (Fig. 3C); whereas, when λ is smaller, say 5×10^{-5}, the average percent difference is higher (about $\pm 1\%$) for the same coverage fraction (Fig. 3B). An alternative way of saying this is that, when $\lambda = 5 \times 10^{-5}$, a coverage fraction of more than 10% (something on the order of 10–20%) would be needed to consistently achieve the same $\pm.5\%$ average percent difference obtained with a coverage fraction of 10% when $\lambda = 5 \times 10^{-4}$. In fact, when λ is very small, say 5×10^{-6}, the coverage fraction will need to exceed about 30% in order to obtain an average percent difference in $\hat{T}$ and T that is consistently within $\pm.5\%$ (Fig. 3A).

Operationally speaking, 10% coverage of a 5000 m × 5000 m field could be achieved in a number of ways; for example, with 100 transects that are no more than 5 m wide (a somewhat typical scenario for UXO site characterization) or with 10 transects that are no more than 50 m wide. Although the cost associated with the physical sampling process would be expected to change, Fig. 3 (A–C) suggests the quality of the estimate (in terms of average percent difference in T and $\hat{T}$) would be essentially the same for any specified λ. However, when λ changes, the quality of the estimate changes unless a corresponding adjustment in coverage fraction is made.

Summarizing, Fig. 3A–C, and others similar to them, can be used to approximate the coverage fraction that is required to achieve a specified degree of closeness (average percent difference) to the true object abundance. The simulation exercise confirms what might have been expected—that, in settings that can be assumed to be nearly homogeneous, a larger coverage fraction would be needed to find a small number of objects (small λ) than to find a large number of objects (large λ).

In terms of the average estimated variance of the $\hat{T}$ obtained with any given sample, Fig. 4A–C demonstrate that precision improves rapidly as coverage fraction increases. For all three fields, average $s_{\hat{T}}^2$ starts to become strongly asymptotic with the X-axis when coverage fraction is about 10%, suggesting that 10% may be a reasonable minimum. Coverage fractions of more than 10% would be required to drive average $s_{\hat{T}}^2$ close to zero, where, again, coverage fraction is not tied to specific values of n and w. Hence, Fig. 4A–C, and others similar to them, can be used to establish the coverage fraction required to achieve a desired level of precision (specific values of precision depend on the value of λ).

As previously suggested, Fig. 5A and B indicate that, for non-homogeneous fields, there is greater inconsistency in the average difference between $\hat{T}$ and T than for the comparison homogeneous field at almost all but the highest values of coverage fraction (Fig. 3B). The observed inconsistency is likely due to the varying degree to which randomly selected transects intersect the area of increased object intensity. Consequently, in the presence of a spatial trend, it may not be possible to estimate T within less than about $\pm 3 - 5\%$, on average, without a comparatively high coverage fraction. Figures 5A and B indicate that a coverage fraction on the order of about 20% (and perhaps more) may be needed.

Further, Fig. 6A and B indicate that larger coverage fractions are needed to estimate T with the same relative level of precision as the estimates obtained for homogeneous fields (Fig. 4B). In contrast to the homogeneous case, precision is also directly dependent on w. In Fig. 6A and B none of the curves ever become asymptotic to the X-axis, even at coverage fractions of 40%. Hence, driving average $s_{\hat{T}}^2$ to zero is somewhat unrealistic and operationally impractical since the required coverage fraction would have to be so large. Also note in this case that, when the coverage fraction is specified, increasing the w requires a corresponding decrease in n, leading to an overall escalation in average $s_{\hat{T}}^2$ (unlike the homogeneous case). For example, in Fig. 6B, the average $s_{\hat{T}}^2$ for 10% sample coverage based on 25 m transects ($n = 20$) is roughly half the average $s_{\hat{T}}^2$ for 10% sample coverage based on 50 m transects ($n = 10$). Therefore, when using wider transects, more of them (higher coverage fraction) are required to achieve the same precision than when using narrower transects.

Two factors interact to produce this situation in the non-homogeneous case—the usual effect of n plus geographic grouping of objects in the field. Small numbers of randomly selected strip transects of width w have the potential to yield rather diverse estimates of total objects, since some of them may intersect high-intensity features (targets) and some may not. In any case, the ultimate result is to impact operational costs both in terms of the required footprint of the survey instrument and the number of transects to be traversed.

These results suggest that, in general, the sampling requirements necessary to produce high precision estimates of total objects are likely to be larger than anticipated for both homogeneous and non-homogeneous Poisson fields under random, unidirectional strip transect sampling, and that, as expected, the requirements for non-homogeneous fields are likely to be larger than for homogeneous fields. Since all "live" situations are expected to be non-homogeneous to some degree, sufficiently large sampling programs must be devised to ensure the resulting estimates are reliable. Unfortunately, the extent of non-homogeneity (or alternatively, the presence of a spatial trend) in the subsurface is generally unknown, so that the ability to choose the correct approach is somewhat restricted unless prior knowledge is available. Treating a field or remediation region as if it is homogeneous when it is not will almost certainly lead to unreliable estimates of object abundance.

7 Summary and Conclusions

The results reported here establish the empirical foundation for applying strip transect sampling to UXO site characterization. Our work indicates that a design-based formulation is a valid approach to the problem and that strip transect sampling can be used to obtain statistically reliable estimates of object abundance for sufficiently large samples. We first conducted a baseline investigation of the approach applied to fields in which the spatial distribution of objects is assumed to be a homogeneous Poisson process. From this investigation we conclude that precision and accuracy of the estimates depends both on λ and the amount of the field encompassed by the sample; and we demonstrate that the two design-related parameters, n and w, can be effectively replaced by a single parameter, coverage fraction. For a representative range of λ, a series of graphs is provided that can be used to approximate the required coverage fraction for a specified level of precision and/or accuracy. For example, a coverage fraction on the order of about 20% would allow object abundance to be estimated within less than 1% of the true amount assuming $\lambda = 5 \times 10^{-5}$ objects/m^2. Whether 20% is a large or small coverage fraction depends on a number of non-sampling issues, not the least of which is cost; and whether 1% is close enough to the true total depends on the physical risk associated with the remediation effort.

We subsequently applied the strip transect sampling methodology to two different non-homogenous Poisson fields, each having a single, but differently shaped, area of increased object intensity, and compared the findings to those for the corresponding homogeneous field. Contrary to the results we obtained for the baseline (homogeneous) case, we found that transect width, in addition to coverage fraction, directly influences the quality of estimates of object abundance. Hence, we conclude that transect width may be a key parameter in sampling protocols designed for settings where the spatial distribution of objects is thought to be non-homogeneous. Within the range of conditions tested, a coverage fraction on the order of about 20% is needed to estimate object abundance within, say, 3–5% of the true amount.

These percentages are consistent for both non-homogeneous fields, suggesting the sampling and estimation procedures are somewhat robust against simple alterations in the shape of the features of increased object intensity. Sites containing multiple target areas of varying sizes and orientations, and having differential object intensities (the most common situation), would be expected to require higher coverage fractions to get within 5% of the true object count.

8 Limitations and Future Work

While this study yields important information about the impact of strip transect sampling, additional work remains to be done, with the greatest focus on applications in non-homogeneous fields. Because we considered only two non-homogeneous fields constructed with the same background value of λ, other fields with high intensity features of varying shapes, sizes, and orientations, as well as multiple features, need to be investigated. In addition, we considered only one field size (5000 m × 5000 m) and only one realization of each field. Further, the results of sampling schemes applied to such fixed-boundary fields are likely to be impacted by edge effects which have been ignored here. In addition, the relative efficiency of strip transect sampling compared to more conventional sampling designs needs additional study. Other sampling strategies, such as systematic sampling, two-stage or sequential sampling (to gain preliminary information about spatial trends), transect aggregation, and selection of transects with unequal width or length could also be pursued. A more direct comparison of homogeneous and non-homogeneous fields could perhaps be devised if they were required to have the same number of objects. However, this consideration would alter the spatial distribution of objects in the non-homogeneous field (the two fields would no longer have the same background intensity), adding a different kind of complexity to the problem. Finally, additional work pertaining to the detection function needs to be undertaken. Our strong assumption of perfect detection is unlikely to be sustained in actual practice for a number of reasons, including measurement error attributed to the measuring device itself, as well as other uncertainties associated with the uniqueness of detecting objects in the subsurface.

Acknowledgments This work was funded by the Strategic Environmental Research and Development Program (SERDP) as project UX-1200. Sandia is a multiprogram laboratory operated by Sandia Corporation, a Lockheed Martin Company, for the United States Department of Energy's National Nuclear Security Administration under contract DE-AC04-94AL85000.

References

Arvanitis LC, Portier KM (1997) Aerial surveys. http://ifasstat.ufl.edu/nrs/AS.htm
Barry SC, Welsh AH (2001) Distance sampling methodology. J Roy Stat Soc B 63: 31–53
Borchers DL, Buckland ST, Zucchini W (2002) Estimating animal abundance: closed populations. Springer-Verlag, London, p 314

Buckland ST, Anderson DR, Burnham KP, Laake JL (1993) Distance sampling: estimating abundance of biological populations. Chapman Hall, London, p 446

Buckland ST, Anderson DR, Burnham KP, Laake JL, Borchers DL, Thomas L (2001) Introduction to distance sampling: estimating abundance of biological populations. Oxford University Press, Oxford, p 432

Burnham KP, Anderson DR, Laake JL (1980) Estimation of density from line transect sampling of biological populations. Wildlife Society, Washington, DC, p 202

Cowling A (1998) Spatial methods for line transect surveys. Biometrics 54: 27–38

de Gruitjer J, Brus D, Bierkens M, Knotters M (2006) Sampling for natural resource monitoring. Springer, Berlin, p 332

de Gruitjer J, ter Braak CJF (1990) Model-free estimation from spatial samples: a reappraisal of classical sampling theory. Math Geol 22: 407–415

Hedley SL, Buckland ST (2004) Spatial models for line transect sampling. J Agric Biol Environ Stat 9: 181–199

Horvitz DG, Thompson DJ (1952) A generalization of sampling without replacement from a finite universe. J Am Stat Assoc 47: 663–685

Jensen JL, Hart JD, Willis BJ (2007) Evaluating proportions of undetected geological events in the case of erroneous identifications. Math Geol 38: 103–112

Laslett GM (1982) Censoring and edge effects in areal and line transect sampling of rock joint traces. Math Geol 14: 125–140

Lohr SL (1999) Sampling: design and analysis. Duxbury Press, Pacific Grove, CA, p 494

Quang PX, Becker EF (1996) Line transect sampling under varying conditions with application to aerial surveys. Ecology 77: 1297–1303

Ripley BD (1981) Spatial statistics. Wiley, New York, p 252

Särndal C-E, Swennson B, Wretman JH (1992) Model assisted survey sampling. Springer-Verlag, New York, p 694

Saito H, McKenna SA (2007) Delineating high-density areas in spatial Poisson fields from strip-transect sampling using indicator geostatistics: application to unexploded ordnance removal. J Environ Manage 84: 71–82

Saito H, McKenna SA, Zimmerman DA, Coburn TC (2005) Geostatistical interpolation of object counts collected from multiple strip transects: ordinary kriging versus finite domain kriging. Stoch Environ Res Risk Assess 19: 71–85

Scheaffer RL, Mendenhall W III, Ott RL (1996) Elementary survey sampling, 5th edn. Duxbury Press, Belmont, CA, p 501

Shieh S-S, Chu J-Z, Jang S-S (2005) An interactive sampling strategy based on information analysis and ordinary kriging for locating hot spot regions. Math Geol 37: 29–48

Stehman SV, Salzer DW (2000) Estimating density from surveys employing unequal-area belt transects. Wetlands 20: 512–519

Thompson SK (1992) Sampling. Wiley, New York, p 343

Thompson SK, Ramsay FL (1987) Detectability functions in observing spatial point processes. Biometrics 43: 355–362

Upton GJG, Fingleton B (1985) Spatial data analysis by example, vol 1: point pattern and quantitative data. Wiley, New York, p 410

Watson AG, Barnes RJ (1995) Infill sampling criteria to locate extremes. Math Geol 27: 589–608

Increasing Resolution in Exploration Biostratigraphy – Part I

F.M. Gradstein, A. Bowman, A. Lugowski and O. Hammer

Reprinted from *Natural Resources Research* DOI: 10.1007/s11053-008-9070-0, when citing this article please use the DOI number.

Abstract Crossplots with Ranking & Scaling (RASC) and Constrained Optimization (CONOP) zonal sequences increase stratigraphic resolution and correlation potential of biozonations. The crossplots reveal which events do and which do not deviate their stratigraphic position from well to well, and how well they track their average stratigraphic level. The methodology solves the problem that conventional fossil zonations do not rank taxa according to the degree of diachroneity of range endpoints in a correlation scheme. Part 1 of this study applies the crossplots method to a North Sea biostratigraphic data, and part 2 to a proprietary dataset from the Gulf of Mexico.

Keywords Quantitative biostratigraphy · crossplot method · ranking and scaling (RASC) · Constrained Optimization (CONOP) · North sea

1 Introduction

Modern biostratigraphy must cope with occurrence data from hundreds of fossil taxa, in thousands of samples derived from many wells or sections in many

---

F.M. Gradstein
Geological Museum, University of Oslo, N-O318 Oslo, Norway,
e-mail: felix.gradstein@nhm.uio.no

A. Bowman
Chevron Energy Technology Co, 14141 SW Freeway, Sugarland, TX 77478, USA,
e-mail: abowman2@bigred.unl.edu

A. Lugowski
Computer Science Department, University of California, Santa Barbara CA 93106-5110, USA,
e-mail: alugowski@gmail.com

O. Hammer
Geological Museum, University of Oslo, N-O318 Oslo, Norway,
e-mail: oyvind.hammer@nhm.uio.no

This study is dedicated to the memory of our friend and colleague Garry Jones, a staunch supporter of quantitative stratigraphic methods in exploration micropaleontology.

geological basins. The observed first or last occurrence of a fossil is called a stratigraphic event; such events routinely are used to correlate wells or outcrop sections. A paleontological/stratigraphic event is its appearance or disappearance in strata that is deemed to have time correlation significance. For stratigraphic purpose and increased resolution we not only recognize and apply first and last occurrence events, but also first common or consistent, acme, last common or last consistent occurrence events. High fidelity of such events means they do not deviate much their stratigraphic position from well to well, and track their average stratigraphic level.

This study outlines a simple method to rank the stratigraphic fidelity of fossil events using zonal crossplots of RASC (Ranking & Scaling) and CONOP (Constrained Optimization) results. To understand the new approach, we have to understand how quantitative methods differ, and how they can be similar. Three methods are prominent for event type data: Graphic Correlation, Constrained Optimization and Ranking & Scaling (see Table 1).

Graphic correlation is a deterministic method employing interpolation of successive (well-)section data in a semi-objective, bivariate plot mode (Shaw, 1964). The method uses order and thickness spacing of events, and operates best with datasets having both first and last occurrences of taxa. The final answer generally is a composite with maximum taxa ranges. It is best suited for small data sets, and requires selection of an initial reference well that is updated by the other ones in succession. The method produces semi-objective answers, and is slow to use. Software programs are out of date. We will not here deal with graphic correlation.

Constrained Optimization, embodied by program CONOP (Kemple et al., 1995; Sadler, 2001; http://sdsugeology.blogspot.com/2007/01/seminar-dr-peter-sadler.html) is a significant improvement on graphic correlation. It operates inverse in that it picks an initial, random solution that is updated in a 'kind of multidimensional graphic correlation' manner, using all sections simultaneously. The key to the method is the quantification of misfit. Once the misfit in the order of first and last appearances of taxa from one local section to next and to the composite is quantified then an optimising algorithm (simulated annealing) searches for the global composite sequence of events that implies a minimal total amount of range extension (penalty) in the individual well sections. The parameters for the optimization procedure include an initial annealing temperature, the number of cooling steps, and the number of trials per step. Data are event order, event cross-over, and thickness spacing; datasets best have both first and last occurrences of taxa and can be small to medium size. This heuristic method is not capable of proof, and answers can be sub-optimal.

Ranking and Scaling is a probabilistic method, embodied by programs RASC & CASC (Agterberg and Gradstein, 1999; Gradstein, 2004; www.rasc.uio.no). The program (version 20) has an interactive data input module, and extensive colour graphics output. The method uses fossil event order in all wells to construct a most likely and average sequence of events. This optimal order is scaled in relative time using either cross-over frequency of all event pairs or their standardized average interval thickness. The method has detailed error and outlier analysis, and several correlation options. RASC & CASC operates on all wells simultaneously, is very

Table 1 Key properties of the three methods Graphic Correlation, Constrained Optimization and Ranking and Scaling which are used to zone and correlate microfossil events in groups of wells or outcrop sections

Graphic Correlation	Constrained Optimization	Ranking & Scaling
Programs GRAPHCOR, STRATCOR	Program CONOP	Programs RASC & CASC
Deterministic method: graphic correlation in bivariate plots (note: STRATCOR program can operate in a probabilistic or deterministic manner)	Mostly a deterministic method; can also simulate probabilistic solutions. Constrained optimisation with simulated annealing and penalty score	Probabilistic method: ranking, scaling normality testing, and automated, most likely graphic correlation with error analysis.
Uses event order and thickness spacing; works best with datasets having both first and last occurrences of taxa	Uses event order, event cross-over, and thickness spacing; datasets best have both first and last occurrences of taxa	Uses event order from well to well, and scores of cross-over from well to well for all event pairs
Best suited for small data sets, but can operate also on large datasets	Processes medium to large data sets	Processes large data sets fast; has data management and data input module
Requires selection of an initial standard section, then section by section comparison with the intermediate composite in repeated rounds	Treats all sections and events simultaneously, and works inverse through iteratively improved 'guesses' about the solution.	Treats all sections and events simultaneously
Line of Correlation (LOC) fitting, in section by section plots, can be partially automated	Multidimensional LOC; automated fitting; can generate several different composites depending on the many run options	Automated execution; generates several scaled optimum sequences per dataset depending on run parameters, and tests to omit 'bad' sections or 'bad' events
Attempts to find maximum stratigraphic range of taxa among the sections	Attempts to find maximum or most common stratigraphic ranges of taxa	Finds average stratigraphic position of first and last occurrence events
Builds a composite of events by interpolation of missing events in successive section by section plots, via the LOC	Uses cumulative misfit and simulated annealing to find either the 'best' or a good multidimensional LOC and composite sequence of events	Uses scores of order relationships to find the most likely order of events, which represents the stratigraphic order found on average among the sections

Table 1 (continued)

Graphic Correlation	Constrained Optimization	Ranking & Scaling
Relative spacing of events is a composite of original event spacing in meters in the sections	Relative spacing of events in the composite is derived from original event spacing in meters or sample levels	Relative spacing of the events in the scaled optimum sequence derived from pair-wise cross-over frequency, or average thickness of event intervals
No automatic correlation of sections; can be used to build time scales	Correlates sections automatically; can be used to build a standard time scale	Automated correlation of sections using isochrons
No error analysis; sensitive to geological reworking and other 'stratigraphic noise', and sensitive to order in which sections are composited during analysis	Numerous numerical tests and graphical analysis of stratigraphic results; finds best break points for assemblage zones	Three tests of stratigraphic normality of sections and events; calculates standard deviation of each event as a function of its stratigraphic scatter in wells
Interactive operation under DOS; graphic displays of scattergrams and best fit lines	Batch operation under Windows; colour graphics displays shows progress of runs	Button operated under Windows; fast batch runs; colour graphics of output and options for interactive graphics editing

fast, handles large and complex data sets, and is relative insensitive to noise. Literature on the method is extensive, and there are many applications, particularly in petroleum basins.

Which of the methods to use depends largely on the type of data employed, the type of answers required, and data input skills. Graphic correlation is easy to use, but subjective and devoid of error analysis. RASC+CONOP are an optimal strategy for insight in high-resolution zonations, with optimized tracking of markers.

Table 1 lists key properties of the three methods that apply to microfossil events in wells.

To understand the new crossplots approach to evaluate and express stratigraphic fidelity, we have to understand how RASC and CONOP results differ, and how they can be similar.

The methods yield the following results (Gradstein et al., 1985; Sadler, 2001):

A. (RASC) Most likely Optimum Sequence of events; this is an ordinal composite standard, where calculated event positions in the composite are averages of all (well) section positions encountered. The composite standard levels show estimates of event variance.
B. (RASC) Scaling of the Optimum Sequence; this is a composite standard with interval zones.
C. (CONOP) Composite standards with display of penalty (misfit) of events, according to two strategies:

1) event positions in the composite are unconstrained, and can move either up or down, not unlike RASC
2) event positions in the composite are maximized, either stratigraphically upwards (for tops) or downwards (for bases), as in conventional graphic correlation seeking total stratigraphic ranges. Using this strategy, events from the lower part of ranges generally move downwards in the composite standard, and events in the upper part of ranges upwards.

Figure 1 graphically illustrates different run strategies using first (lowest) and last (highest) occurrence events of a taxon in eight (well) sections taxa. RASC finds the average first and average last occurrence of an event, and hence constructs an average range; the range endpoints have an uncertainty attached, hence the RASC method is probabilistic. CONOP can be instructed to model either average ranges (unconstrained moves mode) or total ranges (maximizing moves mode); CONOP both mimicks probabilistic methods and models deterministic stratigraphic solutions. For both methods to work well, datasets should hold the stratigraphic occurrences of a reasonable number of taxa (e.g. 50 or more) in 10 or more sections.

Both RASC and CONOP also have a provision in their program runs that event positions are not allowed to move, and are correlated as locally observed. An example of an event that may not be allowed to move (relative to its neighbours) is a distinctive ash layer or log horizon.

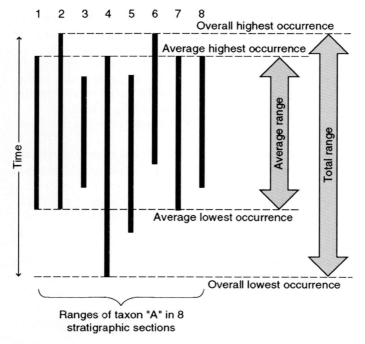

Fig. 1 Graphic illustration of how first (lowest) and last (highest) occurrence events of a taxon in eight (well) sections are treated by respectively Ranking & Scaling (RASC), and by Constrained Optimization (CONOP). RASC finds average first and average last occurrence of events, and hence constructs an average range; range endpoints may have an uncertainty attached; method is probabilistic. CONOP can be instructed to model either average ranges or total ranges

Now we can create stratigraphic cross plots of the probabilistic RASC results (A or B option) with either the probabilistic (option 1) or deterministic (option 2) results using CONOP. If the events in a data set do not deviate much their relative position from well to well a correlation of the RASC Optimum Sequence results with the Optimum Sequence results using CONOP in strategy 1 will show limited or no cross-over of correlation lines. A two way scattergram of the results results in a tight-fitted channel. This approach was also explored by Cooper et al. (2001) for CONOP and RASC results in the Neogene of the Taranaki Basin, New Zealand. The authors calculated high correlation coefficients.

The opposite is true for events in a dataset that deviate their relative stratigraphic positions much from section to section. In this case the average and total stratigraphic ranges differ a lot, and this is readily modelled using CONOP in option 2 mode versus RASC mode. Correlation of RASC and CONOP results will show more misfit, and two way graphs more scatter. The events that misfit more often have above normal variance, and are least useful for tracking their stratigraphic level.

Thus, the crossplot strategy enables well-site paleontologists to predict with confidence what the chances are a particular horizon has been identified, using best marker criteria. Since the methods target all data, not a regional 'mindset', correlation potential of wells is optimized.

2 Results

To test its utility, the crossplot method was applied to two datasets. One well-known datasets is from the North Sea (Gradstein and Bäckström, 1996; Kaminski and Gradstein, 2005) using Last Occurrence (LO) and Last Consistent Occurrence (LCO) events of dinoflagellates and foraminifers. The dataset comes from 30 wells and consists of 1430 occurrences of 88 taxa, and is reported in Part I of this study.

The second dataset, one that Garry Jones helped develop, is from the deep Gulf of Mexico, using LO, LCO, Acme, First Common Occurrence (FCO) and First Occurrence (FO) events of Neogene nannofossils and foraminifers. The Neogene zonation uses 1688 records of 85 events in 13 wells. Both datasets and the cross-plots are outlined below. This dataset is reported in part II of this study.

2.1 North Sea Geology

The North Sea region contain remnants of stratigraphically superimposed sedimentary basins of Late Paleozoic through Cenozoic age, like stacks of incomplete pancackes. The regional history is complex; differential subsidence and repeated uplifts are related to extensive mobilisation of the North Atlantic rift systems. Widespread sealevel rise took place in mid to late Cretaceous time, creating extensive transgression. A Late Cretaceous through Danian chalk blanket formed, originally also covering much of Great Britain and extending across the North Sea, Holland, northern Germany, and Denmark; this 'chalk sea' was 200–400 m deep in places. In the northern part of the North Sea, west of Shetlands and offshore Norway, coeval marine deposition was of a fine-grained terrigenous nature, with marls and shales rather than carbonates.

Deeper water, bathyal sediments, including minor and major gravity flow, siliciclastic wedges, of middle to Late Cretaceous, and of Paleogene age are widespread and contain diversified agglutinated benthic assemblages. In the southern part of the central North Sea, where deep water conditions prevailed into the Miocene, the agglutinated assemblage accordingly extends stratigraphically upwards.

Most diversified, most abundant, and most widespread agglutinated assemblages are found in the fine-grained, deep marine shales that were laid down during the rapid subsidence of the basins in the late Paleocene (Jones, 1988; Gradstein and Bäckström, 1996). This was the time when Paleogene seafloor spreading started in the Atlantic Ocean, north of the Charlie Gibbs fracture zone.

The large scale deposition of volcanic tuffs (Balder Fm) during the earliest Eocene coincides with the eruption of major flood basalts in eastern Greenland and Rockall, at the onset of seafloor spreading in the Norwegian Sea. The ash deposit is a prominent North Sea seismic reflector. Due to the flood-basalt outpourings, the North Sea became restricted, as reflected in the widespread distribution of diatoms, including the now pyritized pillbox *Fenestrella antiqua*, and virtual absence of bottom fauna in the severely dysaerobic basin. Surface water salinity may have been below normal.

A major terrigenous clastic feature in the Central North Sea and West of Shetlands is the presence of large sand bodies intercalated in the upper Paleocene to lower Eocene, considered to be deltaic lobes, deeper water sheet sands and turbidites of late Paleocene to early Eocene age. These deep water sands form the producing horizons in the Forties, Montrose, Frigg, Lomond, Cod and many other oil and gas fields. Studies of the Forties Field Paleocene reservoirs show the presence of typical deep marine fan complexes, with prodeltaic shales and silts passing upward through fine sandstones of the prograding delta slope, which in turn are overlain by coarser sandstones deposited in distributary channels on the delta top.

During the Eocene regional subsidence slowed down, terrigenous clastic supply diminished, and by mid-Cenozoic time the northern North Sea, Viking Graben had filled, but in the more southern North Sea, Central Graben deep marine sedimentation lasted until the middle Miocene. Rapid, basin-wide middle Miocene through Pleistocene basinal subsidence and concommittant uplift of basin edges, led to massive late stage sediment fill deposited in shallow water depths. The cause of this neotectonic phase results may be intensification and re-orientation of compressional intra-plate stresses perpendicular to the basin axis, concommittant with the Alpine orogeny in southern and central Europe. Localised late Miocene sediment condensation was probably caused by the effects of rapid eustatic sealevel lowering in a widespread shallow-marine setting. During the Pleistocene the northern part of the North Sea experienced glacial deposition and erosion.

During the active Paleocene to Eocene phase of North Sea basin subsidence, benthic foraminiferal faunas were markedly different from Central Graben to onshore outcrops. The former were deposited in middle to upper bathyal water depths (<1000 m), whereas onshore deposits were formed in neritic environments (<200 m). Only in the late Paleogene (Oligocene) and Neogene phase of basinal infilling did benthic foraminiferal faunas become gradually more homogeneous over the entire area, although the central part of the North Sea Basin remained deep up to the middle Miocene, as shown by the persistence there of an agglutinated flysch-type fauna, and recurrent incursions of warm-water planktonic foraminifera. Correlations between the onshore and North Sea Basin succession in the Paleocene and Eocene (at least) is best achieved by dinoflagellate cyst biostratigraphy, integrated with the biostratigraphy provided by the calcareous plankton (foraminifera and nannoplankton) and benthic foraminifera, magnetostratigraphy and volcanic ash stratigraphy. This way a correlation network has been established over NW Europe, which serves as the background against which the probabilistic zonation was developed.

2.2 North Sea Biostratigraphy

A total of 30 wells in the Viking and Central Grabens of the North Sea were selected for the zonations, using RASC and CONOP.

The record used for RASC consists of 1360 occurrences of 100 taxa of benthic and planktonic foraminifera, dinoflagellates, miscellaneous microfossils and 6 log markers. Almost all events are last occurrences (LO) in relative time, with two Last Common Occurrences (LCO). Each event occurs in at least 5 out of 30 wells, leaving 107 events; 12 unique events (occurring in fewer than 5 wells) that are useful for age calibration were also inserted. Figure 2 shows the RASC Optimum Sequence with standard deviations; the average standard deviation (solid line) is 1.56; 55 out of 92 events have a standard deviation below average; unique events have no variance calculated. Two taxa, *Reticulophragmoides jarvisi* and *Cystammina pauciloculata* show erratic LO event positions in the wells, resulting in high standard deviations and outlier penalty points; these taxa were deleted from the record.

In Fig. 3 is the RASC Scaled Optimum Sequence with zones assigned subjectively. There are 18 zones and subzones, named NSR 1–13 (North Sea Rasc), of early Paleocene through early Pleistocene age. Large breaks (at events 129, 50, 206, 6, 266 and 23) indicate transitions between natural microfossil sequences, and/or hiatusses. The zones contain 33 agglutinated benthic events (32 × LO and 1 × LCO events) for 32 taxa, 29 of which are described in detail in Kaminski and Gradstein (2005).

Relative to the zonation for offshore northeast Canada (Kaminski and Gradstein 2005, Figs 14 and 15) many more agglutinated foraminiferal taxa are included, in part because the record is more diverse, and in part because of more detailed sampling and more detailed taxonomy. Both late Paleocene and early Oligocene assemblages are markedly diverse, and totally or largely devoid of calcareous benthic and planktonic taxa. One reason for the greater diversity in the North Sea Paleogene record may be less competition from calcareous benthic taxa, relative to the (more fertile and less restricted?) Canadian Atlantic margin.

The total stratigraphic range of taxa may extend younger than the average stratigraphic range, with the result that the average last occurrences displayed in the RASC zonation of Figs. 2 and 3 may be slightly older. On average, event observation in the wells may be closer to the average stratigraphic position than the last occurrence end point of the taxa in a fossil ranges chart that depicts total ranges.

Average LO or LCO events of agglutinated taxa typical for the North Sea Paleogene are:

1. Zone NSR2, late Paleocene: *Ammoanita ingerlisae, A. ruthvenmurrayi, Reticulophragmium pauperum, R. garcilassoi, Spiroplectammina spectabilis* LCO, *Rzehakina minima, Placentammina placenta, Caudammina excelsa*, and *Cystammina sveni*; the latter two have relatively high standard deviations (Fig. 2); *A. ingerlisae, A. ruthvenmurrayi*, together with rare *Conotrochammina voeringensis* are confined to the lower part of the zone.
2. Zone NSR4, early Eocene: *Recurvoidella lamella* and *Spiroplectammina navarroana*. The zone is named after the easily recognisable planktonic *Subbotina*

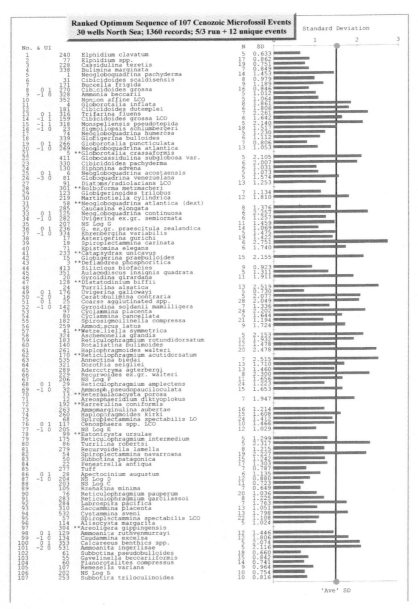

Fig. 2 North Sea Cenozoic Optimum Sequence with standard deviations, calculated with RASC (Ranking & Scaling) program. Record consists of 1360 occurrences of 100 taxa of benthic and planktonic foraminifera, dinoflagellates, miscellaneous microfossils and 6 log markers in 30 wells. Almost all events are last occurrences (LO) in relative time, with two Last Common Occurrences (LCO). Each event occurs in at least 5 out of 30 wells, leaving 107 events; 12 unique events (occurring in fewer than 5 wells) were also inserted. Average standard deviation (solid line) is 1.56; 55 out of 92 events have a standard deviation below average; unique events have no variance calculated

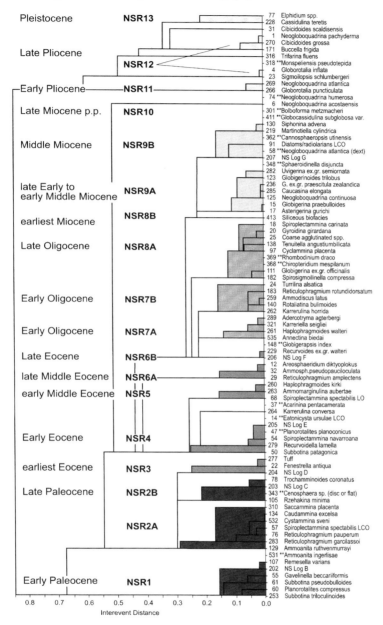

Fig. 3 North Sea Scaled Cenozoic Optimum Sequence, calculated with RASC (Ranking & Scaling) program. Scale is relative, and is derived from cross-over frequency of all Optimum Sequence events. Eighteen NSR (North Sea RASC) zones are recognised of Paleocene through Plio-Pleistocene age (Gradstein and Bäckström, 1996). Large scaling breaks (at events 129, 50, 206, 6, 266 and 23) indicate transitions between natural microfossil sequences, result of hiatuses and /or lithological and facies changes. Zones contain 33 agglutinated benthic foraminiferal events (32 × LO and 1 × LCO events) for 32 taxa, 29 of which are described in Kaminski and Gradstein (2005)

patagonica, which has a circum Atlantic productivity event in the middle part of the early Eocene (see Gradstein et al., 1994, p. 37), when pink marls occur.
3. Zone NSR5, early Middle Eocene: *Spiroplectammina spectabilis* LO, *Ammomarginulina aubertae*, and *Haplophragmoides kirki*; the degree of uncertainty for the LO events of the three taxa is fairly high.
4. Zone NSR6A, late Middle Eocene: *Reticulophragmoides amplectens*, and *Ammosphaeroidina pseudopauciloculata*; the latter event has a fairly high standard deviation; it may be found as young as early Oligocene.
5. Zone NSR7A, early Oligocene: *Annectina biedai, Haplophragmoides walteri, Karreriella seigliei*, and particularly *Adercotryma agterbergi*. The LO events of *H. walteri* and *A. biedai* are often older, and have high standard deviations.
6. Zone NSR7B, early Oligocene: *Ammodiscus latus* and *Reticulophragmium rotundidorsatum*, both with fairly high standard deviations. The cosmopolitan calcareous benthic *Turrilina alsatica* is characteristic for this zone.
7. Zone NSR8, late Oligocene: *Spirosigmoilinella compressa*; rare specimens may be found younger.

Table 2 North Sea exploration wells studied for foraminifers, dinoflagellates, miscellaneous microfossils and log markers

1	Saga	(N)	35/3-1
2	Saga	(N)	35/3-4
3	Hydro	(N)	34/8-1
4	Saga		34/7-1
5	Saga		34/7-2
6	Saga		34/7-4
7	Saga		34/7-15s
8	Saga		34/7-22
9	Saga		34/7-24s
10	Norsk Hydro		31/2-19S
11	Statoil		30/3-1
12	Statoil		30/3-A1
13	Total	(UK)	3/25-1
14	Shell	(UK)	9/23-1
15	Mobil	(UK)	9/13-1
16	Mobil	(UK)	9/13-3A
17	Mobil	(UK)	9/13-5
18	Esso	(N)	16/1-1
19	BP	(UK)	15/20-2
20	Phillips	(UK)	16/17-1
21	Phillips	(UK)	16/29-2
22	BP	(UK)	21/10-1
23	BP	(UK)	21/10-4
24	Mobil	(UK)	21/28-1
25	Shell	(UK)	22/6-1
26	Phillips	(UK)	23/22-1
27	Shell	(UK)	29/3-1
28	Shell	(UK)	30/19-1
29	Saga	(N)	2/2-4
30	Amoco	(N)	2/8-1

Compared to the RASC zonation, the North Sea Optimum Sequence using CONOP is not much different. The data set and Optimum Sequence stack (Table 2) is virtually the same as used with RASC. CONOP was executed in the unconstrained (BOB) mode, event positions in the composite are unconstrained, and can move either up or down, not unlike RASC

Figure 3 is a crossplot of the RASC and CONOP (BOB mode) Optimum Sequences. Spearman's rank correlation is 0.99595 and Kendall' tau 0.95612. One may argue that *G.subglobosa* var., *G.venezuelana* and maybe also *G.trilobus* and *W.symmetrica* are stacked too old with Conop and /or too young with RASC. The first two have high variances; the last one only occurs in 3 wells.

Figure 4 is the same plot, but here CONOP event last occurrences are maximized (socalled DISappearance mode), resulting in looser fit to the RASC Optimum Sequence, although Spearman's rank correlation is still a high 0.98966 and Kendall's tau 0.92671. Subjectively, several stratigraphically conclusions may be drawn. The average last occurrences of *B.frigida*, *M.cylindrica*, *E.elegans*, *R.rotundidorsata*, *C.sveni*, *C.excelsa*, and Calc. benthics spp. locally differ much from the uppermost ranges, as found in some wells (RASC has tables showing in which wells), and RASC variances are also high. The *R.rotundidorsatum* extension is part of a trend with stratigraphically nearby *C.cancellata*, *C.placenta*, Coarse agglutinated spp., *G.girardana*, Silicious biofacies and also *G. praebulloides* locally (Central Graben of the North Sea) extending one or even two zones younger. *C.dutemplei*, *C.subglobosa* var., *G.trilobus* (again) are either stacked too old in CONOP, or too young in RASC. North Sea Log marker G, Log F and Log E are stacked too old in CONOP, since their stratigraphic position in the RASC solution was independently confirmed (see Gradstein and Bäckström, 1996). In this particular CONOP run, the log markers were treated as marker (ASH) horizons, allowing minimal stratigraphic movement; this seems to result in underplotting relative to the DIS events.

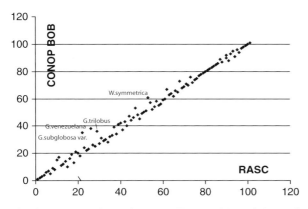

Fig. 4 Crossplot of Ranking & Scaling (RASC) and Constrained Optimization (CONOP, BOB mode) Optimum Sequences. The (stratigraphic) fit is good. Spearman's rank correlation is 0.99595 and Kendall' tau 0.95612. One may argue that *G.subglobosa* var., *G.venezuelana* and maybe also *G.trilobus* and *W.symmetrica* are stacked too old with CONOP and /or too young with RASC. First two taxa have high variances; last one only occurs in 3 wells

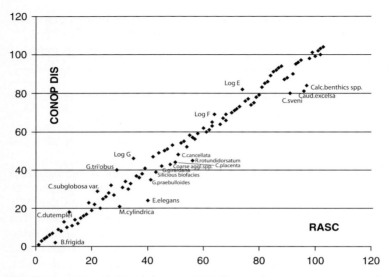

Fig. 5 Same plot as in Fig. 4, but now CONOP event last occurrences are maximized (socalled DISappearance mode), resulting in looser fit to RASC Optimum Sequence, although Spearman's rank correlation is still a high 0.98966 and Kendall's tau 0.92671. For details see text

Although many of the stratigraphic diachroneity conclusions drawn above also could be derived from RASC results alone, the bias is confidence in one method only. Deriving these conclusions from two independent methods strengthens the stratigraphic insight and geologic utility of the data.

References

Agterberg FP, Gradstein FM (1999) The RASC method for ranking and scaling of biostratigraphic events. Earth Sci Rev 46: 1–25

Cooper RA, Crampton JS, Raine I, Gradstein FM, Morgans HEG, Sadler PM, Strong CP, Waghorn D, Wilson GJ (2001) Quantitative biostratigraphy of the Taranaki Basin, New Zealand: a deterministic and probabilistic approach. Am Assoc Petrol Geol Bull 85 (8): 1469–1498

Gradstein FM, Agterberg FP, Brower JC, Schwarzacher W (1985) Quantitative stratigraphy. Reidel Publishing Company, Dordrecht and UNESCO, Paris, p 598

Gradstein FM (2004) Quantitative methods for applied microfossil biostratigraphy. In: Koutsoukos E (ed) Applied stratigraphy. Kluwer, Dordrecht, The Netherlands

Gradstein FM, Bäckström S (1996) Cainozoic biostratigraphy and palaeobathymetry, northern North Sea and Haltenbanken. Norsk Geologisk Tidsskrift 76: 3–32

Gradstein FM, Kaminski MA, Berggren WA, Kristiansen IL, D'Iorio MA (1994) Cenozoic biostratigraphy of the Central North Sea and Labrador Shelf. Micropaleontolgy 40 (Supplement): p 152

Jones GD (1988) A paleoecological model of Late Paleocene "flysch-type" agglutinated foraminifera using the paleoslope transect approach, Viking Graben, North Sea. Abh Geol Bundesansfelt 41: 143–153

Kaminski MA, Gradstein FM (2005) Atlas of Paleogene cosmopolitan deep-water agglutinated foraminifera. Grzybowski Foundation, Special Publication 10, London, p 548

Kemple WG, Sadler PM, Strauss DJ (1995) Extending graphic correlation to many dimensions. Stratigraphic correlation as constrained optimization. Soc Econ Paleo Mineral Spec Pub 53: 65–82

Sadler PM (2001) Constrained optimization approaches to the paleobiological correlation and seriation problems. A user guide and reference manual to the CONOP program family. Unpublished, author released document

Shaw AB (1964) Time in stratigraphy. McGraw-Hill Book Co., New York, p 365

Increasing Resolution in Exploration Biostratigraphy – Part II

A. Bowman, F.M. Gradstein, A. Lugowski and O. Hammer

Reprinted from *Natural Resources Research* DOI: 10.1007/s11053-008-9071-z, when citing this article please use the DOI number.

Keywords Quantitative biostratigraphy · crossplot method · Ranking and Scaling (RASC) · Constrained Optimization (CONOP) · Gulf of Mexico

1 Introduction

Exploration biostratigraphy in the oil and gas industry typically focusses on extinction levels ("tops") of species observed in a sedimentary sequence. Increased biostratigraphic resolution is gained through the investigation of the potential usefulness of other and non-traditional bioevents (first downhole increases, acmes, etc.). The greater the biostratigraphic resolution, the more traceable horizons are created, which allow professional geoscientists to better understand the subsurface geology. Crossplots with Ranking & Scaling (RASC) and Constrained Optimization (CONOP) zonal sequences increase stratigraphic resolution and correlation potential of biozonations. The new method will be demonstrated in the deep Gulf of Mexico, using Neogene petroleum exploration data.

A. Bowman
Chevron Energy Technology Co, 14141 SW Freeway, Sugarland, TX 77478, USA,
e-mail: abowman2@bigred.unl.edu

F.M. Gradstein
Geological Museum, University of Oslo, N-O318 Oslo, Norway,
e-mail: felix.gradstein@nhm.uio.no

A. Lugowski
Computer Science Department, University of California, Santa Barbara CA 93106-5110, USA,
e-mail: alugowski@gmail.com

O. Hammer
Geological Museum, University of Oslo, N-O318 Oslo, Norway,
e-mail: oyvind.hammer@nhm.uio.no

2 Gulf of Mexico

The Gulf of Mexico (GOM) has a rich tradition of petroleum exploration. Many wells have been drilled, producing an extensive micropaleontological database, mostly consisting of calcareous nannofossils and foraminifers (benthic and planktonic).

The Mad Dog Field occupies Blocks 782, 783, and 826 in the Green Canyon exploration area of the deepwater Gulf of Mexico, approximately 200 miles south of New Orleans (Fig. 1). The average water depth of the field area is between 4500 and 6800 feet. The field itself is a faulted, four-way closure located under the Sigsbee Escarpment. Three reservoir sands have been discovered in the lower Miocene, and were termed the DD, EE, and FF. The sands have been interpreted as gravity flow, turbidite sands, and are laterally continuous over several miles. The DD, EE, and FF sands are medium to fine grained, and have about 360 feet of total thickness (Smith et al., 2001). The field was first discovered in 1998, and is currently in the production phase. Production began in January 2005, and total reserves are estimated at 200–450 million barrels oil equivalent.

With the reservoir sands located under salt, and at subsurface depths greater than 20,000 feet, seismic correlation and mapping are difficult. Thus, biostratigraphic data are an essential tool for correlating key sands.

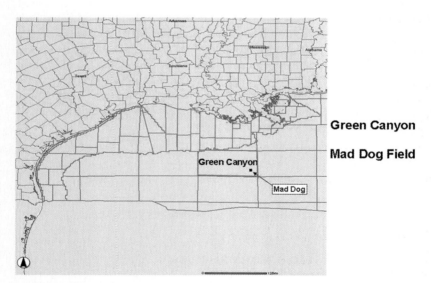

Fig. 1 Mad Dog Field occupies Blocks 782, 783, and 826 in Green Canyon exploration area of deepwater Gulf of Mexico, approximately 200 miles south of New Orleans

3 Methods

Thirteen wells that collectively spanned the GOM deepwater Miocene section were chosen for the study. Wells were chosen based on the quality of biostratigraphic data, and depth of penetration within the Miocene. The key focus was to incorporate as many wells as possible in order to fully represent the proper order of events in the Miocene GOM.

In an effort to increase success with correlations among wells in the Mad Dog Field, the Ranking and Scaling (RASC) and Constrained Optimization (CONOP) methods were applied to arrive at a high-resolution biozonation using as many data as possible. The methods for creating the Optimum Sequence are outlined in Gradstein et al. (2008); here we focus on data preparation for the Mad Dog program runs.

The initial process required evaluating and validating bioevents of hundreds of Miocene age taxa using hard copy range charts (BugCad plots), followed by analysis of the documented bioevents through the biostratigraphic software program Integrated PaleoSystem (IPS). This two stages process yielded the placement of the more common types of bioevents (species range tops and bases), as well as discerning new and useful subordinate bioevents. The RASC result was an optimum stratigraphic order of bioevents, with uncertainty of the reliability of each.

3.1 Defining Bioevents

For RASC analysis to be of use, it is crucial to input quality data. Therefore producing and following "rules" is necessary when documenting various bioevents. Determining the criteria that defines each bioevent is an important step. Uncertainty exists when correlating non-traditional bioevents among wells because biostratigraphers may have different ideas as to how non-traditional bioevents are defined. The rules produced for this study were created based on careful examination of the data, and were modified if they were too complex or illogical. The final set of eight taxonomic-stratigraphic rules were strictly followed during the process of documenting bioevents for the deepwater GOM Miocene wells (Table 1).

3.2 Capturing and Recording Bioevents

Bioevents were captured for calcareous nannofossil and foraminifera species using the BugCad distribution charts and IPS software. A key asset of IPS is the ability to calculate and plot peaks in species abundance and species diversity within a well, therefore making it possible to choose the key bioevents in condensed intervals associated with periods of maximum flooding (within fine-grained lithologies with the best preservation).

Table 1 Eight different rules for stratigraphic bioevents were captured for calcareous nannofossils and foraminifers in thirteen Miocene Gulf of Mexico wells studied

HO = Highest stratigraphic occurrence
cannot be in a sand body

TOP = Extinction level
Based on biostratigraphic interpretation

FDI = First downhole abundance increase
Change from low abundance (1, 2, 6 specimens) to at least 20

FDCO = First downhole common and consistent occurrence
"common" = 50 or greater
> 90 ft of "common" interval
no > 50s above FDCO event
no breaks > 90 ft in beginning of "common" interval (3 continuous occurrences)

LDCO = Last downhole common and consistent occurrence
Analogous to FDCO

ACME = Abundance spike
short and abrupt event; dramatic increase in abundance
if many possibilities, "best" ACME chosen

BASE = Evolutionary first appearance
Based on biostratigraphic interpretation

LO = Lowest stratigraphic occurrence
cannot be in a sand body

Each well had between 32 and 57 bioevents, with the average number of bioevents per well being 46. The number of bioevents in each well is a function of depth of penetration (age), and quality of microfossil preservation. Data from each well was then recorded in ASCII files, listing depth for each specific bioevent and the name of the particular bioevent (ex. "Sphenolithus dissimilis TOP"). Recorded bioevents were then prepared for several iterations of RASC.

3.3 RASC and CONOP Analysis

During RASC runs, the statistical requirements were altered as necessary in order to achieve the most useful output. The main statistical parameter to alter during iterations is the "k value," or the minimum number of wells that a bioevent must occur in. This threshold directly controls the stratigraphic results. As the "k value" increases, the total number of bioevents in the Optimum Sequence decreases. The opposite is obviously true if the "k value" is decreased.

RASC analysis was carried out using threshold k values of 7 and of 6. The "k = 7" RASC run produced 71 total bioevents in the Optimum Sequence, 49 of which had standard deviations below the average of 4.70. The "k = 6" RASC run left 91 bioevents in the Optimum Sequence, with 61 bioevents possessing standard deviations lower than the average (avg. = 4.80).

Increasing Resolution in Exploration Biostratigraphy

The same data were also analyzed with the Constrained Optimization (CONOP) method. CONOP in average event "ranges" (BOB) mode yielded results closely comparable to RASC, with a more diverging sequence using the DIS mode that maximizes event range endpoints.

Figure 2 is a crossplot of the RASC and CONOP (BOB mode) that shows good coherence and low scatter, whereas the crossplot of Fig. 3 with CONOP in DIS mode produced considerable scatter. Outlier events in the upper left quadrant of Fig. 3 are events that vary their BASE or LO much from well to well; outlier events in the lower right quadrant are events that vary their FDI, HO or TOP occurrence much from well to well. Seventeen of these outliers were thus deleted from further RASC runs.

Since a threshold k value equal to 6 produces the most bioevents with minimal change in data quality, and results in the most useful Optimum Sequence, analysis was continued with this run. Firstly, bioevents that had standard deviations greater than the average were omitted from the final Optimum Sequence. Next, all bioevents involving the highest occurrence (HO) and lowest occurrence (LO) of species were deleted from the RASC Optimum Sequence, as they generally created confusion when related to the true extinction top and true evolutionary base of the various species. This step was performed based on knowledge and experience of calcareous nannofossil biostratigraphy.

After the less reliable bioevents were thus omitted, a total of 38 bioevents were retained, of which 34 were calcareous nannofossil events and 4 were foraminifer

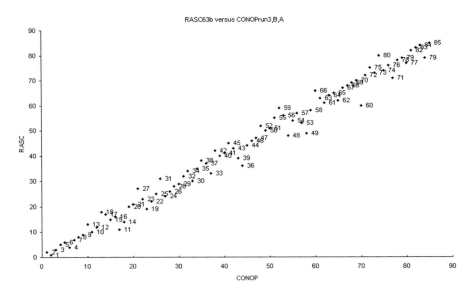

Fig. 2 Crossplot of RASC and CONOP (BOB mode) Optimum Sequences for Miocene Gulf of Mexico. Stratigraphic fit is good, result of close convergence on same Optimum Sequence using both methods

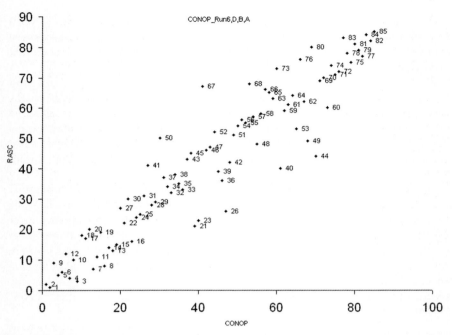

Fig. 3 Same crossplot as in Fig. 2, but now CONOP event last occurrences are maximized (so-called DISappearance mode), resulting in a much looser fit to RASC Optimum Sequence. Outlier events in upper left quadrant are events that vary their BASE or LO much from well to well; outlier events in lower right quadrant are events that vary their FDI, HO or TOP occurrence much from well to well. Seventeen of these outliers were thus deleted from further RASC runs. These are the *too low* events #41-*C.macintyrei* LO, # 50-*R.gelida* BASE, # 67-*D.quinqueramus* LO, # 68-*M.convallis* BASE, # 7-*D.quinqueramus* BASE, #76-*D.surculus* BASE, and # 80-*D.radiatus* BASE; and the *too high* event # 21-*G.dehiscens* FDI, # 23-G.mayeri FDI, # 26-*G.dehiscens* TOP, # 40-*U.peregrina* TOP, # 44-*S.moriformis* HO, # 49-*U.jafari* FDI, # 53-*P.bulloides* HO, and # 60-*O.universa* FDI

events. Table 2 lists the final, curtailed Optimum Sequence, showing the bioevents with standard deviation, and the number of wells in which each occurred.

In an effort to apply the final biozonation created by the Optimum Sequence, four wells within the Mad Dog field were used in this study to refine the biostratigraphic correlation. The names of the wells have been omitted for proprietary concerns. The purple correlation lines are bioevents used from the Optimum Sequence. Most bioevents throughout the entire Miocene have very low crossover frequencies. However, because the most crucial reservoirs in the Mad Dog Field are Middle and Early Miocene age, the focus of the biostratigraphic correlations was directed to these intervals.

Figure 4 illustrates the biostratigraphic correlations for the entire Miocene, created using the bioevents produced from the Optimum Sequence. Figure 5 shows an enlarged portion of the correlation for the Lower to Middle Miocene sedimentary

Table 2 Miocene Optimum Sequence using Ranking and Scaling (RASC) on 13 wells in Gulf of Mexico. RASC run using events that occur in 6 or more wells (k = 6) left 91 bioevents in Optimum Sequence, with 61 bioevents possessing standard deviations lower than average (avg. = 4.80). Bioevents involving highest occurrence (HO) and lowest occurrence (LO) of species were deleted from Optimum Sequence, as they generally created confusion when related to true extinction top and true evolutionary base of various species. This step was performed based on knowledge and experience of calcareous nannofossil biostratigraphy. After less reliable bioevents were omitted, a total of 38 bioevents were retained, of which 34 were calcareous nannofossil events and 4 were foraminifer events. List shows final Miocene Optimum Sequence of bioevents with their standard deviations, and number of wells in which each occurred. Foraminifera events are represented by a numeric sign (#)

Event	S.D. (Avg. = 4.80)	Number of Wells
Discoaster pentaradiatus BASE	3.19	7
Discoaster surculus BASE	3.66	7
Discoaster quinqueramus BASE	2.19	6
Discoaster bollii TOP	1.89	6
Minylitha convallis BASE	2.63	7
Discoaster bollii BASE	2.33	10
Discoaster prepentaradiatus BASE	2.26	6
Discoaster neohamatus BASE	2.12	7
Discoaster brouweri BASE	4.20	7
Catinaster coalitus TOP	1.82	8
Discoaster exilis TOP	1.73	6
Coccolithus miopelagicus TOP	3.10	12
Discoaster musicus TOP	4.44	9
Discoaster sanmiguelensis TOP	3.24	10
Calcidiscus premacintyrei TOP	2.39	9
Globorotalia peripheroacuta TOP#	4.25	9
Cyclicargolithus floridanus TOP	1.74	11
Globorotalia peripheroronda TOP#	4.81	7
Discoaster deflandrei TOP	4.66	11
Sphenolithus heteromorphus TOP	2.62	11
MMR	2.95	8
Cyclicargolithus floridanus FDI	3.90	11
Sphenolithus heteromorphus FDI	2.75	11
Calcidiscus premacintyrei FDI	3.92	6
Discoaster petaliformis TOP	1.97	11
Globorotalia peripheroacuta BASE#	3.51	8
Orbulina universa BASE#	4.11	7
Discoaster petaliformis FDI	2.91	7
Helicosphaera ampliaperta TOP	3.69	10
Discoaster petaliformis BASE	4.09	9
Discoaster deflandrei FDI	1.93	8
Helicosphaera kamptneri-carteri ACME	4.50	9
Sphenolithus heteromorphus ACME	3.58	11
Sphenolithus heteromorphus BASE	3.24	7
Sphenolithus belemnos TOP	2.17	7
Discoaster calculosus TOP	3.63	8
Discoaster deflandrei ACME	4.20	6
Triquetrorhabdulus carinatus TOP	2.27	6

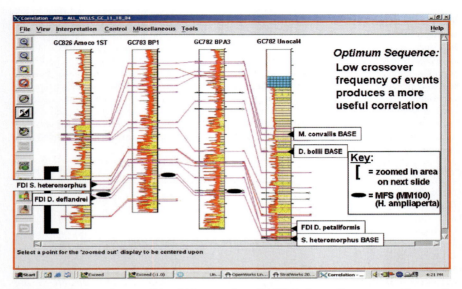

Fig. 4 Application of non-traditional bioevents from RASC Optimum Sequence used in correlation of Mad Dog Field

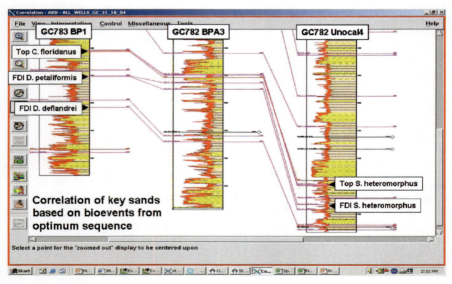

Fig. 5 Zoomed in portion of well correlation for Mad Dog Field. Traditional and non-traditional bioevents from RASC Optimum Sequence are used to improve biostratigraphic correlations

section. The standard and commonly used bioevents (with associated standard deviations), along with the reliable seismic reflector (MMR) are:

Discoaster pentaradiatus BASE (S.D. = 3.19)
Discoaster bollii TOP (S.D. = 1.89)
Minylitha convalis BASE (S.D. = 2.63)
Discoaster bollii BASE (S.D. = 2.33)
Discoaster prepentaradiatus BASE (S.D. = 2.26)
Catinaster coalitus TOP (S.D. = 1.89)
Discoaster exilis TOP (S.D. = 1.73)
Coccolithus miopelagicus TOP (S.D. = 3.1)
Discoaster sanmiguelensis TOP (S.D. = 3.24)
Calcidiscus premacintyrei TOP (S.D. = 2.39)
MMR (Middle Miocene Reflector) (S.D. = 2.95)
Discoaster petaliformis TOP (S.D. = 1.97)
Sphenolithus heteromorphus BASE (S.D. = 3.24)

The most useful non-traditional bioevents produced from this work, and used for correlation of wells in the Mad Dog field include: *Sphenolithus heteromorphus* FDI, *Discoaster deflandrei* FDI, and *Discoaster petaliformis* FDI. The bioevents with lowest standard deviations and the associated Miocene NN Zone (Martini, 1971) are listed below:

NN10: *Discoaster quinqueramus* BASE (S.D. = 2.19)
NN6: *Cyclicargolithus floridanus* TOP (S.D. = 1.74)
NN5: *Sphenolithus heteromorphus* TOP (S.D. = 2.62)
NN3: *Sphenolithus belemnos* TOP (S.D. = 2.17)
NN2: *Triquetrorhabdulus carinatus* TOP (S.D. = 2.27)

4 Conclusions

Biostratigraphic correlations between wells across fields are a crucial tool for mapping key sands. Increasing the biostratigraphic resolution also increases the success in understanding the connectivity and character of important stratigraphic units. The goal of this work was to determine new and reliable bioevents to help build a higher resolution biostratigraphic correlation throughout the Mad Dog Field. The steps followed to achieve this goal were: build criteria (rules) for establishing new bioevents to add biostratigraphic resolution, identify these new bioevents, run the new bioevents through RASC analysis to produce a valid Optimum Sequence, and use the produced Optimum Sequence for correlation of key sands in the Middle and Early Miocene intervals within the Mad Dog Field.

Results show the bioevents incorporated from the Optimum Sequence have low crossover frequencies, therefore allowing for a reliable high resolution biostratigraphic correlation in the Mad Dog Field. This work has provided proven value within the exploration and development realms, and has great potential as a valuable tool not only in the Gulf of Mexico, but worldwide as well.

References

Gradstein FM, Bowman A, Lugowski A, Hammer O (2008) Increasing resolution in exploration biostratigraphy-Part I. In: Bonham-Carter GF, Chenq Q (eds) Forty years of progress in geomathematics. Springer, Heidelberg, pp x–y

Martini E (1971) Standard Tertiary and Quaternary calcareous nannoplankton zonation. In: Farinacci A (ed) Proceedings of second planktonic conference, Roma 1970, Roma, Tecnoscienza, pp 739–785

Smith T, Kenney D, DiMarco M, Poulin M, Trevena A, Steffens G (2001) Geological and geophysical reservoir description at Mad Dog Field – an ultra-deepwater, sub-salt oil discovery in the Gulf of Mexico. In: Abstracts 2001 unocal technology conference, Bangkok, Thailand, March 2001

RASC/CASC: Example of Creative Application of Statistics in Geology

Zhou Di

Reprinted from *Natural Resources Research* DOI: 10.1007/s11053-008-9075-8, when citing this article please use the DOI number.

Abstract RASC/CASC is a computer-based system for quantitative stratigraphic analysis developed by F. Agterberg, F. Gradstein, and their co-workers. The application of the system to the Neogene biostratigraphy of the Pearl River Mouth Basin has demonstrated the advantages of the system. The occurrence data of hundreds of fossils from dozens of wells were analyzed objectively based on established stratigraphic and statistical rules embedded in the system. Through permutation of score matrix the optimum sequence of fossil events was obtained. The calculation of inter-fossil distances allowed the automated biostratigraphic zonation and the age-event correlation. Then the regional geological time table was constructed, and the inter-well chronological correlation and high-resolution subsidence analysis become possible even for wells with incomplete fossil records. Uncertainty at each step was quantified. While all these are important accomplishments in a stratigraphic study, results of the study also helped identifying problems in allocation of fossil events and in dating lithologic divisions.

Keywords RASC/CASC · quantitative stratigraphy · Pearl River Mouth Basin · South China sea

The early development of statistics was largely related to the need for dealing with masses of biological data. In the course of its growth, geology has made more and more contributions to statistics in theory, methodology, as well as in applications. The best known example is the creation of geostatistics which has opened a new area of statistics of regionalized variables. Other examples include the statistics of directional data (Mardia, 1972; Watson, 1983) and compositional data (Aitchison, 1986). The RASC/CASC (RAnking and SCaling / Correlation And Standard error Calculation) in quantitative stratigraphic analysis is another highlight.

Quantitative stratigraphy has been an active field in mathematical geology, applying various statistical and numerical methods (Gradstein et al., 1985; Agterberg, 1990). In October 8–15, 1986, Frits Agterberg led 4 other world leading quantitative stratigraphers to the Jianglin city, SW China, and gave a short course on

Zhou Di
CAS Key Laboratory of Marginal Sea Geology, South China Sea Institute of Oceanology, Chinese Academy of Sciences, Guangzhou 510301, China, e-mail: zhoudiscs@scsio.ac.cn

quantitative stratigraphy. That very successful short course opened the door of quantitative stratigraphy to over 200 Chinese geologists, including me. I was amazed by the theories and methods that were presented to me for the first time. RASC/CASC has attracted me deeply not only by its power in quantifying stratigraphic correlation, but also by its smart and creative way in applying statistics to this particular geological problem. Later I applied this technique together with my student to the biostratigraphic correlation in the Pearl River Mouth (Zhujiangkou) Basin, northern South China Sea (Zhou and Wang, 1994), using the computer programs that were kindly provided by F. Agterberg and F. Gradstein. Based on selected fossil events from multiple wells in the basin, the optimal sequence and biostratigraphic zonation were established, and inter-well correlation was made. Our study provided quantitative evidence that helped to resolve a long-standing debate on the L/M Miocene boundary of the basin. This first RASC/CASC practice in China was successful and won a research award from the China Offshore Oil Company. In this paper I will discuss some highlights and advantages of RASC/CASC according to this experience.

1 The Need of Quantitative Stratigraphy

The Pearl River Mouth Basin (PRMB, i.e. the Zhujiangkou Basin) is a major oil/gas bearing sedimentary basin in northern South China Sea (Fig. 1) with a total area of $\sim$150 000 km^2 bounded by the 1 km sediment isopach. Sediment infilling of the basin consists of syn-rifting continental Paleogene and post-rifting marine Neogene. The total thickness of Cenozoic sediments may exceed 11 km. There were over 60 wells penetrating the sediments by the time of our study, and the general stratigraphic column was established based on bio- and litho-stratigraphic data (Fig. 2). However, there were controversies concerning the location of Paleogene/Neogene boundary, and on the divisions of formations within Paleogene and Neogene (e.g., Qin, 2000). There was a need for a technology that could perform an objective synthesis of as much as possible the existing data in order to resolve the controversy.

Quantitative stratigraphy using RASC/CASC provided the best choice. It could synthesize a large quantity of fossil data automatically and objectively based on established stratigraphic and statistical rules (Gradstein et al., 1985; Agterberg, 1990). The resulting biostratigraphic zonation, composite stratigraphic column, and cross-well correlations are usually more close to reality, because they are less subjective and less subject to random errors than those synthesized qualitatively by human's brain. Thus we decided to apply RASC/CASC to the stratigraphic study of the basin. This was the first application of the method in China.

As the Paleogene strata in the basin are continental and contain only spore and pollen fossils which are more complicated to correlate, our study was focused on the Neogene strata which include 3–6 km thick marine sequences with abundant foraminifera and nannofossils, as well as sporopollens. Out of the 60 wells available at the time, we selected 34 wells for the analysis.

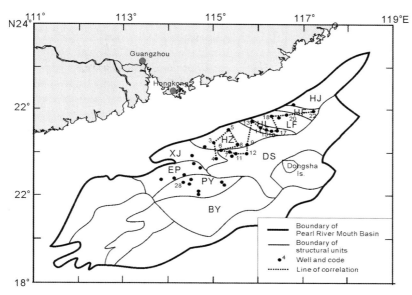

Fig. 1 Map of the Pearl River Mouth Basin, showing major structural units and localities of the wells used in this study. The dotted line connects the wells for the inter-well correlation as shown in Fig. 6. BY- Baiyun Sag; PY-Panyu Low Uplift; EP-Enping Sag; XJ-Xijiang Sag; HZ-Huizhou Sag; HL-Huilu Low Uplift; LF-Lufeng Sag; HF- Haifeng Uplift; HJ-Hanjiang Sag; DS-Dongsha Massif

2 Ranking and Scaling of Events (RASC)

2.1 Definition and Ranking of Event

The first key step of introducing stratigraphic data to statistic analysis is to define "event" (Agterberg, 1990). A biostratigraphic event is defined by a particular stage in a taxon's evolution that may be located uniquely in its time context, as inferred from its position in a rock sequence. The most commonly used events are the last and first appearance (LA and FA), the last and first consistent appearance (FCA and LCA), and the peak appearance (ACME) of a taxon. RASC/CASC allows also the use of non-biostratigraphic events such as a dated volcanic sheet, a dated magnetic reversal, etc. The observed time series of events in the well sections compose the basic data set for the study. In our study, 112 events out of a total of 268 events were retained for the analysis as they are observed in at least 6 wells. These include 55 calcareous nannofossil, 40 foraminifera, and 10 sporopollen events, of which most are LAs with minor FA, FCA, and ACME events.

By definition the order of events should be unique, representing the evolutionary sequence. In reality, however, the uniqueness is rarely observed due to the post-depositional disturbances by organisms, sedimentary hiatus or reworking, tectonic deformation, and magmatic and metamorphic activity, also duo to the bias of fallen well-cuttings and by random errors in sampling and fossil determination. Thus there

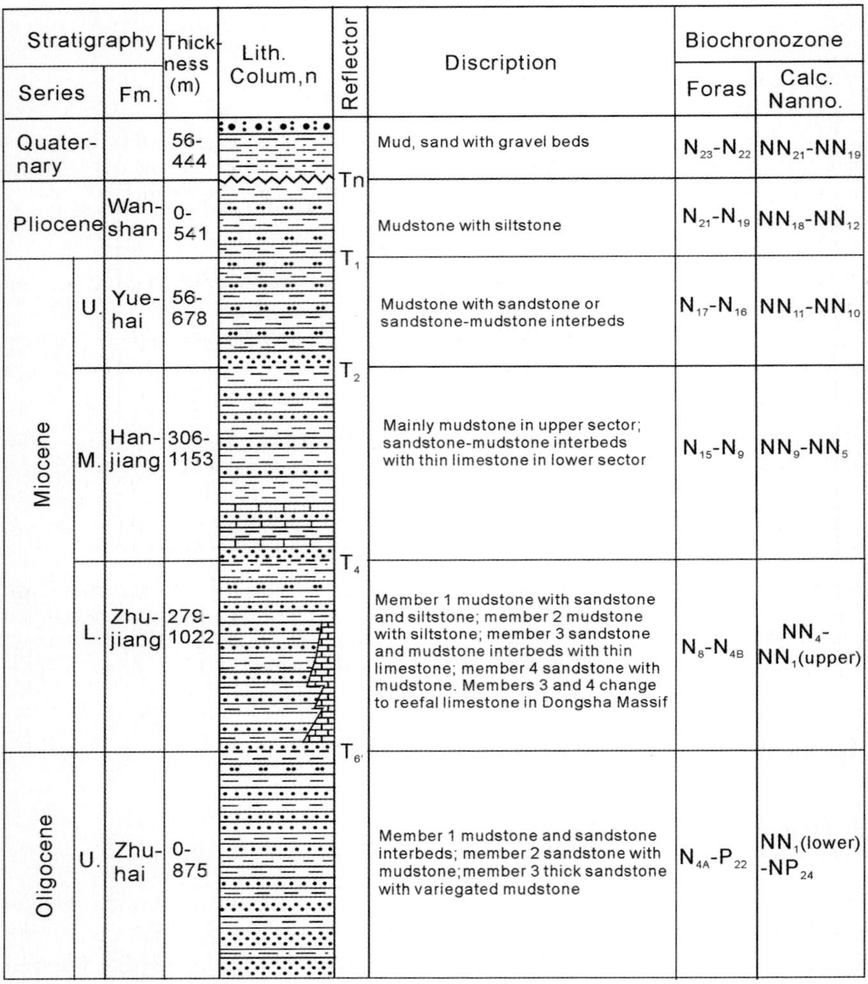

Stratigraphy		Thickness (m)	Lith. Colum,n	Reflector	Discription	Biochronozone	
Series	Fm.					Foras	Calc. Nanno.
Quaternary		56-444		Tn	Mud, sand with gravel beds	N_{23}-N_{22}	NN_{21}-NN_{19}
Pliocene	Wanshan	0-541		T_1	Mudstone with siltstone	N_{21}-N_{19}	NN_{18}-NN_{12}
Miocene	U. Yuehai	56-678		T_2	Mudstone with sandstone or sandstone-mudstone interbeds	N_{17}-N_{16}	NN_{11}-NN_{10}
Miocene	M. Hanjiang	306-1153		T_4	Mainly mudstone in upper sector; sandstone-mudstone interbeds with thin limestone in lower sector	N_{15}-N_9	NN_9-NN_5
Miocene	L. Zhujiang	279-1022		$T_{6'}$	Member 1 mudstone with sandstone and siltstone; member 2 mudstone with siltstone; member 3 sandstone and mudstone interbeds with thin limestone; member 4 sandstone with mudstone. Members 3 and 4 change to reefal limestone in Dongsha Massif	N_8-N_{4B}	NN_4-NN_1(upper)
Oligocene	U. Zhuhai	0-875			Member 1 mudstone and sandstone interbeds; member 2 sandstone with mudstone; member 3 thick sandstone with variegated mudstone	N_{4A}-P_{22}	NN_1(lower) -NP_{24}

Fig. 2 General stratigraphic column for the Pearl River Mouth Basin

is a need to find out what is the optimum sequence (the most-likely sequence) of events based on multi-well observations.

In RASC the problem is solved by the permutation of order-score matrix (S-matrix of Agterberg (1990), p. 147). An order-score matrix is a matrix S_{ij} in which the upper right triangle contains scores of i event occurs above j event, and the lower triangle contains scores of i event occurs below j event (Fig. 3). If events i and j occur at the same level, then the score is halved between the element s_{ij} and s_{ji}. If the initial sequence in the S-matrix is correct, each element in the upper right triangle should be greater than the corresponding element in the lower-left triangle. Otherwise we perform permutation of rows and columns of the S-matrix so that

	A	B	C
A	x	1.5	5.0
B	4.5	x	5.5
C	1.0	0.5	x

	B	A	C
B	x	4.5	5.0
A	1.5	x	5.5
C	0.5	1.0	x

Fig. 3 Example showing the construction and permutation of S-matrix. Left – original matrix containing cross-over scores of events A, B, and C; right – permuted S-matrix where elements in the upper-right triangle are greater than corresponding elements in the lower-left triangle

each element in the upper right triangle is greater than its corresponding element in the lower-left triangle. Thus we obtain the optimum sequence.

A tricky problem in the matrix permutation is the occurrence of "dead cycles" which bring the order of some events back to its original status after several operations. The "dead cycles" occurs more commonly when there are several events appearing on the same level, and/or when the number of occurrence of some event is small. In RASC, a tolerance level (Tol, e.g. Tol = 0.5) is set and both s_{ij} and s_{ji} are set to zero when the difference of corresponding scores s_{ij}–s_{ji} is less than Tol. By adjusting Tol value the "dead cycles" can usually be resolved, and then the optimum sequence is obtained.

RASC provides also the option of inserting unique events into the optimum sequence. Unique events are the events with occurrences less than the tolerance threshold but with high biostratigraphic importance, such as index fossils, marker horizons, etc. The location of a unique event is determined by the average of adjacent events in the wells in which the event appears. By this option we are able to retain age-diagnostic features for analysis.

2.2 Definition of Inter-Fossil Distance and Scaling of Events

After the ranking procedure we obtained the optimum sequence of events, but we still needed to know how close the events are to each other in order to build biostratigraphic zonation. A highlight of RASC is the definition of "inter-fossil distance", also called as the RASC distance, which characterizes the closeness of events based on probability theory and allows a computerized zonation (Gradstein et al., 1985; Agterberg, 1990).

For events A and B, assuming the observed locations X_A and X_B are normally distributed random variables with means of EX_A, EX_B and common variance of σ^2, then the difference $d_{AB} = X_A - X_B$ is normally distributed with mean $\Delta_{AB} = EX_A - EX_B$ and variance $2\sigma^2$. Because the closeness or distance of events is a relative concept, it is set that $\sigma^2 = 0.5$ so that the variance is simplified to 1. Then Δ_{AB} is defined as the inter-fossil distance which can be estimated from the probability A over B in the following way:

Assuming that the probability of A over B equals the frequency of A over B, then

$$P_{AB} = P(d_{AB} \geq 0) = F_{AB} = S_{AB}/N_{AB}$$

where S_{AB} is the element in the upper-right triangle of the ranked S-matrix, and N_{AB} is the number of wells where A and B both appear. Then

$$P(d_{AB} \geq 0) = \Phi(\Delta_{AB}) \text{ and } \Delta_{AB} = \Phi^{-1}(P(d_{AB} \geq 0)),$$

where Φ represents the fractile of the normal distribution in standard form. Thus the inter-fossil distance Δ_{AB} may be found in the table of standard normal distribution when $P(d_{AB} \geq 0)$ is known.

In real data the number of event pairs is usually too small to allow a statistically significant estimate. RASC utilizes indirect estimates to solve the problem, for example, to use the pairs of AC and BC. Such a distance is denoted as $d_{AB.C}$ which can be estimated from d_{AC} and d_{BC} (Agterberg, 1990).

The optimum sequence with inter-fossil distances is called the scaled optimum sequence. From it a dendrogram may be constructed for biostratigraphic zonation. 10 assemblage zones were identified from our dendrogram with inter-zonal distance greater than 0.26, and 15 sub-zones were divided by distance greater than 0.2 (Fig. 4). These zones will be called as RASC zones in this paper.

2.3 Stratigraphic Normality and Uncertainty Analysis in RASC

Stratigraphic normality refers to the degree of correspondence between the individual stratigraphic record and the standard record (the optimum sequence) (Gradstein et al., 1985). A beauty of RASC is that multiple ways to improve the stratigraphic normality and to analyze uncertainty are carefully designed. At each step the input data are screened, and the outputs are statistically tested. Here I briefly describe some examples.

Before the analysis, 3 thresholds have to be set for input data: the minimum number of appearances in wells for an event, the minimum pairs of events in the S-matrix, and the minimum pairs of events in scaling. In our study we used 6, 4, and 4 for these thresholds so that as many events are retained as possible while ensuring the results significant statistically.

While constructing the optimum sequence the uncertainty range of each event in the optimum sequence is also output. The uncertainty range of an event is defined as the upper and lower range that the event has appeared in the studied wells. This provides a quick method for evaluating how firmly an event is positioned in the optimum sequence.

The normality test, based on cumulative inter-fossil distances, is used to find out if some events in individual wells are out of place compared to the optimum sequence. This is achieved by calculating for individual wells the difference in cumulative distance between one event and its adjacent two events (the 1st-order difference) and the difference between the 1st-order differences (the 2nd-order difference). The 2nd-order differences are assumed independent normal variables with the same variance. If the 2nd-order difference of an event in a well exceeds 95 or 99%

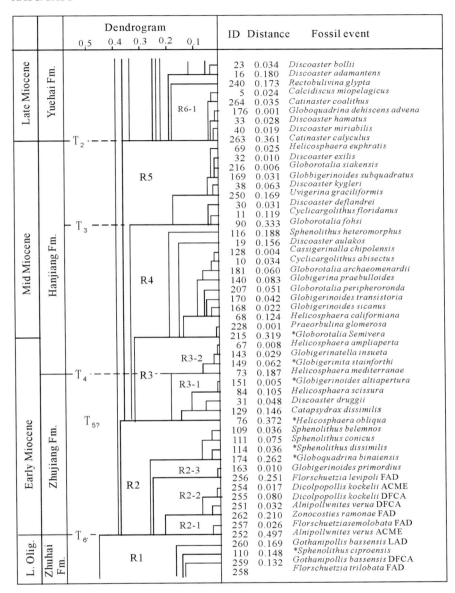

Fig. 4 A segment of the dendrogram of the optimum sequence and biostratigraphic zonation for the PRMB. T denotes seismic interface; R denotes assemblage zones and sub-zones; *- unique event; LAD-last appearance datum; FAD-first appearance datum; DFCA-last consistant appearance datum; ACME-peak appearance datum; data of unspecified events are all LAD

probability limit of a normal distribution, the event in the well is to be considered as out of place with more than 95 or 99% confidence. A close examination may be applied to find out the problems in the stratigraphic location of these events.

3 Correlation and Standard-error Calculation of Multiple Sections (CASC)

3.1 Fitting Distance-Age Curve

The optimum sequence scaled by cumulative inter-fossil distances contains information about not only the sequence of events but also the closeness between events, thus it provides the best basis for event-age correlation. Using the cumulative distance of the optimum sequence as x coordinate, the age as y coordinate, and the events with known age as nodal points, the distance-age curve may be constructed by cubic spline fitting (Fig. 5), and the ages of all events may be read from the curve by their cumulative distances in the optimum sequence. The distance-age curve given by the cubic spline fitting based on cumulative distance is much more close to reality than the curve from linear fitting as used previously.

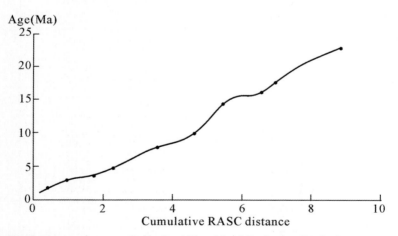

Fig. 5 Distance-age curve for the Pearl River Mouth Basin given by CASC

3.2 Constructing Regional Geological Time Table

With the distance-age curve we can easily do two things: construct the regional time table and perform multi-well correlation.

The geological time table for the PRMB was constructed by correlating the biostratigraphic zonation (the RASC zonation as that in Fig. 4), the global biochronozones (e.g., those of (Martini, 1971) and (Haq et al., 1987)), and the lithologic divisions and seismic reflectors of the basin (Fig. 6). This table is based on the

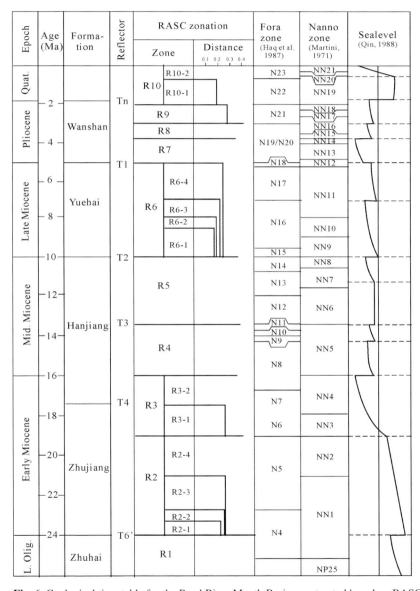

Fig. 6 Geological time table for the Pearl River Mouth Basin constructed based on RASC/CASC output

statistical average of multiple wells and thus more reliable than that constructed based on qualitative synthesis.

From Fig. 6 it is seen that boundaries between the RASC zones correspond quite well with cycles of regional sea-level changes and seismic reflectors. The most important RASC boundary appears between R1 and R2 with inter-fossil distance of 0.48, the highest on the entire sequence. This reflects a dramatic change of environment from dominantly continental to dominantly marine at the Paleogene/Neogene unconformity of $\sim$24 Ma, which was verified by the well #1148 of the ODP leg 148 (Li et al., 2006). The secondarily prominent boundaries are R2/R3, R5/R6, and R8/R9 boundaries. The R2/R3 with distance of 0.37 and age of $\sim$19 Ma is located on top of the "big sandstone member" within the Zhujiang Formation, indicating the end of a low stand period. The R5/R6 with a distance of 0.36 corresponds to the T2 reflector and the boundary between Middle and Upper Miocene. The R8/R9 with a distance of 0.38 occurs in the upper part of the Paleocene Wanshan Formation, corresponding to a large regression with some foraminifera species of N_{21} zone missing.

The RASC zonation in Fig. 6 correlates well with global biochronozones in general but with a couple of exceptions. The last appearance datum (LAD) of *Globorotalia limbata* (event 198) belong to N_{21} zone according to Blow (1979), but here it appears in the R8 zone together with calcareous nannofossils of NN_{16} and NN_{15} zones of Mid. Pliocene. The allocation of this event in the PRMB is questionable. Another problem is the correlation of the events in the N_{15} zone. Blow (1979) used the first appearance datum (FAD) of *Globorotalia acostaensis* to define the top of N_{15} zone and correlated it with the NN_8 zone, dated as late Mid. Miocene. But Haq et al. (1987) correlated N_{15} with NN_9 of early Late Miocene. In the PRMB the LAD of *G. continuosa* (event 184) and *Globoroquadrina dehiscens advena* (event 176) are usually used in stead of the FAD of *G. acostaensis*, as FAD is hard to identify. All three of these events belong to N_{15} (Kennett and Srinivana, 1983). In our study, the two LAD events are associated with events in NN_{10} and NN_9, while the LAD of *G. continuosa* appears between NN_{10} and NN_9, closer to NN_{10} according to its inter-fossil distance. Thus we suggest that the LAD of *G. continuosa* is an event within N_{16}, and the N_{15} zone should be correlated with the lower NN_9. This correlation agrees with that of Haq et al. (1987).

3.3 Inter-Well Chronological Correlations

The distance-age curve (Fig. 5) greatly facilitates the inter-well chronological correlation, especially for the wells with few or no index fossils. For each well an age-depth curve is constructed based on the depths of events in the well, as the age of each event may be read out from the distance-age curve. Then the chronological correlation between the wells is rather straightforward.

Correlations of 17 wells from 4 geological units are shown in Fig. 7. Isochrons at 5, 10, and 24 Ma respectively agree with lower boundaries of the Wanshan, Yuehai, and Zhujiang formations, but the 16 Ma isochron appears in all the wells higher

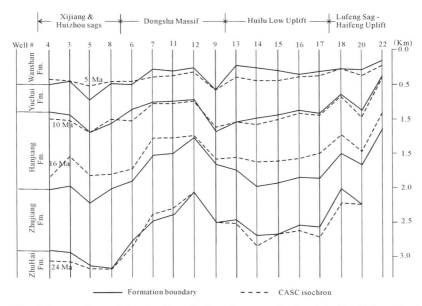

Fig. 7 Inter-well correlation for 11 wells from 4 structural units of the Pearl River Mouth Basin. Numbers on top of each vertical lines are the well codes as shown in Fig. 1

than the lithologic lower boundary of the Hanjiang Formation. The difference is 56–440 m and in average 215 m, larger in depressions and smaller on uplifts. It is not possible that such a large difference is caused by random errors. We suggest that the lower boundary of the Hanjiang Formation is not the boundary between Lower and Mid Miocene, but a lithologic boundary within the Lower Miocene. Our quantitative analysis has brought the long-standing debate on this boundary to a conclusion.

It should be pointed out that in CASC there are also functions to show uncertainties. The uncertainty of the distance-age curve is expressed as SD, the standard deviation in age of the cubic spline fitting. The uncertainties of the isochron depths used for inter-well correlation are represented by error bars whose lengths are the products of SD and the rates of sedimentation at respect depths.

4 High-Resolution Subsidence Analysis

The age-depth curves of the wells given by CASC analysis made possible a high-resolution subsidence analysis. 1D subsidence analysis was performed for Neogene sections in all selected wells from the basin using the program of Stam et al. (1987). Results of the analysis show that the basin was subjected to gross subsidence during the Neogene time, with $1 \sim 2$ km subsidence in total. The subsidence histories are similar within individual depressions or uplifts, but different between depression and uplifts. The amplitude of subsidence increases southwestward in general.

Computed subsidence rates for representative wells from 5 structural divisions are compared in Fig. 8. The subsidence was fast in early Early Miocene and much slower after then. A sharp decrease in subsidence occurred in 19 ~ 20 Ma, corresponding to the top of the R_2 zone in Fig. 6. Differential subsidence was strong at other two times: One occurred at the end of Mid Miocene (~10 Ma, the top of R5),

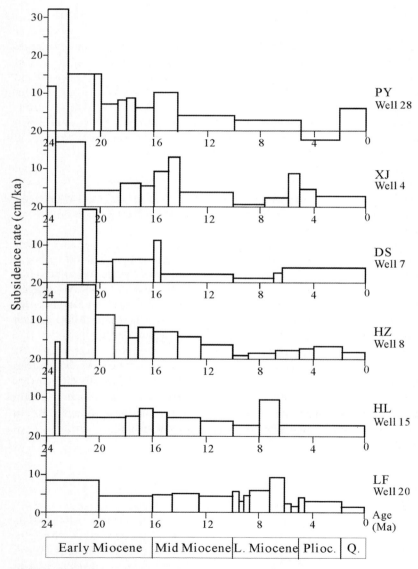

Fig. 8 Rates of tectonic subsidence for selected wells of the Pearl River Mouth Basin. On the right side of each histogram there are notations of structural divisions and well codes, as shown in Fig. 1

when erosion occurred in Dongsha Massif and Huizhou Sag, causing the loss of fossil zones. The other occurred in Pliocene time when erosion was seen in Panyu Low Uplift. These are respectively the first and the second phase of the Dongsha Movement (Chen et al. 2003).

5 Concluding Remarks

By designing and improving RASC/CASC methodology and computer package, F. Agterberg, F. Gradstein, and their co-workers have paved for us a broad road to quantitative stratigraphic analysis. By using RASC/CASC, we have manipulated hundreds of fossil events from dozens of wells from the Pearl River Mouth Basin. The most prominent advance of RASC/CASC over many other procedures of quantitative stratigraphy is that it gives not only the optimum sequence of events, but also quantifies the processes of biostratigraphic zonation and the event-age correlation by means of inter-fossil distance. This facilitates the construction of the geological time table for the basin, and makes the inter-well chronologic correlation and high-resolution subsidence analysis possible even for wells with incomplete fossil records. While all these are important accomplishments in a stratigraphic study, results from RASC/CASC also have helped us in identifying problems in allocation of fossil events and in dating lithologic divisions.

RASC/CASC has impressed me deeply by its comprehensive solutions to quantify each step in biostratigraphic analysis. The procedures in RASC/CASC are very smart and sound statistically. The output of each procedure is accompanied by its uncertainty level. To work with RASC/CASC gives me a feeling of playing with violin: on the background of smooth and pleasant melody once a while there is swift, bright, and resounding sound that brings you happy surprises. I enjoyed the beauty all through the entire practice.

One disadvantage of RASC/CASC is perhaps its complexity, as expressed by the computer program of RASC/CASC in the form which I was worked. This has prevented some of its applications when some workers could not understand its usage in a short time. Fortunately F. Agterberg and his co-workers now published a Windows graphics version of RASC/CASC (RASC version 20, free download from http://www.rasc.uio.no/) which is much easier to work with while retain or improve the original power.

References

Aitchison J (1986) The statistical analysis of compositional data. Chapman & Hall, London
Agterberg FP (1990) Automated stratigraphic correlation. Elsevier, Amsterdam
Blow WH (1979) The cainozoic globigerinida. E.J. Brill, Leiden
Chen C, Shi H, Xu S, Chen X (2003) Conditions for oil and gas reservoirs in Pearl River Mouth (East). Science Press, Beijing (in Chinese)

Gradstein FM, Agterberg FP, Brower JC, Schwarzacher WS (1985) Quantitative stratigraphy. D. Reidel, Dordrecht

Haq BU, Hardenbol J, Vail PR (1987) Chronology of fluctuating sea levels since the Triassic. Science 235: 1156–1167

Kennett JP, Srinivana MS (1983) Neogene planktonic foraminifera, A phylogenetic atlas. Hutchinson Ross Publ. Co., Stroudsburg

Li Q, Jian Z, Li B (2006) Oigocene–miocene planktonic foraminifer biostratigraphy, site 1148, Northern South China Sea. In: Prell WL, Wang P, Blum P, Rea DK, Clemens SC (eds) Proceedings of the ocean drilling program, scientific results, vol 184, pp 1–26

Mardia KV (1972) Statistics of directional date. Academic press, New York

Martini E (1971) Standard Tertiary and Quaternary calcareous nannoplankton zonation. In: Farinacci A (ed) Proceedings of second planktonic conference, vol 2. Rome, pp 739–777

Qin G (2000) Investigation to the stratigraphy and construction of the comprehensive geological columnar section of Cenozoic formation in Pearl River Mouth Basin: China Offshore Oil Gas Geol 14: 21–28 (in Chinese)

Stam B, Gradstein FM, Lioyd P, Gillis D (1987) Algorithms for porosity and subsidence history. Comput Geosci 13: 317–349

Watson GS (1983) Statistics on spheres. John Wiley, New York

Zhou D, Wang P (1994) Neogene bniostratigraphy of Zhujiangkou Basin, South China Sea, a quantitative study. In: Zhou D, Liang Y-B, Zeng C-K (eds) Oceanology of China Seas. Kluwer Academic Publishers, Dordrecht, pp 385–394

Euclidean Distances and Singular Value Decomposition: Useful Tools for Geometric Morphometrics in Biology and Paleontology

James C. Brower

Abstract Although not widely known, euclidean distances between landmarks provide a useful tool for geometric morphometrics in biology and paleontology. The data for each specimen comprise euclidean distances between a set of landmark points. The data matrix may include both size and shape or the distances can be standardized so that only shape is retained. After origin centering, using the average distances, the data are subjected to Singular Value Decomposition (SVD). The distances for selected specimens, representing one or more eigenvectors, can be reconstructed from the SVD and the average distances. Metric multidimensional scaling or principal coordinates provides the coordinates needed to prepare drawings of the specimens modeled. If desired, diagrams of vector displacements of landmarks or transformation grids can be employed to illustrate the changes in shape between pairs of specimens. In shorthand text, this method is termed the Euclidean Distance technique and is indicated by the acronym EDSVD. Case studies on Lower Cambrian trilobites belonging to the genus *Bristolia* illustrate the output from EDSVD. The results of EDSVD are generally similar to the analysis of Procrustes residuals, Bookstein shape coordinates, or the logarithms of euclidean distances by SVD or principal components but they diverge from the thin plate spline technique of Bookstein.

Keywords Geometric morphometrics · landmarks · euclidean distances · singular value decomposition · biology · paleontology · Lower Cambrian trilobites · *Bristolia*

1 Introduction

Geometric morphometrics is the branch of multivariate statistics that deals directly with size and shape. As such, these techniques are of wide and long-ranging interest in biology and paleontology (recent review in Adams et al., 2004). Two

James C. Brower
Department of Earth Sciences, Syracuse University, Syracuse, New York, 13244-1070, USA,
e-mail: karen16@localnet.com

main categories of methods exist: 1. Analysis of outlines, for example with various types of Fourier Analysis and Eigenshapes (reviews in Rohlf and Bookstein, 1990). 2. Analysis of landmarks, which are either homologous points or points of constant topographic reference (see Dryden and Mardia, 1998, p. 3–5; Bookstein, 1991, p. 63–66). The most frequently-used techniques are Bookstein's thin plate spline algorithm (Bookstein, 1991; matrix formulation in Rohlf, 1993; Bookstein, 1996 for recent discussion of the uniform shape component; Zelditch et al., 2004), the application of Principal Components or Singular Value Decomposition to a matrix of point coordinates determined by Bookstein shape coordinates or Procrustes residuals (e. g., MacLeod, 2001; Mitteroecker et al., 2004; Hammer and Harper, 2006, p. 117–122), and by eigenanalysis of a matrix of euclidean distances or their logarithms between landmarks (e. g., MacLeod, 2001; Rao and Suryawanshi, 1996; this paper). Euclidean Distance Matrix Analysis, EDMA, operates on euclidean distances in a different way to estimate the mean shape of a sample or to compare samples of anthropological material (Lele, 1993; Lele and Richtsmeier, 1991, 2001). Outlines and landmarks can be combined as in the extended eigenshapes of MacLeod (1999) or the sliding landmark method of Bookstein (1997) and Bookstein et al. (1999).

It is not widely appreciated that euclidean distances are also a useful tool for geometric morphometrics. In fact, euclidean distances between landmarks exhibit several advantages over superposition techniques for point coordinates (Lele, 1993; Lele and Richtsmeier, 1991, 2001). For example, euclidean distances are not biased by the choice of the coordinate system for the points examined. The size and shape changes revealed by methods such as Procrustes residuals and Bookstein shape coordinates are influenced by the selected coordinate system and the algorithm for superposition of the specimens. This can lead to misleading inferences about the structure in the data set. Euclidean distances allow one to analyze size and shape with respect to identifiable elements of the forms involved. Lele (1993) discusses the problems of estimating the average shape and the variance thereof associated with superposition methods and the advantages of euclidean distances between landmarks in this context. Euclidean distances seem most useful for organisms that lack clear geometrical or biological axes that would provide the basis for a meaningful coordinate system.

The purpose of this paper is to outline one particular method for the analysis of euclidean distances between landmarks by singular value decomposition. Although rather idiosyncratic and not entirely original, the technique combines familiar elements in a somewhat different fashion. The method is useful for the examination of size and shape in various organisms, especially irregular forms like "carpoid" echinoderms with no natural geometrical or biological axes for coordinate systems. In later discussion, the method will be indicated by the acronym EDSVD (for a combination of Euclidean Distances and Singular Value Decomposition) and indicated in short as the Euclidean Distance technique. Several examples are presented to demonstrate that the technique is effective at recovering information about size and shape. The case studies are taken from various species of the Lower Cambrian trilobite *Bristolia*. One study treats the growth of a single population or species. Another example deals with a mixed population consisting of several species. Analyses based on shape and a combination of size and shape are presented, along with comparative results from other algorithms often applied to landmark data.

MacLeod (2001) compared analyses via Bookstein thin plate splines, analysis of Procrustes residuals, and euclidean distances on several data sets. MacLeod's distance matrix procedure is somewhat different from that described here.

2 Statistical Methods

The EDSVD (Euclidean Distance Technique) analysis begins with digitizing p landmarks on n specimens. For planar objects, the data consist of X and Y coordinates, whereas X, Y, and Z coordinates are recorded for three-imensional shapes. Euclidean distances are then calculated between the p landmarks for each of the n specimens. If all distances are retained, each specimen will be represented by p^*p distances. Not all distances are necessary; for example, the distances are symmetrical and $d_{ij} = d_{ji}$, so all of the d_{ij} or d_{ji} can be deleted. All of the p^*p distances are used in this paper. The original data matrix is assembled with the distances in the columns and the specimens located in the rows.

2.1 Size Standardization

The original data matrix includes both size and shape components. However, the data can be standardized so they only reflect shape, with size treated as a separate component; this is the usual practice for geometric morphometrics. Mosimann (1970) demonstrated that an appropriate size variable for multivariate normal linear data equals the sum of the measurements for that particular specimen. This philosophy is followed here and the average non-zero distance for a specimen provides its standard size. Dividing the p^*p distances for a specimen by its standard size produces its shape data. This procedure is repeated for all n specimens to generate a data matrix based on shape. The size information is retained for later analysis. The case studies in this paper employ both types of data matrices, termed X, in later discussion.

Various other size measures are available, such as centroid size (e. g., Dryden and Mardia, 1998, p. 23–26). The comparative analyses outlined later denote that EDSVD yields similar results for any reasonable measure of size.

2.2 Origin Centering

A vector, AVD, containing the average for each of the p^*p distances is computed for the n specimens; this vector represents the average shape or combination of size and shape for the specimens in the data set. The data are origin centered by subtracting this vector from each specimen or row in the data matrix. The intent is to produce a first eigenvector extending through the major axis of the data points rather than connecting the origin of the data space with its mean value (e. g., Davis, 2002, p. 123–158). The origin centered data are designated XC.

2.3 Singular Value Decomposition

The XC data matrix, with the $p*p$ distances in the columns and the n specimens in the rows, is subjected to singular value decomposition SVD (e. g., Davis, 2002, p. 500–508). Typically the distances outnumber the specimens. Hence, the SVD's of this paper are executed as follows. The number of non-zero eigenvalues of XC is limited by the number of distances or specimens, whichever is smaller. Furthermore, the eigenvalues of the crossproduct matrices $XC^{t*}XC$ (distances or R-mode) and $XC^{*}XC^{t}$ (specimens or Q-mode) are identical so one can operate in the smaller mode, here $XC^{*}XC^{t}$. The eigenvectors of $XC^{*}XC^{t}$ are for the specimens and these are normalized to 1.0 to produce EVQ. The eigenvector scores for the variables or distances, $XC^{t*}EVQ$, gives the orthogonal projections of the distances onto the eigenvector axes defined by the specimens. Normalizing these scores to 1.0 gives the eigenvectors for the distances in EVR.

One property of SVD is that it is simple to model the origin-centered data matrix, XC, as a function of one or more eigenvectors by $XC_{est} = EVQ^{*}SV^{*}EVR^{t}$, where SV is a diagonal matrix with the singular value(s) for the eigenvector(s) used in the data models. The data modeled here are origin-centered. To convert XC_{est} into euclidean distances, one must add the AVD vector with the average distances to each specimen in XC_{est} to calculate X_{est}.

Although formulated in terms of SVD, a program for principal components of a covariance matrix can easily be modified for the form of EDSVD discussed herein.

Only the linear version of EDSVD will be investigated in detail here. A logarithmic form, proposed by Rao and Suryawanshi (1996), can be considered as a generalized form of multivariate allometry (see also Dryden and Mardia, 1998, p. 280–287). Comparative results for this algorithm are outlined subsequently in this paper.

2.4 Models of Specimens

Drawings of specimens from the estimated distances allow visualization of the results from the desired eigenvector or eigenvectors. The estimated distances for each specimen are structured into a square $p*p$ matrix of euclidean distances. This matrix is treated by principal coordinates or metric multidimensional scaling, PCOORD (see Green, 1978, p. 411–416). For two dimensional figures, a two axis eigenvector solution contains the coordinates for the drawing. If the PCOORD gives a perfect fit to the data, the third and higher eigenvalues will equal zero. Otherwise, the magnitudes of these eigenvalues reflect the degree of mismatch between the PCOORD and the euclidean distance matrix. For the examples in this paper, the first two PCOORD's account for 99 percent or more of the estimated euclidean distance matrices, so the resulting drawings are clearly satisfactory. A similar situation applies to three-dimensional organisms.

An alternative would be to use non-metric multidimensional scaling as outlined by Carpenter et al. (1996).

The specimen drawings from PCOORD are generally not oriented correctly, because the first two eigenvectors are the orthogonal axes that explain the largest amounts of variance in the matrices of the estimated euclidean distances. The drawings can be rotated analytically or graphically within a drawing program.

Two graphical techniques frequently aid in the interpretation of the drawings, namely vector displacements and transformation grids (see Zelditch et al., 2004). The grids of this paper are constructed with a simple ad hoc procedure. The end-member specimen drawings are rotated into the desired orientation as mentioned previously, and one drawing is designated as reference for a square or rectangular grid. The X and Y coordinates of the second specimen are then contoured on the reference shape with a distance weighted least squares (DWLS) algorithm. Deformation of the grid displays the changes in size and shape between the two specimens. Rather than working with a contouring algorithm as an interpolation function, a thin plate metal spline could be adopted.

2.5 Computer Programs

The specimens were digitized via GraphClick, a shareware program for MacIntosh computers with System 10. Contouring was done by the DWLS algorithm of the statistical package SYSTAT. All other programs were written by the author in the APL programming language.

2.6 Case Studies

Cephalons of the Lower Cambrian trilobite *Bristolia* provide examples to demonstrate the application of Euclidean Distance Matrix technique (EDSVD) to geometric morphometrics. Limitations of preservation dictate that the data for each specimen represent the right or left half of a cephalon. Separate records are obtained for both sides of some well preserved individuals. To facilitate comparisons, all specimens are pictured as right sides of the cephalons. Eighty specimens are available from six described species. Illustrations of the trilobites are taken from Hazzard and Crickmay (1933), Lieberman (1999), Mount (1980), Palmer and Halley (1979), Palmer and Repina (1993), Resser (1928), and Riccio (1952). Note that this paper is only intended to show the power of EDSVD in geometric morphometrics. It is not a definitive examination of *Bristolia*; such a study is currently underway by Webster (e. g., 2002).

The 18 landmarks are pictured in Fig. 1. Most are anatomical landmarks which should be considered as homologous (see Dryden and Mardia, 1998, p. 3–5 and

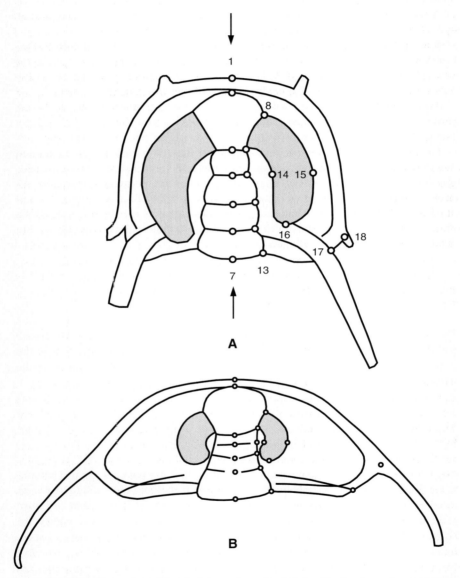

Fig. 1 A, B Juvenile specimen 35 and adult specimen 41 of *B. fragilis* showing points measured, ×33 and ×3.0. Ocular lobes are *shaded; arrows* point out axis of glabella, and selected points are labeled on **A**

Bookstein, 1991, p. 63–66 for classifications of landmarks). Such points include 1–7, the intersections between the central axis and various glabellar features; 8, the junction between the side of the glabella and the anterior margin of the ocular lobe; 9–11, where the various glabellar furrows meet the side of the glabella; 12 and 13, where the occipital ring reaches the side of the glabella; 17, outer part of adgenal spine at the cephalic rim; and 18, center of genal spine along the cephalic rim. Notes about points 9 and 17 are needed. Although the adgenal spine is only present in small specimens, its corresponding position is easily identified in adults (e. g., Palmer and Repina, 1993). Point 9 is the intersection between the side of the glabella and the anterior glabellar suture, and in most specimens the point is at or near where the ocular lobe joins the glabella; if this is not the case the point is projected as seen in Fig. 2F and L–N. Mathematical landmarks are 14–16, which define the inner, outer, and posterior parts of the ocular lobe. The ocular lobe is easily deformed during preservation or by deformation. Consequently Points 14 and 15 are measured in line with one of the glabellar furrows.

Points 1–7 lie on the anterior-posterior axis of the cephalon (Fig. 1). Trilobites have coordinate axes that are geometrically and biologically reasonable. During growth of *Bristolia*, especially the earliest stages, major developmental changes take place along the anterior-posterior axis. Likewise, significant ontogenetic differentiation is observed along the width dimension at right angles to the anterior-posterior axis. Modifications along these axes are significant in the evolution of *Bristolia*. The presence of meaningful coordinate axes facilitates the comparison of methods based on point coordinates or distances between landmarks as noted subsequently.

Drawings of the original specimens referred to in later parts of this paper are in Fig. 2. All drawings are scaled so the cephalon length is constant.

2.7 *Ontogeny of* Bristolia anteros

The growth of *B. anteros* shows the results of EDSVD when a single population is examined. The 17 specimens range from a small Stage 2 immature individual to a large Stage 5 adult animal with cephalon lengths of 0.96 and 7.2 mm, respectively. The growth stages mentioned here are the olenellid trilobite growth stages outlined by Palmer (see Palmer and Halley, 1979; Palmer and Repina, 1993). For the data set based on shape, the first three eigenvalues explain 84.5, 6.1 and 3.2 percent of the variance, listed in the same order (Table 1). The first eigenvector arranges the trilobites in order of progressively decreasing size (Fig. 3). The Spearman rank correlation between the coefficients for the specimens on the first eigenvector and the standard size measure (i. e., the average non-zero distance for the original data) equals −0.91 which is highly significant at the 0.01 probability level. The magnitudes of the Spearman rank correlations for the higher eigenvectors versus size vary from 0.23 to 0.012, none of which differ from a population value of nil at the 0.05 probability level. Hence, I conclude that the first eigenvector extracts all of the shape changes in *B. anteros* that are related to the size of the trilobites. This is consistent

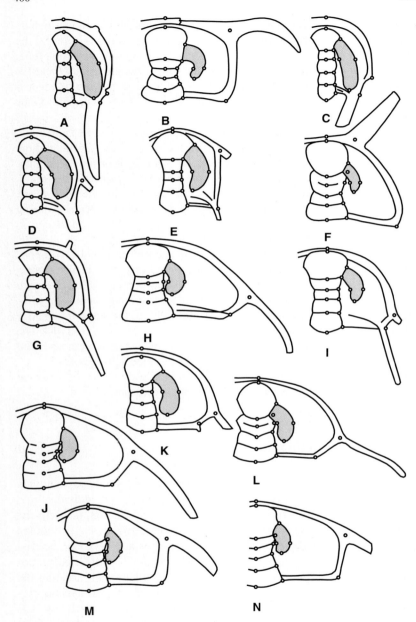

Fig. 2 Original specimens showing growth and variation in six species of *Bristolia*. **A–D** *B. anteros*, Juvenile specimen 1, adult specimen 14, small specimens 4 and 9, ×28, ×4.1, ×22 and ×20. **E, F** Large juvenile specimen 18 and adult specimen 33 of *B. insolens*, ×9.8 and ×2.5. **G, H** Juvenile specimen 35 and adult specimen 41 of *B. fragilis*, ×18 and ×2.4. **I, J** Large juvenile specimen 44 and adult specimen 59 of *B. harringtoni*, ×13 and ×2.0. **K, L** Large juvenile specimen 69 and adult specimen 62 of *B. mohavensis*, ×6.2 and ×2.9. **M, N** Adult specimens 58 and 61 of *B. bristolensis*, ×2.2 and ×1.3. Ocular lobes are *shaded*; if necessary, all glabellar furrows are extended to cover the entire width of the glabella; all specimens are illustrated as right side of cephalon

Table 1 List of percents of variance explained by selected eigenvalues for the trilobite data sets

Eigenvalue number	Ontogeny of *Bristolia anteros* Shape data set	Ontogeny of *Bristolia anteros* Size and shape data set	Growth and variation in six species of *Bristolia* Shape data set	Growth and variation in six species of *Bristolia* Size and shape data set
1	84.5	99.6	41.2	95.9
2	6.15	0.140	34.8	2.49
3	3.16	0.0809	7.21	0.614
4	1.73	0.0705	3.84	0.399
5	1.47	0.0295	3.05	0.149
Total	97.0	100.0	90.0	99.5

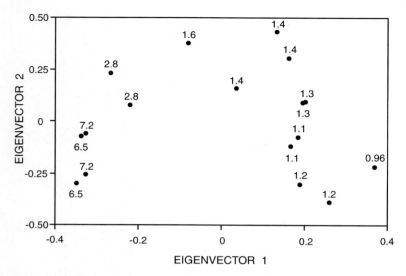

Fig. 3 Plot of first two eigenvectors for growth and variation in specimens of *Bristolia anteros*. The numbers associated with the data points give length of the cephalon to two significant figures

with Mosimann (1970) who showed that only one size variable exists for a single population with a multivariate normal distribution. Furthermore, the shape changes involved in the higher eigenvectors are randomly distributed relative to the size of the animals.

Figure 4 contains drawings of the specimens as modeled by the appropriate eigenvectors, vector displacement diagrams, and the transformation grids. In the model specimens, all of the appropriate points are connected with straight lines, aside from Points 1 and 18 which are joined by a simple arc.

Major shape changes are associated the ontogeny of *B. anteros*. The cephalon and the glabella become relatively wider. Two large scale divergences are the outward migration of the adgenal area along with forward and lateral displacement of the genal spine (Points 17 and 18). The area between the anterior border of the cephalon and the anterior edge of the glabella is shortened (Points 1 and 2). An overall decrease in the size of the ocular lobe involves a large anterior movement of its posterior margin in conjunction with smaller posterior shifts of the other ocular lobe points (Points 8, 9, 14–16). The juvenile glabella is long and slender with parallel sides and roughly equal glabellar lobes. The adult structure becomes wider, centrally constricted, and exhibits regular changes in the orientation of the glabellar lobes and furrows; the size of the anterior glabella lobe expands at the expense of the more posterior lobes (Points 2–7, 8–13). The developmental pattern is clearly displayed by the models of the smallest and a large adult cephalon generated from the first eigenvector, as well as the transformation grid, and vector displacement displacement diagram of Fig. 4. Comparison of the model cephalons with the original trilobites in Fig. 2 denotes that most of the shape information is captured by the first eigenvector.

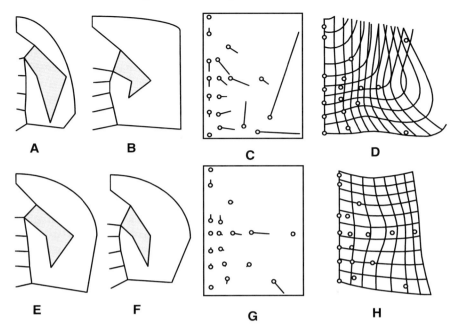

Fig. 4 Growth and variation in *Bristolia anteros* as modeled from Eigenvectors 1 and 2 of shape data set. **A, B** Juvenile specimen 1 and adult specimen 14 reconstructed from Eigenvector 1, ×34 and ×4.9. **C** Vector diagram showing ontogenetic differences in shape from specimen 1 to 14 according to Eigenvector 1; points are for specimen 1, vectors demonstrate changes from corresponding points on specimen 1 to specimen 14; registration is based on points 1 and 7 as constant. **D** Transformation grid for changes from specimen 1 to specimen 14 for Eigenvector 1; points are for specimen 14, contours are for specimen 1; registration as for **C**; reference grid is squares; some grid squares are not closed because of extreme changes in point 18. **E, F** Size independent variation between specimens 9 and 4 as derived from Eigenvector 2, ×24 and ×27. **G** Vector diagram for displacements from specimen 4 to specimen 9 taken from Eigenvector 2; registration as in **C**. **H** Transformation grid for specimens 9 and 4 from Eigenvector 2; points are specimen 9, specimen 4 is contoured; reference grid is squares; registration as in Fig. **C**. Ocular lobes are shaded; points 17 and 18 are connected by a simple arc to give more reality to the drawings; other points are joined by straight lines. Compare with original trilobites in Fig. 2

As mentioned previously, the second eigenvector involves 6.1 percent of the variation that is independent of size (Table 1). Drawings of two end-member specimens, numbers 4 and 9, calculated from the second eigenvector and the average distances are given in Fig. 4 along with a transformation grid and vector displacement plot. The most striking changes consist of lateral shifts in the middle of the ocular lobe, points 14 and 15, and a posterior-lateral movement of the adgenal spine base at point 17. Minor transformations are observed for the anterior glabellar lobe, points 2, 8, and 9, and several other glabellar points. As expected, the differences between the end-member specimens modeled from this eigenvector are much smaller than those of the previous eigenvector.

Furthermore, comparison of the Eigenvector 2 models in Fig. 4 with the original trilobites in Fig. 2 shows only general resemblances. Both situations are caused by the fact that the Eigenvector 2 models are only derived from the mean distances along with 6.1 percent of the variation in the origin centered data matrix (Table 1).

Similar small-scale variation is seen for the third eigenvector, which is linked to 3.2 percent of the variance in the data (Table 1).

An analysis was also done for the ontogeny of *B. anteros* with euclidean distances, which include both size and shape components. As predicted, the first eigenvector represents the size and shape changes that are attributed to growth. However, the first eigenvalue extracts almost all of the variance in the data set, namely 99.6 percent (Table 1). The first eigenvector for the size and shape data arrays the specimens in almost the same order as for the previously discussed shape data, and the Spearman rank correlation for the two vectors comprises 0.924, a value that is highly significant at the 0.01 probability level. I conclude that Eigenvector 1 for both data sets extracts essentially the same information. As with the shape data, the higher eigenvectors of the size and shape data are grouped with minor shape changes that are random with respect to the size of the trilobites (Table 1). All in all, the results from the two data sets exhibit great similarity.

2.8 Growth and Variation in Six Species of Bristolia

The purpose of this example is to treat a series of populations made up of 80 trilobites from six described species. As previously, I will initially outline the results of EDSVD on data consisting of shape information, from which size has been removed. The first three eigenvalues explain 41.2, 34.8, and 7.21 percent of the variance in the data set (Table 1). The first two eigenvectors exhibit significant correlations with the size of the animals. The Spearman rank correlations constitute 0.485 and −0.684, respectively, and both figures differ significantly from a population value of nil at the 0.01 probability level. Clearly, Eigenvectors 1 and 2 extract information that compounds both size and shape. The higher eigenvectors are independent of size; for example, the rank correlation for Eigenvector 3 and the average distances only equals 0.031 which is not significant at any reasonable probability level. Figure 5 illustrates a plot of the first two eigenvectors for the specimens, whereas selected cephalons that are modeled from Eigenvectors 1 and 2 are pictured in Fig. 6. Comparison of the model cephalons with their original counterparts in Fig. 2 indicates that the models recover the major features of the original specimens. However, numerous minor structures and individual variations cannot be seen in the models. Basically, the geometry represented in the eigenvector models is generalized. This reflects the fact that the first two eigenvectors only account for 76 percent of the variance in the origin centered distance data. Twenty-four percent of the total variance remains unexplained.

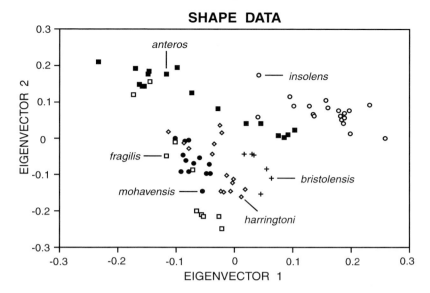

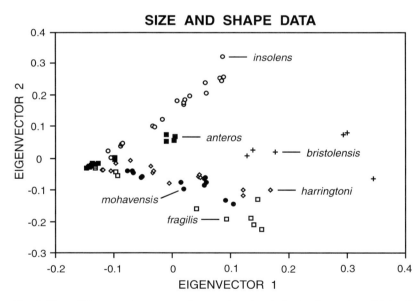

Fig. 5 Plots of first two eigenvectors for growth and variation in six species of *Bristolia* for data sets based on shape and size and shape. The different species are indicated by their trivial names listed on the graphs

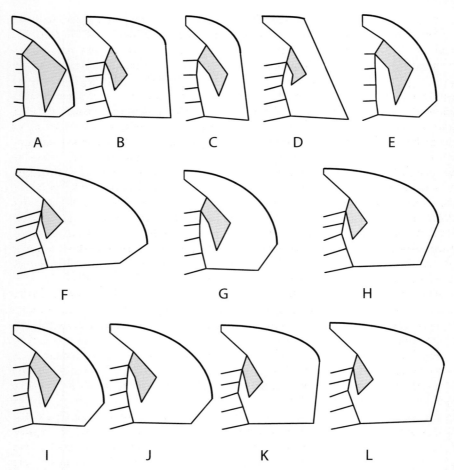

Fig. 6 Growth and variation in six species of *Bristolia* as computed from Eigenvectors 1 and 2 of shape data set. **A, B** Juvenile specimen 1 and adult specimen 14 of *B. anteros*, ×32 and ×4.6. **C, D** Large juvenile specimen 18 and adult specimen 33 of *B. insolens*, ×11 and 2.8. **E, F** Juvenile specimen 35 and adult specimen 41 of *B. fragilis*, ×20 and ×2.7. **G, H** Large juvenile specimen 44 and adult specimen 59 of *B. harringtoni*, ×15 and ×2.2. **I, J** Large juvenile specimen 69 and adult specimen 62 of *B. mohavensis*, ×7.0 and ×3.3. **K, L** Adult specimens 58 and 61 of *B. bristolensis*, ×2.5 and ×1.5. Ocular lobes are *shaded*; points 17 and 18 are connected by a simple arc to give more reality to the drawings; other points are joined by straight lines. Compare with original trilobites in Fig. 2

The specimens and species are systematically distributed in the space defined by the first two eigenvectors (Fig. 5). Inasmuch as the higher eigenvectors do not relate to separation between the species or developmental changes within the taxa, these will not be discussed further. The smallest trilobites, Stage 2 and 3 individuals belonging to *B. anteros* and *B. fragilis* with cephalon lengths varying from

0.96 to 1.5 mm, are concentrated in the upper left area of the plot. Larger and more mature animals are found toward the right side and lower regions of the plot. The individuals of *B. bristolensis* and *B. mohavensis* are all mature. Reasonably complete growth sequences are known for the other species, especially *B. anteros* and *B. fragilis* (Figs. 2, 6). More or less common developmental shape changes consist of shortening the preglabellar field, formation of a glabella with a central constriction and systematic changes in the glabellar furrows, expansion of the anterior glabellar lobe and decrease in relative size of the more posterior lobes, overall widening of the glabella and cephalon, and reduction of the ocular lobe. Many of these changes occur at different rates in the various species. For example, the adults of *B. insolens* are characterized by the smallest ocular lobes relative to the whole cephalon. The growth trajectories of the several species diverge widely with respect to the points of the adgenal and genal spine bases, points 17 and 18 (Fig. 6). For example, *B. fragilis* develops a widely-flaring posterior-lateral cephalic margin with the genal spine located near the posterior. *B. harringtoni* is similar but less exaggerated. As noted previously, the genal spine of *B. anteros* migrates outward and toward the anterior in progressively larger individuals. This trend is even more pronounced in *B. insolens* in which the adult genal spine is close to the anterior part of the glabella.

The different species tend to occupy separate parts of Fig. 5. *B. anteros* and *B. insolens* with their advanced genal spines occupy the upper part of the plot, in which equivalent sized cephalons of the former overly those of the latter. The forms with genal spines closer to the posterior of the cephalon typically fall below those of *B. anteros* and *B. insolens*. *B. harringtoni* and *B. mohavensis* overlap greatly and these two taxa could be conspecific. Specimens of *B. fragilis*, with genal spines near the posterior of the cephalon, cover a narrow region to the left of *B. harringtoni* and *B. mohavensis*, and the large adults of *B. bristolensis* are on the right. The groupings of the species are consistent with their phylogenetic relationships. According to the cladogram for *Bristolia* and allied trilobites shown by Lieberman (1999), *B. fragilis, B. mohavensis, B. harringtoni*, and *B. bristolensis* are closely related and could share a common ancestry. On the other hand, *B. anteros* and *B. insolens* represent separate and more advanced species. Virtually all of the separation between the different species and groups of species can be correlated with five of the euclidean distances associated with the adgenal and genal spine areas and the overall outline of the cephalon, namely $d_{1,17}$, $d_{1,18}$, $d_{2,17}$, $d_{2,18}$, and $d_{17,18}$.

Transformation grids derived from the first two eigenvectors are presented for *B. anteros, B insolens, B. fragilis*, and *B. harringtoni* in Fig. 7. The grids reveal the major changes in the shapes of the four taxa. Although the appearance of the four ontogenetic grids differs considerably, all are modeled by the same functions. This suggests that a common framework, at least to some extent, underlies the ontogenetic trajectories of the four species. Considered with reference to the plot of the first two eigenvectors, the differences between the development of the four species can be related to variations with respect to their starting and ending points on the eigenvector plot (Fig. 5). Both eigenvectors are involved in the changes in the glabella and the ocular lobes. The two eigenvectors dictate differential changes in the outline of the cephalon. The first eigenvector largely relates to the anterior and axial

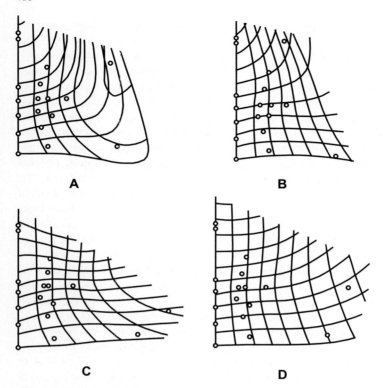

Fig. 7 Transformation grids for species of *Bristolia* in which the ontogenies are reasonably well known; reconstructions are from Eigenvectors 1 and 2 of the shape data set. **A** *B. anteros*, specimens 1 (juvenile contoured) and 14 (adult points). **B** *B. insolens*, specimens 18 (small specimen contoured) and 33 (adult points). **C** *B. fragilis*, specimens 35 (juvenile contoured) and 41 (adult points). **D** *B. harringtoni*, specimens 44 (small trilobite contoured) and 59 (adult points). Registrations are based on points 1 and 7 as constant; reference grids are squares; some grid squares on outside of specimens are not closed. Compare the grids with the original specimens in Fig. 2. The deformations of the square grids show the major growth changes of the species illustrated

or inward migration of the genal spines compared to the adgenal spines in mature specimens of *B. anteros* and especially in *B. insolens*; conversely the widely flaring cephalon of adult *B. fragilis* with its posteriorly placed genal spines is mostly produced by the second eigenvector. These changes are less accentuated in individuals of *B. harringtoni* and *B. mohavensis*.

Analysis of the data including both size and shape produces a complementary picture (Fig. 5). Only the first eigenvector is correlated with the overall size of the animals; in fact, the Spearman rank correlation coefficient rounds to 1.0 at three significant figures. The first eigenvector explains 95.9 percent of the variance in the origin-centered distance matrix (Table 1). Progressively older individuals are associated with larger coefficients on the first eigenvector. Essentially, the first eigenvector extracts the ontogenetic changes that are seen in all of the species (see previous discussion). Although, the second eigenvector tends to separate the various species, this effect only accounts for 2.49 percent of the information in the data set (Table 1). The amount of divergence between the various taxa increases in larger specimens. Adults of the various species are easily distinguished, aside from the overlap between *B. harringtoni* and *B. mohavensis*. Conversely, the juveniles of all forms are identical or nearly so. The relative arrangement of the species is similar for both data sets, but the two eigenvector plots are rotated relative to each other (Fig. 5). However, differences between juveniles are more clear in the shape data than in the data set based on both size and shape. This is because the variance within juveniles is very small in the data for size and shape, but these animals become more variable when size is removed from the data set.

3 Comparison with Other Methods

Results for the various data sets and methods are compared as follows. Ordinations for the first two and the first seven eigenvectors, normalized to their corresponding eigenvalues, are computed. These serve for calculating euclidean distances between all specimens in each data set. A matrix of Pearsonian correlations displays the relationships between the euclidean distances for the different data sets which are illustrated by Unweighted-Pair-Group-Method (UPGM) dendrograms. Plots of the first several eigenvectors are inspected to provide a qualitative estimate of the similarities of the interpretations that would be obtained from each of the data sets being compared. Data on the six species of *Bristolia* and the growth of *B. anteros* will be annotated.

As mentioned previously, various measures could be employed for size standardization of the euclidean distances to generate a shape data set; the following are tested: the average euclidean distance as used here; length of the cephalon, which is a typical morphometric variable; the square root of centroid size for Points 1–7 along the axis of the glabella; and the square root of centroid size for all 18 points. The ranges of the correlations comparing the analyses for the different size parameters are 0.981–0.998 for *Bristolia anteros* and 0.952–0.999 for the six species of *Bristolia*. The correlations obtained for solutions based on two and seven eigenvectors are almost the same. Likewise, the plots of the eigenvectors for the different size measures are highly similar. I conclude that EDSVD for shape data is not greatly sensitive to the size parameter for standardizing the distances. The obvious caveat is that a reasonable size variable must be used.

Ordinations for similar techniques are also treated. The four distance methods are EDSVD for shape data (D1), EDSVD for data including both size and shape (D2), logarithms of non-zero euclidean distances that are standardized for size (logD1) and logarithms of non-zero euclidean distances including both size and shape components (logD2) as advocated by Rao and Suryawanshi (1996). [For the logD1 analysis, the original distances for each specimen were standardized by division with their geometric means. The data set comprises the logarithms of these standardized distances.] Three Procrustes superposition algorithms are also evaluated. Here, the points for the origin centering, size standardization if done, and rotations are Numbers 1–7 along the central axis of the cephalon and glabella. This generates a coordinate system for all specimens with one axis that parallels the length of the cephalon. All superpositions are based on least squares and are subjected to Singular Value Decomposition (SVD). The methods consist of: Procrustes residuals in which size is removed from the data (P1; see MacLeod, 2001 for a similar approach); Procrustes residuals where size is retained in the data (P2); and the "Procrustes form analysis" of Mitteroecker et al., 2004); the last scheme takes the P1 data and adds another column with the logarithm of centroid size (P3). Lastly Bookstein Shape Coordinates (SC) were determined with Points 1 and 7 forming the baseline.

As outlined earlier, the data set for *Bristolia anteros* is comparatively simple inasmuch as a single species is represented. The results of all methods are strikingly similar, and the correlation coefficients for all pairs of techniques range from 0.957 to over 0.999 with little difference between the correlations for two versus seven eigenvectors. The dominant theme is growth and the associated changes in size and shape. This pattern is extracted by the first eigenvector which explains from 74.3 to 99.6 percent of the variance. Nearly identical shape changes are linked to the first eigenvector in all cases; as one would expect, the specimens are also arranged in largely the same order along all first eigenvectors. Variations in shape that are independent of size and age are associated with smaller amounts of variance in the data, and these changes are portrayed by the higher eigenvectors. The second and third eigenvectors of all analyses generally resemble one another, both with respect to shapes and some or all of the end-member specimens. However, the higher eigenvectors of the various analyses differ to a greater extent. Nevertheless, the same basic interpretation would be obtained from all analyses on this data set.

The data for the genus *Bristolia* are far more complex. Six species are represented along with much greater changes in size and shape. Here, the different data sets begin to produce divergent results. As with the previous comparisons, the correlations for the two- and seven eigenvector solutions show only miniscule differences (Table 2). The dendrogram in Fig. 8 contains two clusters, one with the analyses for shape data (D1, log D1, P1, P3, and SC) and the other with the analyses for both size and shape (D2, P2, and logD2). Clearly, the nature of the *Bristolia* data sets overrides the technique applied to them. Within the dendrogram, two joins between pairs of analyses have mutually exclusive correlations, namely P1 and SC at 0.992, and P2 and D2 at 0.963. The ordinations within each pair of methods are almost the same for the first several eigenvectors although they may be rotated, flipped, differentially expanded or compressed relative to one another. Slightly less similar plots are noted for the analyses of P3 and D1 and those of P1 and SC with correlations spanning an interval of 0.929–0.949. Distinctly lower correlations from 0.871 to 0.748 are obtained for logD1 against the other analyses in the shape cluster; however, these values exceed all correlations between logD1 and any analysis for the size and shape data sets. The output from logD2 is somewhat intermediate between

Euclidean Distances and Singular Value Decomposition 411

Table 2 Matrix of Pearson correlation coefficients between analyses of the data set on six species of the Lower Cambrian trilobite *Bristolia* for various methods. Acronyms for the different methods are: D1 – EDSVD for shape data, D2 – EDSVD for size and shape data, logD1 – logarithms of non-zero euclidean distances that are standardized for size, logD2 – logarithms of non-zero euclidean distances including both size and shape components, P1 – Procrustes residuals in which size is removed from the data, P2 – Procrustes residuals where size is retained in the data, and P3 – "Procrustes form analysis." SC – Bookstein shape coordinates. Data above and below diagonal are for seven and two eigenvectors, respectively

	P1	P2	logD1	logD2	P3	D1	D2	SC
P1	1.0	0.531	0.748	0.480	0.949	0.929	0.401	0.992
P2	0.540	1.0	0.470	0.723	0.675	0.543	0.963	0.524
logD1	0.890	0.549	1.0	0.709	0.833	0.871	0.456	0.769
logD2	0.494	0.744	0.696	1.0	0.724	0.628	0.814	0.488
P3	0.950	0.686	0.934	0.729	1.0	0.944	0.598	0.946
D1	0.930	0.562	0.987	0.634	0.943	1.0	0.464	0.937
D2	0.413	0.969	0.486	0.817	0.606	0.476	1.0	0.399
SC	0.996	0.538	0.899	0.503	0.950	0.937	0.414	1.0

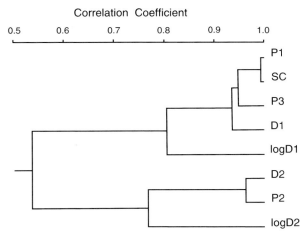

Fig. 8 Dendrogram showing similarities between analyses of the data set on six species of the Lower Cambrian trilobite *Bristolia* for various methods. Clustering is by the unweighted-pair-group-method on a matrix of Pearson correlation coefficients. Acronyms on the dendrogram are: D1 – EDSVD for shape data, D2 – EDSVD for size and shape data, logD1 – logarithms of non-zero euclidean distances that are standardized for size, logD2 – logarithms of non-zero euclidean distances including both size and shape components, P1 – Procrustes residuals in which size is removed from the data, P2 – Procrustes residuals where size is retained in the data, and P3 – "Procrustes form analysis." SC – Bookstein shape coordinates

that for the size and shape analyses of D2 and P2 and the five analyses on the shape data; the correlations between logD2 and D2 and P2 equal 0.814 and 0.723, whereas those for logD2 versus the analyses in the shape cluster vary from 0.724 to 0.480.

Analyses D2 and P2 are for data retaining both size and shape. The first eigenvector accounts for 96.2 and 76.3 percent of the variance and it relates to changes in size and shape that are common to all species. The adults of the some of the taxa begin to separate along the second axis, but the juveniles are more similar and they are not differentiated along this axis (e. g., Fig. 5). The analyses for the shape data are more somewhat diffuse and the first two eigenvectors are both correlated with size and size-linked changes in shape. Percents of variance attributed to the first two eigenvectors range from 44.7 to 55.6 and 18.6 to 35.9, respectively. Although *B. harringtoni* and *B. mohavensis* overlap considerably, specimens of the other species generally tend to be more or less distinct at most growth stages. However, the logD1 analysis exhibits more overlap between the taxa and the size groups of the larger specimens than any of the other analyses of the data sets based on shape. As noted previously, the logD2 analysis falls between the shape set of analyses and the typical analyses involving both size and shape, largely because the smallest specimens are characterized by greater dispersion than those of P2 and D2. An earlier version of this *Bristolia* data set was treated with the thin plate spline algorithm of Bookstein (1991). This analysis differs from the previous methods because more local changes of shape received greater emphasis (see also MacLeod, 2001). Consequently, a quantitative comparison is not presented for the thin plate spline.

4 Statistical Inference

Although I visualize EDSVD as primarily a structural technique, it should be possible to draw inferences about the underlying population or populations from which the data are drawn. Rohlf (2000b) generated shape space models for triangles with metrics that are commonly applied in geometric morphomerics; of those shape spaces, the ones applicable to the methods of this paper consist of euclidean distances as in Euclidean Distance Matrix Analysis (EDMA; see Lele, 1993; Lele and Richtsmeier, 1991, 2001), logarithms of euclidean distances that are standardized for size (Rao and Suryawanshi, 1996), and Kendall's shape spaces. Rohlf (2000a) modeled the statistical power of tests for comparing different samples of triangles, tetrahedrons, and a simple figure with six landmarks. This approach was extended to estimating mean shapes by Rohlf in 2003. If these experiments can be extrapolated to real data, they suggest that conventional statistical tests using the classical distributions may not provide satisfactory results for euclidean distances and their logarithms. Instead, one should rely on bootstrapping and permutation methods (Rohlf, 2000a, b). Bootstrapping and permutation techniques for geometric morphometrics are reviewed by Zelditch et al. (2004). For the examples outlined here, the significance tests with Spearman rank correlations are all supported by bootstrapping the data.

5 Summary

The Euclidean Distance-Singular Value Decomposition technique (EDSVD) is effective at recovering patterns of changes in size and shape for geometric morphometrics, as demonstrated by two case studies drawn from the Lower Cambrian trilobite *Bristolia*. Although not shown here, the results of EDSVD are generally similar to the analysis of Procrustes residuals, Bookstein shape coordinates, and logarithms of euclidean distances by singular value decomposition or principal components but they diverge from the thin plate spline technique of Bookstein. These findings are the same as those of MacLeod (2001).

EDSVD exhibits several advantages over other methods for analyzing landmarks in geometric morphometrics. The results of EDSVD are not biased by the choice of coordinate systems. However, EDSVD is certainly affected by the choice of the distances to be studied. The thin plate spline method and the analysis of Bookstein coordinates or Procrustes residuals are usually performed on data from which size has been removed. EDSVD can operate on either shape data or on information that combines elements of size and shape. EDSVD is founded on concepts that are familiar to biologists and paleontologists, namely the analysis of euclidean distances and relatively direct matrices of crossproducts or covariances by eigenanalysis. Readers should observe that EDSVD is not constrained to operate within Kendall's shape space (see Dryden and Mardia, 1998; Zelditch et al., 2004).

EDSVD is primarily a structural technique for the examination of changes in size and shape in a particular data set. Statistical inferences about the underlying population or populations are probably best carried out with bootstrapping and permutation methods (review in Zelditch et al., 2004).

Acknowledgments Ralph Chapman suggested that I resurrect my long-dormant interest in geometric morphometrics, and he also indicated that the Lower Cambrian trilobite *Bristolia* would provide a suitable example for study. Comments by two anonymous reviewers on an earlier version of this paper are appreciated. Graeme Bonham-Carter and Qiuming Cheng invited me to participate in this volume in honor of Frits Agterberg. Most importantly, I celebrate my good friend Frits Agterberg for his inspirational, wide-ranging, long, and distinguished career in mathematical geology.

References

Adams DC, Rohlf FJ, Slice DE (2004) Geometric morphometrics: ten years of progress following the "Revolution." Ital J Zool 71: 5–16

Bookstein FL (1991) Morphometric tools for landmark data. Cambridge University Press, Cambridge, p 435

Bookstein FL (1996) Standard formula for the uniform shape component in landmark data. In: Marcus LF, Corti M, Loy A, Naylor GJP, Slice DE (eds) Advances in morphometrics. Plenum Press, New York and London, pp 153–168

Bookstein FL (1997) Landmark methods for forms without landmarks. Med Image Anal 1: 225–243

Bookstein FL, Schafer K, Prossinger H, Seidler H, Fieder M, Stringer C, Weber GW, Arsuaga J-L, Slice DE, Rohlf FJ, Recheis W, Mariam AJ, Marcus LF (1999) Comparing frontal cranial profiles in archaic and modern *Homo* by morphometric analysis. Anat Rec (New Anatomist) 257: 217–224

Carpenter KE, Sommer HJ III, Marcus LF (1996) Converting truss interlandmark distances to Cartesian coordinates. In: Marcus LF, Corti M, Loy A, Naylor GJP, Slice DE (eds) Advances in morphometrics. Plenum Press, New York and London, pp 103–111

Davis JC (2002) Statistics and data analysis in geology, 3rd edn. John Wiley & Sons, New York, p 638

Dryden IL, Mardia KV (1998) Statistical shape analysis. John Wiley & Sons, New York, p 347

Green PE (1978) Analyzing multivariate data. The Dryden Press, Hinsdale, IL, p 519

Hammer O, Harper DAT (2006) Paleontological data analysis. Blackwell Publishing Ltd., Oxford, p 351

Hazzard JC, Crickmay CH (1933) Notes on the Cambrian Rocks of the eastern Mohave Desert, California. Univ Calif Publ, Bull Dept Geol Sci 23: 57–80

Lele SR (1993) Euclidean distance matrix analysis: estimation of mean form and form differences. Math Geol 25: 573–602

Lele SR, Richtsmeier JT (1991) Euclidean distance matrix analysis: a coordinate-free approach for comparing biological shapes using landmark data. Am J Phys Anthropol 86: 415–427

Lele SR, Richtsmeier JT (2001) An invariant approach to statistical analysis of shapes. Chapman & Hall/CRC, New York, p 308

Lieberman BS (1999) Systematic revision of the Olenelloidea (Trilobita, Cambrian). Bulletin 45, Peabody Museum of Natural History, Yale University, p 150

MacLeod N (1999) Generalizing and extending the eigenshape method of shape visualization and analysis. Paleobiology 25: 107–138

MacLeod N (2001) Landmarks, localization, and the use of morphometrics in phylogenetic analysis. In: Adrain JM, Edgecombe GD, Lieberman BS (eds) Fossils, phylogeny, and form, an analytical approach. Kluwer Academic/Plenum Publishers, New York, pp 197–233

Mitteroecker P, Gunz P, Bernhard M, Schaefer K, Bookstein FL (2004) Comparison of cranial ontogenetic trajectories among great apes and humans. J Hum Evol 46: 679–698

Mosimann JE (1970) Size allometry: size and shape variables with characterizations of the log-normal and generalized gamma distributions. J Am Stat Assoc 65: 930–945

Mount JD (1980) Characteristics of early Cambrian Faunas from eastern San Bernardino County, California. South Calif Paleontol Soc Spec Publ 2: 19–29

Palmer AR, Halley RB (1979) Physical stratigraphy and trilobite biostratigraphy of the Carrara Formation (Lower and Middle Cambrian) in the southern Great Basin. United States Geological Survey Professional Paper 1047, p 131

Palmer AR, Repina LD (1993) Through a glass darkly: taxonomy, phylogeny, and biostratigraphy of the Olenellina. The University of Kansas Paleontological Contributions, New Series, Number 3, p 35

Rao CR, Suryawanshi S (1996) Statistical analysis of shape of objects based on landmark data. Proc Natl Acad Sci USA 93: 12132–12136

Resser CH (1928) Cambrian fossils from the Mohave Desert. Smithsonian Misc Colln 81: 1–13

Riccio JF (1952) The lower Cambrian Olenellidae of the southern Marble Mountains, California. Bull South Calif Acad Sci 51: 25–49

Rohlf FJ (1993) Relative warp analysis and an example of its application to mosquito wings. In: Marcus LF, Bello E, García-Valdecasas A (eds) Contributions to morphometrics, vol 8. Monografias del Museo Nacional de Ciencias Naturales, Madrid, Spain, pp 131–159

Rohlf FJ (2000a) Statistical power comparisons among alternative morphometric methods. Am J Phys Anthropol 111: 463–478

Rohlf FJ (2000b) On the use of shape spaces to compare morphometric methods. Hystrix 11: 9–25

Rohlf FJ (2003) Bias and error in estimates of mean shape in geometric morphometrics. J Hum Evol 44: 665–683

Rohlf FJ, Bookstein FL (eds) (1990) In: Proceedings of the Michigan morphometrics workshop. Special Publication Number 2, University of Michigan Museum of Zoology, Ann Arbor, MI, p 380

Webster M (2002) Stratigraphic trends in morphology: the evolution of *Bristolia* (Trilobita, Cambrian). Geol Soc Am Abstr Prog 34 (2): A14

Zelditch ML, Swiderski DL, Sheets HD, Fink WL (2004) Geometric morphometrics for biologists: a primer. Elsevier Academic Press, New York, p 443

A Note on Seasonal Variation in Radiolarian Abundance

Richard A. Reyment, Isao Motoyama, Miyuki Ota and Yuichiro Tanaka

Reprinted from *Natural Resources Research* DOI: 10.1007/s11053-008-9064-y, when citing this article please use the DOI number.

Abstract A statistical analysis of two consecutive sequences of observations on radiolarian abundances in the western North Pacific, by methods appropriate to data on the simplex (i.e. compositional data) show that although the overall graphical presentations of the frequencies appear similar there are in effect, substantial differences, particular in the earlier part of each of the series. The results of the multivariate analyses are used for identifying those species that contribute most to the analysis. A brief guide to the mathematical properties of compositional data is given.

Keywords Compositional data · simplex · principal component analysis · radiolaria · paleoecology

1 Introduction

Marine biologists and marine micropaleontologists are often concerned with two aspects, productivity and composition, of plankton (palaeo)ecology in pelagic realms. Productivity is represented by "flux" (i.e., abundance per unit area per unit time) observed in terms of sediment- trap experiments; composition is assessed as percentage abundance of species. Vertical fluxes of Radiolaria from the surface to the depths have been investigated by several workers in order to acquire an

---

Richard A. Reyment
Section for Palaeozoology, Natural History Museum, Box 50007, 10405, Stockholm, Sweden,
e-mail: richard.reyment@nrm.se

Isao Motoyama
Department of Earth Evolution Sciences, University of Tsukuba, Tsukuba 305-8572, Japan,
e-mail: isaomoto@sakura.cc.tsukuba.ac.jp

Miyuki Ota
Department of Earth Evolution Sciences, University of Tsukuba, Tsukuba 305-8572, Japan

Yuichiro Tanaka
Institute of Geology and Geoinformation, Geological Survey of Japan, National Institute of Advanced Industrial Science and Technology, Higashi 1-1-1, Tsukuba 305-8567, Japan

understanding of production rate and sinking processes (e.g., Takahashi and Honjo, 1983; Takahashi, 1987; Abelmann, 1992; Boltovskoy et al., 1993). These studies have revealed that dissolution of sinking polycystine radiolarians in the water column is not significant before settling on the seafloor and that modern radiolarian fluxes in the high-latitude and equatorial oceans are associated with clear seasonal signals. For the other plankton groups, Kuroyanagi et al. (2002), Eguchi et al. (2003) and Mohiuddin et al. (2005) have reported apparent seasonality in faunal composition of planktonic foraminifers based on sediment- trap experiments in the North Pacific. Based on these observations, Nakato et al. (2005) and Motoyama et al. (2005) considered seasonal variability in radiolarian composition as well as fluxes at the studied station in a temperate marine climatic zone where environmental seasonality is clear. They found, however, insignificant seasonal variation at the family-level composition, although there was seasonal variability in total radiolarian abundance. The Radiolaria are a very diversified planktonic group composed of some 500 living species in the world ocean which is why the analyses have been made at the family-level, instead of at the species-level; this approach turned out to be time-consuming for the preliminary reports by Nakato et al. (2005) and Motoyama et al. (2005). We now have a species-level radiolarian abundance data-set from a two-year observational period at a sediment-trap station in the Northwest Pacific which permit analysis of their seasonal variation or seasonal stability in reasonable detail.

In the present note we apply standard methods of compositional multivariate statistical analysis to frequency data for radiolarians observed over two consecutive years. The compositional mode for analysing the data is necessary inasmuch as abundances are characterized by each row of an array of observations summing to a constant. Constrained data of this kind are said to lie in simplex space, a subspace of full space. Full-space methods of multivariate statistical analysis cannot be applied to constrained data-sets without appropriate modifications. Although not immediately obvious, flux-data are also constrained in that the observations derive from observations on a fixed area.

We document here substantial seasonal variations in the radiolarian faunal composition. The statistical analysis suggests that this variation can be ascribed to a small number of species and that many of species incorporated in the analysis do not show marked seasonal variability.

2 Material and Methods

Particularly for shell-bearing microplankton, the sediment-trap method is a powerful tool for fixed-point collection of series in time over several days to years (Honjo and Doherty, 1988). Time-series sediment traps (coned-shaped traps with 21-cup collectors and a collection area of $0.5\,\mathrm{m}^2$) were used to collect sinking particles at Station WCT-2 ($39°$N, $147°$E; trap depth, $\sim 1500\,\mathrm{m}$) in the Northwest Pacific (Fig. 1; Table 1). Forty samples were collected over a period of 1 year and 10 months between November 18th 1997 and August 9th, 1999 (Tables 1 and 2). Sampling

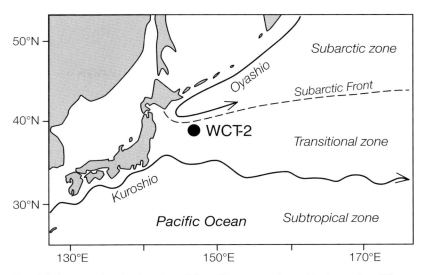

Fig. 1 Index map showing location of the sediment trap site and hydrography of the upper layer of the northwestern Pacific Ocean

intervals for each sample were 13 days for the 20 samples set up by Cruise NH97, and 6 days for 1 sample and 18 days for 19 samples by Cruise NH98. Samples were fixed with formalin to create a 3% solution buffered with sodium borate. The samples were wet-sieved through a 1 mm mesh and a 63 µm mesh to remove large and small particles. Each sample < 1 mm, > 63 µm was split into several aliquots (1/16–1/512) using a high precision rotary splitter. Each aliquot was treated with HCl to remove calcareous components, followed by rinsing with a 63 µm mesh, and then filtered through Millipore filters (0.45 µm pore size) using a vacuum pump. After filtering, the filters were air dried and then mounted on the glass slides with Canada Balsam. All specimens in a slide were counted under a transmitted light microscope. More than 500 polycystine species were encountered during the investigation. Of these species, 31 were selected for statistical analysis. These species constitute 70–85% of the total polycystine assemblage in each sample. The taxonomic information is summarized in Tables 3 and 4.

The total of 40 observational sets was divided into two equally-sized samples. The first set comprises 20 observations made over ten months from November 1997

Table 1 Locations, mooring depths, duration of sediment traps at Station WCT-2

Location	Seafloor depth (m)	Duration	Trap depth (m)
39°00.1'N, 146°59.7'E	5356	18 November 1997–5 August 1998	1371
39°01.0'N, 147°00.1'E	5322	26 August 1998–9 August 1999	1586

Table 2 Samples from Station WCT-2

No.	WCT-2 Sample	Trap cup Opened	Closed	Duration (days)
	Cruise NH97			
1	39N U01	18 Nov. '97	1 Dec. '97	13
2	39N U02	1 Dec. '97	14 Dec. '97	13
3	39N U03	14 Dec. '97	27 Dec. '97	13
4	39N U04	27 Dec. '97	9 Jan. '98	13
5	39N U05	9 Jan. '98	22 Jan. '98	13
6	39N U06	22 Jan. '98	4 Feb. '98	13
7	39N U07	4 Feb. '98	17 Feb. '98	13
8	39N U08	17 Feb. '98	2 Mar. '98	13
9	39N U09	2 Mar. '98	15 Mar. '98	13
10	39N U10	15 Mar. '98	28 Mar. '98	13
11	39N U11	28 Mar. '98	10 Apr. '98	13
12	39N U12	10 Apr. '98	23 Apr. '98	13
13	39N U13	23 Apr. '98	6 May. '98	13
14	39N U14	6 May. '98	19 May. '98	13
15	39N U15	19 May. '98	1 Jun. '98	13
16	39N U16	1 Jun. '98	14 Jun. '98	13
17	39N U17	14 Jun. '98	27 Jun. '98	13
18	39N U18	27 Jun. '98	10 Jul. '98	13
19	39N U19	10 Jul. '98	23 Jul. '98	13
20	39N U20	23 Jul. '98	5 Aug. '98	13
	Cruise NH98			
21	39N U01	26 Aug. '98	1 Sep. '98	6
22	39N U02	1 Sep. '98	19 Sep. '98	18
23	39N U03	19 Sep. '98	7 Oct. '98	18
24	39N U04	7 Oct. '98	25 Oct. '98	18
25	39N U05	25 Oct. '98	12 Nov. '98	18
26	39N U06	12 Nov. '98	30 Nov. '98	18
27	39N U07	30 Nov. '98	18 Dec. '98	18
28	39N U08	18 Dec. '98	5 Jan. '99	18
29	39N U09	5 Jan. '99	23 Jan. '99	18
30	39N U10	23 Jan. '99	10 Feb. '99	18
31	39N U11	10 Feb. '99	28 Feb. '99	18
32	39N U12	28 Feb. '99	18 Mar. '99	18
33	39N U13	18 Mar. '99	5 Apr. '99	18
34	39N U14	5 Apr. '99	23 Apr. '99	18
35	39N U15	23 Apr. '99	11 May. '99	18
36	39N U16	11 May. '99	29 May. '99	18
37	39N U17	29 May. '99	16 Jun. '99	18
38	39N U18	16 Jun. '99	4 Jul. '99	18
39	39N U19	4 Jul. '99	22 Jul. '99	18
40	39N U20	22 Jul. '99	9 Aug. '99	18

A Note on Seasonal Variation in Radiolarian Abundance

Table 3 Taxonomic list. Key to Fig. 6

Sample no	1	2	3	4	5	6	7	8	9	10	11	12	13	14	15	16	17	18	19
1 *Actinomma medianum* Nigrini	19.7	285.5	167.4	167.4	29.5	344.6	295.4	246.2	295.4	423.4	728.6	315.1	68.9	49.2	68.9	226.5	128.0	29.5	167.4
2 *Cladococcus cervicornis* Haeckel	29.5	98.5	108.3	246.2	68.9	39.4	49.2	68.9	118.2	49.2	78.8	68.9	59.1	98.5	39.4	177.2	137.8	246.2	502.2
3 *Rhizoplegma boreale* (Cleve)	14.8	246.2	98.5	98.5	39.4	128.0	236.3	275.7	384.0	324.9	334.8	137.8	29.5	98.5	59.1	88.6	88.6	98.5	68.9
4 *Stylochlamydium venustum* (Bailey)	7.4	29.5	9.8	29.5	9.8	39.4	49.2	29.5	98.5	167.4	226.5	275.7	68.9	226.5	39.4	196.9	374.2	19.7	206.8
5 *Stylodictya validispina* Joergensen	12.3	39.4	19.7	9.8	19.7	19.7	59.1	39.4	29.5	49.2	196.9	187.1	39.4	29.5	0.0	98.5	39.4	19.7	29.5
6 *Spongotrochus glacialis* Popofsky	12.3	285.5	78.8	118.2	39.4	167.4	167.4	187.1	216.6	147.7	344.6	167.4	59.1	177.2	29.5	216.6	137.8	49.2	206.8
7 *Tetrapyle octacantha* Mueller	73.8	384.0	78.8	137.8	29.5	147.7	118.2	108.3	157.5	128.0	226.5	118.2	0.0	49.2	0.0	118.2	29.5	19.7	19.7
8 *Larcopyle buetschlii* Dreyer	27.1	265.8	147.7	137.8	29.5	196.9	305.2	187.1	295.4	374.2	866.5	492.3	98.5	285.5	68.9	256.0	118.2	29.5	167.4
9 *Tholospira cervicornis* Haeckel	51.7	295.4	118.2	118.2	19.7	49.2	108.3	108.3	39.4	177.2	275.7	128.0	49.2	147.7	39.4	128.0	118.2	19.7	137.8
10 *Cladoscenium ancoratum* Haeckel	19.7	265.8	147.7	108.3	49.2	167.4	177.2	147.7	324.9	246.2	285.5	206.8	128.0	236.3	98.5	167.4	167.4	118.2	256.0
11 *Gonosphaera primordialis* Joergensen	17.2	128.0	9.8	39.4	0.0	19.7	88.6	39.4	39.4	78.8	167.4	59.1	9.8	68.9	39.4	108.3	128.0	59.1	118.2
12 *Plagiacantha arachnoids* Claparede	0.0	128.0	118.2	187.1	118.2	128.0	324.9	315.1	521.8	246.2	521.8	236.3	59.1	137.8	49.2	285.5	206.8	118.2	187.1
13 *Plagiacantha panarium* Dumitrica	9.8	59.1	9.8	29.5	9.8	108.3	137.8	128.0	610.5	275.7	1378.5	787.7	610.5	2727.4	1939.7	10289.2	5996.3	1033.8	856.6
14 *Pseudocubus obeliscus* Haeckel	9.8	571.1	157.5	108.3	39.4	216.6	295.4	196.9	620.3	443.1	1772.3	935.4	246.2	1201.2	295.4	2569.8	768.0	246.2	128.0
15 *Semantis gracilis* Popofsky?	0.0	59.1	177.2	29.5	9.8	49.2	29.5	0.0	68.9	19.7	39.4	19.7	9.8	0.0	0.0	39.4	29.5	29.5	19.7
16 *Antarctissa* sp. 1	0.0	0.0	0.0	0.0	0.0	19.7	78.8	137.8	226.5	236.3	620.3	403.7	39.4	364.3	9.8	216.6	78.8	59.1	49.2
17 *Arachnocorys circumtexta* Haeckel	0.0	68.9	39.4	9.8	39.4	29.5	0.0	9.8	29.5	9.8	29.5	0.0	147.7	0.0	0.0	9.8	78.8	19.7	246.2
18 *Lithomelissa setosa* Joergensen	4.9	137.8	108.3	59.1	19.7	98.5	128.0	206.8	984.6	462.8	2825.8	1664.0	147.7	669.5	344.6	886.2	344.6	29.5	147.7
19 *Lithomelissa thoracites* Haeckel	22.2	295.4	118.2	49.2	49.2	137.8	157.5	128.0	305.2	374.2	590.8	315.1	49.2	157.5	118.2	196.9	108.3	19.7	98.5
20 *Lophophaena* cf. *capito* Ehrenberg	2.5	0.0	9.8	0.0	0.0	0.0	39.4	0.0	68.9	128.0	334.8	285.5	78.8	246.2	19.7	256.0	137.8	59.1	147.7
21 *Peridium longispinum* Joergensen	4.9	728.6	265.8	226.5	88.6	364.3	443.1	226.5	521.8	226.5	433.2	354.5	68.9	423.4	177.2	374.2	177.2	88.6	167.4
22 *Peridium spinipes* Haeckel	27.1	541.5	295.4	78.8	9.8	177.2	295.4	275.7	334.8	256.0	1024.0	620.3	29.5	275.7	118.2	147.7	108.3	78.8	108.3
23 *Peromelissa phalacra* Haeckel	9.8	98.5	68.9	49.2	49.2	147.7	137.8	9.8	118.2	147.7	512.0	177.2	39.4	167.4	19.7	98.5	68.9	0.0	19.7
24 *Phormacantha hystrix* (Joergensen)	54.2	1063.4	797.5	384.0	246.2	1171.7	817.2	354.5	1191.4	640.0	2235.1	1191.4	403.7	955.1	462.8	1772.3	1211.1	738.5	1132.3
25 *Plectacantha oikiskos* Joergensen	4.9	600.6	797.5	364.3	128.0	226.5	393.8	433.2	964.9	521.8	905.8	866.5	344.6	521.8	344.6	580.9	571.1	512.0	758.2
26 *Plectacantha trichoides* Joergensen	9.8	177.2	108.3	39.4	19.7	59.1	78.8	98.5	157.5	108.3	128.0	59.1	29.5	49.2	9.8	128.0	265.8	108.3	531.7
27 *Pseudodictyophimus gracilipes* (Bailey)	41.8	531.7	137.8	167.4	108.3	305.2	305.2	324.9	452.9	354.5	580.9	502.2	108.3	344.6	49.2	384.0	285.5	78.8	403.7
28 *Lophophaeninae* sp. A8	2.5	0.0	9.8	9.8	9.8	39.4	19.7	68.9	59.1	39.4	118.2	49.2	9.8	29.5	0.0	29.5	9.8	0.0	0.0
29 *Plagiacanthinae* sp. F	0.0	0.0	0.0	0.0	0.0	0.0	0.0	19.7	147.7	167.4	157.5	206.8	39.4	78.8	9.8	59.1	49.2	19.7	0.0
30 *Eucecryphalus* sp. A	12.3	157.5	29.5	59.1	59.1	147.7	206.8	137.8	108.3	68.9	196.9	29.5	0.0	19.7	19.7	19.7	29.5	0.0	0.0
31 *Pterocorys zancleus* (Mueller)	27.1	128.0	59.1	19.7	19.7	68.9	9.8	59.1	49.2	19.7	88.6	68.9	59.1	59.1	9.8	78.8	49.2	0.0	39.4
32 other polycystins	706.5	7079.4	3042.5	2264.6	758.2	3446.2	4352.0	3308.3	5376.0	5385.8	9511.4	6222.8	1309.5	4381.5	1201.2	4706.5	4046.8	1171.7	4056.6
total polycystins	1235.7	14749.5	7345.2	5376.0	2116.9	8260.9	9905.2	7876.9	14916.9	12297.8	27736.6	17152.0	4292.9	14276.9	5671.4	24910.8	16086.6	5120.0	10978.5

	20	21	22	23	24	25	26	27	28	29	30	31	32	33	34	35	36	37	38	39	40
1	49.2	32.0	170.7	85.3	99.6	56.9	85.3	455.1	483.6	426.7	170.7	568.9	483.6	85.3	113.8	170.7	284.4	199.1	284.4	312.9	4.7
2	128.0	96.0	483.6	256.0	184.9	227.6	28.4	455.1	2218.7	512.0	341.3	256.0	170.7	28.4	71.1	227.6	426.7	199.1	284.4	483.6	11.9
3	88.6	74.7	227.6	369.8	213.3	398.2	142.2	85.3	369.8	170.7	227.6	625.8	682.7	284.4	142.2	199.1	426.7	199.1	227.6	113.8	14.2
4	295.4	64.0	142.2	56.9	56.9	85.3	56.9	85.3	56.9	170.7	170.7	227.6	113.8	56.9	128.0	142.2	170.7	170.7	56.9	28.4	16.6
5	9.8	0.0	56.9	0.0	56.9	56.9	0.0	142.2	170.7	85.3	28.4	85.3	56.9	56.9	156.1	56.9	85.3	28.4	341.3	28.4	7.1
6	98.5	96.0	113.8	284.4	0.0	56.9	85.3	398.2	625.8	426.7	227.6	398.2	312.9	398.2	71.1	199.1	312.9	142.2	341.3	85.3	14.2
7	59.1	10.7	56.9	85.3	42.7	113.8	256.0	455.1	654.2	369.8	113.8	341.3	113.8	85.3	0.0	85.3	56.9	56.9	56.9	142.2	4.7
8	68.9	74.7	256.0	227.6	85.3	56.9	341.3	369.8	512.0	455.1	170.7	341.3	455.1	227.6	128.0	199.1	654.2	369.8	284.4	227.6	16.6
9	59.1	42.7	85.3	284.4	56.9	28.4	142.2	426.7	512.0	398.2	170.7	256.0	113.8	113.8	28.4	85.3	142.2	199.1	113.8	256.0	4.7
10	265.8	74.7	284.4	199.1	142.2	625.8	256.0	512.0	540.4	398.2	142.2	654.2	170.7	28.4	113.8	199.1	426.7	341.3	398.2	398.2	14.2
11	88.6	0.0	0.0	85.3	113.8	56.9	113.8	28.4	28.4	0.0	28.4	113.8	28.4	28.4	56.9	85.3	113.8	113.8	113.8	0.0	0.0
12	118.2	53.3	85.3	56.9	113.8	113.8	227.6	170.7	910.2	483.6	540.4	739.6	768.0	227.6	113.8	85.3	28.4	199.1	113.8	85.3	4.7
13	315.1	42.7	28.4	142.2	14.2	28.4	0.0	512.0	2560.0	1109.3	654.2	1166.2	2560.0	3128.9	1322.7	5091.6	8021.3	2702.2	2389.3	682.7	4.7
14	29.5	21.3	0.0	170.7	71.1	85.3	56.9	682.7	8818.	369.8	113.8	142.2	455.1	1877.3	398.2	2588.4	1934.2	1223.1	2787.6	625.8	26.1
15	0.0	0.0	0.0	28.4	14.2	28.4	0.0	0.0	28.4	56.9	28.4	0.0	28.4	28.4	71.1	28.4	56.9	85.3	227.6	85.3	9.5
16	78.8	0.0	85.3	142.2	0.0	28.4	142.2	199.1	426.7	398.2	199.1	256.0	512.0	426.7	71.1	483.6	568.9	142.2	170.7	113.8	4.7
17	98.5	0.0	170.7	369.8	28.4	227.6	56.9	142.2	170.7	85.3	0.0	56.9	28.4	28.4	0.0	0.0	0.0	0.0	0.0	0.0	0.0
18	29.5	42.7	142.2	369.8	28.4	85.3	0.0	1080.9	1223.1	455.1	170.7	512.0	768.0	1621.3	199.1	1194.7	2275.6	568.9	1479.1	455.1	40.3
19	88.6	10.7	142.2	170.7	113.8	142.2	170.7	512.0	455.1	170.7	170.7	113.8	113.8	483.6	28.4	369.8	312.9	256.0	56.9	199.1	11.9
20	98.5	64.0	227.6	28.4	0.0	0.0	0.0	85.3	28.4	28.4	0.0	56.9	113.8	56.9	14.2	56.9	0.0	0.0	0.0	0.0	0.0
21	187.1	170.7	312.9	455.1	270.2	682.7	625.8	1194.7	1564.4	768.0	455.1	1052.4	796.4	711.1	327.1	824.9	796.4	568.9	1194.7	739.6	26.1
22	98.5	53.3	199.1	284.4	142.2	28.4	142.2	682.7	6542.	682.7	28.4	369.8	28.4	568.9	113.8	483.6	369.8	483.6	455.1	483.6	14.2
23	19.7	0.0	28.4	113.8	42.7	56.9	85.3	341.3	256.0	170.7	56.9	85.3	85.3	85.3	14.2	56.9	113.8	0.0	0.0	28.4	0.0
24	738.5	192.0	1280.0	1564.4	583.1	1109.3	881.8	2759.1	3555.6	2104.9	1080.9	2474.7	2673.8	3100.4	640.0	2247.1	3697.8	2417.8	3925.3	2133.3	97.2
25	423.4	192.0	910.2	597.3	611.6	1223.1	455.1	967.1	2161.8	1934.2	1223.1	4494.2	3185.8	1621.3	1720.9	1792.0	3612.4	2161.8	3640.9	2247.1	49.8
26	393.8	85.3	284.4	227.6	113.8	113.8	85.3	85.3	199.1	170.7	85.3	256.0	256.0	56.9	28.4	85.3	284.4	426.7	284.4	426.7	2.4
27	344.6	128.0	796.4	142.2	128.0	170.7	369.8	369.8	853.3	199.1	341.3	483.6	227.6	85.3	128.0	170.7	512.0	398.2	341.3	455.1	7.1
28	0.0	21.3	28.4	85.3	0.0	0.0	199.1	0.0	113.8	142.2	28.4	113.8	28.4	28.4	99.6	0.0	85.3	113.8	56.9	142.2	2.4
29	0.0	0.0	85.3	0.0	0.0	28.4	28.4	0.0	0.0	0.0	28.4	85.3	85.3	28.4	0.0	369.8	938.7	426.7	170.7	0.0	2.4
30	0.0	10.7	56.9	142.2	14.2	170.7	28.4	56.9	256.0	341.3	199.1	312.9	199.1	56.9	99.6	28.4	28.4	170.7	56.9	85.3	4.7
31	59.1	21.3	0.0	28.4	28.4	28.4	28.4	113.8	142.2	85.3	0.0	284.4	28.4	28.4	0.0	170.7	0.0	170.7	0.0	0.0	4.7
32	7436.3	1696.0	4551.1	4494.2	1792.0	4152.9	3982.2	7509.3	9756.4	6030.2	3214.2	7196.4	4693.3	4323.6	1479.1	3413.3	6001.8	4721.8	6314.7	4209.8	310.5
	7768.6	3370.7	11292.4	11264.0	5276.4	10268.4	9073.8	21048.9	32369.8	19200.0	10353.8	24120.9	20508.4	19968.0	7736.9	21191.1	32739.6	19256.9	25827.6	15274.7	732.4

A Note on Seasonal Variation in Radiolarian Abundance

Table 4 Radiolarian abundance data at Station WCT-2

Sample no.	1	2	3	4	5	6	7	8	9	10	11	12	13	14	15	16	17	18	19
1 Actinomma medianum Nigrini	1.6	1.9	2.3	3.1	1.4	4.2	3.0	3.1	2.0	3.4	2.6	1.8	1.6	0.3	1.2	0.9	0.8	0.6	1.5
2 Cladococcus cervicornis Haeckel	2.4	0.7	1.5	4.6	3.3	0.5	0.5	0.9	0.8	0.4	0.3	0.4	1.4	0.7	0.7	0.7	0.9	4.8	4.6
3 Rhizoplegma boreale (Cleve)	1.2	1.7	1.3	1.8	1.9	1.5	2.4	3.5	2.6	2.6	1.2	0.8	0.7	0.7	1.0	0.4	0.6	1.9	0.6
4 Stylochlamydium venustum (Bailey)	0.6	0.2	0.1	0.5	0.5	0.5	0.5	0.4	0.7	1.4	0.8	1.6	1.6	1.6	0.7	0.8	2.3	0.4	1.9
5 Stylodictya validispina Joergensen	1.0	0.3	0.3	0.2	0.9	0.2	0.6	0.5	0.2	0.4	0.7	1.1	0.9	0.2	0.0	0.4	0.2	0.4	0.3
6 Spongotrochus glacialis Popofsky	1.0	1.9	1.1	2.2	1.9	2.0	1.7	2.4	1.5	1.2	1.2	1.0	1.4	1.2	0.5	0.9	0.9	1.0	1.9
7 Tetrapyle octacantha Mueller	6.0	2.6	1.1	2.6	1.4	1.8	1.2	1.4	1.1	1.0	0.8	0.7	0.0	0.3	0.0	0.5	0.2	0.4	0.2
8 Larcopyle buetschlii Dreyer	2.2	1.8	2.0	2.6	1.4	2.4	3.1	2.4	2.0	3.0	3.1	2.9	2.3	2.0	1.2	1.0	0.7	0.6	1.5
9 Tholospira cervicornis Haeckel	4.2	2.0	1.6	2.2	0.9	0.6	1.1	1.4	0.3	1.4	1.0	0.7	1.1	1.0	0.7	0.5	0.7	0.6	1.3
10 Cladoscenium ancoratum Haeckel	1.6	1.8	2.0	2.0	2.3	2.0	1.8	1.9	2.2	2.0	1.0	1.2	3.0	1.7	1.7	0.7	1.0	2.3	2.3
11 Gonosphaera primordialis Joergensen	1.4	0.9	0.1	0.7	0.0	0.2	0.9	0.5	0.3	0.6	0.6	0.3	0.2	0.5	0.7	0.4	0.8	1.2	1.1
12 Plagiacantha arachnoids Claparede	0.0	0.9	1.6	3.5	5.6	1.5	3.3	4.0	3.5	2.0	1.9	1.4	1.4	1.0	0.9	1.1	1.3	2.3	1.7
13 Plagiacantha pananum Dumtrica	0.8	0.4	0.1	0.5	0.5	1.3	1.4	1.6	4.1	2.2	5.0	4.6	14.2	19.1	34.2	41.3	37.3	20.2	7.8
14 Pseudocubus obeliscus Haeckel	0.8	3.9	2.1	2.0	1.9	2.6	3.0	2.5	4.2	3.6	6.4	5.5	5.7	8.4	5.2	10.3	4.8	4.8	1.2
15 Semartis gracilis Popofsky?	0.0	0.4	2.4	0.5	0.5	0.6	0.3	0.0	0.5	0.2	0.1	0.1	0.2	0.0	0.0	0.2	0.2	0.6	0.2
16 Antarctissa sp. 1	0.0	0.5	0.0	0.0	1.9	0.2	0.8	1.8	1.5	1.9	2.2	2.4	0.9	2.6	0.2	0.9	0.5	1.2	0.4
17 Arachnocorys circumtexta Haeckel	0.0	0.0	0.5	0.2	0.0	0.4	0.0	0.1	0.2	0.1	0.1	0.0	0.0	0.0	0.0	0.0	0.5	0.4	2.2
18 Lithomelissa setosa Joergensen	0.4	0.9	1.5	1.1	0.9	1.2	1.3	2.6	6.6	3.8	10.2	9.7	3.4	4.7	6.1	3.6	2.1	0.6	1.3
19 Lithomelissa thoracites Haeckel	1.8	2.0	1.9	1.5	2.3	1.7	1.6	1.6	2.0	3.0	2.1	1.8	1.1	1.1	2.1	0.8	0.7	0.4	0.9
20 Lophophaena cf. capito Ehrenberg	0.2	0.0	0.0	0.0	0.0	0.0	0.4	0.0	0.5	1.0	1.2	1.7	1.8	1.7	0.3	1.0	0.9	1.2	1.3
21 Peridium longispinum Joergensen	0.4	4.9	3.6	4.2	4.2	4.4	4.5	2.9	3.5	1.8	1.6	2.1	1.6	3.0	3.1	1.5	1.1	1.7	1.5
22 Peridium spinipes Haeckel	2.2	3.7	4.0	1.5	0.5	2.1	3.0	3.5	2.2	2.1	3.7	3.6	0.7	1.9	2.1	0.6	0.7	1.5	1.0
23 Peromelissa phalacra Haeckel	0.8	0.7	0.9	0.9	2.3	1.8	1.4	0.1	0.8	1.2	1.8	1.0	0.9	1.2	0.3	0.4	0.4	0.6	0.2
24 Phormacantha hystrix (Joergensen)	4.4	7.2	10.9	7.1	11.6	14.2	8.3	4.5	8.0	5.2	8.1	6.9	9.4	6.7	8.2	7.1	7.5	14.4	10.3
25 Plectacantha oikiskos Joergensen	0.4	4.1	10.9	6.8	6.0	2.7	4.0	5.5	6.5	4.2	3.3	5.1	8.0	3.7	6.1	2.3	3.5	10.0	6.9
26 Plectacantha trichoides Joergensen	0.8	1.2	1.5	0.7	0.9	0.7	0.8	1.3	1.1	0.9	0.5	0.3	0.7	0.3	0.2	0.5	1.7	2.1	4.8
27 Pseudodictyophimus gracilipes (Bailey)	3.4	3.6	1.9	3.1	5.1	3.7	3.1	4.1	3.0	2.9	2.1	2.9	2.5	2.4	0.9	1.5	1.8	1.5	3.7
28 Lophophaeninae sp. A8	0.2	0.0	0.1	0.2	0.5	0.5	0.2	0.9	0.4	0.3	0.4	0.3	0.2	0.2	0.0	0.1	0.1	0.0	0.0
29 Plagiacanthinae sp. F	0.0	0.0	0.0	0.0	0.0	0.0	0.0	0.3	1.0	1.4	0.6	1.2	0.9	0.6	0.2	0.2	0.3	0.4	0.0
30 Eucecryphalus sp. A	1.0	1.1	0.4	1.1	2.8	1.8	2.1	1.8	0.7	0.6	0.7	0.2	1.1	0.1	0.2	0.1	0.2	0.0	0.0
31 Pterocorys zancleus (Mueller)	2.2	0.9	0.8	0.4	0.9	0.8	0.1	0.8	0.3	0.2	0.3	0.4	1.4	0.4	0.2	0.3	0.3	0.0	0.4
32 other polycystins	57.2	48.0	41.4	42.1	35.8	41.7	43.9	42.0	36.0	43.8	34.3	36.3	30.5	30.7	21.2	18.9	25.2	22.9	37.0

	20	21	22	23	24	25	26	27	28	29	30	31	32	33	34	35	36	37	38	39	40
1	0.6	0.9	1.5	0.8	1.9	0.6	0.9	2.2	1.5	2.2	1.6	2.4	2.4	0.4	1.5	0.8	0.9	1.0	1.1	2.0	0.6
2	1.6	2.8	4.3	2.3	3.5	2.2	0.3	2.2	6.9	2.7	3.3	1.1	0.8	0.1	0.9	1.1	1.3	1.0	1.1	3.2	1.6
3	1.1	2.2	2.0	3.3	4.0	3.9	1.6	0.4	1.1	0.9	2.2	2.6	3.3	1.4	1.8	0.9	1.3	1.0	0.9	0.7	1.9
4	3.8	1.9	1.3	0.5	1.1	0.8	0.6	0.8	0.2	0.9	1.6	0.9	1.1	0.3	1.7	0.7	0.5	0.9	0.2	0.2	2.3
5	0.1	0.0	0.5	0.0	0.3	0.6	0.0	0.7	0.5	0.4	0.3	0.4	0.6	0.3	2.0	0.3	0.3	0.1	0.0	0.2	1.0
6	1.3	2.8	1.0	2.5	1.1	0.6	0.9	1.9	1.9	2.2	2.2	1.7	1.5	0.3	0.9	0.9	1.0	0.7	1.3	0.6	1.9
7	0.8	0.3	0.5	0.8	0.8	1.1	2.8	2.2	2.0	1.9	1.1	1.4	0.6	0.4	0.0	0.4	0.2	0.3	0.2	0.6	0.6
8	0.9	2.2	2.3	2.0	1.6	0.6	3.8	1.8	1.6	2.4	1.6	1.4	2.2	1.1	1.7	0.9	2.0	1.9	1.1	1.5	2.3
9	0.8	1.3	0.8	2.5	1.1	0.3	1.6	2.0	1.6	2.1	1.6	1.1	0.6	0.6	0.4	0.4	0.4	1.0	0.4	1.7	0.6
10	3.4	2.2	2.5	1.8	2.7	6.1	2.8	2.4	1.7	2.1	1.4	2.7	0.8	0.1	1.5	0.9	1.3	1.8	1.5	2.6	1.9
11	1.1	0.0	0.0	0.8	2.2	0.6	1.3	0.1	0.1	0.0	0.3	0.5	0.1	0.1	0.7	0.4	0.3	0.6	0.4	0.0	0.0
12	1.5	1.6	0.8	0.5	2.2	1.1	2.5	0.8	2.8	2.5	5.2	3.1	3.7	1.1	1.5	0.4	0.1	1.0	0.4	0.6	0.0
13	4.1	1.3	0.3	1.3	0.3	0.3	0.0	2.4	7.9	5.8	6.3	4.8	12.5	15.7	17.1	24.0	24.5	14.0	9.3	4.5	0.6
14	0.4	0.6	0.0	1.5	1.3	0.8	0.6	3.2	2.7	1.9	1.1	0.6	2.2	9.4	5.1	12.2	5.9	6.4	10.8	4.1	3.6
15	0.0	0.0	0.0	0.3	0.3	0.3	0.0	0.4	0.1	0.3	0.3	0.4	0.1	0.1	0.0	0.1	0.2	0.4	0.9	0.6	1.3
16	1.0	0.0	0.8	1.3	0.0	0.3	1.6	0.9	1.3	2.1	1.9	1.1	2.5	2.1	0.9	2.3	1.7	0.7	0.7	0.7	0.6
17	1.3	0.0	1.5	3.3	0.0	2.2	0.6	0.7	0.5	0.4	0.0	0.2	0.1	0.1	0.0	0.0	0.0	0.0	0.0	0.0	0.0
18	0.4	1.3	1.3	0.8	0.5	0.8	0.0	5.1	3.8	2.4	1.6	2.1	3.7	8.1	2.6	5.6	7.0	3.0	5.7	3.0	5.5
19	1.1	0.3	1.3	1.5	1.1	1.4	1.9	2.4	1.4	0.9	1.1	0.5	0.6	2.4	0.4	1.7	1.0	1.3	0.2	1.3	1.6
20	1.3	1.9	2.0	0.3	2.2	0.0	0.0	0.4	0.1	0.1	0.0	0.2	0.3	0.3	0.2	0.3	0.0	0.0	0.0	0.0	0.0
21	2.4	5.1	2.8	4.0	5.1	6.6	6.9	5.7	4.8	4.0	4.4	4.4	3.9	3.6	4.2	3.9	2.4	3.0	4.6	4.8	3.6
22	1.3	1.6	1.8	2.5	2.7	0.3	1.6	3.2	2.0	3.6	0.3	1.5	0.1	2.8	1.5	2.3	1.1	2.5	1.8	3.2	1.9
23	0.3	0.0	0.3	1.0	0.8	0.6	0.9	1.6	0.8	0.9	0.5	0.4	0.4	0.4	0.2	0.3	0.3	0.0	0.0	0.2	0.0
24	9.5	5.7	11.3	13.9	11.1	10.8	9.7	13.1	11.0	11.0	10.4	10.3	13.0	15.5	8.3	10.6	11.3	12.6	15.2	14.0	13.3
25	5.4	5.7	8.1	5.3	11.6	11.9	5.0	4.6	6.7	10.1	11.8	18.6	15.5	8.1	22.2	8.5	11.0	11.2	14.1	14.7	6.8
26	5.1	2.5	2.5	2.0	2.2	1.1	0.9	0.4	0.6	0.9	0.8	1.1	1.2	0.3	0.4	0.4	0.9	2.2	1.1	2.8	0.3
27	4.4	3.8	7.1	1.3	2.4	1.7	4.1	1.8	2.6	1.0	3.3	2.0	1.1	0.4	1.7	0.8	1.6	2.1	1.3	3.0	1.0
28	0.0	0.6	0.3	0.8	0.0	0.0	2.2	0.0	0.4	0.7	0.3	0.5	0.1	0.1	0.0	0.0	0.3	0.6	0.2	0.9	0.3
29	0.0	0.3	0.8	0.0	0.3	0.3	0.3	0.3	0.0	0.0	0.3	0.4	0.4	0.1	1.3	1.7	2.9	2.2	0.7	0.0	0.3
30	0.0	0.3	0.5	1.3	0.3	1.7	0.3	0.5	0.8	1.8	1.9	1.3	1.0	0.3	0.4	0.1	0.1	0.9	0.2	0.6	0.6
31	0.8	0.6	0.0	0.3	0.5	0.3	0.3	0.5	0.4	0.4	0.0	1.2	0.1	0.1	0.0	0.8	0.0	0.9	0.0	0.0	0.6
32	44.2	50.3	40.3	39.9	34.0	40.4	43.9	35.7	30.1	31.4	31.0	29.8	22.9	21.7	19.1	16.1	18.3	24.5	24.4	27.6	42.4

to August 1998. The second set comprises 20 observations made over the period of twelve months ranging from August 1998 to August 1999.

2.1 Oceanographic Setting

Surface-waters in the Northwest Pacific off Japan can be classified into three zones, subtropical, transitional and subarctic zones (Fig. 1). Our observation site, Station WCT-2, was located at the transitional water-mass between the warm Kuroshio Current and the cool Oyashio current. The sea surface temperature ranges from 9.7 to 22.7° C during the period encompassed by the observations (Reynolds and Smith, 1994).

3 Brief Description of the Statistical Properties of Data on the Simplex

Compositional data-analysis is little known among marine biologists and marine geologists. For this reason, we have deemed it necessary, to include a short account of the basic principles that apply.

One of the most common types of observations occurring in applied geology concerns compositions, a significant aspect of which is that the data are in the form of frequencies, proportions or percentages, and all of which have the common property that the rows of the data-matrix sum to a constant. This may not seem to be much of an obstacle but there is indeed a geometrical stumbling block involved that may be severe enough as to distort, or even invalidate, an analysis (Aitchison, 1986; Reyment and Savazzi, 1999). Geometrically, compositions lie in simplex space, a subset of full space.

The study of compositions is essentially concerned with the relative magnitudes of "ingredients" and not their absolute values such as is the case for, say, measurements on a skull. These ingredients are not variables in the accepted sense of that term in statistics, but *parts*. What justifies this distinction? Consider any compositional vector **x** with non-negative elements

$$x_1 + x_2 + \ldots x_D = 1 \quad (1)$$

This vector is subject to the "unit-sum constraint", which means that the composition **x** is composed of D parts summing to 1. The components of (1) cannot be independent because they are constrained to sum to the same value.

The characteristic features of a compositional data-set are:

(a) Each row of the $N \times D$ data-matrix corresponds to a single object (in the present connexion, a biological population or equivalent).
(b) Each column of the data-matrix represents the frequency of a single part.

(c) Each row of the data-matrix sums to 1 (for proportions), respectively, 100 (for percentages).
(d) Correlations fluctuate erratically when one or more of the parts is/are removed from the data matrix (or a new part is added) because of the mathematical necessity of re-establishing the constant row-sum.
(e) Each entry in the data-matrix is non-negative.

Property (d) provides part of the key to understanding the complexity of compositional data. Correlations computed for "normal" data-matrices are invariant to the number of variables included. If you delete one or more variables from a set of measurements (array) on some anatomical feature, this has no effect on the correlations between the remaining variables. Deleting a part does, however, change correlations between all remaining parts. For example, removing the proportion of CaO from a chemical array of values does influence all other oxides in an unpredictable way, row by row, each of which will after the deletion have a row-sum differing generally from all other rows of the array and which must be restored to the constant-sum state. A multivariate statistical analysis performed on an "unadjusted" compositional data-set is seldom useful.

Is all of this a recent discovery? Not at all, and in fact the problem of spurious correlation is perhaps one of the oldest in biometry, but probably the one that is least observed in practice. The founding father of Biometry, Karl Pearson, wrote in 1897 in his essay on the mathematical foundations to the theory of evolution that a form of spurious correlation may arise when indices are used in the measurement of organisms. The modern theory of compositional data-analysis is due to Aitchison (1983, 1986, 1997). Important expansions of the geometrical aspects of theory are given in Egozcue and Pawlowsky-Glahn (2007).

Subcompositions: The question is often put by people unfamiliar with the geometrical properties of data lying in simplex space as to why it is not acceptable just to delete uninteresting parts from a data-set. The formation of a subcomposition is not merely a matter of deleting a part from each composition. If this is done, the entire balance in the data is disturbed in an unpredictable manner. If S is any subset of the parts $1,\ldots,D$ of a D-part composition $\mathbf{x}$, and $\mathbf{x}_S$ is the subvector formed from the corresponding components of $\mathbf{x}$, then $C(\mathbf{x}_S)$ is called the subcomposition of the parts S (Aitchison, 1986, p. 196). The significance of this can be seen from the following exemplification for 5 parts from which parts 1, 4 and 5 are selected to form a subcomposition, where C is known as the closure operator (Aitchison, 1986, p. 31, 34).

$$(s_1, s_2, s_3) = C(x_1, x_4, x_5)$$

Geometrically, this is a transformation from the original sample space $\mathbf{S}^4$ to a new simplex $\mathbf{S}^2$. An important property of compositional data, and one that overrules the "leave-one-out" manipulation, is that the ratio of any two components must be the same as the ratio of the corresponding two components in the full, original composition. Hence,

$$s_i/s_j = x_i/x_j \qquad (2)$$

which is the fundamental attribute of "preserved ratio relationships".

3.1 The Concepts of Covariance and Correlation in Simplex Space

1. The problem of negative bias. A correlation coefficient computed between two parts is not free to range over the interval $(-1, +1)$. Thus, in the case of two parts, say alleles A and B of the AB0 relationship of serology,
$$\text{Corr}(x_1, x_2) = -1$$
and the product-moment correlation is constrained to taking a specified value.
2. There is no relationship between the product-moment correlations of a subcomposition and those of the full composition. As the dimensionality of a subcomposition is decreased, so do the crude covariances/correlations fluctuate in sign, which is an outcome of the incoherency of the product-moment correlation coefficient in simplex space (Aitchison, 1997).
3. The concept of null correlation in reference to simplex space does not have the same meaning with respect to independence as is the case for full-space data. Futile attempts have been made in the past in geochemistry and analytical chemistry to define a zero correlation in simplex space.
4. The concept of perturbations within the simplex is another fundamental property of compositional data. A perturbation with the original composition **x** is operated upon by the perturbing vector **u** to form a perturbed composition $\mathbf{X} = \mathbf{u}°\mathbf{x}$. This is familiar to mathematical geneticists as the relationships of genotypes before and after selection (Edwards, 2000, Chap. 2).

The logical necessities of scale-invariance, subcompositional coherence and perturbation as fundamental operations in the simplex led Aitchison (1986, 1997) to adopt certain log-ratio forms of defining patterns of compositional variability. These are compatible with the *additive logistic normal class* of distributions on the simplex. One example is the set of final divisor log-ratios

$$y_i = \log(x_i/x_D)(i = 1, \ldots, D-1) \qquad (3)$$

3.2 Log-Ratio Covariance-Matrices

Three log-ratio covariance matrices are available for constrained multivariate analysis, the variation matrix, the log-ratio covariance matrix and the centred log-ratio covariance matrix (Aitchison, 1986; Reyment and Savazzi, 1999). It is the latter variant that is employed in the present analysis. The centred log-ratio covariance matrix is defined as the symmetric treatment of all D parts of a vector of compositions achieved by the manipulation using the geometric mean of all D components as a divisor. For a D-part composition, the centred log-ratio covariance matrix of the D-dimensional random vector

$$\mathbf{z} = \log\{\mathbf{x}/g(\mathbf{x})\}$$

where $g(\mathbf{x}) = (x_1, \ldots, x_D)^{1/D}$ is the geometric mean of the parts, is

$$\Gamma = \text{cov}\left[\log\left(x_i/g(\mathbf{x})\right),\ \log(x_j/g(\mathbf{x}))\right] \tag{4}$$

This matrix is the one that is interpretationally most useful for many multivariate analogues of full-space statistics. It is easy to explain in that it is symmetric with respect to all parts. The drawback is that this matrix is singular and hence does not possess a "normal" inverse and, where relevant, requires a generalized matrix inverse. Another, important drawback is that if one attempts to work with a full composition, and another person works with a subcomposition, special care is called for in interpreting the respective results. This is because the correlation between two centre log-ratio transformed parts is not the same when considered within a composition or within a subcomposition.

3.3 Log-Contrast Principal Component Analysis and Principal Coordinate Analysis

For the purposes of the present exposition, the multivariate method chosen is the widely used one of principal component analysis (Aitchison, 1983; Aitchison, 1986, p. 190) well known from many spheres of quantitative biology. The covariance matrix used as input is that of (4), the centred log-contrast covariance matrix. A log-contrast of a D-part composition $\mathbf{x}$ is defined as any log-linear combination $\mathbf{a}'\log\mathbf{x}$ with

$$a_1 + \ldots + a_D = 0$$

The principal component analysis follows from the reduction of a centred log-ratio covariance matrix in the usual manner by finding the latent roots and vectors satisfying.

$$(\Gamma - \lambda_i \mathbf{I})\mathbf{a}_i = \mathbf{0} \tag{5}$$

Log-contrast principal coordinate analysis is interpretable as the Q-mode dual of log-contrast principal component analysis as expressed by (5). The computations were made using the program *prcrd.exe* given in Reyment and Savazzi (1999, p. 140). In principal coordinate analysis, the data matrix represents distances between D points as defined as follows.

$$\sum_{i=1}^{D}\left(\left[\log(x_{1i}/g(\mathbf{x}_1)) - \log(x_{2i}/g(\mathbf{x}_2))\right]^2\right)^{1/2}$$

4 The Statistical Analysis

The first step in the analysis is a log-contrast principal coordinate analysis of the full data-set consisting of 40 observational time-points on which frequencies of 31 species were recorded. The plot of the first two constrained axes for the first set of

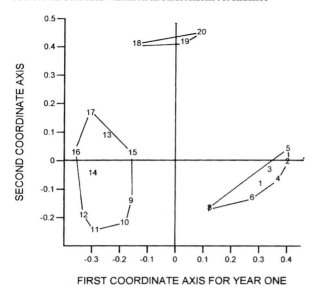

Fig. 2 Plot of first and second log-contrast principal coordinate axes for Set 1

20 observational vectors shows that the points do not form a homogeneous cluster (Fig. 2). There are three distinct groupings.

1. Observations 1–8 form a group
2. Observations 9–17 form a second group
3. Observations 18–20 form a third group.

A significant detail arising from this graph is that the clusters are homogeneous in that constituents of each group are located sequentially in time. The three clusters are well separated from each other.

The plot of the first two constrained principal coordinate axes for the second set of 20 observational vectors (Fig. 3) differs from the first group (Fig. 2) in being less markedly differentiated. The three subdivisions obtained for the graph of the constrained principal coordinate axes for the first group occur again, with one time-point displaced (the ninth). The conclusion suggested here is that notwithstanding the difference in spread of the points in each of the three groupings for both years, there is evidence of an ecological factor (temperature?) that seems to be controlling the abundance of species.

Turning now to the constrained principal coordinates graph for the pooled samples ($N = 40$), the location of the points are fairly compatible (Fig. 4), such that the first eight, respectively nine, samples fall in the right-hand quadrants (without overlap) and the second set of samples in the left-hand quadrants, with some overlap. The last three samples for both sets are located close to each other.

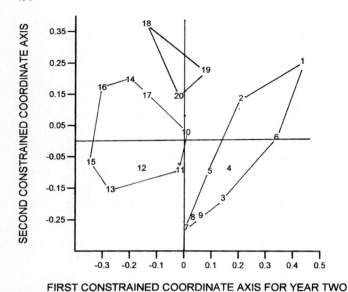

Fig. 3 First and second log-contrast principal coordinate axes for Set 2

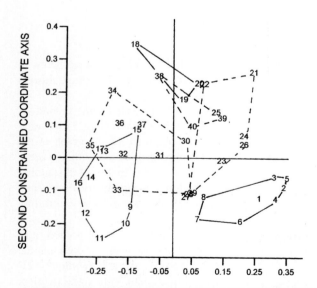

Fig. 4 The first log-contrast principal coordinate plotted against time-points over two consecutive years. The convex hulls for the first set are shown by complete lines, those for the second set by dashed lines

A Note on Seasonal Variation in Radiolarian Abundance 431

The second step in the analysis is to examine the time-sequence expressed by all samples as represented by the first coordinate axis where the elements of the first vector of coordinates is plotted against time. The graph (Fig. 5) for the two years has an oscillatory shape without obvious outlying points, thus suggesting that there is a systematic pattern of fluctuations in frequencies such that it could seem logical to assume that the order was almost identical for both series.

Closer inspection shows, however, that the two sequences are not identical, notwithstanding that they resemble each other. The following differences occur. The sinusoidal appearance of the curve for the first year (denoted "1") is more pronounced than for the curve for the second year (denoted "2"). For the first year, the initial gradient is a descent, whereas that for the second year is an ascent. After this initial divergence, the agreement in the curves begins to increase until, over the last third of the sequences, when the shapes agree more closely. Part of the difference could be due to the time-spread in the data and hence a lag in seasonal cycles.

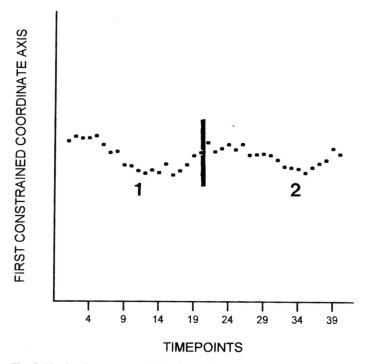

Fig. 5 The first log-contrast principal coordinate score plotted against time-point for the first (denoted "1") and second year (denoted "2") (N = 40). The vertical bar marks the end the series for the first set and the beginning of the series for the second set

4.1 Contrasting Samples

An appropriate procedure for comparing and contrasting the two samples is that of constrained discriminant function analysis (Aitchison, 1986, p. 177; Reyment and Savazzi, 1999, p. 168). The plot of the discriminant scores (Fig. 6) discloses that there is complete separation between the two groups of observations. A check for misidentifications of observations showed that none of them classifies wrongly. The method of discriminant functions can also be used to ascertain, approximately, which components are responsible for most of the separation which is simply done by inspecting the standardized discriminant coefficients. Examination of the coefficients suggests that most of the separation between sets is due to species 2, 4, 10, 12, 13, 16, 21, with less important support from species 1, 5, 8, 9, 23, 26, 27, and 28 (the key to these species is given in Table 3.

This result, if valid under repeated sampling, could be a significant feature in the study of the temporal distribution of radiolarians.

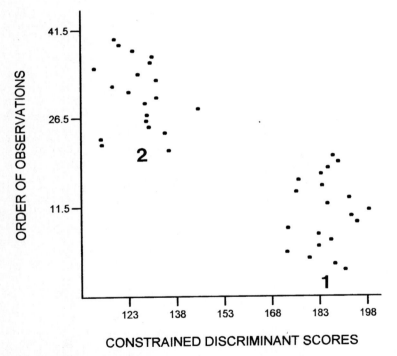

Fig. 6 The constrained discriminant function scores for the two sets of observations (first year in right half (denoted "1"), second year in left half of the graph (denoted "2")

5 Conclusions

The present brief study suggests that some species of radiolarians react differently from season to season, whereas others may be less affected by environmental conditions, as disclosed by the constrained discriminant function analysis of the data. Moreover, over the course of a particular season, constellations of species are formed which are expressed by heterogeneity in frequency patterns. An interesting outcome is that these patterns are consistent, as far as can be understood from the limited data-set available. There is evidence that an ecological factor such as temperature could be influencing the abundance of some of the species. The results presented here are to be regarded as preliminary.

Acknowledgments Samples used in this study were collected from cruises NH 97 and NH 98 of a project assigned to the Kansai Environmental Engineering Center Co. Ltd. by the New Energy and Industrial Technology Development Organization (NEDO). This study was partly supported by a Grant-in-Aid for Scientific Research No. 15540446 from the Ministry of Education, Culture, Sports, Science and Technology of Japan. Reyment's work was supported by the Japanese Society for the Promotion of Science (J. S. P. S.) and the Royal Swedish Academy of Sciences. A referee is thanked for a constructive report which greatly enhanced the value of the paper.

References

Abelmann A (1992) Radiolarian flux in Antarctic waters (Drake Passage, Powell Basin, Bransfield Strait). Polar Biol 12: 357–372

Aitchison J (1983) Principal component analysis of compositional data. Biometrika 70: 57–63

Aitchison J (1986) The statistical analysis of compositional data. Chapman and Hall, London, xv + 416 pp

Aitchison J (1997) One hour course in compositional data-analysis, or compositional data-analysis is easy. In: Pawlowsky-Glahn V (ed) Proceedings of the 1997 annual conference of the international association for mathematical geology, Universitat Politècnica de Catalunya, Barcelona, pp 3–35

Boltovskoy D, Alder VA, Abelmann A (1993) Annual flux of radiolaria and other shelled plankters in the eastern equatorial Atlantic at 853 m: seasonal variations and polycystine species-specific responses. Deep Sea Res I 40: 1863–1895

Edwards AWF (2000) Foundations of mathematical genetics, 2nd edn. Cambridge University Press, Cambridge, p 121

Egozcue JJ, Pawlowsky-Glahn V (2007) Simplicial geometry for compositional data. Geological Society of London, Special Publication SP264: 145–159

Eguchi NO, Ujiié H, Kawahata H, Taira A (2003) Seasonal variations in planktonic foraminifera at three sediment traps in the subarctic, transition and subtropical zones of the central North Pacific Ocean. Mar Micropaleontol 48: 149–163

Honjo S, Doherty KW (1988) Large aperture time-series sediment traps; design objective, construction and application. Deep Sea Res 35: 133–149

Kuroyanagi A, Kawahata H, Nishi H, Honda M (2002) Seasonal changes in planktonic foraminifera in the northwestern North Pacific Ocean: sediment trap experiments from subarctic and subtropical gyres. Deep Sea Res II 49: 5627–5645

Mohiuddin MM, Nishimura A, Tanaka Y (2005) Seasonal succession, vertical distribution, and dissolution of planktonic foraminifera along the Subarctic Front: implications for paleoceanographic reconstruction in the northwestern Pacific. Mar Micropaleontol 55: 129–156

Motoyama I, Ota M, Kokushou T, Tanaka Y (2005) Seasonal changes in fluxes and assemblages of radiolarians collected by sediment trap experiments in the northwestern Pacific: a family-level analysis. J Geol Soc Jpn 111 (7): 404–416 (in Japanese with English abstract)

Nakato A, Motoyama I, Kawahata H (2005) Seasonal and latitudinal changes of radiolarian sinking population in sediment trap samples from the central North Pacific. Bull Geol Surv Jpn 56: 225–236 (in Japanese with English abstract)

Reyment RA, Savazzi E (1999) Aspects of multivariate statistical analysis in geology. Elsevier Science B. V., Amsterdam, x + 285 pp

Reynolds R, Smith TM (1994) Improved global sea surface temperature analyses. J Climatol **7**: 929–948 http://ingrid.ldep.columbia.edu/SOURCES/.IGOSS/.nmc/.weekly/

Takahashi K (1987) Radiolarian flux and seasonality: climatic and El Nino response in the subarctic Pacific, 1982–1984. Global Biogeochem Cycles 1: 213–231

Takahashi K, Honjo S (1983) Radiolarian skeletons: size, weight, sinking speed, and residence time in tropical pelagic oceans. Deep-Sea Res 30: 543–568

Application of Markov Mean First-Passage Time Statistics to Sedimentary Successions: A Pennsylvanian Case-Study from the Illinois Basin

John H. Doveton

Abstract Markov chains have been widely used in the statistical analysis of sedimentary successions for the discrimination of non-random transitions between lithologies and the examination of potential cyclic patterns. A lithologic transition probability matrix can also be transformed to a matrix of mean first-passage times that represent the expected number of steps for first occurrences between lithologies or facies. These passage times capture the statistics of multiple transition paths and provide a metric of distances that can be used for comparisons between sections separated stratigraphically or geographically. In addition, systematic long-term elements can be differentiated from the short-term model provided by the limited memory contained within a transition probability matrix. A case-study example is described that demonstrates the use of mean first-passage times to show changes in the interplay of deltaic and marine facies, based on transition matrices from type sections of Pennsylvanian formations in the Illinois Basin.

Keywords Markov chains · mean first passage times · stratigraphy · Illinois Basin

1 Introduction

Markov chain analysis has been applied for many years as a simple and powerful method for the discrimination of potential patterns of lithological ordering in sedimentary sequences. Following the pioneer paper of Vistelius (1949), numerous stratigraphic applications have been published including Allegre (1964), Carr et al. (1966), Krumbein (1967), Gingerich (1969), Read (1969), Doveton (1971), and Ethier (1975). For the most part, these studies were concerned with attempts to isolate systematic transition patterns that could then be interpreted in terms of sedimentological processes or cyclic phenomena. Statistical tests (typically chi-square)

John H. Doveton
Kansas Geological Survey, University of Kansas, Lawrence, Kansas 66047, USA,
e-mail: doveton@kgs.ku.edu

are commonly applied to the null hypothesis of a random process to verify whether a Markov property is significant and establish the structure of the repetitive pattern.

The fundamental Markovian descriptor of a sequence is the transition probability matrix, P, which describes a first-order Markov chain. In general applications, observations of the state of a process are taken at equal increments of time. In the case of stratigraphic data, the state of events (lithologies, facies or other characteristics) is recorded at equal increments of depth. The substitution of depth for time simply changes the reference framework from time to space.

The off-diagonal elements of the transition probability matrix capture the transition characteristics between states and will be the same for any sampling interval finer than the thinnest bed. The transition probabilities of a state to itself occur on the main diagonal of the transition probability matrix. They dictate the statistics of the distribution of thicknesses of each state and will vary with the length of the interval used. The mean thickness, m, and its variance, v, of the state i are easily computed from the equations:

$$m = \frac{1}{(1 - p_{ii})}$$

and

$$v = \frac{p_{ii}}{(1 - p_{ii})^2}$$

where the units are in number of interval spacings (Kemeny and Snell, 1960). The equations of these parameters show that thicknesses implied by Markov transition probabilities follow a geometric distribution (Krumbein and Dacey, 1969). This becomes an important consideration if the Markov transition probability matrix is used to model a synthetic stratigraphic succession. Studies of sedimentary bed thicknesses have generally and empirically concluded that they are lognormally distributed (see e.g. Pettijohn, 1957; Potter and Siever, 1955).

Patterns in the ordering of lithologies can be examined in a modified model, the embedded Markov chain (Gingerich, 1969; Doveton, 1971). Rather than make observations at fixed intervals, the observation matrix is used to record only transitions between states. Transitions of a state to itself are now precluded and the main diagonal of the transition probability matrix has zero values. Each step of the Markov chain marks the occurrence of a lithology state that differs from states in the adjacent steps.

2 Markovian Comparisons of Stratigraphic Sections Separated in Space or Time

Basin analysis studies commonly attempt to characterize changes in sedimentation pattern between successive formations or as lateral facies changes across a basin. Transition probability matrices represent statistical summaries of sequence character drawn from outcrops or core and so are useful data for these studies. However, a geological insightful comparison of one probability matrix with another must be considered carefully. A common practice of differentiating transitions that occur either more often or less often than independent events is through the computation of a difference matrix that contrasts observed probabilities with those expected for a random model (see e.g. Gingerich, 1969). A similar methodology could be used in comparing transition probability matrices for successions separated in time or space. However, observed differences would be constrained to individual transitions and especially sensitive to small sample size or non-stationarity in long successions.

An alternative measure that can be used in section comparison is that of mean first-passage time statistics (Doveton and Duff, 1984). The mean first-passage time expresses the average number of events that occurs after leaving a state before the same state is reentered. The expected passage time statistics can be calculated from the transition probability matrix as a matrix, **M**, of mean values. A fundamental matrix, **Z**, may be defined as:

$$\mathbf{Z} = (\mathbf{I} - \mathbf{P} + \mathbf{A})^{-1}$$

where **I** is an identity matrix, **P** is the transition probability matrix, and **A** is the independent events probability matrix. Then:

$$\mathbf{M} = (\mathbf{I} - \mathbf{Z} + \mathbf{E}\mathbf{Z}_\mathrm{d})\mathbf{D}$$

where **E** is an $m \times m$ matrix of unit values; $\mathbf{Z}_\mathrm{d}$ is the diagonal matrix of **Z**; **D** is a diagonal matrix whose elements are the reciprocals of the independent-events probability matrix; and m is the number of states. An example of a mean first passage time matrix computed from a transition tally matrix is shown in Table 1. Interested

Table 1 Transition tally and mean first passage time matrices compiled from outcrops of the Mattoon Formation in Illinois. Key: A = non-marine shale; S = sandstone; M = marine shale; Y = underclay; C = coal; L = limestone

(a) Transition tally matrix

	A	S	M	Y	C	L
A	0	16	3	2	3	3
S	9	0	0	12	1	0
M	7	1	0	0	0	1
Y	0	1	1	0	13	2
C	5	5	5	0	0	2
L	6	0	0	3	0	0

(b) Mean first passage-time matrix

	A	S	M	Y	C	L
A	3.7	2.6	10.1	4.4	4.8	10.5
S	3.0	4.6	10.6	3.1	4.2	11.1
M	1.6	3.5	11.2	5.2	5.8	10.3
Y	3.5	4.3	9.3	5.9	2.2	10.0
C	2.6	3.3	8.4	5.2	5.9	10.3
L	2.2	4.2	10.8	4.0	5.0	11.2

readers are referred to Kemeny and Snell (1960) for additional details on the derivation of these matrix algebra relationships. These passage times are an improvement on individual transition probabilities because they summarize the statistics of multiple transition paths and provide a metric of distances that can be graphed and assessed visually.

If applied to an embedded model (no transitions of a state to itself) then the mean first-passage times are the Markovian prediction of the average number of lithologies that intervene between successive occurrences of any given lithological state. These measures are particularly useful when applied in successions believed to have been deposited as cyclic sequences. If one of the states can be identified with the initiation or conclusion of a cycle, then the state forms a reference marker and the passage time statistics of this state are Markovian descriptor of the cycle. The method is demonstrated in the following case-study of Pennsylvanian formations in the Illinois Basin.

3 A Pennsylvanian Case-Study from the Illinois Basin Using Mean First-Passage Times

Duff (1974) analyzed the lithological succession in type and reference sections of the Pennsylvanian Kewanee and McLeansboro Groups of Illinois described by Kosanke et al (1960) and Smith and Smith (1967). Transition probability matrices were separately calculated from tally matrix counts of transitions between reference lithologies of coal (C), shale (A), sandstone (S), underclay (Y), limestone (L), and marine shale (M) for the Spoon, Carbondale, Modesto, Bond, and Mattoon formations. The location of the measured sections are shown on the map of Fig. 1. Mean first-passage time matrices were computed for each of the five formations that expresses the average separation measured in numbers of intervening lithologies predicted on the basis of first-order transition probabilities. A graphic representation of all the passage-time elements taken simultaneously would be impossible to draw because they collectively represent a complex network of interlithology distances rather than measurements from a fixed origin. However, profiles may be drafted with respect to a specific reference lithology chosen as a local datum. The reference lithology should be identified with an event that is diagnostic of the initiation (or termination) of a repetitive sequence, which in the Pennsylvanian strata of Illinois is termed a "cyclothem". Although Weller (1956) and other American authors traditionally have favored the choice of sandstone, European authors generally have defined coal-bearing cyclothems as bounded by underclays or coals. Clearly, the appropriate reference lithology must be dictated by the process that created the cyclothem, whether it is autocyclic (the interplay of sedimentation mechanisms of a deltaic complex prograding onto a marine platform), allocyclic (tectonic or eustatic changes in sea-level), or a combination of processes.

Two alternative graphic representations of mean first-passage time patterns are shown in Fig. 2 (using underclay as the datum lithology) and Fig. 3 (using marine

A Pennsylvanian Case-Study from the Illinois Basin

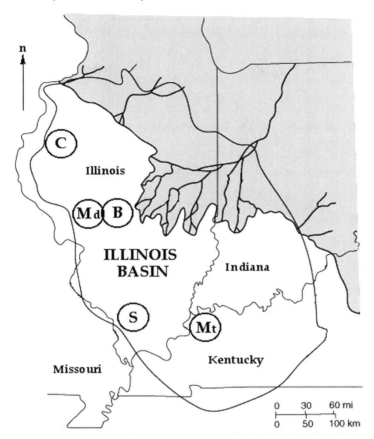

Fig. 1 Map and generalized paleogeography of the Illinois Basin during Pennsylvanian time (modified from Jacobson, 2000) indexed with locations of type and reference sections used in this study: S = Spoon Formation; C = Carbondale Formation; Md = Modesto Formation; B = Bond Formation; Mt = Mattoon Formation

shale as the datum lithology). The structure of both diagrams shows essentially the same features, as would be expected for statistics computed from the entire transition probability matrix rather than individual transition probabilities. The formations are ordered in time from left to right from the oldest (Spoon Formation) to the youngest (Mattoon Formation) and some generalized conclusions on facies changes can be drawn from Fig. 2, which are also reflected in Fig. 3. There is a decline in mean first-passage time from the underclay to marine shale moving upwards from the Spoon to the Bond, which then increases in the Mattoon. The passage time to limestones broadly parallels that of the marine shale and together, suggest an increasing marine influence, which reaches a maximum in the Bond Formation. Sandstone mean first-passage times are related inversely with marine shale times, indicating a contrast of coarse clastic influx in deltaic environments with fine-grained

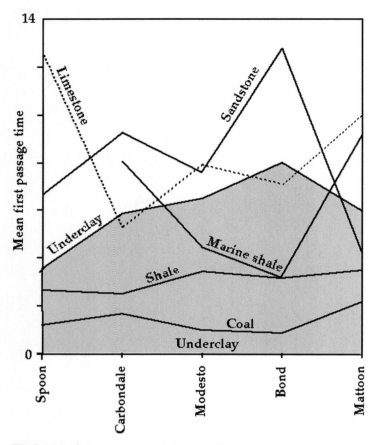

Fig. 2 Mean first-passage time profile using Underclay as the reference datum for Pennsylvanian formations in the Illinois Basin computed from transition probability matrices of type and reference sections

marine sediments. The interplay between freshwater and marine sediments may also be marked by the increase of underclay recurrence times to a maximum in the Bond Formation, implying the occurrence of cyclothems with the greatest number of lithologies.

Although the mean first-passage time profiles are based on type and reference sections, the transition statistics are calculated from relatively limited samples and so should be related to regional studies that have drawn on more extensive outcrop and core observations. Jacobsen (2000) summarized the changes in lithological composition of the Pennsylvanian succession in the Illinois Basin as a lithology distribution chart which he related to changes in depositional environment tied to relative coastline position. The chart is redrawn in Fig. 4 for the Spoon to Mattoon Formations sequence and shows a strong concordance with the mean first-passage time profiles. Coals have their maximum development in the Carbondale Formation

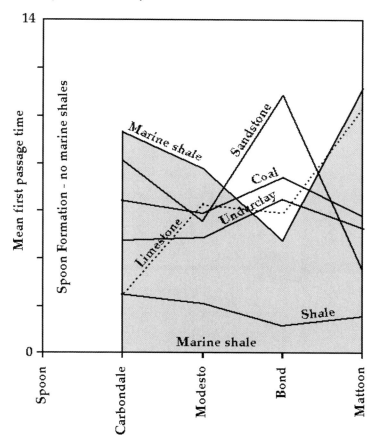

Fig. 3 Mean first-passage time profile using Marine Shale as the reference datum for Pennsylvanian formations in the Illinois Basin computed from transition probability matrices of type and reference sections

and marine limestones become dominant in the Bond Formation with a concomitant decline in sandstones. While mean first-passage times based on limited sections are no substitute for regional assessments based on a larger data-base, the confirmation of the patterns shown on the passage-time profiles is encouragement that the technique could be usefully extended to the assessment of stratigraphic changes as a tool in basin analysis.

4 Discussion

Markov chain analysis has been used extensively in studies that attempt to extract statistically significant repetitive patterns of lithology. The mean first- passage time is the average spacing that separates the occurrence of any two lithologic states. As

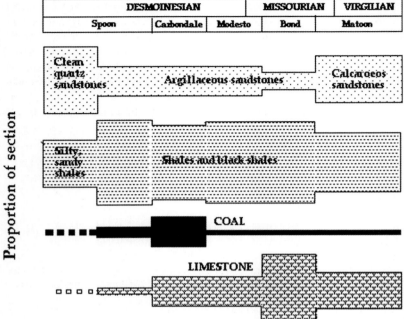

Fig. 4 Lithology distribution chart summarizing the changes in lithological composition of the Pennsylvanian succession in the Illinois Basin (modified from Jacobsen, 2000)

such, it is a useful metric of distance that is computed from the entire transition probability matrix and has potential in basin analysis studies to compare sequences separated stratigraphically or geographically. If the transition of a lithology to itself is counted, then the distance measures will be in units that match the sampling interval. However, the distances will include thicknesses of lithologies computed as expectations of an exponential distribution that may be an unsatisfactory match with natural, possibly lognormal, distributions. The use of a embedded Markov chain model removes the complicating factor of lithology thickness characterization to an assessment of lithology sequence, in which case the mean first-passage time is the average number of occurrences of lithologies between two (or the same) given lithologies.

Results from a study of Pennsylvanian formations in the Illinois Basin demonstrate the practicality of mean first-passage time statistics as comparative measures between sections that can be interpreted in terms of sedimentation changes in time. By extension, the same methodology could be used in comparison of coeval sections in basin analysis studies of lateral facies changes. Ultimately, the mean first-passage time is calculated from the statistics of the adjacency of lithologies, so that only short-term variation is contained in the prediction of long term changes. However, it is suggested that the short-term variability captures much of the localized authigenic sedimentation products, as contrasted with more regional allogenic features introduced by tectonism and eustatic sea-level changes. Although many of sedimentary sequence characteristics of the Pennsylvanian coal-bearing sequences can be attributed to local controls, sea-level changes are now widely accepted to be driven principally by ice-cap changes in an icehouse world. Clearly, the sedimentation products of these long-term mechanisms would be excluded from a Markov model based on short-memory lithologic transitions. However, by the same token, long-term predictions from a Markov model that is probably dominated by authogenic mechanisms could be applied as a template to differentiate allogenic features generated by tectonism and eustatic sea-level change.

References

Allegre C (1964) Vers une logique mathematique des series sedimentaires. Bull Soc Geologique Fr 7 (6): 214–218
Carr DD, Horowitz A, Hrarbar SV, Ridge KF, Rooney R, Straw WT, Webb W, Potter PE (1966) Stratigraphic sections, bedding sequences, and random processes. Science 28 (3753): 89–110
Doveton JH (1971) An application of Markov chain analysis to the Ayrshire coal measures succession. Scott J Geol 7 (1): 11–27
Doveton JH, Duff PMcLD (1984) Passage-time characteristics of Pennsylvanian sequences in Illinois. Ninth Int Carboniferous Congr 3: 599–604
Ethier VG (1975) Application of Markov analysis to the Banff Formation (Mississippian), Alberta. Math Geol 7 (1): 47–61
Gingerich PD (1969) Markov analysis of cyclic alluvial sediments. J Sediment Petrol 39 (1): 330–332
Jacobsen RJ (2000) Depositional history of the Pennsylvanian rocks in Illinois. Illinois Geological Survey Geonote 2, p 7
Kemeny JG, Snell JL (1960) Finite Markov chains. Van Nostrand, Princeton, p 210
Kosanke RM, Simon JA, Wanless HR, Willman HB (1960) Classification of the Pennsylvanian strata of Illinois. Illinois Geological Survey Report of Investigations 214, p 84
Krumbein WC (1967) FORTRAN IV computer programs for Markov chain experiments in geology. Kans Geol Surv Comput Contribut 13: 38
Krumbein WC, Dacey MF (1969) Markov chains and embedded Markov chains in geology. Math Geol 1 (2): 79–96
Pettijohn FP (1957) Sedimentary rocks: Harper and Bros., New York, p 718
Potter PE, Siever R (1955) A comparative study of upper Chester and lower Pennsylvanian stratigraphic variability. J Geol 63 (5): 429–451
Read WA (1969) Analysis and simulation of Namurian sediments in central Scotland using a Markov process model. Math Geol 1 (2): 199–219
Smith WH, Smith GE (1967) Description of late Pennsylvanian strata from deep diamond drill cores in the southern part of the Illinois Basin. Illinois Geological Survey Circular 411
Vistelius AB (1949) On the question of the mechanism of formation of strata. Dokl Akad Nauk SSSR 65: 191–194
Weller JM (1956) Argument for diastrophic control of late Paleozoic cyclothems. Am Assoc Pet Geol Bull 40 (1): 17–50

The Beta Distribution for Categorical Variables at Different Support

Clayton V. Deutsch and Zhou Lan

Abstract Categorical variables are commonly encountered in mathematical geology. The categories may represent facies, rock types, soil types or some other discrete variable. These categories are mutually exclusive at the small data support; however, they become mixed as scale increases; the facies indicators become proportions at larger support. Geostatistical simulation at a fixed support requires a model for the probability distribution at the support being considered. The first and second order moments of the univariate and bivariate scale-dependent facies proportion distribution may be predicted by well established scaling laws. The shape of the scale-dependent multivariate distribution of facies proportions may be modeled by the ordinary Beta distribution. This has widespread application in geostatistical modeling of geological sites.

Keywords Geostatistics · simulation · categorical data · Beta distribution · facies analysis

1 Introduction

Suppose in a three-dimensional space Ω, there exist K facies categories $S_1, S_2, \ldots, S_K$. Each point scale location $\mathbf{u}_\alpha$ in the space corresponds to a unique facies category, that is, a set of indicator variables $I(\mathbf{u}_\alpha, k)$ $(k = 1, 2, \ldots, K)$, such that $I(\mathbf{u}_\alpha, k) = 1$ when the facies category at $\mathbf{u}_\alpha$ is S_k and $I(\mathbf{u}_\alpha, k) = 0$ otherwise. The proportion of category S_k is obtained by scaling up the facies categories over a neighborhood v_α of location $\mathbf{u}_\alpha$:

Clayton V. Deutsch
Centre for Computational Geostatistics (CCG), Department of Civil and Environmental Engineering, University of Alberta, Alberta, T6G 2W2, Canada, e-mail: cdeutsch@ualberta.ca

Zhou Lan
Centre for Computational Geostatistics (CCG), Department of Civil and Environmental Engineering, University of Alberta, Alberta, T6G 2W2, Canada

$$p_v(\mathbf{u}_\alpha, k) = \frac{1}{v} \int_{v_\alpha} I(\mathbf{u}, k) d\mathbf{u}, \quad k = 1, 2, \ldots, K. \tag{1}$$

The proportions $p_v(\mathbf{u}, k)$ are volume-dependent, that is, they depend on the support of the neighborhood v. For many years, the scaling laws governing the changes in mean, variance, covariance, and variograms of random variables have been studied. Some of the important works include: (1) Journel and Huijbregts (1978) develop a series of theoretical concepts and theorems for volume scaling, (2) Isaaks and Srivastava (1989) present the scaling laws with a practical case study, and (3) Deutsch and Frykman (1999) discuss variogram modeling at different volumetric support as well as sequential simulation based on multiscale data. The following ideas are of particular importance in understanding the volume dependent distribution of categorical variables. The ideas in this paper build on the work of Haas and coworkers (Haas and Formery, 2002; Biver et al., 2002).

Given two different volumetric support v and V, there are three important concepts: dispersion variance $D^2(v,V)$, average variogram $\overline{\gamma}(v,V)$ and mean covariance $\overline{C}(v,V)$ which are defined as (Journel and Huijbregts, 1978):

$$D^2(v,V) = E[m_v - m_V]^2 \tag{2}$$

$$\overline{\gamma}(v,V) = \frac{1}{Vv} \int_V \int_v \gamma(\mathbf{y} - \mathbf{y}') d\mathbf{y}' d\mathbf{y} \tag{3}$$

$$\overline{C}(v,V) = \frac{1}{Vv} \int_V \int_v C(\mathbf{y} - \mathbf{y}') d\mathbf{y}' d\mathbf{y} \tag{4}$$

where m_v, m_V are average values at the support of scale v and V respectively. The average variogram and average covariance are in fact the average values of, respectively, the point variogram $\gamma(\mathbf{h})$ and covariance $C(\mathbf{h})$, where one extremity of the distance vector $\mathbf{h}$ falls in the volume of V and the other extremity independently falls in the domain v. The definitions in (2) through (4) and, indeed, in the remainder of this paper are based on an assumption of second order stationarity. The following results are well known (Journel and Huijbregts, 1978; Kupfersberger, 1998):

$$D^2(v,\Omega) = D^2(V,\Omega) + D^2(v,V) \quad (v \subset V \subset \Omega) \tag{5}$$

$$\sigma^2(v,V) = \overline{C}(v,v) - \overline{C}(V,V) = \overline{\gamma}(V,V) - \overline{\gamma}(v,v) \tag{6}$$

$\sigma^2(\bullet, \bullet) = 0$ and $\overline{\gamma}(\Omega, \Omega) = D^2(\bullet, \Omega)$. The symbol "$\bullet$" denotes the support of the point data. These results are useful in understanding how variances change with changes in support. For scaled up variable Z_v, volumetric support variogram $\gamma_v(\mathbf{h})$ is defined as:

$$2\gamma_v(\mathbf{h}) = E\{[Z_v(\mathbf{u}) - Z_v(\mathbf{u} + \mathbf{h})]^2\} \text{ where } Z_v(\mathbf{u}) = \frac{1}{v} \int_v Z(\mathbf{y} + \mathbf{u}) d\mathbf{y} \tag{7}$$

Journel and Huijbregts (1978) showed that $\gamma_v(\mathbf{h})$ could be expressed as:

$$2\gamma_v(\mathbf{h}) = 2\overline{\gamma}[v(\mathbf{u}), v(\mathbf{u}+\mathbf{h})] - \overline{\gamma}[v(\mathbf{u}), v(\mathbf{u})] - \overline{\gamma}[v(\mathbf{u}+\mathbf{h}), v(\mathbf{u}+\mathbf{h})] \quad (8)$$

Assuming second order stationarity, we have

$$\overline{\gamma}[v(\mathbf{u}), v(\mathbf{u})] = \overline{\gamma}[v(\mathbf{u}+\mathbf{h}), v(\mathbf{u}+\mathbf{h})] = \overline{\gamma}(v, v) \quad (9)$$

and thus

$$\gamma_v(\mathbf{h}) = \overline{\gamma}[v(\mathbf{u}), v(\mathbf{u}+\mathbf{h})] - \overline{\gamma}(v, v) \quad (10)$$

For a large vector distance $\mathbf{h}$ compared with the size of v, the value of $\overline{\gamma}[v(\mathbf{u}), v(\mathbf{u}+\mathbf{h})]$ will be close to the value of $\gamma(\mathbf{h})$ causing $\gamma_v(\mathbf{h}) \simeq \gamma(\mathbf{h}) - \overline{\gamma}(v, v)$. The variogram model at arbitrary support v is defined as:

$$\gamma_v(\mathbf{h}) = C_v^0 + \sum_{i=1}^{nst} C_v^i \gamma^i(\mathbf{h}) \quad (11)$$

where $\gamma^i(\mathbf{h})$ represents the ith nested structure and nst the total number of nested structures. C_v^0 denotes the nugget effect and C_v^i the variance contribution of the ith nested structure. The sum of variance contributions equals the dispersion variance, that is:

$$D^2(v, \Omega) = C_v^0 + \sum_{i=1}^{nst} C_v^i \quad (12)$$

with Ω the volume of the region of interest. The range of the variogram increases for larger volumes. In particular for a change from volume support v to volume support V, the range increases as:

$$a_V = a_v + (|V| - |v|) \quad (13)$$

Depending on the shape of the large volume V, the range may increase in some directions and stay the same in other directions. The purely random component, the nugget effect, decreases with an inverse relationship of the volume:

$$C_V^0 = C_v^0 \cdot \frac{|v|}{|V|} \quad (14)$$

The decreases in variance contribution as support increases is determined by the average variogram $\overline{\Gamma}$ calculated from the nested structure Γ^i, that is:

$$C_V^i = C_v^i \cdot \frac{1 - \overline{\Gamma}(V, V, \mathbf{a}^i)}{1 - \overline{\Gamma}(v, v, \mathbf{a}^i)} \quad (15)$$

These scaling laws relate to the first and second order moments of the univariate and bivariate distribution. The actual shape of the distribution is of particular concern in geostatistical applications. There are analytical change-of-shape models for continuous variables, but there are no published references on the multivariate multiscale distribution of facies proportions. The univariate distribution for each indicator variable changes from a bimodal distribution of ones and zeros at point support to the global average at large support, see Fig. 1. Predicting the transition

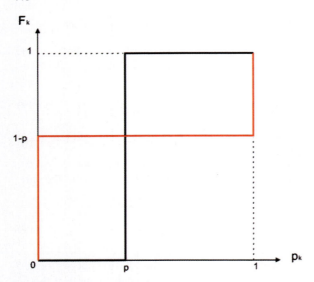

Fig. 1 CDF of p_k for cases $v = 0$ (*in red*) and $v = \infty$ (*in black*)

between a bimodal to a unimodal distribution as volume increases is of interest for the multiscale modeling of categorical variables.

2 Multiscale Distribution of Facies Proportions

Let the facies proportion for category S_k be denoted p_k. Consider n elementary volumes discretizing a block. Each elementary volume is one category. Let $x_k = \sum_{\alpha=1}^{n} I(\mathbf{u}_\alpha; k)$, the number of points where S_k occur within a certain block $v(\mathbf{u}_\alpha)$ satisfies the following condition: $\sum_{k=1}^{K} x_k = n$, where n the entire number of grid nodes in the block and $\sum_{k=1}^{K} p_k = 1$. This suggests a multinomial distribution of variables $X_k, k = 1, 2, \ldots, K$, that is (Kotz et al. 2000)

$$p[X_1 = x_1, \ldots, X_K = x_K] = \frac{n!}{x_1! x_2! \ldots x_K!} p_1^{x_1} p_2^{x_2} \cdots p_K^{x_K} \quad (16)$$

This distribution, however, does not fit actual multiscale facies distributions. An important assumption of the binomial distribution is that the indicator variables $I(\mathbf{u}_\alpha; k)$ and $I(\mathbf{u}_\beta; k)$ at two locations are independent from each other for any distinct locations $\mathbf{u}_\alpha$ and $\mathbf{u}_\beta$. The assumption of independence is violated by virtually all geological data.

Consider the Beta distribution. The Beta distribution for a random variable X within a closed interval [0,1] and has the probability density functions (pdf):

$$f(x) = \frac{\Gamma(\alpha + \beta)}{\Gamma(\alpha)\Gamma(\beta)} x^{\alpha-1} (1-x)^{\beta-1} \quad 0 \le x \le 1 \quad (17)$$

where the Gamma function is defined as $\Gamma(z) = \int_0^\infty t^{z-1} e^{-t} dt$. The probability distribution is completely determined by α and β. Based on the known expected values (global mean p_k) and variances $Var_v(p_k)$, the parameters α and β are:

$$\alpha = p_k \left[\frac{p_k(1-p_k)}{Var_v(P_k)} - 1 \right] \text{ and } \beta = (1-p_k) \left[\frac{p_k(1-p_k)}{Var_v(P_k)} - 1 \right]. \quad (18)$$

Furthermore, let Ω be the region of interest, and $D_k^2()$ be the dispersion variances of proportion p_k based on a volumetric support, we have:

$$\frac{p_k(1-p_k)}{Var_v(p_k)} - 1 = \frac{D_k^2(\bullet,\Omega)}{D_k^2(v,\Omega)} - 1 = \frac{D_k^2(\bullet,\Omega)}{D_k^2(\bullet,\Omega) - D_k^2(\bullet,v)} - 1$$

$$= \frac{D_k^2(\bullet,v)}{D_k^2(\bullet,\Omega) - D_k^2(\bullet,v)} = \frac{1}{\theta - 1} \quad (19)$$

Where $\theta = \frac{D_k^2(\bullet,\Omega)}{D_k^2(\bullet,v)} = \frac{\tilde{p}_k(1-\tilde{p}_k)}{D_k^2(\bullet,v)}$, that is: $\alpha = \frac{\tilde{p}_k}{\theta-1}$ and $\beta = \frac{1-\tilde{p}_k}{\theta-1}$. The expected values and variances are:

$$E_v[p_k] = \frac{\alpha}{\alpha+\beta}, \; Var_v[p_k] = \frac{\alpha\beta}{(\alpha+\beta)^2(\alpha+\beta+1)} \quad (20)$$

Figure 2 gives a view of a 3-D facies model that was scaled to different resolutions for empirical observation of multiscale distributions of facies proportions. Figure 3 compares sample distributions from an arbitrary facies model and fitted

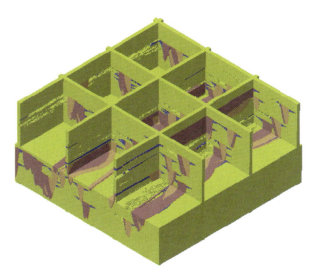

Fig. 2 Block diagram of one facies model that was upscaled to different sizes to investigate the support-dependent nature of facies proportion distributions

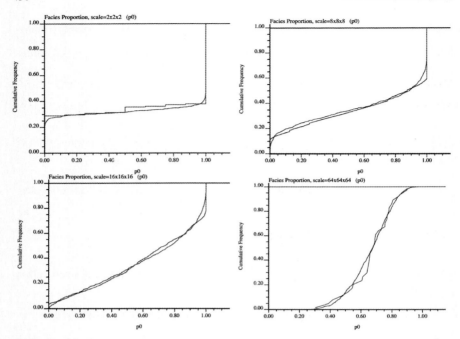

Fig. 3 Fit of experimental support-dependent facies proportion distributions with the Beta distribution

Beta distributions. The Beta simulated realizations give good reproductions of the marginal sample distributions at most supports and the facies categories. The frequencies of extreme proportion values (around 0 and 1) are partially over-estimated, making the simulated CDF curves over-smooth at both ends. When $Var_v(p_k) \to 0$ for any k, both $\alpha, \beta \to \infty$, the distribution appears normal. Some other training images were tested and reasonable fits were obtained provided the global proportion is not less than 0.01 or greater than 0.99.

To proceed with multiscale facies modeling we require a multivariate distribution and not simply K univariate distributions.

3 Dirichlet Distribution for Multivariate Multiscale Distirbutions

For K facies categories $S_1, \ldots, S_K$ in the area of interest, the proportions $(p_1, \ldots, p_K)$ will fall on a hyperplane determined by $p_1 + \ldots + p_K = 1$. The marginal distribution for each facies proportion might be fitted by the Beta distribution. This suggests using the Dirichlet Distribution, which is a generalized Beta distribution to model the joint distribution of $p_1, \ldots, p_K$.

A Dirichlet distribution (Johnson and Kotz, 2000) is defined for n random variables $X_1, X_2, \ldots, X_n$ that have the joint probability density function (joint-pdf):

The Beta Distribution for Categorical Variables at Different Support

$$f(\mathbf{x}; \boldsymbol{\alpha}) = \frac{\Gamma\left(\sum_{i=1}^{n} \alpha_i\right)}{\prod_{i=1}^{n} \Gamma(\alpha_i)} \prod_{i=1}^{n} x_i^{\alpha_i - 1} \quad (21)$$

where $x_1, x_2, \ldots, x_n \in [0,1]$ and $\sum_{i=1}^{n} x_i = 1$. $\alpha_1, \alpha_2, \ldots, \alpha_n$ are shape parameters. The expected values and variances of the random variables are:

$$E[X_i] = \frac{\alpha_i}{\sum_{i=1}^{n} \alpha_i}, \text{ and } Var[X_i] = \frac{\alpha_i \left[\sum_{j=1}^{n} \alpha_j - \alpha_i\right]}{\left[\sum_{i=1}^{n} \alpha_i\right]^2 \left[\sum_{i=1}^{n} \alpha_i + 1\right]} \quad (22)$$

It can be shown that the marginal distributions are all Beta. Taking into consideration the constraint $\sum_{i=1}^{n} x_i = 1$, only $n-1$ variables are free and $x_n = 1 - \sum_{i=1}^{n-1} x_i$. The joint-pdf for the Dirichlet distribution becomes:

$$f(x_1, x_2, \ldots, x_{n-1}; \boldsymbol{\alpha}) = \frac{\Gamma\left(\sum_{i=1}^{n} \alpha_i\right)}{\prod_{i=1}^{n} \Gamma(\alpha_i)} \left[\prod_{i=1}^{n-1} x_i^{\alpha_i - 1}\right] \cdot \left(1 - \sum_{i=1}^{n-1} x_i\right)^{\alpha_n - 1} \quad (23)$$

The joint pdf of facies proportion $p_0, p_1, \ldots, p_{K-1}$, can then be fitted as:

$$f(p_0, p_1, \ldots, p_{K-2}; \boldsymbol{\alpha}) = \frac{\Gamma\left(\sum_{k=0}^{K-1} \alpha_k\right)}{\prod_{k=0}^{K-1} \Gamma(\alpha_k)} \left[\prod_{k=0}^{K-2} p_k^{\alpha_k - 1}\right] \cdot \left[1 - \sum_{k=0}^{K-2} p_k\right]^{\alpha_{K-1} - 1} \quad (24)$$

In case of only two facies categories, it reduces to a standard Beta distribution and parameters α and β are uniquely determined by the mean and variance of either one of the two facies proportions. In cases of more than two facies categories, the following conditions should be satisfied:

$$E[X_i] = \frac{\alpha_i}{\sum_{i=1}^{n} \alpha_i}, \text{ and } Var[X_i]$$

$$= \frac{\alpha_i \left[\sum_{j=1}^{n} \alpha_j - \alpha_i\right]}{\left[\sum_{i=1}^{n} \alpha_i\right]^2 \left[\sum_{i=1}^{n} \alpha_i + 1\right]}, \quad i = 1, 2, \ldots, n \quad (25)$$

Here we have n variables, $\alpha_1, \alpha_2, \ldots, \alpha_n$, to be determined, satisfying $2n$ constraints. One possible approach is to use the separate expected values, but the variance of the most important category. From the constraint on the mean, we obtain:

$$\alpha_i = E[X_i] \cdot \sum_{j=1}^{n} \alpha_j = E[X_i] \cdot \upsilon \text{ for all } i = 1, 2, \ldots, n \quad (26)$$

Denote the sum of the α's as υ. Substitute this into the equation for $Var[x_d]$ where x_d is the selected important category:

$$\upsilon = \frac{E[x_d] \cdot (1 - E[x_d])}{Var[x_d]} \text{ and } \alpha_i = E[x_i] \cdot \upsilon \text{ for all } i = 1, 2, \ldots, n. \quad (27)$$

A number of examples were considered. The shapes of the marginal distributions were approximately reproduced, particularly for the case of smaller supports. The mean of each facies category is reproduced. In the output distributions, the variances for facies proportion p_0 were reproduced at different supports of volumetric support, while the variances for other facies categories were over estimated to different extents. For p_1 and p_2, the estimated variances were close to the real levels, but for p_4 and p_5 (and the proportions of more facies if present), the variances were overestimated and the shapes of distributions deviate from experimental observations.

4 Ordinary Beta Multivariate Multiscale Distribution

One possible solution to the problem of variances in Dirichlet distrbution that do not match experimental results lies in the use of the generalized Beta distribution introduced by Mauldon (1959). Mauldon defined an integral transformation (ϕ_β) of the distribution of n random variables $x_1, x_2, \ldots, x_n$ with joint CDF $F(x_1, x_2, \ldots, x_n)$:

$$\phi_\beta(t; a_1, \ldots, a_n) = E\left[(t - \sum_{j=1}^{n} a_j x_j)^{-\beta}\right]$$
$$= \int_{-\infty}^{\infty} \int_{-\infty}^{\infty} \cdots \int_{-\infty}^{\infty} \left(t - \sum_{j=1}^{n} a_j x_j\right)^{-\beta} dF(x_1, \ldots, x_n) \quad (28)$$

and defined $x_1, x_2, \ldots, x_n$ as a n − dimensional Beta distribution when there exist parameters c_{ij} and β_i ($i = 1, 2, \ldots, r$) such that

$$\phi_\beta = \prod_{i=1}^{r} \left(t - \sum_{j=1}^{n} a_j c_{ij}\right)^{-\beta_i} \text{ where } \beta = \sum_{i=1}^{r} \beta_i \quad (29)$$

The c_{ij} parameters form a coordinate matrix. Mauldon showed that when the coordinate matrix is a unit matrix (with all $c_{ij} = 1$), and $X_1, \ldots, X_n$ fall within $(0,1)$ and $x_1 + \ldots + x_n = 1$, the joint pdf has the form:

$$f(x_1, \ldots x_n) = \frac{\Gamma(\beta)}{\prod \Gamma(\beta_i)} \prod_{j=1}^{n} x_j^{\beta_j - 1} \quad (30)$$

Mauldon called this a Basic Beta distribution. Note that it is in fact the Dirichlet distribution as we discussed above. Mauldon also showed that any n − dimensional Beta distributions can be obtained by $\mathbf{y} = M\mathbf{x}$ from basic Beta distributed variables $X_1, \ldots, X_n$ through M. Mauldon called this the Ordinary Beta distribution.

The result from Mauldon is helpful in solving our problem. Let $p_1, p_2, \ldots, p_K$ be the K facies proportions. The joint distribution can be modeled by $\mathbf{p} = M\mathbf{x}$ where $\mathbf{x}$ follows a Dirichlet distribution with:

$$E[X_i] = \frac{\beta_i}{\beta} \text{ and } Var[X_i] = \frac{\beta_i(\beta - \beta_i)}{\beta^2(\beta + 1)} \quad (31)$$

and
$$E[\mathbf{p}] = M \cdot E[\mathbf{x}], \quad COV[\mathbf{p}] = M \cdot COV[\mathbf{x}] \cdot M^T \qquad (32)$$

where $COV[\mathbf{x}]$ denotes the covariance matrix of variable vector $\mathbf{x}$ and M^T. By solving the preceding system of equations we can find M and parameters β_i which make $p_1, \ldots, p_K$ reproduce the correct population means, variances and covariances. Specifically, applying a diagonal matrix:

$$\mathbf{M} = \begin{pmatrix} a_{11} & 0 & \cdots & 0 \\ 0 & a_{22} & \cdots & 0 \\ \vdots & \vdots & \ddots & \vdots \\ 0 & 0 & \cdots & a_{KK} \end{pmatrix} \qquad (33)$$

$\mathbf{x} = (x_1, x_2, \ldots, x_K)$ form a K – dimensional basic Beta (Dirichlet) distribution with parameters β_i ($i = 1, 2, \ldots, K$), we obtain:

$$\begin{cases} E[p_i] = \dfrac{a_{ii}\beta_i}{\beta} \\ Var[p_i] = a_{ii} \cdot \dfrac{\beta_i(\beta - \beta_i)}{\beta^2(\beta + 1)} & i = 1, 2, \ldots, K \\ \beta = \beta_1 + \beta_2 + \ldots + \beta_K \end{cases} \qquad (34)$$

Or, equivalently:

$$\begin{cases} a_{ii} = \dfrac{\beta \cdot Var[p_i] + Var[p_i] + (E[p_i])^2}{E[p_i]} \\ \beta_i = \dfrac{\beta \cdot E[p_i]}{a_{ii}} & i = 1, 2, \ldots, K \\ \beta = \beta_1 + \beta_2 + \ldots + \beta_K \end{cases} \qquad (35)$$

This system is solved for a_{ii} and β_i and we obtain the distribution of $\mathbf{p} = M\mathbf{x}$ that will reproduce both the target means and variances. One problem is that the values of $p_1, p_2, \ldots, p_K$ may not be in [0,1] or sum to one. This difficulty can be avoided by setting:

$$p_i^* = \frac{p_i}{\sum_{j=1}^{K} p_j} \quad i = 1, 2, \ldots, K \qquad (36)$$

This was implemented and the fit to experimental multivariate multiscale facies proportion distributions was very good. One major problem that may occur in ordinary Beta distribution fitting lies in the roots of β, β_i's or a_{ii}. To solve the system of equations it is necessary to solve Kth degree polynomial equations. Fortunately, our observations suggest that all the non-real roots and most of the non-positive roots occur in those extreme situations where the expected values $E[p_k]$ for certain k's in $1, 2, \ldots, K$ are greater than 0.99 or less than 0.01. Note that for all $0 \leq p_k \leq 1$, we have:

$$Var[p_k] = E[p_k^2] - (E[p_k])^2 \leq E[p_k] - (E[p_k])^2 \leq E[p_k]. \qquad (37)$$

10^9 pairs of uniformly distributed random vectors $(\mathbf{u},\mathbf{v})$, 5-dimensional or 4-dimensional, were drawn such that $0.01 < \mathbf{u}(i) < 0.99$ and $0.0005 < \mathbf{v}(i) < \mathbf{u}(i)$, treated respectively as $E[\mathbf{p}]$ and $Var[\mathbf{p}]$. Real roots occurred in all the cases and positive roots occurred in more than 97.5% of the cases. In the event all real roots were negative, we could slightly reduce the required variances and obtain positive roots. The problem of negative roots is solved by reducing the maximum of the target variances.

5 Joint PDF for Ordinary Beta Distribution

The parametric joint pdf for the Ordinary Beta distribution can be derived by applying change of variables theorem. In the transformation $\mathbf{p} = M\mathbf{x}$, where $\mathbf{x}$ follows a Dirichlet distribution with joint pdf:

$$f_{\mathbf{x}}(x_1,\ldots,x_K) = \frac{\Gamma(\beta)}{\prod \Gamma(\beta_i)} \prod_{j=1}^{K} x_j^{\beta_j - 1} \tag{38}$$

and M an invertible matrix with inverse $M^{-1} = [b_{ij}]$, $i, j = 1, 2, \ldots, K$ we have

$$\mathbf{x} = M^{-1}\mathbf{p} \tag{39}$$

or equivalently $x_i(\mathbf{p}) = \sum_{j=1}^{K} b_{ij} p_j$ $(i, j = 1, 2, \ldots, K)$. The Jacobian of the transformation is:

$$J = [s_{ij}], \text{ where } s_{ij} = \frac{\partial x_i(\mathbf{p})}{\partial p_j} = b_{ij} \; (j = 1, 2, \ldots, K) \tag{40}$$

That is: $J = M^{-1}$. Denoting the determinant of the Jacobian J as $\det(J)$ and its absolute value as $|\det(J)|$, applying the integral transformation theorem, we obtain the joint pdf for $\mathbf{p}$ as:

$$\begin{aligned} f_{\mathbf{p}}(p_1,\ldots,p_K) &= f_{\mathbf{x}}[x_1(\mathbf{p}),\ldots,x_K(\mathbf{p})] \cdot |\det(J)| \\ &= |\det(M^{-1})| \cdot \frac{\Gamma(\beta)}{\prod \Gamma(\beta_i)} \cdot \prod_{j=1}^{K} \left(\sum_{k=1}^{K} b_{jk} p_k \right)^{\beta_j - 1} \end{aligned} \tag{41}$$

If M is diagonal:

$$\mathbf{M} = \begin{pmatrix} a_{11} & 0 & \cdots & 0 \\ 0 & a_{22} & \cdots & 0 \\ \vdots & \vdots & \ddots & \vdots \\ 0 & 0 & \cdots & a_{KK} \end{pmatrix} \tag{42}$$

where, as previously discussed, $a_{ii} > 0$ for all $i = 1, 2, \ldots, K$, the joint pdf $f_{\mathbf{p}}$ can be simplified to:

$$f_{\mathbf{p}}(p_1,\ldots,p_K) = \frac{\Gamma(\beta)}{\prod_{i=1}^{K}[a_{ii}\cdot\Gamma(\beta_i)]}\cdot\prod_{j=1}^{K}\left(\frac{p_j}{a_{jj}}\right)^{\beta_j-1}$$
$$= \frac{\Gamma(\beta)}{\prod_{i=1}^{K}\left[a_{ii}^{\beta_i}\cdot\Gamma(\beta_i)\right]}\cdot\prod_{j=1}^{K}(p_j)^{\beta_j-1} \tag{43}$$

and joint CDF is:

$$F_{\mathbf{p}}(p_1,\ldots,p_K) = \int_0^{p_1}\ldots\int_0^{p_K}f_{\mathbf{p}}(t_1,\ldots,t_K)dt_1\ldots dt_K$$
$$= \frac{\Gamma(\beta)}{\prod_{i=1}^{K}\left[a_{ii}^{\beta_i}\cdot\beta_i\cdot\Gamma(\beta_i)\right]}\cdot\prod_{j=1}^{K}(p_j)^{\beta_j} \tag{44}$$

This parametric form is particularly useful to simulate categorical variables at a non-point support. The simulated outcomes are not discrete categories (unless the support is the point support). Point support data could be combined with data at any other support in block kriging to simulate at an arbitrary support. The lack of a reasonable multivariate distribution for multiscale categorical variable proportions has been a limitation in geostatistics.

6 Conclusions

The distribution of large support proportions of categorical variables is complex. At a small support, proportion of a facies category occurs as a set of discrete values. When the support increases, a continuous distribution is observed. Means of the scaled up facies proportions are independent of the support. The variance of facies proportions will decrease as the support increases. The changes in variance and variograms can be predicted with the variogram and covariance models. Both the marginal distribution of proportion for each facies category and the joint (multivariate) distribution of a full set of facies categories depend on the means and variances of all categories. They depend on the volumetric support. Beta distributions and Ordinary Beta distributions are manageable for the modeling the marginal and joint distribution, respectively, of the facies proportions based on the means and variances at different support.

References

Biver P, Haas A, Bacquet C (2002) Uncertainties in facies proportion estimation II: application to geostatistical simulation of facies and assessment of volumetric uncertainties. Math Geol 34 (6): 703–714

Deutsch CV, Frykman P (1999) Geostatistical scaling laws applied to core and log data. Society of Petroleum Engineers Paper Number 56822

Haas A, Formery P (2002) Uncertainties in facies proportion estimation I. theoretical framework: the Dirichlet distribution. Math Geol 34 (6): 679–702
Isaaks EH, Srivastava RM (1989) An introduction to applied geostatistics. Oxford University Press, New York, p 561
Journel AG, Huijbregts CJ (1978) Mining geostatistics. Academic Press, New York, p 600
Kotz S, Balakrishnan N, Johnson NL (2000) Continuous multivariate distributions: models and applications, 2nd edn. Wiley Interscience, New York, p 752
Kupfersberger H, Deutsch CV, Journel AG (1998) Deriving constraints on small-scale variograms due to variograms of large-scale data. Math Geol 30 (7): 837–852
Mauldon JG (1959) A generalization of the beta-distribution. Ann Math Stat 30 (2): 509–520

Enhancement of Seafloor Maps for Mecklenburg Bay, Baltic Sea, Using Proxy Variables

Ricardo A. Olea, Bernd Bobertz, Jan Harff and Rudolf Endler

Abstract A common situation in the earth sciences is the unfortunate availability of data in inverse abundance to needs. The geostatistical method of cokriging is used here to address the situation of lack of measurement for geophysical and geotechnical characteristics of the seafloor of the Baltic, where, in contrast, there is abundant information about the bathymetry and the granulometry of the seafloor sediments. New maps for porosity, bulk density, grain density, p-wave velocity, acoustic impedance, and critical shear stress velocity show consistency among themselves and good agreement with old maps prepared with direct measurements, but covering about one third of the total area of interest. These maps show seafloor properties controlled by the effect of erosion and redeposition of glacial tills that have resulted in the accumulation of sediments of decreasing coarseness from coastal areas to a depocenter in the middle of an embayment.

Keywords Baltic Sea · seafloor · geotechnics · cokriging

Ricardo A. Olea
Leibniz Institute for Baltic Sea Research Warnemünde (IOW), 18119 Rostock, Germany; Presently at: United States Geological Survey, 12201 Sunrise Valley Drive, Mail Stop 956, Reston, VA, 20192, USA, e-mail: olea@usgs.gov

Bernd Bobertz
Leibniz Institute for Baltic Sea Research Warnemünde (IOW), 18119 Rostock, Germany; Presently at: Institut für Geographie und Geologie, Ernst-Moritz-Arndt-Universität Greifswald, Friedrich-Ludwig-Jahn-Strasse 16, 17487 Greifswald, Germany, e-mail: bobertz@uni-greifswald.de

Jan Harff
Leibniz Institute for Baltic Sea Research Warnemünde (IOW), 18119 Rostock, Germany,
e-mail: jan.harff@io-warnemuende.de

Rudolf Endler
Leibniz Institute for Baltic Sea Research Warnemünde Baltic Sea Research Institute (IOW), 18119 Rostock, Germany, e-mail: rudolf.endler@io-warnemuende.de

1 Introduction

Physical characteristics of seabeds are becoming increasingly important for modeling of sediment transport and solving geotechnical problems, such as planning for dredging and dumping of sediments. Lange and Jäger (1984) manually prepared a series of maps partly covering Mecklenburg Bay, northern Germany (Fig. 1). The seabed porosity map in Fig. 2 is part of their series. This reproduction is provided as a general reference and it is not expected to be fully readable given the severe reduction in size from an original paper copy. Two contour lines have been made more prominent for better comparison to the new maps. For a posting of the porosity data, see Fig. 4.

The Lange and Jäger sample extends mostly along the central part of the bay, where the seafloor consists almost exclusively of fine sediments, thus providing a geographically restricted and biased sample relative to the entire bay. Although there have been interest and need to know the seafloor properties of the bay, there was no additional sampling until 2006, when a traverse comprising 20 short cores plus two more stations were sampled. All new measurements mostly overlap with the old sampling. Thus lack of systematic surveys of geophysical parameters across the entire Mecklenburg Bay persists.

While geophysical characteristics of the Baltic Sea remain poorly sampled, there has been a generous acquisition of bathymetry and granulometry over the years. Bobertz et al. (2008) took advantage of secondary information to prepare a bed roughness map applying regionalized classification. In the present study we rely on the completely different estimation method of cokriging because of its capability of

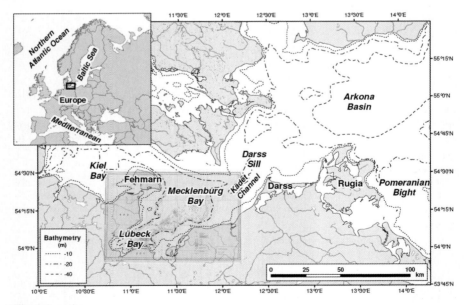

Fig. 1 Location map

Enhancement of Seafloor Maps for Mecklenburg Bay 459

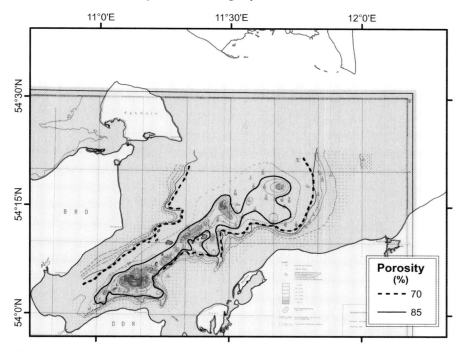

Fig. 2 Map of seafloor porosity in Mecklenburg Bay after Lange and Jäger (1984). Considering that the original map is unreadable after reduction to page size, two of the most relevant contour lines have been retouched. In addition, the study area has been expanded to the one in this contribution

not only handling multivariate estimation, but also of providing the standard error of the results. The new suit of maps includes porosity, grain size density, bulk density, p-wave velocity, acoustic impedance, and critical shear stress velocity. Lange and Jäger (1984) had prepared maps for all these attributes except for the last two, all extending over exactly the same area as the porosity map in Fig. 2.

The objectives of this publication are to release the new maps and to illustrate that proper mathematical modeling is not always possible when employing default approaches and parameters, but requires paying careful attention to details. As application of cokriging is not restricted by the physical nature of the attributes, it is expected that this study will serve to others as a guide to model other attributes, not necessarily geothechnical seafloor properties.

2 Methodology

Geostatistics offers the cokriging method for the estimation of several attributes sampled within the same geographical region (e.g. Wackernagel, 1995; Goovaerts, 1997; Olea, 1999). There are several forms of the method, of which simple cokriging is the one best suited to the characteristics of our sampling.

The notation is more compact and analogous to that for the estimation of a single attribute if one uses vectorial and matrix notation. Let

$$\mathbf{Z}(\mathbf{s}_i) = [Z_1(\mathbf{s}_i) \quad \ldots \quad Z_k(\mathbf{s}_i) \quad \ldots \quad Z_p(\mathbf{s}_i)]' \tag{1}$$

denote p attributes sampled in the same domain and let

$$\mathbf{s}_i = [easting_i \quad northing_i]' \tag{2}$$

denote geographical location at $\mathbf{s}_i$ for a two-dimensional sampling. Then, for a subset of n observations, the simple cokriging estimate for all p attributes $\mathbf{Z}^*(\mathbf{s}_0)$ is:

$$\mathbf{Z}^*(\mathbf{s}_0) = \mathbf{m} + \sum_{i=1}^{n} \mathbf{\Lambda}_i (\mathbf{Z}(\mathbf{s}_i) - \mathbf{m}) \tag{3}$$

where $\mathbf{m}$ is a vector with the mean for each one of the p attributes and $\mathbf{\Lambda}_i$ is a square coefficient matrix of order p with λ_{jk}^i being the weight for sampling site $\mathbf{s}_i$ using n measurements for attributes j and k.

$$\mathbf{m} = [m_1 \quad \ldots \quad m_k \quad \ldots \quad m_p]' \tag{4}$$

$$\mathbf{\Lambda}_i = \begin{bmatrix} \lambda_{11}^i & \lambda_{12}^i & \ldots & \lambda_{1p}^i \\ \lambda_{21}^i & \lambda_{22}^i & \ldots & \lambda_{2p}^i \\ \vdots & \vdots & \vdots & \vdots \\ \lambda_{p1}^i & \lambda_{p2}^i & \ldots & \lambda_{pp}^i \end{bmatrix} \tag{5}$$

Cokriging provides weights λ_{jk}^i such that they minimize the error in a mean square sense. If one knows $\mathbf{m}$ and has measurements for some of the p attributes at some n sites around the estimation location $\mathbf{s}_0$, then the weights $\mathbf{\Lambda}_i$ to calculate $\mathbf{Z}^*(\mathbf{s}_0)$ come from the solution of the following linear system of equations:

$$\left. \begin{array}{l} \sum_{i=1}^{n} \mathbf{\Lambda}_1' \mathbf{Cov}(\mathbf{s}_1, \mathbf{s}_i) = \mathbf{Cov}(\mathbf{s}_1, \mathbf{s}_0) \\ \sum_{i=1}^{n} \mathbf{\Lambda}_2' \mathbf{Cov}(\mathbf{s}_2, \mathbf{s}_i) = \mathbf{Cov}(\mathbf{s}_2, \mathbf{s}_0) \\ \ldots\ldots\ldots\ldots\ldots\ldots\ldots\ldots\ldots\ldots\ldots\ldots \\ \sum_{i=1}^{n} \mathbf{\Lambda}_p' \mathbf{Cov}(\mathbf{s}_p, \mathbf{s}_i) = \mathbf{Cov}(\mathbf{s}_p, \mathbf{s}_0) \end{array} \right\} \tag{6}$$

where $\mathbf{Cov}(\mathbf{Z}(\mathbf{s}), \mathbf{Z}(\mathbf{s}+\mathbf{h}))$ is the following matrix of covariances:

$$\mathbf{Cov}(\mathbf{Z}(\mathbf{s}), \mathbf{Z}(\mathbf{s}+\mathbf{h})) = \begin{bmatrix} Cov_{11}(Z_1(\mathbf{s}), Z_1(\mathbf{s}+\mathbf{h})) & \vdots & Cov_{1p}(Z_1(\mathbf{s}), Z_p(\mathbf{s}+\mathbf{h})) \\ \vdots & \vdots & \vdots \\ Cov_{p1}(Z_p(\mathbf{s}), Z_1(\mathbf{s}+\mathbf{h})) & \vdots & Cov_{pp}(Z_p(\mathbf{s}), Z_p(\mathbf{s}+\mathbf{h})) \end{bmatrix} \tag{7}$$

As a byproduct of the minimization problem, cokriging also provides the standard error, $\sigma^*(\mathbf{s})$, which is nothing but the value of the square root of the objective function in the minimization problem:

$$\sigma^*(\mathbf{s}_0) = \sqrt{\text{Trace}\left(\mathbf{Cov}(0) - \sum_{i=1}^{n}\Lambda_i\mathbf{Cov}(\mathbf{s}_i,\mathbf{s}_0)\right)} \tag{8}$$

By repeating the calculations at multiple locations—most commonly at the nodes of a regular grid—one can display the results in the form of maps. For the preparation of the grids, we used program cok3db in software GSLIB (Deutsch and Journel, 1992) modified to allow estimation for means other than zero and also for the possibility of solving systems of equations not including primary data, solely secondary data. The GSLIB grids were transformed into plain ACSII files with one column per parameter using a text editor. Afterwards they were imported into the ArcGIS (ESRI, 2006) as event themes. Contour maps are made with the Geostatistical Analyst Extension incorporating first order local polynomial interpolation.

Calculation of the pair $(\mathbf{Z}^*(\mathbf{s}_0), \sigma^*(\mathbf{s}_0))$ requires knowledge of the covariances for all attributes and crosscovariances for all possible combinations of attributes. Estimation of covariances and crosscovariances is customarily done in terms of semivariograms and crosssemivariograms, which are related through:

$$\left.\begin{array}{l}\text{Cov}_{jk}(\mathbf{h}) = \gamma_{jk}(\infty) - \gamma_{jk}(\mathbf{h}) \\ \gamma_{jk}(\mathbf{h}) = \dfrac{1}{2}\text{E}[(Z_j(\mathbf{s}) - Z_j(\mathbf{s}+\mathbf{h}))(Z_k(\mathbf{s}) - Z_k(\mathbf{s}+\mathbf{h}))]\end{array}\right\} \tag{9}$$

Note that the modeling of the crosssemivariogram requires that both attributes be measured at the same locations. The semivariogram is a special case of the crosssemivariogram for $j = k$.

To assure that the square of the standard error is positive, in practice one uses negative definite models fitted to the estimated points rather than a tabulation of estimated points. The most direct approach is to use linear coregionalization models, which are of the form:

$$\mathbf{\Gamma}(\mathbf{h}) = \sum_{l=1}^{L} B_l \gamma_l(\mathbf{h}) \tag{10}$$

For $\mathbf{\Gamma}(\mathbf{h})$ to be negative definite, the determinant, all minor determinants, and all eigenvalues of every coefficient matrix B_l, must be positive. An example of linear coregionalization model for two attributes would be:

$$\mathbf{\Gamma}(\mathbf{h}) = \begin{bmatrix} 2 & 1 \\ 1 & 3 \end{bmatrix} + \begin{bmatrix} 4 & 2 \\ 2 & 3 \end{bmatrix}\left(1 - e^{-\frac{3\mathbf{h}}{200}}\right) + \begin{bmatrix} 1 & 1 \\ 1 & 2 \end{bmatrix}\left(\frac{3}{2}\frac{\mathbf{h}}{150} - \frac{1}{2}\left(\frac{\mathbf{h}}{150}\right)^3\right), \tag{11}$$

which means, in this case, that the crosssemivariogram is:

$$\gamma_{12}(\mathbf{h}) = 1 + 2\left(1 - e^{-\frac{3\mathbf{h}}{200}}\right) + \left(\frac{3}{2}\frac{\mathbf{h}}{150} - \frac{1}{2}\left(\frac{\mathbf{h}}{150}\right)^3\right). \tag{12}$$

A weighted least square program (Jian et al., 1996; Olea 2006) was used for the modeling and GSLIB (Deutsch and Journel, 1998) was employed in the estimation of experimental points and display of results, as well as for the preparation of histograms and cumulative distributions.

Note that there are no special assumptions about the nature of the attributes involved in a cokriging estimation, thus the method is applicable to any attribute with a geographical distribution. Note also, that according to (3), one can estimate all attributes in one step. The downsides are larger systems of equations, and above all, more crosssemivariograms to model. For that reason we solved cokriging systems separately for the various geophysical attributes of interest, thus avoiding modeling of crosssemivariograms among all geotechnical attributes of interest, and having smaller and more stable systems of equations. Additionally, the geotechnical data are all collocated, situation that does not result in better estimates when cokriging each primary variables individually instead of in one large system of equations.

Given the repetitious nature of the applications of the same approach to several primary variables, while rendering all maps, we limit detailed exposition to the preparation of the porosity map only.

3 Data

Figure 3 shows the composite sample considered in our study, comprising four different sources:

- The 97 stations from the Lange and Jäger study with measurements for a variety of geophysical seafloor attributes including porosity, p-wave velocity, bulk density, and grain density.
- A complete survey of 22 stations taken in 2006, all but two along a traverse in the central part of the Mecklenburg Bay. This survey also has measurements for porosity plus bulk density, grain density, and p-wave velocity.
- Data from a granulometric IOW database comprising granulometric data and water depth (Bobertz and Harff, 2004).
- Values of water depth taken from an IOW bathymetric database (Seifert et al., 2001) employed in the estimations. These measurements were taken only in areas not covered by the other three sources.

Figure 4 posts all measurements for seafloor porosity, the primary variable that we will use to illustrate the methodology. Figure 5 is a different display of the same data, this time to show the relative abundance of values. There is a clear dominance of high values.

Although the 1984 map collection includes granulometric information, it is for the same stations where there are values for porosity. For proxy variables to be of use to enhance mapping a primary attribute, the proxy variables must be measured at sites other than the primary variable, hopefully at a higher sampling density. In addition, advances in instrumentation and laboratory analysis after 1984 allow today

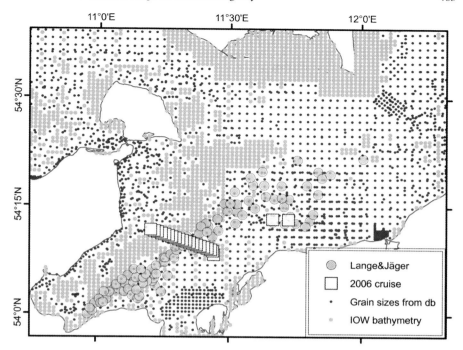

Fig. 3 Posting of all sampling locations

obtaining more accurate and diverse measurements. Figures 6 and 7 show values of median for the grain size of seafloor sediments that have been acquired over the years in the Mecklenburg Bay area, which are part of a larger database covering the whole German Baltic (Bobertz and Harff, 2004).

Another seafloor attribute that has been measured even more densely than the granulometry is water depth. Stations in the Lange and Jäger study do not include bathymetry measurements; they prepared a hand-drawn map with isolines. Instead of using their map, we assigned the closest grid node from an IOW bathymetry dataset (Seifert et al., 2001). For areas without granulometry values, we used the grid nodes themselves. Figures 8 and 9 summarize the fluctuations of the bathymetry values that we considered in this study.

4 Structural Analysis

We have seen that use of cokriging requires semivariogram and crosssemivariogram models. While the estimation method itself does not need to have both the primary and the secondary variables measured at the same station, we have seen above that modeling of the cross semivariograms does it. As an approximation, we

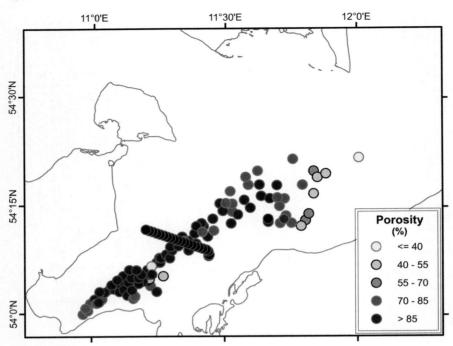

Fig. 4 Posting of the values in Fig. 2 and those from a 2006 cruise

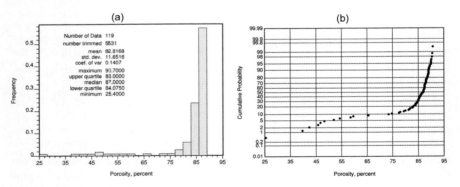

Fig. 5 Distribution of seafloor porosity data; (**a**) histogram; (**b**) cumulative distribution

assigned to the Lange and Jäger porosity stations the median of the closest modern granulometric station (Fig. 6). Figure 10 shows the range in distances. This approximation was used only for the cross semivariogram modeling, not for the estimation.

The next illustration, Fig. 11, was prepared for the purpose of exploring the dependencies between all seabed attributes we intend to use in the preparation of the new porosity map. Figure 11a shows the correlation between porosity and grain size median after eliminating 17 pairs. Those pairs clearly deviated from the negative

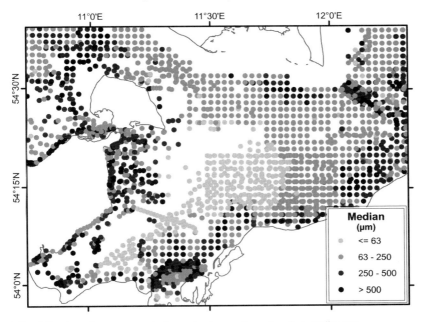

Fig. 6 Posting of measurements for grain size median of seafloor sediments

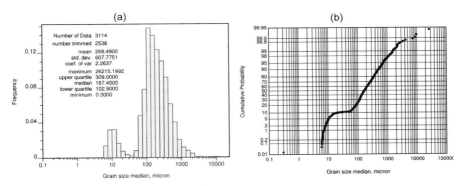

Fig. 7 Data distribution for grain size median; (**a**) histogram; (**b**) cumulative distribution

correlation that was expected according to the physics of sedimentology (e.g., Hamilton, 1972, p. 641). All discarded pairs but one had high median values corresponding to porosity values above 80% and several of the same pairs had large distances between the paired stations, some between 1750 and 2250 m.

Confirming the experience gained in the Baltic Sea (Endler, 2008), Fig. 11a and b show better correlation of porosity with granulometry than with water depth. Note that, except for scatterplot 11d, these findings are based only on those stations displayed in Fig. 4, as there are no porosity measurements for the rest of the bay. In that sense, scatterplots 11a–c are geographically biased toward an area rich in specific type of sediments.

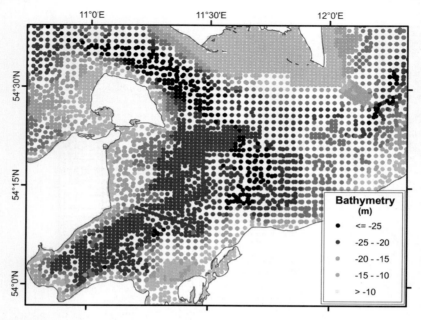

Fig. 8 Water depth at Mecklenburg Bay

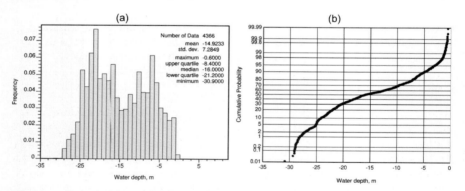

Fig. 9 Data distribution for water depth; (**a**) histogram; (**b**) cumulative distribution

The scatterplots in Fig. 11 show correlations strong enough to go into the modeling of semivariograms and crosssemivariograms for all possible pairs of variables. The resulting experimental values and best fitting models are in Fig. 12. Note the models are for normal scores of the original observations, which are dimensionless. There are at least two reasons for utilizing normal scores:

- Although geostatistical estimation methods are general enough to work with data following any distribution, they perform better under conditions of normality. As seen in the distributions above, all three attributes strongly deviate from normality.

Enhancement of Seafloor Maps for Mecklenburg Bay

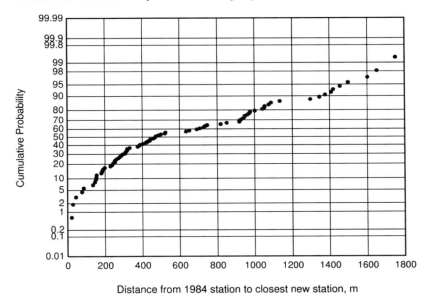

Fig. 10 Distance from Lange and Jäger station to closest station in the IOW granulometric database

- Cokriging does not provide a direct way to force the results to fall into a given range. Porosities, for example, must vary between 0 and 100%. Normal score transformation provides an indirect way to properly constrain the estimates. Cokriging in the original sampling space resulted in some negative porosities as well as values above 100%.

There is a large discrepancy in sample size and geographical distribution of the data used to prepare the models in Fig. 12. Models 12a, d, and e used solely the 119 stations in the central part of the bay (Fig. 4). The other models are based on close to 20 times more values from an almost exhaustive sample of granulometry and depth extending to every corner of the bay. Both the difference in sample size and sample area have a lot to do with the difference in styles among the models in Fig. 12. Figure 13 shows modeling for the semivariogram of normal scores of water depth using the sample covering all the study area and only those from the 2006 survey plus the Lange and Jäger data. The discrepancy is even more significant working in the original sampling space instead of normal scores.

The discrepancies resulting from the different sizes and areal extension of the samples are mostly real as the semivariogram for the entire bay does not have to be the same as those for the central sector. It is also well known that the influence of outliers grows as the sample size reduces, resulting in more erratic fluctuations as those along the sill of model 13b.

The cokriging method employed for the mapping requires that all semivariograms and cross semivariograms must be of the same type and with the same range, which forces some approximations. In addition, the determinants for both the nugget

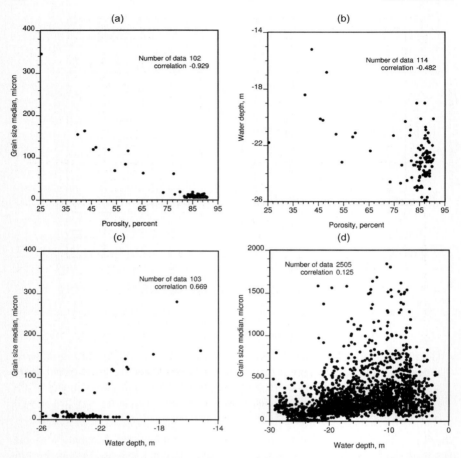

Fig. 11 Correlations between the three attributes considered in the preparation of the porosity map. The first three scatterplots were prepared using only information in the central part of the Mecklenburg Bay from the 1984 report and the 2006 cruise. The last scatterplot used all available information, as displayed in Figs. 6 and 8

and sill matrices must be positive and all eigenvalues must be positive, which sometimes requires additional tuning of the parameters to satisfy these conditions. That was not our case. The linear coregionalization model of our choice was:

$$\begin{bmatrix} 0.20 & & \\ -0.03 & 0.22 & \\ -0.01 & 0.01 & 0.01 \end{bmatrix} + \begin{bmatrix} 0.82 & & \\ -0.61 & 0.75 & \\ -0.44 & 0.64 & 1.05 \end{bmatrix} \text{Sph}(25). \qquad (13)$$

This notation means, for example, that the cross semivariogram between attribute 1 (porosity) and 2 (grain size median) is:

Enhancement of Seafloor Maps for Mecklenburg Bay

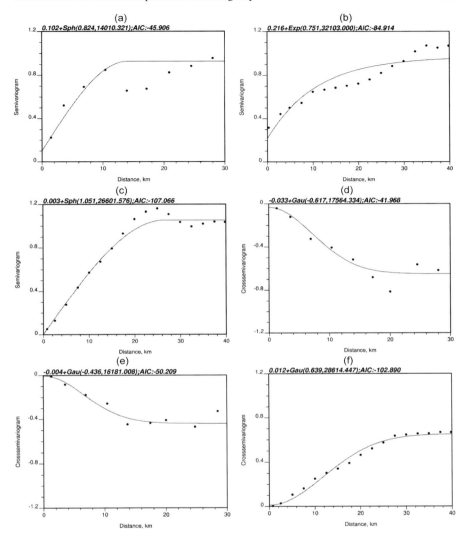

Fig. 12 Experimental values (*dots*) and best models for spatial correlation (*solid lines*) of normal scores of seafloor attributes; (**a**) porosity semivariogram (**b**) semivariogram for grain size median at seabed; (**c**) water depth semivariogram; (**d**) crosssemivariogram between porosity and grain size median; (**e**) crosssemivariogram between porosity and water depth; (**f**) crosssemivariogram between grain size median and water depth

$$\gamma_{1,2}(\mathbf{h}) = -0.03 - 0.61 \left(\frac{3}{2} \frac{\mathbf{h}}{25} - \frac{1}{2} \left(\frac{\mathbf{h}}{25} \right)^3 \right). \quad (14)$$

The selection of these models was done considering that the parameters of a model based on a larger sample size is more representative of reality, and that the

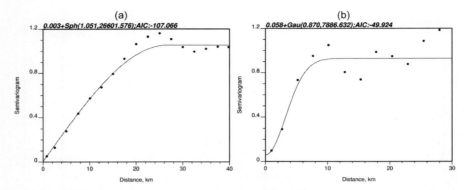

Fig. 13 Experimental semivariogram (*dots*) and models (*continuous line*) for normal scores of water depth using two difference data sources; (**a**) the same semivariogram as that in Fig. 12c employing all available data; (**b**) estimates and model using solely the 2006 survey plus the Lange and Jäger values

increase rate of spherical models is intermediate between Gaussian and exponential for short distances.

5 Estimation

Figure 14 shows a kriging map obtained using only the seabed porosity information (Fig. 4) and its semivariogram (Fig. 12a). Kriging is a limiting form of cokriging for only one attribute. Figure 14 covers about the same contoured area as Fig. 2. The kriging map has a tendency to exaggerate the extension of the high porosity area, which is in agreement with a well known poor performance of geostatistical methods in extrapolating. These drawbacks prompted the use of proxy variables.

Figure 15 renders the new simple cokriging map using both the seabed porosity information in Fig. 4, the proxy information for grain size median and water depth, and the models in (15). Figure 16 is the associated standard error map. The new map is in agreement with the original map in Fig. 2 within the limited area where the old contouring is supported by data.

Equation (15) is a slightly modified version of (13),

$$\begin{bmatrix} 0.20 \\ -0.03 \ 0.22 \\ -0.01 \ 0.01 \ 0.01 \end{bmatrix} + \begin{bmatrix} 0.82 \\ -0.41 \ 0.75 \\ -0.55 \ 0.64 \ 1.05 \end{bmatrix} \text{Sph}(25), \quad (15)$$

modification that was particularly important for obtaining also satisfactory grain densities (Fig. 17) jointly using the bulk density (Fig. 18) and porosity maps. These adjustments were done by trial and error, in a fashion analogous to the history matching in oil reservoir simulation (Caers, 2005, Chap. 4).

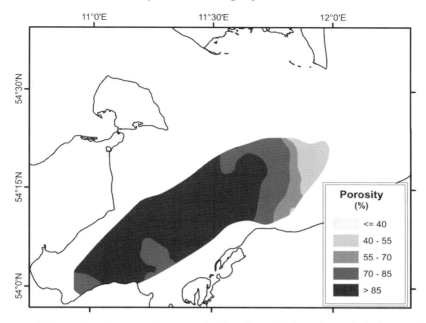

Fig. 14 Map of seafloor porosity prepared with ordinary kriging using exclusively porosity data

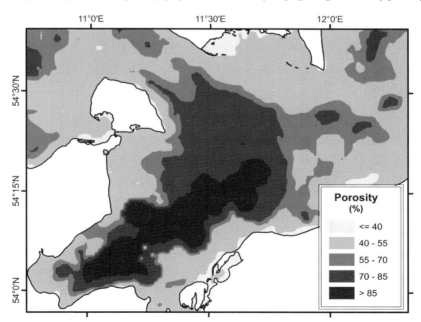

Fig. 15 Simple cokriging map of seafloor porosity using grain size median and water depth as proxy variables. Estimation was done utilizing normal scores that were then backtransformed to porosity space

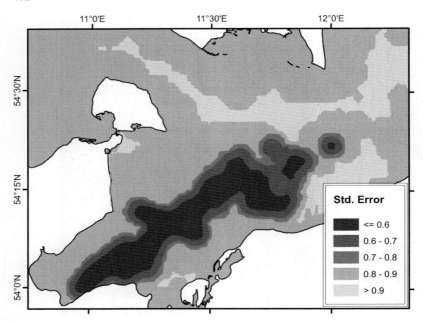

Fig. 16 Simple cokriging standard error map of seafloor porosity for the normal scores in Fig. 15

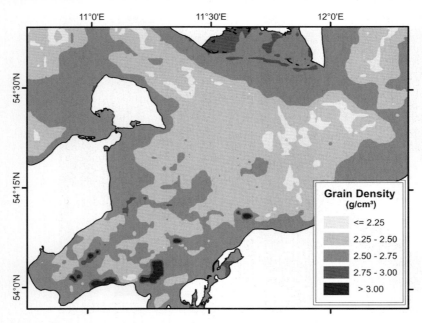

Fig. 17 Simple cokriging map for grain density employing water depth and grain size median as proxy variables

Enhancement of Seafloor Maps for Mecklenburg Bay

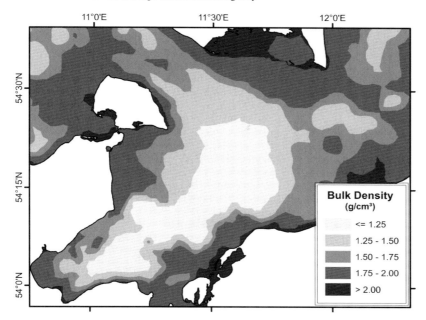

Fig. 18 Simple cokriging map for bulk density with depth and grain size median as proxy variables

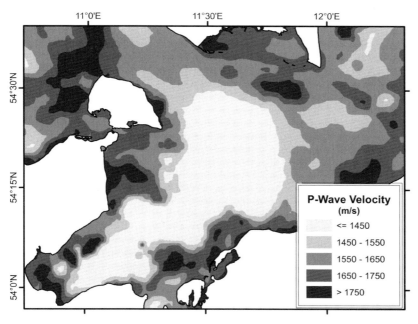

Fig. 19 Simple cokriging map for p-wave velocity using water depth and grain size median as proxy variables

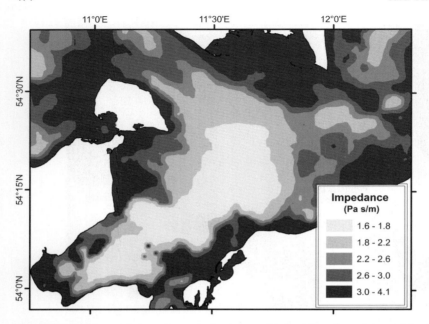

Fig. 20 Map for acoustic impedance obtained as a grid-to-grid product of the nodal values in Figs. 18 and 19

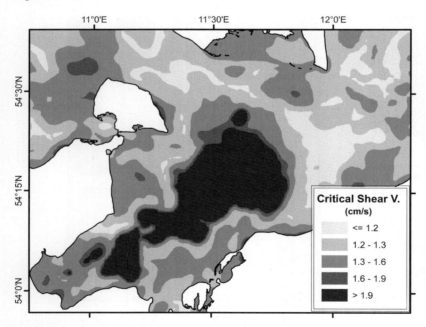

Fig. 21 Map for critical shear stress velocity obtained as transformation of the map in Fig. 22 according to the relationship displayed in Fig. 23

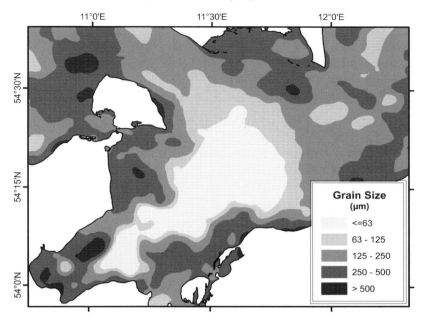

Fig. 22 Simple cokriging map of median grain size using water depth as proxy variable

The successful use of the porosity map in generating a map of a dependable attribute should be considered an additional validation for the results in Figs. 15–16.

Following the general methodology outlined above and discussed below, the complete study also included preparation of the map in Fig. 19. The acoustic

Table 1 Transformation used to convert grain size median into critical shear stress velocity. If d is the grain size median in meters, the expression for $27 < 10^6 \, d \leq 148$ is the Hjulström equation and the one above 148 microns is the Shields equation. The critical shear stress velocity is in cm/s

$$V_s = \begin{cases} 3.0, & \text{if } 10^6 \, d \leq 27 \\ 0.14\sqrt{g \cdot \rho' \cdot d} + 0.735\dfrac{\upsilon}{d}, & \text{if } 27 < 10^6 \, d \leq 148 \\ \sqrt{\theta_{cr} \cdot g \cdot \rho' \cdot d}, & \text{if } 148 < 10^6 \, d \end{cases}$$

$$\rho' = \frac{\rho_s - \rho_w}{\rho_w}$$

$$D_* = \left(\frac{g \cdot \rho'}{\upsilon^2}\right)^{\frac{1}{3}} \cdot d$$

$$\theta_{cr} = \frac{0.3}{1 + 1.2 \cdot D_*} + 0.055 \cdot (1 - e^{-0.02 \cdot D_*})$$

$$\rho_s = 2650 \text{ kg/m}^3$$
$$\rho_w = 1018 \text{ kg/m}^3$$
$$g = 9.81 \text{ m/s}^2$$
$$\upsilon = 10^{-6} \text{ m}^2/\text{s}$$

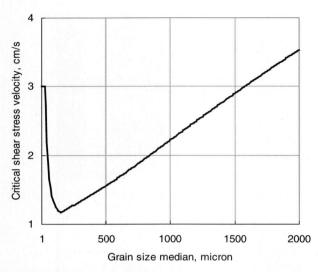

Fig. 23 Display of the transformation in Table 1

impedance map in Fig. 20 was obtained multiplying the grid for p-wave velocity (Fig. 19) by the one of bulk density (Fig. 18).

The map for critical shear stress velocity in Fig. 21 is a nonlinear transformation of the grid values in Fig. 22 according to the expression and Table 1 that appears in graphical form in Fig. 23.

6 Discussion

Mathematical formulations are commonly too simple to account for the numerous special situations that arise in practice. Such formulations are still capable of generating valid results provided special care is taken to handle the peculiarities not considered in the formulation of the methods. In Mecklenburg Bay, first we had the problem that cokriging does not have a direct way to restrict results to a given interval. In our case, for example, cokriging in the sampling space produced values of porosities less than zero and larger than 100%, and bulk densities below the density of water of 1 g/cc. To solve that problem, we resorted to the indirect procedure of converting the data to normal scores, doing the estimation in normal score space, and backtransforming the results.

Secondly, our study involves preferential sampling. Our opinion and that of other IOW geologists are that the distribution of seafloor porosity in Fig. 5 is not representative of the distribution for the entire Mecklenburg Bay (Endler, 2008). Figure 5 is the result of preferential sampling of the finest fraction along the deepest part of the central bay. In straight simple cokriging estimation of normal scores, the practice is to employ zero as the mean for the normal scores for all attributes because for unbiased samples, the normal score of the true mean is indeed zero.

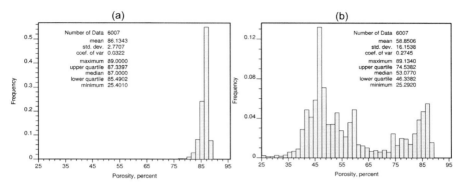

Fig. 24 Histogram of estimated seafloor porosity values; (**a**) using $\mathbf{m} = [0 \ \ 0 \ \ 0]'$; (**b**) taking $\mathbf{m} = [-1.3 \ \ -0.6 \ \ -0.3]'$

In our case, the seafloor porosity mean of 82.8% (Fig. 5a) is higher than the true mean for the entire bay. Figure 24a gives the histogram resulting after backtransformation of estimates of normal scores with $\mathbf{m} = [0 \ \ 0 \ \ 0]'$, which is basically a reproduction of the data histogram of Fig. 5a. Figure 24b is the histogram obtained with $\mathbf{m} = [-1.3 \ \ -0.6 \ \ -0.3]'$ after minimizing the discrepancy in the standardized distributions for grain size median and 100 minus porosity (Fig. 25). The values in the second histogram vary within the range of the data and the histogram

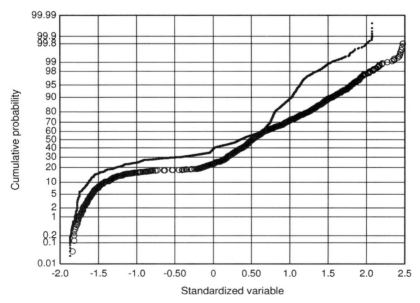

Fig. 25 Standardized cumulative distributions. The small bullets denote cumulative frequencies for the estimated values of 100 minus porosity values in Fig. 15 and the large open circles those for grain size median in Fig. 6

is bimodal, like the grain size median histogram in Fig. 7a. Remember that grain size median and porosity are negatively correlated (Figs. 11a and 12d). In our opinion, Fig. 24b is a good rendition of reality, which could also have been obtained through the less parsimonious approaches of postulating a normal score distribution, assuming a backtransformation different from that supported by the data, or both.

Cokriging automatically gives more importance to the primary variable when there are both primary and secondary information at the same station, to the point that it is almost irrelevant to consider or ignore secondary information collocated with primary information. There is no similar screening among secondary information, even though a similar discrimination is here necessary for the prediction of an attribute such as seafloor porosity in the Baltic Sea. According to Forster et al. (2003), granulometry is a more reliable and consistent attribute than water depth for predicting seafloor porosities. For that reason, water depth was employed as a proxy variable only at those locations more than 1 km away from stations with neither porosity nor grain size median measurements. This manual discarding of depth values was done only for the estimation. We used all values in the preparation of the models in Fig. 12.

The relationship between seafloor porosity and water depth depends to a large extent on the strength of bottom current too. So, for example, for the same water depth in an embayment like the Lübeck Bay in the southwest, seabed porosity and bathymetry relate differently than along a channel, such as the Kadet to the northeast (Fig. 1). Considering that the correlation between water depth and porosity was established based primarily on the fine sediment of the quiet seafloor of Lübeck Bay and central Mecklenburg Bay, the water depth values along the strong current channels in the northern part of the study area were not used in the estimation even in the absence of all other information. Use of depth values along the channels results in too high values of porosity (Forster et al., 2003). Fortunately, there was good coverage of granulometric values, so the problem was satisfactorily solved by relying only on the granulometry, providing one last lesson: use proxy information only when contributes to enhance the estimation; quality of modeling is not always proportional to the quantity of data. Another possibility could have been to consider a third proxy variable properly discriminating, such as average velocity of bottom currents.

Harff et al. (2004) postulate that the main sediment sources in our study area are glacial tills. Our maps complement and improve previous results that in general show seafloor properties that are consistent with sediment sequences ranging from coarse coastal sediments to fine grain muds at the center in Lübeck Bay and at the center of Mecklenburg Bay. This distribution is controlled by a combination of erosional forces of currents near the coast and the clockwise rotation of the currents in Mecklenburg Bay, which is in good agreement with the analysis of sediment types and modeling of marine bottom currents. Further research includes utilization of our maps for additional studies of sediment dynamics based on the hydrographic modeling recently developed by Seifert et al. (2008).

7 Conclusions

While exclusive use of direct measurements of porosity, density, and p-wave velocity of the seafloor along the central part of Mecklenburg Bay allows limited computer mapping with inaccuracies in the extrapolation, consideration of grain size median and water depth has allowed mapping of six seafloor geophysical and geotechnical properties within the entire Mecklenburg Bay showing agreement with early mapping and the experts' expectations.

Routine application of geostatistics without paying attention to the physics of the attributes and peculiarities in the sampling may result in unrealistic estimates. In the case of seabed porosity in Mecklenburg Bay:

- Working on original sampling space resulted in some seafloor porosity estimates less than zero and others larger than 100%. This problem was eliminated by employing normal score transformations and backtransformation to porosity space.
- Because of preferential sampling, use of zero mean for the normal scores of all attributes in simple cokriging estimation produces unrealistic values. Consideration of pseudomeans resulted in the correct scaling.
- Cokriging strongly discriminates in favor of measurement of the same attribute being estimated, but does not do something similar among the other variables. Considering that granulometry is a better predictor of seafloor porosity than depth, depth values were eliminated from the sample in the presence of collocated grain size median measurements.
- Limited sampling forced investigation of the crosssemivariogram porosity-water depth only along the southwestern part of the bay and not along the deep parts of channels to the north, where fortunately there was good coverage of the more informative grain size median. Considering that the geologists expect different correlations for these two areas, we did not use depth measurements along the channels in the northern part of the bay, a perfectly sound step under the circumstances.

The seafloor porosity map was also successfully used to prepare a seafloor grain density map in combination with a seafloor bulk density map. Additional satisfactory modeling included maps for p-wave velocity, acoustic impedance, and critical shear stress velocity.

The new series of maps confirm and expands previous results and can be used to continue advancing the geology of the Baltic Sea.

Acknowledgments Work by the two most senior authors was supported by a grant from the Forschungsanstalt der Bundeswehr für Wasserschall und Geophysik (FWG) obtained by Jan Harff. The authors are particularly indebted to Thomas Wever from the FWG for a faithful and productive collaboration. We benefited from a round of discussions we had with Dieter Lange and Wolfgang Jäger, the authors of the original 1984 study, and from comments from two anonymous reviewers.

References

Bobertz B, Harff J (2004) Sediment facies and hydrodynamic setting: a study in the south western Baltic Sea. Ocean Dyn 54 (1): 39–48

Bobertz B, Harff J, Bohling B (2008) Parameterisation of clastic sediments including benthic structures. J Mar Syst 73 (3–4), in press

Caers J (2005) Petroleum geostatistics. Society of Petroleum Engineers, Richardson, TX, p 88

Deutsch CV, Journel AG (1992) GSLIB: Geostatistical Software Library and User's Guide. Oxford University Press, New York, p 340, 1998 2nd edn, p 384

Endler R (2008) Sediment physical properties of the DYNAS study area. J Mar Syst 73 (3–4), in press

ESRI (2006) ArcGIS Desktop 8.2. http://www.esri.com

Forster S, Bobertz B, Bohling B (2003) Permeability of sands in the coastal areas of the southern Baltic Sea: Mapping a grain-size related sediment property. Aquat Geochem 9: 171–190

Goovaerts P (1997) Geostatistics for natural resources evaluation. Oxford University Press, New York, p 483

Hamilton EL (1972) Compressional wave attenuation in marine sediments. Geophysics 37 (4): 620–646

Harff J, Bobertz B, Lemke W, Granitzki K, Wehner K (2004) Sand and gravel deposits in the southwestern Baltic Sea, their utilization and sustainable development. Z Angew Geologie, Special Volume No. 2: 111–124

Jian X, Olea RA, Yu Y-S (1996) Semivariogram modeling by weighted least squares. Comput Geosci 22 (4): 387–397

Lange D, Jäger W (1984) Geologische und petrophysikalische Untersuchungen an spätglazialen und holozänen Sedimenten im Bereich der südlichen Beltsee unter besonderer Berücksichtigung des Festlandsockels der Deutschen Demokratischen Republik. Institut für Meereskunde, Rostock, unpublished combined postdoctoral thesis and doctoral thesis

Olea RA (1999) Geostatistics for engineers and earth scientists. Kluwer Academic Publishers, Norwell, MA, p 303

Olea RA (2006) Erratum to "Semivariogram modeling by weighted least squares". Comput Geosci 32 (3): 419

Seifert T, Fennel W, Kuhrts C (2008) High resolution model studies of transport of sedimentary material in the south-western Baltic. J Mar Syst 73 (3–4), in press

Seifert T, Tauber F, Kayser B (2001) A high resolution spherical grid topography of the Baltic Sea, 2nd edn. Baltic Sea Science Congress, Stockholm 25–29 November 2001, Poster 147, www.io-warnemuende.de/iowtopo

Wackernagel H (1995) Multivariate geostatistics. Springer-Verlag, Berlin, p 256

Statistical Analysis of Physiographic and Structural Directional Data in the U.S. Midcontinent (Kansas)

Daniel F. Merriam and John C. Davis

Abstract Kansas, located in the stable U.S. Midcontinent, exhibits a variety of physiographic and structural directional features. The direction of river valleys, fractures including joints and faults, lineaments, anticlinal axes, and geophysical anomalies are analyzed to determine their relation to each other and what could be interpreted from these data about the structure and structural development of Kansas. The direction of each feature was measured on surface or subsurface maps and statistically analyzed and compared. Although each measured property has slightly different directions, several trends are recognized; in general, three directions are dominant: northeast, east-northeast, and northwest.

Keywords Trends · orientation data · statistics · rose diagrams · joints · fractures · faults · plains-type folds (anticlines) · geophysical anomalies · topographic features · lineaments

1 Introduction

Directional data are of concern to geologists because they give clues into understanding the geology and geologic development of an area or region. In the flatland of the U.S. Midcontinent these data are all the more important to understand the geologic history of the region. Surface features exhibited by the physiography are traces of features in the subsurface, thus recognizing features on the surface may give hints as to those present in the subsurface (Merriam, 2005). Subsurface features usually are interpreted from contour maps based on well data or geophysical data.

D.F. Merriam
University of Kansas, 1930 Constant Avenue, Campus West, Lawrence, Kansas 66047, e-mail: dmerriam@kgs.ku.edu

J.C. Davis
DAVCON, Box 353, Baldwin City, Kansas 66006-0353, e-mail: john.davis5@mchsi.com

Table 1 Definition of terms as employed here (Jackson, 1997)

Feature	Definition
Fracture	a general term for any surface within a material across which there is no cohesion, e.g. a crack. Fracture includes cracks, joints, and faults.
Joint	a planar fracture, crack, or parting in a rock, without displacement; often occurs with parallel joints to form part of a joint set.
Fault	a discrete surface or zone of discrete surfaces separating two rock masses across which one mass has slid past the other.
Linear	arranged in a line or lines; pertaining to the linelike character of some object or objects; not recommended as a synonym for lineament.
Lineament	an extensive linear surface feature; a linear topographic feature of regional extent that is believed to reflect crustal structure.

Faults, joints, lineaments and the axes of folds are oriented features that do not have a sense of direction. As a consequence, circular histograms (rose diagrams) of orientation measurements are symmetrical and it is only necessary to show half of the diagram. All of the rose diagrams in the following section were made using the same conventions; the class width is 5° with an origin at 0° (north), the areas of the petals are proportional to frequency, and the diagrams are scaled so the radius of the largest class is a constant. Alternative conventions are discussed by Davis (2002). The diagrams were made using VectorRose (Zippi, 2000).

Directional data may differ with depth or age. Geologic features may change orientation slightly or axes of structures may migrate laterally depending on the time of formation and subsequent history. Salisbury and Merriam (1984) noted that for upstate New York and Ontario/Quebec (Canada) lineaments in younger rocks are dependent on (1) rock type; (2) thickness of overlying units; (3) distance from major structural features; and (4) time and continuity of promulgation of the patterns. We wanted to see if these findings also were pertinent for the U.S. Stable Continental Interior and Kansas in particular. Some terms are defined here for clarity (Table 1).

We have collected a variety of directional data from different sources to see if they can provide some insight into the geologic history of Kansas. The data have been analyzed by standard statistical techniques (Davis, 2002).

2 Previous Work

The fracture system in Precambrian and younger rocks has been of interest for many years. Most of this work in Kansas is either general for a large area or specific for a small area. Most of the previous works were qualitative or semiqualitative and interpretations were based on visual aspects and comparisons (Fig. 1). In 1988, Johnsgard produced an excellent masters thesis including a comprehensive summary of

Statistical Analysis of Physiographic 483

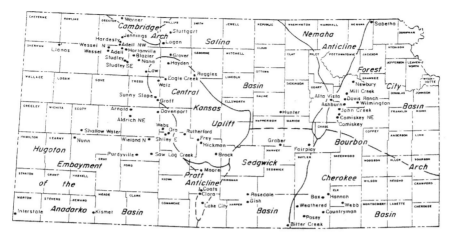

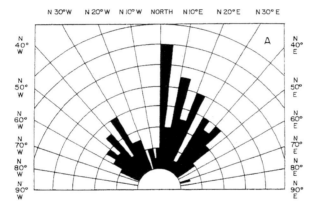

Fig. 1 Location of anticlines in relation to major structural elements in Kansas (Merriam, 1963). Rose diagram of orientation of anticlines. Note northeast, northwest, and north-south dominant trends

all the work up to that time with references. A brief extract of his compilation of joint directions is included here (Table 2).

Johnsgard noted that the Permo-Pennsylvanian joint sets were relative consistent, but different from the joint sets in the Cretaceous. He also suggested that the drainage system was controlled, at least partly, by fracture patterns. He summarized his and previous work on lineaments and on geologic and geophysical anomaly trends. He concluded that the general directional pattern of N10°–30°E and N40°–50°W represented the Precambrian basement fracture system. If his interpretation is correct, then directions of joints in younger rocks seemingly are different (Table 2).

Table 2 Summary of joint directions recorded in Kansas (Johnsgard, 1988)

Age and Place (county)	Direction		Source
Cretaceous Niobrara	N2°–4°E	N10°–14°W	Bass (1926)
Cretaceous Ft. Hays (Ellis)	N25°E	N70°W	Neuhauser (1986)
Permian ls (Marshall)	N76°E	N15°W	Nelson (1952)
Lower Permian ls	N72°E	N25°W	Neff (1949)
Lower Permian ls (Nemaha)	N85°E	N35°W	DuBois (1978)
Lower Permian ls (Morris)	N45°E	N52°W	Macfarlane (1979)
Lower Permian ls (Wabausee)	N65°E	N21°W	Eccles (1981)
Permo-Penn (S-C KS)	N60°E	N35°W	Ward (1968)
Upper Penn ls (Wilson)	N55°E	N35°W	Wagner (1961)
Upper Penn (Douglas.)	N63°E	N41°W; N2°W	Jefferis (1969)
Pennsylvanian ls (NE KS)	N60°E; N3°E	N35°W	Stewart (1967)

Johnsgard stated that the orientation data for Kansas linear features '...implies a similar genetic cause.' Most of his findings had been proposed previously, but he made a nice compilation and summary.

3 The Data

The data used here were measured on published maps (Table 3). The original data were interpreted by their respective author, and it is presumed they collected and recorded them correctly. The accuracy is assumed to give approximate values sufficient for geologic interpretations.

Our data consist of measured angles of the trend in mapped features. All of the data sources were published geological or geophysical maps except the major drainage features which were compiled from the 1:500,000 topographic map of Kansas. Geographic locations of some of the data sets are shown in Fig. 2.

3.1 Baars and Others 1995 U.S. Megalineament

Don Baars and colleagues interpreted the pattern of deformation in Precambrian rocks in the conterminous United States. They noted that orthogonal basement blocks were separated by fault zones striking northeast and northwest. This orthogonal pattern also was recognized by Watney et al. (1997) in the sedimentary interval overlying the crystalline basement. We measured orientations of the megafeatures in Kansas on the Baars et al. (1995) map of the U.S. and plotted the directions. Although the features exhibit the northeast and northwest trends obvious on the map, they are more dispersed than expected. This dispersion may be the result of the sparse amount of data.

Statistical Analysis of Physiographic 485

Table 3 Data sets used in study. Directional properties measured from published maps (source indicated)

Lineaments
U.S. Megalineaments (Baars et al., 1995)
U.S. Precambrian basement map (Sims et al., 2005)
Kansas Landsat eastern (McCauley et al., 1978)
Kansas Landsat western (McCauley, 1988)
Western Kansas Landsat (Zeller et al., 1976)
Kansas Photolineaments northeastern (Johnsgard, 1988)
Kansas Photolineaments north-central (Cooley, 1984)
Kansas topographic interpretation southeast (Merriam and Sorensen, 1982)
Structure
Kansas (anticlines) (Jewett, 1951 extended by Merriam, 1963)
Kansas structural study (anticlines/faults) southeast (Merriam and Förster, 1996)
Kansas (faults) (Cole, 1962)
Kansas (faults) (Merriam, 1963)
Kansas (joints) southeast (Ward, 1968)
Drainage
Kansas Topographic Map (1:500,000)
    Drainage east of Nemaha Ridge
    Drainage along the Nemaha Ridge
    Drainage between Nemaha Ridge and High Plains
    Drainage on the High Plains
Geophysics
Magnetic anomalies (Yarger, 1983)
    Eastern Kansas
    Western Kansas
Gravity lineaments (Lam, 1987)
    MCR, Midcontinent Rift
    sCP, southern Central Plains orogeny
    SGR, Southern Granite-Rhyolite province

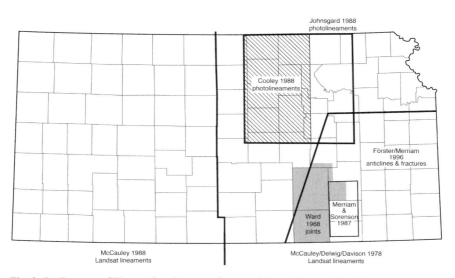

Fig. 2 Outline map of Kansas showing area of some of the studies mentioned

3.2 U.S. Precambrian Basement Map (Sims et al., 2005)

Sims et al. (2005) presented a preliminary Precambrian basement structure map of the continental United States based on an interpretation of geologic and aeromagnetic features. The map is a mass of data with faults overlain on a color coded aeromagnetic base.

3.3 Landsat Lineaments (Eastern Kansas), McCauley et al. (1978)

As part of his doctoral dissertation, Jim McCauley compiled the lineaments in eastern Kansas at a scale of 1:500,000 (McCauley, Dellwig, and Davison, 1978). In 1988 McCauley compiled lineaments for western Kansas. In the eastern area the bedrock is largely in Permo-Pennsylvanian sediments (limestone, sandstone, and shale) and in the western area the bedrock consists of Cretaceous and Tertiary sediments (shale and siltstone), which are less indurated. Lineaments are oriented in northeast, northwest, and north–south sets.

3.4 Western Kansas Landsat Lineaments (McCauley, 1988)

The lineaments for western Kansas were compiled by Jim McCauley ten years later. The orientations are slightly different than those noted in the east part of the state probably because of the difference in age and type of rock exposed on the surface.

3.5 Central Great Plains Landsat Lineaments of Zeller et al. (1976)

In much the same area that McCauley identified lineaments, Zeller and colleagues determined that there was a close spatial relationship between the occurrence of uranium in the subsurface and regional geomorphic lineaments. The direction of their measured lineaments were N58°E and N53°W, which visually are different from those measured by McCauley for western Kansas, but this was not surprising considering the circumstances and objectives.

3.6 Johnsgard's Photolinements (1988) in North-Central Kansas

As part of a University of Kansas thesis, Johnsgard mapped the fracture pattern in north-central Kansas to determine its relation to hydrogen soil gas anomalies over the Midcontinent Rift System.

3.7 Landsat Photolinements of Cooley (1984)

Johnsgard (1988) reproduced a portion of a map of Landsat photolineaments originally produced by M.E. Cooley of the USGS for an area in northeastern Kansas. The orientation of lineaments is northeast ($\approx$N35°E) and northwest ($\approx$N45°W) with a strong north-south component. According to Johnsgard (p. 69) '...it is fairly obvious that not every single one of the photolineaments ...mapped corresponds to an actual geologic or structural feature.'

3.8 Merriam and Sorensen's (1982) Topographic Trends

Merriam and Sorensen, in a study of the Howard Limestone (Pennsylvanian) in southeastern Kansas, noted linements as interpreted from topographic features. Again, the trends are northeast ($\approx$N45°E) and northwest ($\approx$N35°W), but the directional properties are not as prominent as on other maps.

3.9 Statewide Structures Catalogued by Jewett (1951)

In 1951 Mark Jewett catalogued the major structures in Kansas. Jewett's list was modified and extended by Merriam in 1963. Although the data are limited and slightly biased to structures trending northeast, the summary does hint of the importance of major northeast- and northwest-trending features in the state.

3.10 Merriam and Förster's (1996) Structural Study

In an effort to determine the influence of the Precambrian basement on the development of plains-type folds (anticlines), Merriam and Förster (1996) measured fractures in the basement and orientation of anticlinal axes in the overlying Paleozoic sediments in southeastern Kansas. They determined the two patterns are similar with major orientations northeast and northwest; the northwest trend is dominant. Thus the anticlines developed in the sedimentary section closely reflect the basement structural grain, supporting the supposition of basement control.

3.11 Cole's 1962 Precambrian Faults

Virgil Cole showed several faults on his contour map of the buried Precambrian surface in Kansas. These faults were interpreted from well data and to some extent from geophysical investigations, and they show the northeast and northwest alignment as other structural maps. There is a prominent north-northeast trend, a northwest trend, and a lesser trend almost due north. Unfortunately, the analysis is based on only 45 faults.

3.12 Merriam's 1963 Statewide Faults

Merriam's (1963) extended fault map of Kansas in 1963 showed additional faults not included on Cole's map of 1962. Merriam noted that the fault trends statewide were not as prominent as the anticlinal features but revealed much the same pattern as the folds. He concluded that 'Faulting as now recognized [in 1963] has played a much greater role in the development of Kansas structure than was heretofore understood.'

3.13 Merriam's 2003 Analysis of Surface Faults

In a study of the Worden Fault in southwestern Douglas County in eastern Kansas, Merriam (2003) compared the Worden to other known faults exposed and mapped on the surface in the eastern part of the state. He determined the average direction of those surface faults is N25°E and N35°W. The trend of the Worden is N30°E.

3.14 Merriam's 1963 Study of Anticlines in Kansas

For his *The Geologic History of Kansas*, Merriam (1963) tabulated data on 75 anticlinal oil fields in Kansas. The northeast and northwest trends are apparent and a north-south trend is prominent. The data were tabulated and plotted on a rose diagram (Fig. 1).

3.15 Ward's Joint Patterns (1968)

John Ward (1968) made a study of the joint patterns in the Permo-Pennsylvanian sedimentary section in an area in south-central Kansas on and adjoining the buried Nemaha Anticline. Ward determined two major joint sets – northeast (N60°E) and northwest (N35°W). He noted most of the joints were vertical and probably related to and contemporaneous with the Nemaha Anticline (post lower Permian and pre-lower Cretaceous).

3.16 Topography and Drainage Patterns (Kansas 1:500,000 Topographic Map)

Trends in surface drainage trends are diverse probably because streams and rivers follow the course of least resistance, which may be a prominent fault zone or fracture system. These features may have slightly different directions depending on the location, type of feature, the exposed rocks, and fracture pattern. The most obvious example is the major Arkansas River which extends in a zig-zag pattern across the state from west to east through Garden City, Dodge City, Great Bend, Hutchinson, and Wichita.

3.17 Magnetic Anomaly Trends From Yarger (1983)

Trends in magnetic anomalies can be interpreted from the aeromagnetic contour map of Yarger (1983). Aeromagnetic anomalies depend on the material causing the anomaly, subsequent deformation, and metamorphic history. In Kansas the magnetic field strongly reflects the Precambrian basement (see Van Schmus and others, 1993, for more information on the Precambrian of the Midcontinent).

3.18 Lam's 1987 Gravity Alignments

Gravity anomalies in Kansas seemingly are more diffuse than magnetic anomalies. There are three obvious trends, however; northeast (N35°E), north–south (N2°W), and northwest (N45°W).

4 The Analyses

There are several sources of error in collecting data for a study of this type. There is observational error in measurement, for example, determining the properly oriented framework in which to measure the angles. Some judgments must be made in the selection of features to measure, which can bias the results, and there is operator error including mis-measuring, mis-recording, etc. We have taken precautions to keep these problems to a minimum.

The numerical analysis of the directional data was made using standard statistical techniques as described in Davis (2002) and Watson (1966, 1970, 1983).

5 Comparing Two Samples of Directional Data

To compare two samples, we used Watson's U^2 test, a nonparametric procedure that detects any type of difference between the samples being compared. The test is a variant of the Kolmogorov-Smirnov test of the equality of two distributions, adapted by Watson (1961) for circular data. These procedures compare the two distributions in cumulative form, by finding the sum of the squared differences between the distributions at each sample point.

If the two samples consist of n observations in the first and m in the second, the total number of differences between the two distributions is $n + m = N$. The differences are $d_1, d_2, \ldots, d_N$ with mean $\bar{d} = \sum_1^N d_k/N$. Watson's U^2 test statistic is

$$U^2 = \frac{nm}{N^2} \sum_{k=1}^{N} (d_k - \bar{d})^2$$

The test statistic is compared to critical values tabulated by Zar (1998) and by Batschelet (1981). The null hypothesis that the two distributions were derived from the same parent (i.e., are "equal") is rejected if the test statistic exceeds the critical value at the selected level of significance. For values of n and m greater than about 15, the critical values are essentially independent of the sizes of the two samples. Critical values for three widely used significance levels and large samples such as those in this study, are:

$$U^2_{(0.10)} = 0.152, \quad U^2_{(0.05)} = 0.187, \quad U^2_{(0.01)} = 0.268$$

The VectorRose software (Zippi, 2000) used by us calculates Watson's U^2 statistic for any two selected data sets.

Figure 3 illustrates the manner in which Watson's U^2 statistic reflects differences between two cumulative distributions of directional data. Figure 3a compares the directions of axes of 46 anticlines in Kansas contoured by D. F. Merriam (dashed line) to the directions of axes of 50 anticlines in the southeast corner of the state (solid line). Watson's U^2 statistic is 0.6346, which greatly exceeds the critical value and leads to rejection of the hypothesis that the axial directions are drawn from the same population. Figure 3b compares cumulative distributions for axial directions of 46 anticlines in Kansas contoured by D. F. Merriam (dashed line) and axial directions of 56 Kansas anticlines contoured by other geologists (solid line). Watson's U^2 statistic is 0.06248, which is less than the critical value, so there is no evidence of any difference between the two sets of observations.

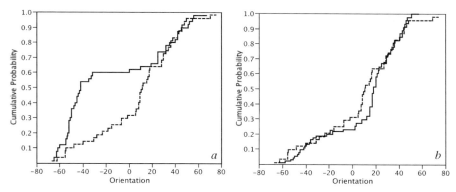

Fig. 3 Cumulative distributions of anticlinal axis orientations: *a*, dashed line—axes of anticlines in Kansas contoured by Merriam (1963); solid line—axes of anticlines in southeastern Kansas. Watson's $U^2 = 0.0625$; distributions are significantly different. *b*, Dashed line—axes of anticlines in Kansas contoured by Merriam; solid line—axes of anticlines in Kansas contoured by others. Watson's $U^2 = 0.0625$; the differences between the distributions are not significant

6 Comparisons

6.1 Differences

Megalineament observations by Baars et al. (1995) are not significantly different from those measured by Sims et al. (2005), but both data sets are small (Fig. 4). The northeast–northwest 'regionalization blocks' of Watney et al. (1997) dominate the pattern on a smaller scale although the data set is very small; their directions noted were N44°E, N47°W, and N–S.

Landsat lineament data compiled by Jim McCauley (1988) and McCauley et al. (1978), show a significant difference between eastern Kansas and the western part of the state (Fig. 5). The lineaments in eastern Kansas are mainly developed in Permo-Pennsylvanian rocks whereas western Kansas is blanketed by Cretaceous and Tertiary sediments.

The Kansas fault data set of Cole and the one of Merriam deviate significantly. The Merriam/Forster comparison of anticlines and faults are visually similar, but statistically the two data sets differ significantly (Fig. 6).

6.2 Similarities

Anticlinal data contoured by Merriam (1963) and contour maps by other geologists published in the Kansas Geological Society Oil and Gas Fields (volumes I, II, III, IV, and V) were analyzed separately to determine if there was a bias in contouring by different geologists. It was determined that there was no difference in the interpretation by contouring by different geologists, a reassuring finding (Fig. 7).

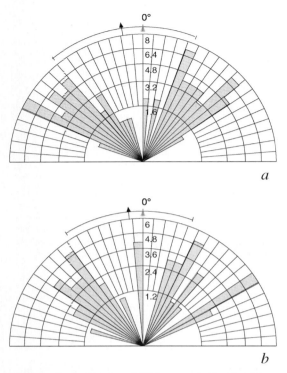

Fig. 4 a, Rose diagram of 77 large-scale Precambrian structural features in continental USA mapped by Barrs et al. (1995) with a mean direction of 351.2°: **b,** Orientation of 51 megalineaments mapped in continental US by Sims, Saltus, and Anderson (2005); mean direction is 357.1°. Distributions are not significantly different ($U^2 = 0.0925$)

The topographic data set was partitioned and analyzed separately to determine if there was any different by geographic area or age of bedrock. The state was partitioned into four parts: (1) east of the Nemaha Anticline; (2) on the Nemaha; (3) between the Nemaha and the High Plains and (4) on the High Plains (Table 3). Area (1), (2), and (3) are in Permo-Pennsylvanian consolidated rocks, where the drainage is expressed differently than in (4), which is mainly in softer less consolidated Cretaceous, Tertiary, and younger sediments. So, the drainage pattern reflects the bedrock in the different physiographic provinces.

The differences in trends in Merriam and Sorensen's topographic data and the orientation of rivers in eastern Kansas are not statistically significant (Fig. 8) adding weight to the concept that river direction is controlled in part by the rock fracture system. Faults mapped by Cole (1962) and those noted by Merriam in 1963 essentially have the same trend. There is an indication (by offset) that movement on the northwest-trending faults is younger than on the northwest-trending faults.

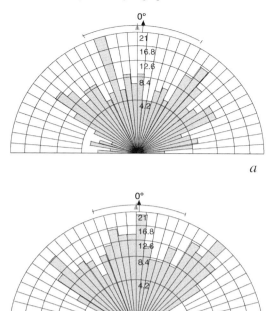

Fig. 5 a, Rose diagram of 246 lineations measured on Landsat images of western Kansas (McCauley, 1988). Mean direction is 2.3°; **b**, Orientation of 304 lineations measured on Landsat images in eastern Kansas (McCauley et al., 1978). Mean direction is 0.4°. Distributions are statistically different at a significance level of 0.01 ($U^2 = 0.3153$)

7 Relation of Features

Lam (1987) plotted the orientations of gravity anomalies east and west of the Abilene Anticline. There is a noticeable difference in the dominant trend of the anomalies. East of the anticline, the major direction is northwest and west it is northeast.

The magnetic trends mimic structural trends, but the gravity anomalies are different and related to the Precambrian basement provinces (Table 4). This is not surprising considering each province is composed of different a rock type: MRC is consists of metasediments and intrusive gabbros; SCP is mostly gneissic granites, and the sCP is a rhyolite. In general the age of the provinces decrease in a southerly direction as expected with in accretionary terrane as proposed by Carlson and Treves (2005). Any test statistic test greater than $U^2 = 0.187$ indicates the two distributions being compared are significantly different. Only the SCP Province (granite-rhyolite province) is *not* significantly different between the eastern and western part.

The significant difference in the two parts of the sCP east and west of the Midcontinent Rift System is not understood geologically; a closer inspection of the distinction between the two parts is warranted.

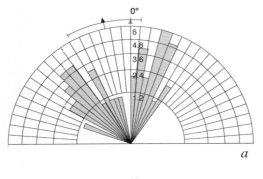

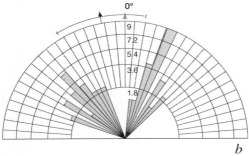

Fig. 6 a, Rose diagram of 45 faults in the Precambrian of Kansas mapped by Cole (1962). Mean direction is 347.9°; **b**, Orientation of 44 faults mapped in Kansas by Merriam (1963). Mean direction is 354.4°. Distributions are not significantly different ($U^2 = 0.1175$)

8 The Results

8.1 Previous Conclusions

Previous authors have reached various conclusions about the relationships among fractures, structures, and lineament features in Kansas.

Don Baars et al. (1995) concluded that the oldest fault zones in the central part of the North American continent are Precambrian, about 2.6 Ga.in age and these zones were reactivated periodically.

White (1990) concluded that for Permian rocks in east-central Kansas (1) stream morphology is controlled by the joint system and (2) major lineaments in this part of Kansas are related to the Humboldt fault zone, which in turn exerts an influence on the minor structures. White concluded that the fault system probably developed as a result of the Ouachita Orogeny in Pennsylvanian time.

Gary Stewart (1967) in an investigation of joints in Pennsylvanian rocks in northeastern Kansas determined there are two set of joints – tension and shear with the shear joints being the older of the two sets.

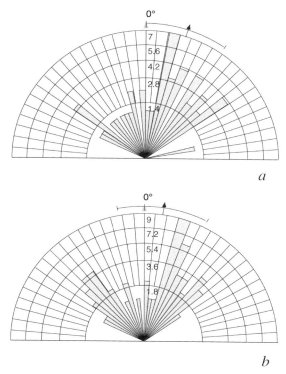

Fig. 7 a, Rose diagram of axes of 56 anticlines in Kansas contoured by Merriam (1963). Mean direction is 17.1°; **b**, axes of 46 anticlines in Kansas contoured by other geologists and published by the Kansas Geological Society in Kansas oil and gas fields. Mean direction is 1.3°. Distributions are not significantly different ($U^2 = 0.0625$). These same data are shown in cumulative form in Fig. 1

Jon Callen (1985) in a study of lineaments in northeastern Kansas, determined the lineaments were correlated with drainage patterns and that basement rocks influenced the surface expression of lineations.

Ken Neuhauser (1986) studied joints in the Cretaceous of western Kansas and determined they were shear fractures forming a conjugate system that probably formed in the late Cretaceous–early Tertiary as a result of the Laramide Revolution.

Table 4 Statistical results of gravity data vs the Precambrian terrane. MRC is Midcontinent Rift System; sCP is southern Central Province; and SCP is granite-rhyolite province

MRS vs sCP = $U^2 = 0.714$
MRS vs SCP = $U^2 = 0.436$
sCP vs SCP = $U^2 - 0.223$
SCP (east) vs SCP (west) = $U^2 = 0.114$
sCP (east) vs sCP (west) = $U^2 = 0.321$

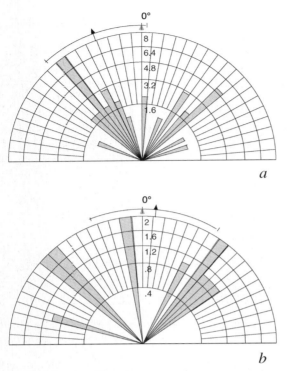

Fig. 8 a, Rose diagram of 36 topographic trends (Merriam and Sorensen, 1982) with a mean direction is 339.7°; **b**, Orientation of 13 major river segments in eastern Kansas from the 1:500,000 topographic map of the state. Mean direction is 2.8°. Distributions are not significantly different ($U^2 = 0.1121$)

8.2 Current Understanding

Our study indicates that most of the fractures, lineaments, drainage trends, and geophysical anomalies are related to some extent. The fracture pattern was initiated in the Precambrian and propagated up through the overlying sedimentary section mainly during structural movement in adjacent areas. The lineaments, topographic features, and drainage patterns are surface expressions of this fracture/fault pattern, although expressed slightly differently in different rock types. The drainage pattern is controlled to some extent by the fracture pattern with the Arkansas River as a good example of a major drainage feature reflecting the fracture pattern.

Undeniably, the Precambrian crystalline basement plays a big role in development of the forms and patterns in the overlying sedimentary blanket in the U.S. Midcontinent. Many of the Precambrian features promulgate through the overlying cover and are expressed on the present-day surface. The orientation and trend of directional properties of geological features, for example structure, topography, or drainage, may be expressed essentially the same except for local variations, which may be important geologically.

Geologists with different objectives and backgrounds may contour a feature somewhat differently, although the major trends in their maps may be much the same. Thus, different investigators using the same data set may arrive at different conclusions that reflect their interest and purpose. A statistical analysis assures that the same or similar results are obtained even by different investigators using the same data; it also quantifies and substantiates the visual analysis of the investigator. Usually, a geologist is a good identifier of trends and can recognize patterns in the data, but that recognition is enhanced and substantiated with use of quantitative techniques.

References

Baars DL, Thomas WA, Drahovzal JA, Gerhard LC (1995) Preliminary investigations of basement tectonic fabric of the conterminous USA. In: Ojakangas RW, others (eds), Basement tectonics: Kluwer Acad. Publ., The Netherlands, pp 149–158

Bass NW (1926) Geology of Hamilton County. In: Geologic investigations in western Kansas. Kansas Geol Survey Bull 11: 53–83

Batschelet E (1981) Circular statistics in biology. Academic Press, London, p 371

Callen JM (1985) Lineament azimuth trend analysis in the Forest City Basin (Kansas). The Compass 62 (2), pp 92–96

Carlson MP, Treves SB (2005) The Elk Creek carbonatitie, southeast Nebraska – an overview. Nat Resour Res 14 (1): 39–45

Cole VB (1962) Configuration map on top of Precambrian basement rocks in Kansas. Kansas Geol Survey, Oil and Gas Invest 26, map

Cooley ME (1984) Linear features determined from Landsat imagery in western Kansas: U.S. Geol. Survey, Open-File Rept. 84–0241, 1 sheet, scale: 1:500,000

Davis JC (2002) Statistics and data analysis in geology 3rd edn. John Wiley & Sons, New York, p 638

DuBois SM (1978) The origin of surface lineaments in Nemaha County, Kansas: unpubl. masters thesis, University of Kansas, p 37

Eccles KR (1981) Structural geology and recent seismicity of southeast Riley, eastern Geary, and western Wabaunsee counties: unpubl. masters thesis, Kansas State University, p 86

Jackson JA (1997) Glossary of geology, 4th edn. Am. Geol. Inst., Alexandria, Virginia, p 769

Jefferis LH (1969) An evaluation of radar imagery for structural analysis in gently deformed strata: a study in northeast Kansas: unpubl. masters thesis, University of Kansas, Lawrence, p 32

Jewett JM (1951) Geologic structures in Kansas: Kansas Geol Survey Bull 90: 105–172, pt. 6

Johnsgard SK (1988) The fracture pattern of north-central Kansas and its relation to hydrogen soil gas anomalies over the Midcontinent Rift System: unpubl. masters thesis, University of Kansas, Lawrence, p 112

Lam C-K (1987) Interpretation of state-wide gravity survey of Kansas: unpubl. doctoral dissertation, University of Kansas, p 213 (Kansas Geol. Survey Open-File Rept. 87–1)

Macfarlane PA (1979) Geologic constraints on land use in northeastern Morris County, Kansas: unpubl. masters thesis, University of Kansas, Lawrence, p 117

McCauley JR (1988) Landsat lineament map of western Kansas: Kansas Geol. Survey, Open-file Rept. 88-20, 1 sheet, scale: 1:1,000,000

McCauley JR, Dellwig LF, Davison EC (1978) Landsat lineaments of eastern Kansas: Kansas Geol Survey, Map M-11, scale: 1:500,000

Merriam DF (1963) The geologic history of Kansas. Kansas Geol Survey Bull 162: 317
Merriam DF (2003) Reinterpretation and reflections on the importance of the Worden Fault in Douglas County, Kansas. Kansas Acad Science 106 (1–2): 11–16
Merriam DF (2005) Surface expression of buried geologic features in Kansas, or a practical example of the metaphor: the princess and the pea. Kansas Acad Science Trans 108 (3–4): 121–129
Merriam DF, Förster A (1996) Precambrian basement control 'on plains-type folds' (compactional features) in the Midcontinent region, USA. In: Oncken O, Janssen C (eds) Basement Tectonics 11. Kluwer Academic Publishers, Dordrecht, pp 149–166
Merriam DF, Sorensen CE (1982) Geology of the Howard Limestone (Wabaunsee Group, Virgilian Stage, Pennsylvanian) in southeastern Kansas: Kansas Geol Society 34th Field Conf. Guidebook. 1–29
Neff AW (1949) A study of the fracture patterns of Riley County, Kansas: unpubl. masters thesis, Kansas State University, p 47
Nelson PD (1952) The reflection of the basement complex in the surface structures of the Marshall-Riley County area of Kansas: unpubl. masters thesis, Kansas State University, p 73
Neuhauser KR (1986) Joint patterns in the Fort Hays Limestone (Cretaceous) of Ellis County, Kansas. Kansas Acad Sci Trans 89 (3–4): 102–109
Salisbury AC, Merriam DF (1984) Relation of Precambrian basement lineaments to overlying Paleozoic lineaments in New York and Ontario and Quebec (Canada): The Compass 61 (2): 77–84
Sims PK, Saltus RW, Anderson ED (2005) Preliminary Precambrian basement structure map of the continental United States; an interpretation of geologic and aeromagnetic data: U.S. Geol Survey, Open-File Rept., 31 p., 1 sheet
Stewart GF (1967) Jointing in Upper Pennsylvanian limestones in northeastern Kansas. Kansas Geol Survey Bull 187: 17–19, pt. 1
Van Schmus WR, et al. (1993) Transcontinental Proterozoic provinces, in Precambrian: conterminous U.S.: Geol Soc America, The Geology of North America C-2: 171–334
Wagner HC (1961) Geology of the Altoona Quadrangle, Kansas: U.S. Geol Survey, Geol. Quad. Map GQ-149, 1 sheet, scale: 62,500
Ward JR (1968) A study of the joint patterns in gently dipping sedimentary rocks of south-central Kansas. Kansas Geol Survey Bull 191: 23, pt. 2
Watney WL, Davis JC, Olea RA, Harff J, Bohling GC (1997) Modeling of sediment accommodation realms by regionalized classification: Geowissenschaften, 15 (1): 28–33
Watson GS (1961) Goodness-of-fit tests on a circle. Biometrika 43: 109–114
Watson GS (1966) The statistics of orientation data. J Geology 74 (5): 786–797, pt. 2
Watson GS (1970) Orientation statistics in the earth sciences: Geol. Inst. Uppsala Bull 2 (9): 73–89
Watson GS (1983) Statistics on spheres: John Wiley & Sons, New York, p 238
White DC (1990) Lineament study of stream patterns in a portion of east-central Kansas: unpubl. masters thesis, Emporia State University, p 57
Yarger HL (1983) Regional interpretation of Kansas aeromagnetic data. Kansas Geol. Survey, Geophys. Ser. 1, p 35
Zar JH (1998) Biostatistical Analysis, 4th edn. Prentice-Hall, Upper Saddle River, NJ, p 929
Zeller EJ, Dreschhoff G, Angino E, Holdoway K, Hakes W, Jayaprakash G, Crisler K (1976) Potential uranium host rocks and structures in the Central Great Plains. Kansas Geol. Survey, Geol. Ser. 2, p 59
Zippi PA (2000) VectorRose 3.02 Manual: PAZ Software, Garland, TX, p 28

Cross-Wavelet Analysis: A Tool for Detection of Relationships Between Paleoclimate Proxy Records

Andreas Prokoph and Hafida El Bilali

Reprinted from *Mathematical Geosciences* DOI: 10.1007/s11004-008-9170-8, when citing this article please use the DOI number.

Abstract Cross-wavelet transform (XWT) is proposed as a data analysis technique for geological time-series. XWT permits the detection of cross-magnitude, phase differences (= lag time), non-stationarity, and coherency between signals from different paleoclimate records that may exhibit large stratigraphic uncertainties and noise levels. The approach presented herein utilizes a continuous XWT technique with Morlet-wavelet as the mother function, allows for variable scaling factors for time and scale sampling, and the automatic extraction of the most significant periodic signals. XWT and cross-spectral analysis is applied on computer generated time-series as well as two independently sampled proxy records (CO_2 content approximated from plant cuticles and paleotemperature derived from $\delta^{18}O$ from marine fossil carbonate) of the last 290 Ma. The influence of nonstationarities in the paleoclimate records that are introduced by stratigraphic uncertainties were a particular focus of this study. The XWT outputs of the computer-models indicate that a potential causal relationship can be distorted if different geological time-scale and/or large stratigraphic uncertainties have been used. XWT detect strong cross-amplitudes (~ 200 ppm•‰) between the CO_2 and $\delta^{18}O$ record in the 20–50 Myr waveband, however, fluctuating phase differences prevent a statistical conclusion on causal relationship at this waveband.

Keywords Cyclicity · time-series · stratigraphic uncertainty

Andreas Prokoph
SPEEDSTAT, 19 Langstrom Crescent, Ottawa, Ontario, K1G 5J5, Canada,
e-mail: aprokocon@aol.com

Hafida El Bilali
Department of Earth Sciences and Ottawa-Carleton Geoscience Centre, Carleton University, Ottawa, Ontario, K1S 5B6, Canada, e-mail: hafida.el@sympatico.ca

1 Introduction

This study aims to enhance the possibilities for detection and quantification of relationships between paleoclimate- proxy data with wide stratigraphic uncertainties. Previous studies on this topic relied on cross-spectral analysis, regression, or correlation techniques. In addition, visual comparison of plots of such records is commonly used and widely accepted (Crowley and Berner, 2001; Royer, 2006; Shaviv and Veizer, 2003). Correlation and trend analysis can determine the significance of relationships between non-stationary time-series (i.e. signal amplitude and wavelengths are not constant through time). However, these methods may not detect correlations between sinusoidal signals of the same wavelength in two records, if these signals are phase shifted. If the phase shift approaches $\Phi = \pi/2$ then both time-series appear uncorrelated. Cross-correlation and cross-spectral analysis can detect such phase shifts, but only as average values, and are not able to represent non-stationarities in the signals. On the other hand, cross-wavelet analysis permits detection, extraction and reconstruction of relationships between two non-stationary signals simultaneously in frequency (or scale) and time (or location) (e.g., Grinsted et al., 2004) (Fig. 1). Cross-spectral analysis is a bivariate version of spectral analysis for comparison of two datasets (e.g., Davis, 2002). Similarly cross-wavelet transform can be considered as a bivariate extension to wavelet transform (WT), which has been developed as a tool to filter, exam, and extract nonstationary signals in time-series and images (Morlet et al., 1982). Software for continuous wavelet

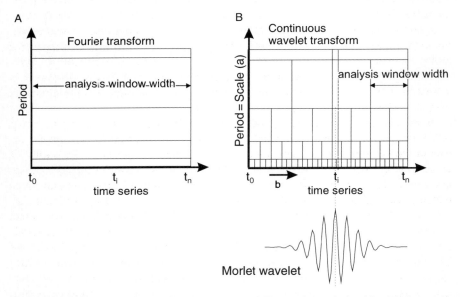

Fig. 1 Analysis windows for (**A**) Fourier analysis and (**B**) wavelet analysis with a Morlet wavelet of scale a centered at time $t_i = t_0 + b$ (modified after Prokoph and Agterberg 1999)

transform (CWT) and discrete wavelet transform (DWT) is readily available (e.g., MATLAB toolbox) and textbooks explain their mathematical-statistical background and complexity in detail (e.g., Kaiser, 1993). A particular good reference on the use of wavelet analysis for geological, climatological and geophysical time-series is given by Torrence and Compo (1998). In the context of this study, the comparison of wavelet transform to spectral analysis methods by Rioul and Vetterli (1991) is recommended.

Cross-wavelet analysis (XWT) has a shorter history than wavelet analysis and its application to geological and climatological time-series is still limited (Jevrejeva et al., 2003; Jury et al., 2002; Rigozo et al., 2004; Valet, 2003). Nevertheless, the use of XWT in these studies provided already new insights in the temporal-changing relationships between El Nino, ice condition or tree growth, or magnetic paleointensity and remnant magnetization. Torrence and Webster (1999) and Grinsted et al. (2004) provide detailed explanations of wavelet squared coherency and cross-wavelet phase angle using Morlet wavelet based on continuous cross-wavelet transform. They use of Monte Carlo Simulations to provide frequency-specific probability distribution (Global wavelet spectrum) that can be tested against by wavelet coefficients. However, significance testing of the cross-wavelet spectrum is not often used, because the assumption of an underlying probability distribution is not given in potentially nonstationary signals (Maraun and Kurths, 2004).

Cross-wavelet transform is applied in order to investigate a paleoclimate-specific issue: the uncertainty in the geological time-scales due to stratigraphic error that add non-stationarities to the paleoclimate records. The distortion of geological time-scales to the real time is a result of propagating uncertainties and errors in radioactive decay constants, accuracy of the tie-ages (e.g. Fish-Cayon-Tuff), biostratigraphic zonation of samples taken, and others (e.g. Gradstein et al., 2004). Some of these uncertainties can be reduced by the use of probability techniques for global Phanerozoic time-scales (Agterberg, 1994). However, it is difficult to eliminate stratigraphic uncertainties in the correlation of time-scale that are restricted to Boreal or Tethyal records, or stratigraphic uncertainties in high-resolution records over short time-intervals (Gradstein et al., 2004).

In the present paper, detecting and reconstructing time-distorted signals using XWT based on computer-simulated time-series is demonstrated. Systematic errors in the geological time-scale are computer-simulated. Moreover, the ability of cross-wavelet analysis to detect these errors is demonstrated. Further, the consequences of these time-scale errors on the determination of causal relationships between geological records dated from different location or using different dating methods (e.g., biostratigraphy based on microfossil, macrofossils, or stable isotope curves) is indicated. In contrast, the common method of assigning stratigraphic errors to tie-point ages and to interpolate between these tie-points is not investigated in this study. The atmospheric CO_2 record derived from plant cuticles (Retallack, 2002) and the tropical sea-surface temperature record derived from oxygen isotope data (Prokoph et al., 2008) are investigated for waveband-dependent relationships.

2 Cross-Wavelet Analysis Method

The wavelet coefficients W of a time series $x(s)$ are calculated by a simple convolution (Chao and Naito, 1995):

$$W_\psi(a,b) = \left(\frac{1}{a}\right) \int x(s) \psi\left(\frac{s-b}{a}\right) ds \tag{1}$$

where ψ is the mother wavelet; the parameter a is the scale factor that determines the characteristic frequency or wavelength; and b represents the shift of the wavelet over $x(s)$. The wavelet coefficient was scaled by $1/a$ which represents wavelet amplitudes while most other applications use $1/\sqrt{a}$ to calculate the modulus or variance of the signals (e.g., Chao and Naito, 1995; Grinsted et al., 2004). The cross-wavelet spectrum of two series $x(s)$ and $y(s)$ is defined by (e.g., Jury et al., 2002)

$$W_{xy}(a,b) = W_x(a,b) W_y^*(a,b) \tag{2}$$

where $W_x(a,b)$ and $W_y(a,b)$ are the CWT of $x(s)$ and $y(s)$ respectively, where* denotes the complex conjugate. In the present application, the modulus of the cross-wavelet transform $|W_{xy}(a,b)|$ represent the cross-amplitudes of $x(t)$ and $y(t)$. A continuous wavelet transform (CWT) is utilized, with the Morlet wavelet as the mother function (Morlet et al., 1982) because it reflects the gradual changes in paleoclimate records (Appenzeller et al., 1998; Bolton et al., 1995; Prokoph et al., 2004). The shifted and scaled Morlet mother wavelet is defined as (Morlet et al., 1982)

$$\psi^l_{a,b}(s) = \pi^{-\frac{1}{4}} (al)^{-\frac{1}{2}} e^{-i2\pi \frac{1}{a}(s-b)} e^{-\frac{1}{2}(\frac{s-b}{al})^2} \tag{3}$$

The parameter l modifies wavelet transform bandwidth resolution either in favor of time or in favor of frequency, because according to the uncertainty principle $\Delta a \Delta b \geq 1/4\pi$ there is always a trade-off between frequency and location resolution. Thus, the bandwidth resolution Δa for wavelet transform varies with $\Delta a = \frac{a\sqrt{2}}{4\pi l}$, and a location resolution $\Delta b = \frac{al}{\sqrt{2}}$. Here, the parameter $l = 10$ is selected for all analyses, which give sufficiently precise results in resolving depth and frequency, respectively (e.g., Prokoph and Barthelmes, 1996).

The cross-wavelet coefficients are subject to edge effects because the wavelet is not completely localized in time (Fig. 1). The boundary of edge effects forms a wavelength dependent curve for significantly edge-effect free wavelet coefficients that is called the cone of influence (Torrence and Compo, 1998). Here, we chose a cone of influence that preserves > 80% of the original signal is chosen, located above the stripped lines in the XWT-scalogram. The first-order trend is removed before application of XWT, eliminating superimposed lag-1 autocorrelative red noise (Mann and Lees, 1996). For computation, the integral in the wavelet transform has to be modified by using the trapezoidal rule for unevenly sampled points to evaluate XWT, and this provides $W^*_{xy}l(a,b)$. In this study the interpolated $W^*l(a,b)$ can be graphically visualized with appropriate shades of gray. The phase difference between the two time-series is defined (Jury et al., 2002) by

$$\Phi(b) = \tan^{-1} \frac{\int_{a1}^{a2} \text{Im}(W_{xy}(a,b))da}{\int_{a1}^{a2} \text{Re}(W_{xy}(a,b))da} \quad (4)$$

with b corresponding to the time lag b. *Im* and *Re* indicate the imaginary and real parts, respectively.

3 XWT on Computer Simulated Paleoclimate Proxy Time Series

A set of time-series was generated with particular emphasis on potential (a) stratigraphic errors and uncertainties and (b) phase-shifted response of one paleoclimate feature (e.g., global sea surface temperature) to a forcing factor (e.g., atmospheric CO_2) represented by another proxy record. Previous XWT studies (e.g., Grinsted et al., 2004) analyzed the coherency of wavebands in detail but considered the time (or distance) scale as error-free. Signals are simplified as sine waves that are superimposed by high-frequency noise, so as to emphasize the effects of the stratigraphic errors. The detection and extraction of more complex signals using CWT or XWT are described in Prokoph and Barthelmes (1996), Prokoph and Patterson (2004), Torrence and Webster (1999), and Valet (2003).

The stratigraphic error in this study is defined as

$$\varepsilon = t - t' \quad (5)$$

with t representing time and t' the geological time-scale used. For simulations, paleoclimate signals are defined as x_i in a geological record $y(t)$

$$y(t) = \Sigma sin x_i(t) + e(t), \quad (6)$$

with $e(t)$ summarising intrinsic environmental, sampling and instrumental high-frequency errors.

Two sets of simulations are generated, assuming (1) a stratigraphic error gradually increasing with geological age and (2) a cyclic fluctuating stratigraphic error. This results in two different geological time scales

$$t'_1 = t + 20\cos(2\pi t/800) + 20 + \varepsilon_1 \quad (7)$$

and

$$t'_2 = t + 20\cos(2\pi t/200) + 20 + \varepsilon_2 \quad (8)$$

with t in Ma and ε_1 and ε_2 (-1 Ma $< \varepsilon < 1$ Ma) representing small stratigraphic errors. The length of the simulated record is set to $t = 543$ Ma (Fig. 2), resembling the duration of the Phanerozoic in the GTS2004 time-scale (Gradstein et al., 2004). The simulated paleoclimate forcing record $x_1(t)$ is set to

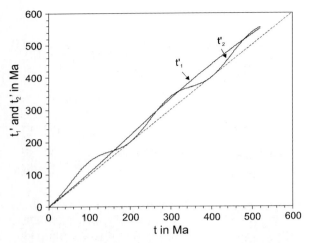

Fig. 2 Comparison of distorted ages used for modeling (t'_1 and t'_2) to undistorted age t (dotted line). See text for equations of models. Note that the derived time-scale appears up to 20 Ma older than the original time-scale (ie. 560 Ma instead of 542 Ma)

$$x_1(t) = cos(2\pi t/35) + 2sin(2\pi t/140) + \varepsilon_3 \quad (9)$$

and the simulated paleoclimate response record $x_2(t)$ is defined as

$$x_2(t) = cos(2\pi t/35 + \phi) + \varepsilon_4 \quad (10)$$

with ε_3 and ε_4 representing random noise $(-4 < \varepsilon_3 < 4)$ and $(-4 < \varepsilon_4 < 4)$ of four times the amplitude of the signal $cos(2\pi t/35)$. Both "forced" and "response" function are unitless in these simulations. A phase shift ϕ is set to $\pi - 0.8 = 2.34$ radians in (10) to simulate non-correlation of the cosine signals in records x_1 and x_2.

Three models with distorted time-scales $t-> t_1'$ (Model 1) or $t-> t_2'$ (Models 2,3) are constructed from (7) to (10) and re-sampled to time intervals $\Delta t' = 1\ Myr$ (Fig. 3A)

$$Model_1 = x_1(t'_1) \quad (11)$$

$$Model_2 = x_1(t'_2) \quad (12)$$

and

$$Model_3 = x_2(t'_2) \quad (13)$$

XWT of Model 1 with Model 3 enhances the cross-amplitude of the correlated 35 Myr from the background of noise and non-correlated 140 Myr signals. The cross-amplitude increased to ~40 Myr, because the stratigraphic error increased the time-scale from 543 to 555 Ma (Fig. 2), but this wavelength remains almost constant (Figs. 3B, 4A). The expected cross-amplitude a^2 of $1 = a\ cos(2\pi t/35)^* a\ cos(2\pi t/35 + \phi)$ with $a = 1$ (9 and 10) is reduced due to the edge effect following

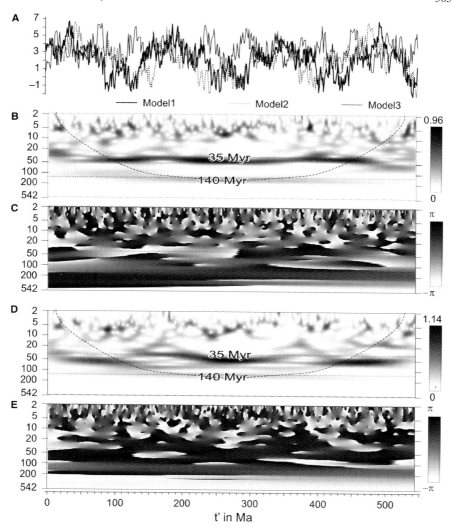

Fig. 3 Cross-wavelet analysis of computer-generated models of paleoclimate-related time-series using Morlet-wavelength. (A) Models 1, 2, and 3. Note that Model 1 is respective time-scale t'_1 and Models 2 and 3 are respective time-scale t'_2. (B) Wavelet spectrum showing the cross-amplitude of signals of Model 1 versus Model 3. Stripped line delineates the 10% signal reduction due to edge effect (see text for details). Dotted lines indicate the 140 Myr and 35 Myr signals that are embedded in the original time-series at time t'. Cross-amplitude scale on right side. (C) Phase-shift spectrum of Model 1 versus Model 3. Scale on right side. Note that the grey scale change of the ∼35 Myr signal at ∼290 Ma marks the transition from phase difference of π to $-\pi$ between the two models in the 35 Myr signal. (D) Wavelet spectrum showing the cross-amplitude of signals of Model 2 vs Model 3. Stripped line delineates the 10% signal reduction due to edge effect (see text for details). Dotted lines indicate the 140 and 35 Myr signals that are embedded in the original time-series at time t. Cross-amplitude scale on right side. Note the large temporal variability of the originally 35 Myr-signal. (E) Phase-shift spectrum of Model 1 vs. Model 3. Scale on right side

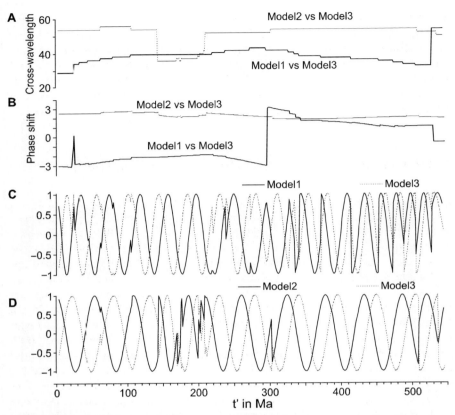

Fig. 4 Digital output of (**A**) the dominant wavelength (ie. with highest cross-amplitude). (**B**) Temporal changes in the phase shift of the most dominant signal (ie. 35 Myr). Note the more stable and more accurate representation of the 35 Myr signal in Model 1 vs. Model 3, but the more accurate representation of the phase shift and "cause-response" order in Model 2 vs. Model 3. (**C**) and (**D**) Reconstruction of original 35 Myr signal in time-scale t' of Model 1 vs. Model 3 and Model 2 vs. Model 3 based on output of cross-wavelength, cross-amplitude and phase-shifts (see Fig. 3 and Fig. 4A,B). Note that the initial phase of the models at time t_0 can be derived from wavelet analysis. Note also the artificial spikes in the reconstructed signals due to the interaction of high-frequency noise with the waveband necessary to extract the ∼35 Myr signal

the cone of influence to a maximum of 0.96 (middle of record) and 0.5 at the top and bottom of the record. The phase spectrum indicates however instability in phase shift in this wavelength, however, including a pronounced phase jump at ∼280 Ma (Figs. 3C, 4B). In contrast, XWT of Model 2 with Model 3 shows high variability (30–55 Myr) from the expected 35 Myr wavelength (Figs. 3D, 4A), but the phase spectrum correctly reconstructs the phase shift $\phi = 2.34$ radians (Figs. 3E, 4B).

The original signals can be partially reconstructed based on the extracted correlative wavelengths (Fig. 4A), cross-amplitude (Fig. 3B,D), and phase shift (Fig. 4B).

A complete reconstruction of the decomposed record in time-scale t' requires the knowledge of the phase at $t = 0$ and the wavelet amplitude of the correlative wavelength of at least one of the two records. These additional parameters can be extracted by using CWT with the same mother wavelet (ie Morlet) and parameter l as used in XWT. Figure 4 shows that the recomposed correlative $\sim$35 Myr-cycle of Model 2 vs. Model 3 reconstructs the original phase shift much more precisely than the Model 1 vs. Model 3 analysis.

4 XWT of Observational Paleoclimate Proxy Time Series

The updated tropical $\delta^{18}O$ record of seawater carbonate (Prokoph et al., 2008) was selected as proxy for global paleotemperature, because this record has been proven to well represent the Phanerozoic paleoclimate (Veizer et al., 1999). Nevertheless, this record has its limits as paleoclimate proxy, because (1) it cannot separate the seawater pH changes from water temperature changes, (2) has an unexplained linear long-term trend, and (3) is not spatially and temporal consistently sampled (e.g., Zeebe, 2001; Royer et al., 2004). The atmospheric carbon dioxide concentration record is derived from fossil plant cuticles of the last 290 Ma (Retallack, 2002). Fluctuations in the atmospheric CO_2 concentration are considered as the major cause of global climate fluctuations throughout the Phanerozoic (Royer, 2006). Both records have been transferred into time-scale GTS2004 (Gradstein et al., 2004) and have been cut at 290 Ma to permit sufficient sampling density, and the linear trend of the $\delta^{18}O$ record has been removed (Fig. 5A). The stratigraphic 1σ uncertainty of the $\delta^{18}O$ record is estimated to be, on average $< 1\%$ of the sample mean age (Prokoph et al., 2008), but with $\sim$3 Myr is much higher in the CO_2 proxy record. The often high and variable stratigraphic uncertainties in the raw CO_2 proxy occasionally led to offsets of peaks in the raw data to peaks in the Gaussian filtered CO_2 proxy record (see Fig. 1A).

XWT of both records indicate consistently high cross-amplitude of up to 530 ppm•‰ in the $\sim$20–50 Myr waveband (Figs. 5B, 6A). Moreover, no other waveband is characterized by cross-amplitudes of > 100 ppm•‰. In addition, the $\sim$140 Myr cycle of $\delta^{18}O$ is missing in the CO_2 proxy record. In the 20–50 Myr waveband, the phase differences shift gradually through the time interval from 290 Ma to $\sim$35 Ma (Figs 5C, 6B). The abrupt gray-scale change shows a phase shift at $\sim$35 Ma from π to $-\pi$ (Fig. 5C) that most likely indicates a gradual shift from $-181°$ to $-179°$ rather than a large phase jump. Figure 6C shows the simplified reconstruction of the CO_2 proxy with $\delta^{18}O$ record relationship. The $\delta^{18}O$ reconstruction has been inverted to simulate the potential temperature changes, because all transfer functions suggest a predominantly inverse linear relationship between $\delta^{18}O$ and temperature (e.g., Faure, 1998). It should be noted that stratigraphic uncertainty in the CO_2 record increases rapidly from 1$\sigma < 1$ Myr in the last 30 Ma to > 3 Myr between 30 Ma and 50 Ma.

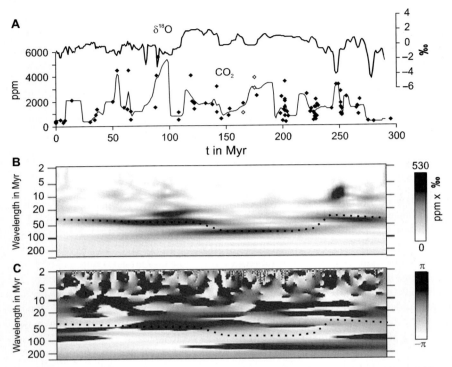

Fig. 5 (**A**) Paleoclimate proxy time-series for the last 290 Ma. original data (*diamonds*) and Gaussian filtered (*solid line*) of CO_2 (in ppm) derived from fossil cuticule (Retallack, 2002) and detrended and Gaussian filtered $\delta^{18}O$ record (in ‰) of marine tropical carbonate of brachiopods, belemnites and planktic foraminifera (Prokoph et al., 2008). (**B**) Wavelet spectrum using Morlet-wavelength showing the cross-amplitude of signals of CO_2 vs. $\delta^{18}O$. Dotted line traces the strongest wavelength (∼25–65 Myr) through time. (**C**) Phase-shift spectrum of CO_2 vs. $\delta^{18}O$. Dotted line traces the strongest wavelength (∼25–65 Myr). Scale on right side

5 Conclusions

XWT can be an efficient tool to detect coherent, phase-shifted cyclicity between paleoclimate-related records. Correlative non-stationary signals with characteristic phase shifts, cross-amplitude and wavelength can be distinguished from non-correlative signals or noise that only occur in one of the compared records. The stratigraphic error in the geological records provides a challenge for interpretation. The XWT of simulated records show that consistency of the time-scales used is essential to reconstruct causal relationship. Moreover stratigraphic uncertainty introduces non-stationarities in the records that question the ability to define confidence intervals on the coherency derived from XWT (Torrence and Webster, 1999) on geological records. Nevertheless, XWT is still able to trace non-stationary signals better than the classical approach of cross-spectral analysis, which cannot distinguish the spectral peaks as belonging to different time intervals (Fig. 7A,B). In

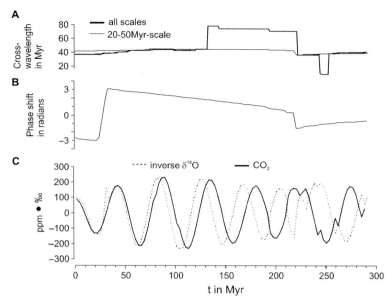

Fig. 6 Digital output of (**A**) the dominant wavelength of CO_2 vs. $\delta^{18}O$ (ie. with highest cross-amplitude) of all wavebands (*bold line*) and 20–50 Myr scale (*solid line*). (**B**) Temporal changes in the phase shift of the 20–50 Myr signal. (**C**) Reconstruction of 20–50 Myr signals CO_2 and $\delta^{18}O$ based on output of cross-wavelength, cross-amplitude and phase-shifts using a common cross-amplitude (see Fig. 5). The phase at t_0 and individual amplitude can be reconstructed using wavelet analysis (Prokoph and Patterson 2004)

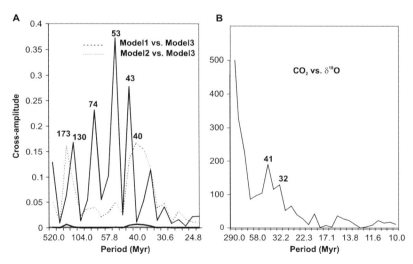

Fig. 7 Cross-amplitudes of cross-spectral analysis from (**A**) the Models and (**B**) the paleoclimate records. Note that spectral analysis suggests ~130 Myr and ~173 Myr signals in all models, and cannot detect that the 32–74 Myr spectral peaks are due to temporal fluctuations in the wavelength of a single ~35 Myr signal

addition, spectral analysis suggests significant cross-amplitudes for signals that occur only in one of the records (see ~ 130–173 Myr cycle in Fig. 7A). However, it is strongly recommended that paleoclimate records selected for statistical tests of their temporal correlations are transformed in a time-scale that can be applied to both of them. This may be difficult if terrestrial and marine records are compared. Alternatively, time-scales can be reconstructed using CWT based on the assumption that the wavelength of the most significant signal is stable and well-known, such as varves or tree-rings for one year (e.g., Prokoph and Patterson 2004).

Acknowledgments We thank G. Retallack for providing a digital version of his published data, including details on their stratigraphic uncertainty.

References

Agterberg FP (1994) Estimation of the geological time scale. Math Geol 26: 857–876
Appenzeller C, Stocker TF, Anklin M (1998) North Atlantic oscillation dynamics recorded in Greenland ice cores. Science 282: 446–449
Bolton EW, Maasch KA, Lilly JM (1995) A wavelet analysis of Plio-Pleistocene climate indicators: A new view of periodicity evolution. Geophys Res Let 22: 2753–2756
Chao BF, Naito I (1995) Wavelet analysis provides a new tool for studying Earth's rotation. EOS 76: 161, 164–165
Crowley TJ, Berner RA (2001) CO_2 and climate change. Science 292: 870–872
Davis JC (2002) Statistics and data analysis in geology, 3rd edn. Wiley, New York, p 637
Faure G (1998) Principles and applications of geochemistry. Prentice Hall, Englewood Cliffs, NJ
Gradstein F, Ogg J, Smith A (2004) A geologic time scale 2004. Cambridge University Press, Cambridge, UK
Grinsted A, Moore JC, Jevrejva S (2004) Application of the cross wavelet transform and wavelet coherence to geophysical time series. Nonlinear Proc Geoph 11: 561–566
Jevrejeva S, Moore JC, Grinsted A (2003) Influence of the Arctic Oscillation and El Nino-Southern Oscillation (ENSO) on ice conditions in the Baltic Sea: the wavelet approach. J Geophys Res 108 (D21): 4677–4687
Jury MR, Enfield DB, Mélice J (2002) Tropical monsoons around Africa: stability of El Niño–Southern Oscillation associations and links with continental climate. J Geophys Res 107 (C10): 3151–3167
Kaiser G (1993) A friendly guide to wavelet. Birkenhaeuser, Basel, Switzerland, p 300
Mann ME, Lees JM (1996) Robust estimation of background noise and signal detection in climatic time-series. Clim Change 33: 409–445
Maraun D, Kurths J (2004) Cross-wavelet analysis: Significance testing and pitfalls. Nonlinear Proc Geophys 11: 505–514
Morlet J, Arehs G, Fourgeau I, Giard D (1982) Wave propagation and sampling theory. Geophysics 47: 203
Prokoph A, Barthelmes F (1996) Detection of nonstationarities in geological time series: Wavelet transform of chaotic and cyclic sequences. Comp Geosci 10: 1097–1108
Prokoph A, Patterson RT (2004) From depth-scale to time-scale. Transforming of sediment image color data into high-resolution time-series. In: Francus P (ed) Image analysis, sediments and paleoenvironments. Dev in Paleoenviron Res Series 7, Springer, Dordrecht, pp 143–164
Prokoph A, Rampino MR, El Bilali H (2004) Periodic components in the diversity of calcareous plankton and geological events over the past 230 Myr. Palaeogeogr Palaeoclimatol Palaeoecol 207: 105–125

Prokoph A, Schields G, Veizer J (2008) Compilation and time-series analysis of a marine carbonate $\delta^{18}O$, $\delta^{13}C$, $^{87}Sr/^{86}Sr$ and $\delta^{34}S$ database through Earth history. Earth Sci Rev 87 (3–4): 113–134

Retallack GJ (2002) Carbon dioxide and climate over the past 300 million years. In: Gröcke DR, Kucera M (eds) Understanding climate change. Proxies, chronology and ocean-atmosphere interactions. Phil Trans Royal Soc London Series A 360, pp 659–674

Rigozo NR, Nordemann DJR, Echer E, Vieira LEA (2004) ENSO influence on tree ring data from Chile and Brazil. Geofisica Internat 43: 87–294

Rioul O, Vetterli M (1991) Wavelets and signal processing. IEEE Special Magazine 14–38

Royer DL (2006) CO_2-forced climate thresholds during the Phanerozoic. Geochim Cosmochim Acta 70: 5665–5675

Royer DL, Berner RA, Montanez IP, Tabor NJ, Beerling DJ (2004) CO_2 as a primary driver of Phanerozoic climate. GSA Today 14 (3): 4–10

Shaviv NJ, Veizer J (2003) Celestial driver of Phanerozoic climate? GSA Today 13: 4–10

Torrence C, Compo GP (1998) A Practical guide to wavelet analysis. Bull Amer Meteor Soc 79: 61–78

Torrence C, Webster PG (1999) Interdecadal changes in the ENSO-Monsoon System. J Climate 12: 2679–2690

Valet J-P (2003) Time variation in the geomagnetic intensity. Rev Geophys 41: 1–48

Veizer J, Ala D, Azmy K, Bruckschen P, Buhl D, Bruhn F, Carden GAF, Diener A, Ebneth S, Goddéris Y, Jasper T, Korte C, Pawellek F, Podlaha OG, Strauss H (1999) $^{87}Sr/^{86}Sr$, $\delta^{13}C$ and $\delta^{18}O$ evolution of Phanerozoic seawater. Chem Geol 161: 59–88

Zeebe RE (2001) Seawater pH and isotopic paleotemperatures of Cretaceous oceans. Palaeogeogr Palaeoclimatol Palaeoecol 170: 49–57

On Correlational Properties for Volcanic Earthquakes Associated with Asamayama (Japan), 1983–2005

Richard A. Reyment

Reprinted from *Natural Resources Research* DOI: 10.1007/s11053-008-9065-x, when citing this article please use the DOI number.

Abstract To a first approximation, earthquakes directly associated with volcanic activity may be studied as point stochastic processes. The earthquakes associated with B-type (movements located at 1 km or shallower) eruptive activity in the caldera of Asamayama differ in correlational properties from concurrent deep-seated seismic activity (A–type, located deeper than approximately 1 km). A-type activity occurs either in the form of independently distributed intervals between events or as dependently distributed intervals which are most appropriately analysed in contiguous subsamples ("windows"). The cross- correlations between the magnitudes of A-type earthquakes and depth of events for three periods from 1983 to 2005 may be of significance for interpreting aspects of the volcanic history of Asamayama. The lag-1 serial correlation coefficient for the A-type sequence from 1983 to 1990 is not significantly different from zero. In the case of the sets for 1991–2002 and 2003–2005, the coefficients are small but not zero. The difference is in part, at least, probably due to the well known confounding effect of trending as opposed to true serial correlation between successive events. The serial correlation coefficient for the B-type crater-sequence is not significant. The novel aspect of the present study concerns the relationship between depth of A-type earth movements and magnitude of associated shocks.

Keywords A-Type and B-Type volcanic earthquakes · Asamayama · cross-correlation · serial correlation · trend

1 Introduction

Asama is one of the scientifically recognized 108 active volcanoes of Japan. It is situated on the border between Nagano and Gumma Prefectures (Honshu), near the town of Karuizawa: latitude 36°24′23″N and longitude 138°31′23″E, with an

---

Richard A. Reyment
Section for Palaeozoology, Natural History Museum, Box 50007, 10405, Stockholm, Sweden,
e-mail: richard.reyment@nrm.se

elevation of 2568 metres, and 140 km NW distant of Tokyo. It lies near the junction of the Izu-Marianas and NE Japanese volcanic arcs. Growth of a dacitic shield-volcano was accompanied by pumiceous pyroclastic flows and by the growth of the Ko-Asama lava-dome (*ko* means "child") on the eastern flank. Maekake, capped by the Kama pyroclastic cone, forms the present summit of the volcano.

Being one of the most active volcanoes of Japan, Asama has been the object of continuous scientific study for several decades. Available, largely anecdotal, early records for Asama date from before the tenth century. A major outbreak took place in 1783 with a considerable loss of life, which tragic event is commemorated in a museum at Kambara near Naganohara town. Geologically, Asama is a stratovolcano (= composite volcano) with two craters situated on the remnants of an older volcano the origin of which is estimated at some 20000 years BP (Bout and Derriau, 1966). Japan encompasses several volcanic regions that have evolved through denser crust from the Pacific Plate subducting between less dense crust to the west. Most Japanese volcanoes extrude silica-rich rocks in the form of highly viscous lava (andesite and dacite), and as a consequence thereof are prone to high-energy explosive eruptions, such as is the case for Asama (Hashimoto, 1991). A critical listing of reported eruptions of Asama, including discredited reports, can be found on Professor Y. Hayakawa's website, to wit, *www.edu.gunma-a.ac.jp/hayakawa/volcanoes/asama/asamasiryo/table106*. (NB *"siryo"/"shiryo"* means "data").

The most recent major eruption dates from September 1, 2004. It was characterized by several vulcanian episodes and a continuous explosive phase of strombolian type. This phase of activity continued for several weeks, extending marginally into 2005 and producing major ashfalls. A previous major phase of eruptive activity took place in 1973, the course of which is similar to, but not identical with the most recent event, not least in respect of the chemical properties of extruded lava. A concise account of the last recorded event is given by Nakada and others (2005), who concluded that a swarm of A-type quakes accompanied the intrusion of magma into the deep-seated reservoir dyke. The first application of point processes to volcanic earthquakes was made by Reyment (1969). Applications have increased rapidly since then, particularly with regard to tectonic earthquakes (Okada, 1999). It is here emphasized that tectonic earthquakes cannot be interpreted in the same manner as those arising from volcanic activity The two categories are not geologically equatable, which is seldom, if ever, understood by investigators lacking a sound footing in the Geosciences.

2 Origin of the Data

The data used in the present account were recorded digitally over the interval from 1983 to 2005 by the Japanese Meteorological Agency (JMA), Karuizawa, and made available through the kind intermediary of Professor Yuji Yagi (Tsukuba University). The coordinates of the hypocentre are latitude = 36.156 − 36.656;

longitude = 138.273 − 138.773. The maximum depth recorded by the JMA for the A-type activity was reported to be 40 km.

In considering volcanic earthquakes a distinction is made between deep-seated shocks, and those that are located within the immediate vicinity of the caldera. On an originally ad hoc basis, quakes that occur at a depth in excess of 1 km are referred to as A-type earthquakes and those that occur at or near the surface of the caldera, as B-type earthquakes. This distinction was introduced by Minakami (1935). Despite the seemingly arbitrary nature of the classification, it was based on astute observation and has proven useful; it has been adopted almost universally by vulcanologists. A-type earthquakes are generally less than magnitude 6 on the Richter logarithmic scale. B-type volcanic earthquakes originate usually in, and adjacent to, active craters at very shallow depths, normally < 1 km. The insightful classification of Minakami (1935) is now on a digitally registered footing. A-type volcanic quakes are high-frequency in nature, whereas B-type events are low frequency (Iguchi et al., 2002). These differences show up clearly in the digital recordings. Iguchi et al., 2002), reporting on the 1996 episode at Satsuma-Iwojima, added a C-type (referred to as monochromatic) variety. In the case of Asama, the "magma chamber" is considered by geophysicists of the JMA to be more in the nature of a dyke than an actual chamber, as has been deduced from geophysical evidence (Nakada et al., 2005). It is partly the energy involved in emptying and replenishing the reservoir that is considered to bring about the A-type movements. The lava formed in the reservoir is a hypersthene-olivine andesite, which being less mobile and more viscous than basalts, causes the explosive outbursts triggered by built-up gas-pressure. Although not manifestly relevant in the present connexion, I wish to mention results of Oike (1977) who, studied the effect of extensive rainfall for triggering minor, non tectonic earthquakes in N.S.W., Australia.

An additional seismic category associated with active volcanoes is that of vague trembling which often takes place continuously thus giving the impression to the heedful of continuous, fuzzy unrest. For statistical purposes only the A- and B-types of activity are of interest.

The question has been posed as to whether the lavas extruded during the two most recent episodes (1973, 2004) are identical in chemical composition. A compositional principal coordinate analysis of the published data, encompassing six analyses (Nakada et al., 2005), is summarized in Table 1. The coordinates for the major elements do not overlap for the two data-sets. In the case of the minor elements, the two sets are much more widely separated. On the basis of the admittedly small data-sets available, the two lavas differ chemically.

Latter (1980) observed that volcanoes and earthquakes tend to have corresponding mondial distributions. This kind of relationship would seem to be amenable to study by the method of linear intensity models proposed by Ogata and Akaike, 1982). The A- and B- types studied herein derive from the data obtained by the JMA and the Tokyo University Earthquake Centre. Recent statistical results by Oikawa and others (2005) imply that the precise determination of the hypocentre for volcanic earthquakes, including B-type episodes, may be more complicated than hitherto conceived. No attempt is made nor implied at presenting a physical model for the

Table 1 Compositional principal coordinate analyses for the lavas extruded in 1973 and 2004

Major elements (Si Ti Al Fe Mn Mg Ca Na K P)

specimen	first three compositional principal coordinates		
1973 eruption			
1	0.152	−0.086	0.104
2	0.192	0.506	−0.120
3	0.115	0.051	0.203
2004 eruption			
4	0.124	−0.049	−0.117
5	0.141	−0.220	−0.142
6	0.139	−0.227	−0.151

Minor elements (Ba Co Cr Cu Nb Sc V Ni Zn Rb Zr Sr Y)

1973 eruption			
1	0.230	−0.188	0.105
2	0.180	−0.215	−0.260
3	0.244	−0.181	0.203
2004 eruption			
4	0.076	0.126	−0.161
5	0.082	0.253	0.039
6	0.091	0.282	0.047

Data from Nakada et al. (2005)

mechanisms underlying the activity of the volcano. In this connexion I refer the reader to Ogata (1982) and the informative review of the application of point processes in vulcanology by the same author (Ogata, 1999).

It should be mentioned that there is disorderliness in the definition of A- and B-types in vulcanology. For Wedge, Young and McKendrick (1994) the two categories denote types of lava exuded by Etna. Tsugura et al. (1997) use a hybridized definition, to wit, B-type events as opposed to "volcanic tremor episodes". Amma-Miyasaka and Nakagawa (2003) use the terms A-type and B-type for crystalline varieties in lava. Tsuruga et al. (1997, p. 337) speak of three B-types and four "volcanic tremor episodes". These postulated variants have yet to receive wider acceptance.

3 Findings

Reyment (1969) applied the theory of point processes (Cox and Lewis, 1966) to the recorded activity of some selected volcanoes, including the Japanese volcanoes Asama, Aso and Kirishima. All three volcanoes display trend in the rate of occurrence of eruptions over time. In the case of Asama, a possible tendency towards pseudo-cyclicity was noted; this observation defied further evaluation because of the small data-base of uneven accuracy. Reyment (1976) considered further the occurrence of volcanic earthquakes for Asama. The main result of that study was to confirm that there is fluctuational trending in the rate of occurrence of earthquakes which is of varying degree and direction. A check made on the present material confirmed these earlier results.

The course of action I use for investigating the statistical properties of geological sequence-data is as follows:

1. The times between events are examined graphically for evidence of trend, and other patterns, such as cyclicity, for contiguous, ad hoc defined, sub-samples (i.e. "windows" of the observations).
2. If the data are judged to be trend-free, then it may be assumed that the series of events is stationary. This implies that the marginal distributions of the Xi's (= the time-intervals between events) are identical.
3. Check for significant serial correlation between the successive Xi's. If the intervals are not correlated, this may be taken as an indication that the times between events are independent and identically distributed, with unknown distribution $F(x)$. Such a sequence represents some kind of a renewal process. The Poisson process is a special case of a renewal process (Cox and Lewis, 1966, p. 206), where

$$F(x) = prob(X \leq x) = 1 - e^{\lambda x}.$$

4. In the absence of significant trend in rates of occurrence the sequence is tested for agreement with a Poisson process and, in continuation, for a model based on a renewal process (Cox and Lewis, 1966).

The times to events (the earthquakes), $\{T_{(i)}\}$, or times between successive events, $\{Xi\}$, completely distinguish the process. Hence

$$0 < T_{(1)} < T_{(2)} < T_{(3)} < \ldots$$

An equivalent expression of a series of events may be made in terms of the counting process, Nt, to wit, the number of events taking place in the interval (0,t}. Nt is a continuous time-parametric stochastic process. the sample functions of which are "jump-functions". The analysis of the interval process, $\{Xi\}$ is essentially the analysis of a time-series composed of positive random variables; the analysis of the counting process Nt lacks a counterpart in the usual analysis of time-series (Cox and Lewis, 1966).

The series of events generated by volcanic earthquakes is geologically more complicated than the structure of the typical point process in that a sequence is bivariate with respect to magnitude of the shock and location in time. An even more complicated relationship can be considered when the location of the epicentre for each shock is included in the analysis; however, for the JMA data such information is not generally available. For present purposes, the events are taken to be distinguished only by where they occur in time.

4 Number of Earthquakes Per Year

Figure 1 displays the yearly pattern of A-type earthquakes from 1983 to 2004. There is a general tendency for the number to increase over time up to 2002 with one outlying value in 1986. The outlier might possibly indicate a period of stepped up activity, but I have not been able to find an unequivocal record of a possible cause. The frequency of A-type events decreases over the period leading up to the eruption of 2004. Comparison with Fig. 2 for B-type earthquakes does not show much agreement in frequencies. The frequency pattern for B-type shocks is one of activity increasing as a function of time from 1997 onwards. Outbreak-sequences for B-type events contain longer periods characterized by the absence of seismic activity. It seems that there has been a building up of factors over a relatively long period (here from 1997 to 2004) culminating in a fully developed crater-eruption.

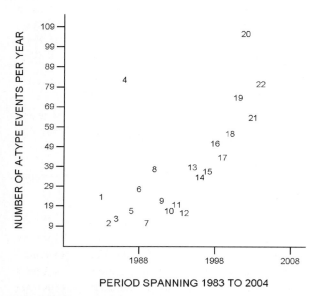

Fig. 1 The pattern of number of A-type volcanic earthquakes from 1983 to 2004. Data extracted from JMA-files. The plotted points are numbered in order to facilitate comparison

5 Testing for Trend in the Rate of Occurrence of Events

A suitable test for assessing the presence of significant trend in a sequence of events can be constructed from the times to events $\{T_i\}$. A simple model for the rate of occurrence of events, $\lambda(t)$, is

Correlations for Asama Volcanic Earthquakes

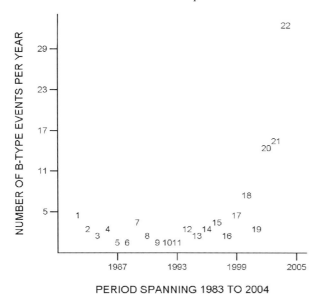

Fig. 2 The pattern of number of B-type volcanic earthquakes from 1983 to 2004. Data extracted from JMA files. The plotted points are numbered in order to facilitate comparison

$$\lambda(t) = \exp(\alpha + \beta t) = \lambda \exp(\beta t) \text{ for } t > 0 \text{ and } \lambda > 0.$$

A test for $\hat{\beta} = 0$, against $\hat{\beta} \neq 0$ is provided by the statistic

$$U = \sum ((T_i/t_0) - n/2))/(n/12)^{1/2}$$

for a series observed for a fixed time-interval $(0, t_0]$ associated with n events. This statistic converges rapidly to a normal unit variable (Cramér, 1946, p. 245; Cox and Lewis, 1966, p. 47). For values of $|U| > 1.96$ (the 5% level of significance) the hypothesis $\hat{\beta} \neq 0$ is rejected. Cox and Lewis (1966) noted, that this is a very conservative reading and that many practitioners tend to accept a value that somewhat exceeds the imposed limit. Lewis and Robinson (1974) took up the question of identifying trend in a modulated renewal process and general point processes. They proposed a modification of U suitable for non-renewal processes, this being the ratio $U/C(X)$, where $C(X)$ denotes the estimated coefficient of variation. Point processes that are over-dispersed with respect to the Poisson process pose a problem of interpretation in distinguishing between trends *s. str.* and the presence of long intervals in a sequence; such data occur in the records of volcanic earthquakes. Vere-Jones and Davies (1966, p. 258) concluded that for large-scale, deep-seated seismicity in New Zealand (i.e. non-volcanic) a Poisson approximation may be a reasonable working hypothesis, even if all theoretical requirements are not met.

The analysis summarized in Table 2 shows that periods marked by statistically significant trending alternate with periods without significant trending. In two cases,

Table 2 Trend-results for windows of consecutive A-type earthquakes and final suite of B-type earthquakes

Subsample	U	$\hat{C}(X)$	$U/\hat{C}(X)$	N
1983–1987	3.39**	1.80	1.88	130
1988–1991	0.30	1.17	NA	84
1992–1995	2.45*	1.24	1.98*	81
1996–1998	2.91*	1.08	2.69*	91
1999–2000	0.47	1.19	NA	84
2001	4.09**	1.59	2.57*	69
2002	2.62*	1.01	1.64	92
2003–2005	1.60	1.00	NA	104
2004	1.40	1.04	NA	51
B-type earthquakes				
2004–2005	5.10***	2.07	2.46**	72

U denotes the Cramér statistic, $\hat{C}/(X)$ the coefficient of variation. NA = not applicable. The asterisks denote the usual manner of indicating statistical significance (5, 1 and 0.1%).

the Lewis-Robinson adjustment, $U/\hat{C}(X)$, yields a value that falls short of the 95% confidence level. This is all the more interesting since the type-A earthquakes for the same period do not display significant trend. The conclusion here would seem to be that the type-A sequence is of recurrent nature, possibly related to adjustments in the magma reservoir. The B-type earthquakes reflect a separate physical process, perforce linked to the A-type activity but not directly controlled by it.

6 Analysis of Correlation Properties

In the present connexion, serial correlation and cross-correlation, are studied for the currently available information for the activity of Asama up to March 2005. The methods of calculation of these correlations are uncomplicated and are to be found in any introductory textbook on time-series analysis, for example, Kendall (1973). Quenouille (1952) gives a very good introduction to the problem posed by interpreting serial correlations The statistical analysis proceeds by the following steps.

If the intervals are significantly correlated, this may either be an expression of a causal link from tremor to tremor, or a general indication of trend due to geotectonic factors. Quenouille (1952) paid careful attention to the difficult problem of separating genuine serial correlation from trending and in so doing noted that in many cases, this could not be satisfactorily achieved.

(The significance of the serial coefficient is ascertained by computing $\hat{\rho}_1\sqrt{(N-1)}$ which has, approximately, a unit normal distribution for large values of N.)

Compute the cross-correlations for magnitude of each seismic event and its registered depth. Plot the sequence of values thus obtained against the time-axis.

The yearly record of A-type earthquakes from 1983 to 2005 as registered by the JMA indicates that there is a general tendency for the number of events to increase over time up to 2002 with one outlying value for 1986. The frequency of A-type events decreases over the period leading up to the eruption of 2004. Comparison with the data for B-type earthquakes does not show much agreement in frequencies inasmuch as their frequency pattern indicates activity increasing as a function of time from 1997 onwards.

7 Correlation Between Magnitude and Depth of Earthquakes

7.1 The A-Type Earthquakes

The method of cross-correlation was applied to the observations recorded by the JMA for depth and magnitude of the volcanic events. The series of observations were considered in relation to the times to three eruptive episodes as recognized and registered by the JMA, to wit, the events of 1990 (N = 148), 2002–2003 (N = 385) and 2004 (N = 72).

Sequence 1983–1990. This sequence encompasses 148 observational pairs. The value of the cross-correlation for the 137th observation is 0.410; the entire set of correlations have an average value of 0.455 (Table 3) which suggests for this section of the sequence that the depth of the A-type earthquake is significantly related to the energy released. This relationship is illustrated in Fig. 3. The eruptive event is succeeded by a slight drop in the cross-correlation values (marked by an arrow in the figure). N.B. The "time-axis" in this and analogous figures represents time-ordered events and not absolute time.

Sequence 1991–2003. This sequence encompasses 385 observational pairs. The value of the cross-correlation for the 384th observation is 0.556; the entire set of correlations have an average value of 0.679. The eruptive event is followed by a slight decline in the cross-correlation values. The relationship is illustrated in Fig. 4, and the location of the eruptive event marked by an arrow.

Sequence 2003–2005. This sequence contains 72 observational pairs. The cross-correlations fall off slowly but rise slightly during the course of the strong eruptive phase. In this respect, the relationship between depth and magnitude of the shocks differs from the two foregoing phases. These phases are shown in the graphs (Figs. 3 and 4) to be quite similar in that both are "hump-shaped" prior to the eruptive event.

Table 3 Means and standard deviations for A-type cross-correlations

Interval	Mean	Standard Deviation	N
1983–1990	0.455	0.0281	148
1991–2000	0.679	0.7170	385
2001–2005	0.525	0.0531	72

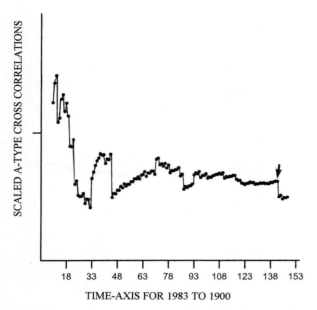

Fig. 3 A-type cross-correlations for 1983–1990. The arrow denotes the minor eruptive event that terminates this sequence. There are several jumps in the course of the graph

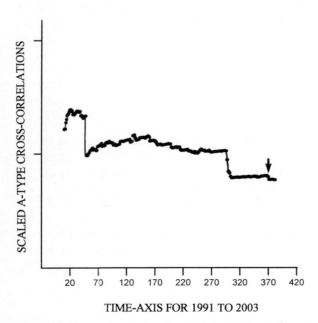

Fig. 4 A-type cross-correlations for 1991–2003. The arrow marks the minor eruptive event that terminates this sequence

Fig. 5 A-type cross-correlations for 2003–2005. The arrow marks the major eruptive event of September 2004

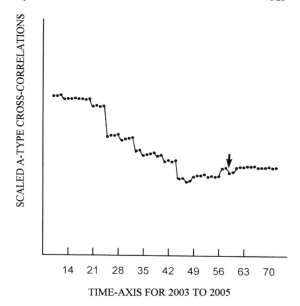

This feature is lacking in the third sequence studied. The average cross-correlation for these values is 0.525. The relationship is illustrated in Fig. 5 in which the arrow denotes the beginning of the eruption of September 2004.

7.2 The B-Type Earthquakes

The sequence of observations for the B-type quakes yield a cross-correlation pattern that differs significantly from that of the A-type observations. Figure 6 represents the relationships over the period 1997 to the first two months of 2005. The cross-correlations increase in importance from a low initial level, rising gradually to a maximum attained on September 1st when the crater eruption commenced. Eruptive activity continued at a high level until around September 11th, after which it declined. It seems as though the relationship between energy and depth has undergone a regular build-up over time until the major eruption of September 2004. It can be surmised that the strength of correlation will possibly decline to the level held around 1997.

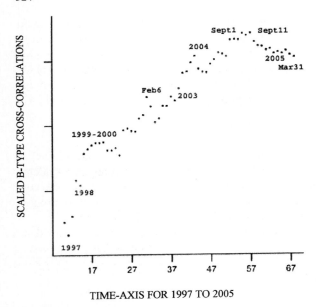

Fig. 6 B-type cross-correlations for 1997–2005. Events for February 6th 2003 and the time-span for September 1st to September 11th are indicated. March 11th 2005 denotes the termination of the observational sequence available for study. The eruptive event of February 2003 does not stand out, whereas the September 2004 phase is readily identifiable

8 Serial Correlations

A summary account of the occurrence of significant serial correlation coefficients for a lag of 1 for the sub-samples is presented in Table 4. These results suggest that for the first set (1983 to the 1990 eruptive event), the lag-1 serial correlations for magnitude are not significantly correlated. The value of 0.246 for the second set is not high, but nevertheless attains statistical significance. The serial correlation for the third set (2003–2005) is likewise significant. The lag-1 serial

Table 4 Lag 1 serial correlations for magnitudes for the three A-type sequences and one B-type sequence

N	Magnitude	Sequence
A-type events		
148	0.007	1991–1990
385	0.246	1991–2002
72	0.393	2003–2005
B-type events		
68	−0.032	1997–2005

correlation coefficients for the B-type events do not differ significantly from nought. The rather unusual result for the first series of values leading up to the first eruptive event considered here was explored using a simple device proposed by Quenouille (1952, 1956) and advocated by Kendall (1973) for diminishing eventual bias in r, the serial correlation coefficient. This involves dividing the sequence into two equal parts, $r(1)$ and $r(2)$, and then computing an unbiassed serial correlation, R.

$$R = 2r - 1/2(r(1) + r(2))$$

The result (Table 4) does not alter, in any significant manner, the original result. The conclusion indicated is that for the first series the values are independently distributed. The second and third series are, on the other hand, not independently distributed. A possible explanation of this differentiation could be that trending reflects the building up of energy which starts from a status of zero trend.

The B-type serial correlation coefficients are indicative of serial independence and hence an event at time t does not affect the manifestation of that taking place at time $t+1$. The series analysed here starts in 1997. Prior to this, B-type registrations are rare in the JMA records.

9 Conclusions

The cross-correlation study of the magnitudes of the type A volcanic earthquakes and the focal depths of the earthquakes is informative. The relationships exposed in the sub-samples fluctuate between gradation as a function of time to perturbed "stop-start" steplike patterns. It can be expected that as further digitally acquired data are obtained, with more sophisticated recording equipment, a fully implemented multi-dimensional approach would become feasible. The results for the cross-correlations indicate that the association between depth and magnitude of earthquakes tends to be relatively high, and somewhat stronger outside an eruptive phase. However, there is no unequivocal evidence that the depth-energy cross-correlations can be directly connected to the events registered for the crater (the B-type shocks).

The sequence of magnitudes leading up to the first eruption encompassed in the data studied can be accepted as being independently distributed. The second and third sets are, however, not independently distributed due in part, at least, to the effects of externally operating trend. The cause of the trending can be reasonably expected to lie with the deep-seated geological processes leading up to the important eruptive phase of September 2004. The non-significant serial correlation for the B-type events would seem to speak for a "decoupling" of the crater-bound seismic events from the deep-seated events.

During the course of this research I began to doubt, as the results unfolded, that a point-process model might not be the best way of studying volcanic activity. When I introduced point-process modelling into the analysis of volcanic eruptions (Reyment, 1969) it seemed that this was a practically useful approach with a rich store of well understood methods. However, in retrospect, the stochastic processes encompassed by the theory of point processes are relatively straightforward, even uncomplicated, in relation to what goes on in the Earth, where forces, that we still are far from understanding, thrust and parry in never-ending jousts. A general approach to the analytical study of volcanic activity could well be made to include a more geodynamic element in the statistical modelling.

A factor that is easily overlooked by mathematical statisticians lacking geological expertise is that the environment of a classic point-process is stable. In the geological world, this condition is seldom met. The logical approach to the study of on-again-off-again sequences in time in a geological environment is to proceed via the analysis of "windows", whereby the properties of succeeding sections of the series are analysed separately, and then compared (I was introduced to this technique many years ago by the late Dr. G. Hill, CSIRO, Australia). Clearly, there can be no compelling reason to support a total analysis of a process that fluctuates in response to random influences.

9.1 Lexical Note

Why does the volcano studied here appear in print as Asama and as Asamayama? The word "yama" is one way of denoting "mountain". YAMA is the native Japanese reading of the Chinese idiograph SAN. Both readings are employed for the English concept of "mountain". However, the readings are not interchangeable once they have become established. For example, the Japanese romanized rendition of what in non-Japanese literature often is known as *Fujiyama* is in reality *Fujisan*. The Japanese romanized word for volcano is *kazan*, being composed of the Chinese idiograph for fire (ka) and the idiograph for mountain (zan) – the s-sound goes to z after a in Japanese.

Acknowledgments This work was supported by grants from the Japanese Council for the Promotion of Science and the Royal Swedish Academy of Sciences. Professor Yuji Yagi, Tsukuba University, provided the digitally obtained data for the period 1983–2005. The Japanese Meteorological Agency (JMA) and in particular Dr. Shinichiro Matsuda, director of the observatory, and Dr. Yoshihiro Ueda, generously made its resources available during my visits. Professor Kazuyoshi Endo, the sponsor of my visit to the Department of Geological Sciences at Tsukuba University is thanked for devoting much time and dedication to the furtherance of my project. Special thanks go to the two referees for constructive criticism, all of which I found useful.

References

Amina-Myasaka, MR, Nakagawa M (2003) Evolution of deeper basaltic and shallower andesitic magmas during the AD 1409–1983 eruptions of Miyakejime Volcano, Izu-Mariana Arc: inferences from temporal variation of mineral composition in crystal clots. J Petrol 44: 2113–2138

Bout P, Derriau M (1966) Recherches sur les volcans explosifs du Japon: Mémoires et Documents, Editions du CNRS de France, fasc. 4

Cox RR, Lewis, PAW (1966) The statistical analysis of series of events. Methuen Monographs on applied probability and statistics, pp 285

Cramér H (1946) Mathematical methods of statistics, Princeton Mathematical Series, Princeton University Press, Princeton, NJ, pp 575

Hashimoto M (ed) (1991) Geology of Japan. Developments in earth and planetary sciences, Terra Scientific Publication Company, Tokyo, pp 249

Iguchi M, Saito E, Nishi Y, Tameguri T (2002) Evaluation of recent activity at Satsuma-Iwojima – felt earthquake on June 8, 1996. Earth, Planets, Space 54: 187–195

Latter J (1980) Volcanoes in New Zealand. Alpha, November (1980), pp 1–6

Lewis PAW, Robinson DW (1974) Testing for monotone trend in a modulated renewal process: reliability and biometry. S. I. A. M., Philadelphia, pp 163–182

Minakami T (1935) The explosion activities of volcano Asama in 1935 (Part 1). Bulletin of the Earthquake Research Institute, University of Tokyo, Tokya, vol 13, pp 629–644

Nakada S, Yoshimoto M, Koyama E, Tsuji H, Urabe T (2005) Comparative study of the 2004 eruption with old eruptions at Asama volcano and the activity evaluation. Earthquake Research Institute, University of Tokyo, Tokya, pp 1–19

Ogata Y (1999) Seismicity analysis through point-process modelling: a review. Pure Applied Geophys 155: 471–507

Ogata Y, Akaike H (1982) On linear intensity models for mixed double-stochastic Poisson and self-exciting point processes. J R Stat Soc B 44: 102–107

Oikawa J, Ido Y, Tsugi H (2005) Precise hypocenter determination for volcanic earthquakes at Asama volcano, Japan. American Geophysical Union (abstract) #V21C-0618

Oike K (1977) On the relation between rainfall and the occurrence of earthquakes. Bulletin of the Disaster Prevention Research Institute 20 (B1): 35–45

Quenouille MH (1952) Associated measurements: Butterworths Scientific Publications, London, p 242

Quenouille MH (1956) Notes on bias in estimation. Biometrika 43: 353–360

Reyment RA (1969) Statistical analysis of some volcanologic data regarded as series of point events. Pure and Applied Geophysics 74: 57–77

Reyment RA (1976) Analysis of volcanic earthquakes of Asamayama (Japan). Symposium 5, International Geological Congress (1976). Springer Verlag, New York, pp 87–95

Tsuruga K, Yomogida K, Honda S, Ito H, Ohminato T, Kawakatsu H (1997) Spatial and temporal variation of volcanic earthquakes at Sakurajima Volcano, Japan. J Volcanol Geotherm Res 75: 337–358

Wedge G, Young PAV, McKendrick IJ (1994) Mapping lava flow hazards using computer simulation. J Geophys Res 99: 489–504

Crosscorrelation of Sea Levels

Joseph E. Robinson

Abstract Crosscorrelation has been used to determine those factors that occur in common within related strings of geological data. Knowledge of these factors allows the isolation of events that either affect all the data sets or are confined to an individual data string. The crosscorrelation can be expressed in either the time or the frequency domain. However, in the frequency domain, amplitude and phase of the input functions can be calculated and utilized as separate spectra. This makes it possible to combine amplitude and phase spectra from different data sets and then calculate inverse transforms that create new time displays.

The amplitude spectrum from the crosscorrelation of related data sets, in conjunction with those for the original data, allow the isolation of components common to all from those that are set-specific, while retaining the original phase spectra for each location. Inverse transforms can then display the revised spectra as time functions containing only the desired frequency components with their new amplitudes and the original phase at the location where the data was recorded. Sea level data from harbors on the Baltic sea illustrate an example of revised sea level measurements separated from local variations that affect the collection locations, yet with the sea components in the correct phase at the desired location.

Keywords Fourier analysis · time series · sea level · crosscorrelation

1 Introduction

Correlation, locating the position of similar features in geological data sets where the data consists of two or more sequences of numeric values, invites the application of mathematical techniques. One method that has been successful in both geology and geophysics is crosscorrelation in the time domain between digital data strings, where a maximum computed value over a definite interval can indicate the best alignment (e.g. Robinson, 1978). The crosscorrelation function cross multiplies the

---

Joseph E. Robinson
Syracuse University, Syracuse, NY 13210, USA, e-mail: joerobinson1@peoplepc.com

values and thus retains and amplifies those data set components that are in common (e.g. Lee, 1960).

Crosscorrelation can be carried out in either the time domain or the frequency domain. The time sequence results of both operations are identical and early papers by Anstey (1964, 1965) provide a basic description, Cooley and Tukey (1965) describe a computation algorithm and Davis (1973, 1986, 2002) provides both descriptions and programs. The main difference between the time and the frequency domain is that phase, which describes the relative position of the individual data components, is fully contained in the time domain crosscorrelation. In the frequency domain, on the other hand, phase can be considered as a separate function. Transforming a data sequence to the frequency domain produces two displays, an amplitude function and a frequency function. In crosscorrelation the original amplitude functions are multiplied whereas the phase functions are added. The advantage of the frequency domain is that either of the frequency functions can be manipulated independently of other spectra to display new and often interesting results. The correlated amplitude spectra (or its square root) can be used as a model to modify or replace any of the original computed amplitudes. The inverse transform of the altered amplitude spectra, combined with the appropriate phase spectra can provide a new time display containing only selected features as related to a specific collection station.

2 Application

Analysis of the effects of global warming and change in world sea levels requires knowledge of the relation between the level of the sea and of the land it abuts. The historical measurement of eustatic changes of sea levels, (e.g. Rossiter, 1967; Dott, 1992), has in most situations, been dependent on sequences of water levels measured in harbors with the land as the base level. In other areas, the sea has been the base level for determining land movements. Both can be important in evaluating the effect of both present and projected global changes.

The longest known continuous sea level record is from the harbor at Stockholm, Sweden where annual measurements date from 1774 (Ekman, 1988). Recorded and published sea level elevations in the Stockholm harbor strongly reflect glacial rebound of the Fennoscandia peninsula (Ekman, 1991, 1996; Bergsten, 1954). However, included in the data are any changes in the actual Baltic sea levels plus local watershed anomalies and harbor changes. The harbor water levels were considered the base line for measuring the extent and amplitude of the rebound without considering possible changes in the sea itself. The sea level was considered to be constant.

During the period 1901 through 1950 there is a notable lowering of the measured sea level elevations at Stockholm (Fig. 1A). Conversely, published sea levels taken for the same years in the harbor at Weismar, Germany show a modest rise in elevation (Fig. 2A). Rossiter (1967) and Ekman (1996) suggest that the land surface along the southern Baltic coast is stable or, perhaps slowly subsiding. Data from other harbors and shore lines substantiate their conclusions on the uplift and tilting

Crosscorrelation of Sea Levels

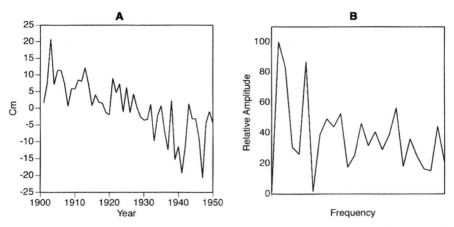

Fig. 1 (**A**) Observed sea levels in harbor at Stockholm, Sweden for years 1901–1950. (**B**) Amplitude spectrum of Fourier transform from Stockholm sea level data

of the Baltic basin. However the elevation of the Baltic Sea may also have been subject to change. Crosscorrelation analysis can aid in evaluating the relationship.

The Baltic Sea is connected by the Kattegat and Skagerrak to the North Sea and hence to the Atlantic Ocean. Other oceans and seas thus exert an influence on the open Baltic sea. The harbor sea level measurements are not, as individual sets of data, a portrayal of sea level change in the open Baltic sea, they certainly reflect major changes in land elevation. However, the land elevation changes tend to be localized and differ from place to place over the entire Baltic Sea basin. The elevation change decreases from north to south (e.g. Ekman, 1991). Water levels over

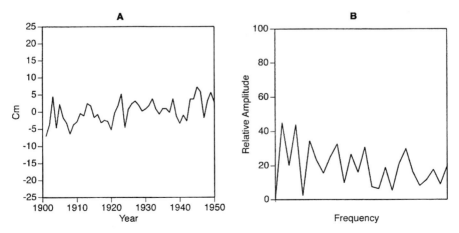

Fig. 2 (**A**) Observed sea levels in harbor at Wismar, Germany, from 1901 to 1950. (**B**) Amplitude spectrum of Fourier transform from Wismar sea level data

the same interval of years may also differ when data sets are collected from separate localities, even from harbors on the same body of water. The size and orientation of the Baltic Sea and its narrow North Sea connection suggests that open sea elevation changes may be expected to affect the entire body of water in a relatively uniform manner, but over an extended period of time. Phase, related to the actual time when changes occur, can therefore be expected to differ with location. Tides are a short term example of this phenomenon.

The crosscorrelation of digital water level data from two or more locations for the same period of time contain only those features that are common to all data sets, regardless of the phase of individual events. Although the basic data are in the time domain. Fourier transforms of the crosscorrelated power series produce frequency spectra where the individual components can be combined with the original spectra to produce new spectra displaying specific attributes. The power amplitiude spectrum displays components that are common to all data sets. The phase spectrum is location specific and a measure of how the various amplitude components are related in position when creating a new time domain data sequence.

Because the power spectrum contains only those amplitudes that are common to all the input data sets, it can be divided into new amplitude spectra to retain or eliminate local anomalies from the original amplitude spectrum at any of the collection locations. When combined with the appropriate phase spectrum, the inverse transform produces the desired new time function. Also, the appropriate approach to roots of the power amplitude spectrum results in new spectra that can be combined with any of the phase spectra to create new time functions related to each location for a measure of open sea levels. Where multiple data series are involved, evaluation and computation is easier if first carried out on data pairs. The reduced new data can undergo additional processing where necessary.

Figures 1B and 2B illustrate the amplitude spectra of the Stockholm and Wismar sea levels computed from published sea level values for the years 1900–1950. Simple multiplication of the corresponding individual amplitudes produces the cross power spectrum and eliminates those unique frequency components that are confined to a single harbor. This new spectrum can be used in determining revised spectra for any of the original sampling locations. An inverse transformation of the new spectrum produces a new sequence of water level values at the desired location. The Baltic Sea with its long history of water level data is an excellent test basin for techniques designed to distinguish between land and sea movements.

On the assumption that changes in sea level may be relatively uniform over broad areas of open ocean basins, their shared components have been retained for the Stockholm data to construct a revised sea level elevation sequence for the Stockholm harbor (Fig. 3). The new values can be considered as an estimate of Baltic sea levels measured immediately offshore from the Stockholm harbor. Elevation change appears less than in the original data and fluctuations have slightly less amplitude. However, there is not an absolute scale so there is no real measure of accuracy. Certainly a great deal of the local noise has been removed. However the effect of basin uplift or tilting is still strongly in evidence. This example is based on amplitude and phase at Stockholm: it is location specific and does not represent the

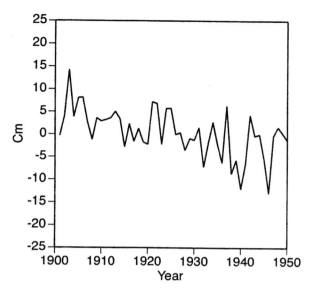

Fig. 3 Revised sea levels for Baltic Sea in vicinity of Stockholm

Baltic Sea as a whole. Analysis of additional data sets from other well separated collection stations would improve the open Baltic sea level estimate and could lead to a new chapter in the Baltic basin story and suggest applications to other seas and oceans.

3 Conclusions

Crosscorrelation in the frequency domain provides a technique for separating specific signals from unwanted noise. In this treatment, it can aid the separation and identification of local harbor effects and land movements as well as eustatic changes. Because both land and sea are subject to change in relative elevation over time, it is important to distinguish both and quantify the effect. Geological inference may also be a help, particularly where there are a number of measurements over a large area. If it is likely that open ocean elevations are more consistent than land elevations, continuity of elevation may be a criterion for separation. Therefore, sea level measurements collected from a number of stations situated over a large area can be analyzed for systematic change and may be combined for optimum resolution. Historic sea level measurements from within harbors can be recalculated to provide better baseline data for comparison with modern sea level measurements. Crosscorrelation methods can aid in calculating more accurate sea level data necessary for the evaluation of recent sea level changes and their relation to past and future trend projections.

References

Anstey NA (1964) Correlaton techniques – a review. Geophys Prospecting 12: 366–382
Anstey NA (1965) Wiggles. J Can Soc Explor Geophysicists 1 (1): 13–43
Bergsten F (1954) The land uplift in Sweden from the evidence of the old water marks. Geografiska Annaler 36: 81–111
Cooley JW, Tukey SW (1965) An algorithm for the machine calculation of complex Fourier series. Math Comput 19: 297–301
Davis JC (1973) Statistics and data analysis in geology, 1st edn. John Wiley & Sons, Inc., New York, p 550
Davis JC (1986) Statistics and data analysis in geology, 2nd edn. John Wiley & Sons, Inc., New York, p 550
Davis JC (2002) Statistics and data analysis in geology, 3rd edn. John Wiley & Sons, Inc., New York, p 550
Dott RH Jr. (1992) An introduction to the ups and downs of Eustasy. Geol Soc America Memoir 180
Ekman M (1988) The world's longest continued series of sea level observations. PAGEOPH 127 (1): 73–77
Ekman M (1991) A concise history of postglacial land uplift research (from its beginning to 1950). Terra Nova 3: 358–365
Ekman M (1996) A consistent map of the post-glacial uplift of Fennoscandia. Terra Nova 8: 1958–1965
Lee YT (1960) Statistical theory of communication. John Wiley & Sons, New York, p 509
Robinson JE (1978) Pitfalls in automatic lithostratigraphic correlation. Comput Geosci 4 (3): 273–275
Rossiter JR (1967) An analysis of annual sea level variations in European waters. Geophys Soc 12: 259–299

Diversion of Flooding Rivers to Residual Mining Open Pits

Ian Lerche

Reprinted from *Natural Resources Research* DOI: 10.1007/s11053-008-9066-9, when citing this article please use the DOI number.

Abstract Massive flooding in East Germany in the summer of 2002 was first alleviated, and then exacerbated, by diversion of the river flood waters into residual open pits, the legacy of lignite mining. The pits at first contained the flood waters but, once filled to capacity, leaked precipitously, causing massive flooding in the flat lands around the pits. This paper examines the problem of constructing quantitative models for assessments of fill, bypass and leakage from such floodwater containing pits. Emergency management teams can then generate quickly not only many different scenarios to help with immediate flood control options and fall-back positions, but can also investigate long term planning that can then be undertaken to estimate better the consequences of permitting such diversion. While the models developed are simple, and are numerically implemented in easy to use spreadsheet format, they have the advantage of guiding directions of flood assessment control and consequent results. The illustrative numerical examples show how one can quickly use such quantitative models to obtain patterns of flooding relevant to situations of sustained, torrential rainfall and subsequent river overflow.

Keywords Simulation model · flood control · emergency management

1 Introduction

One of the more prevalent problems in low-lying land areas is the cyclic flooding that occurs when major rivers flood due to torrential and sustained rain. Such a major georisk event occurred in the summer of 2002 when the Elbe River in Germany, fed by almost continual rains for many weeks, overflowed its banks along most of the length of the river. The damage to the German economy and to the livelihoods of many thousands of people was significant (MZ, 2002).

Ian Lerche
Institut fuer Geophysik und Geologie, Universitaet Leipzig, Talstrasse 35, 04103 Leipzig, Germany, e-mail: lercheian@yahoo.com

In the area of East Germany, through which the Elbe flows along with major tributaries, there are numerous open pits, the residual legacy of the lignite mining from the old German Democratic Republic. These pits range in size from a few acres to several square miles across and in depth to almost 100 m. in particular places, with an average open-pit depth of about 60 m. During severe river flooding there is overflow of the flooding river into the open pits, which then fill to capacity, and often to overcapacity so that secondary flooding from the pit overflow takes place. In addition, the open pits are not completely isolated from their neighboring rock formations, which range from clays and shales, through to the sands and carbonates of the Tertiary. Thus there can be, and often is, leakage into the subsurface of the river overflow waters that found their way into the open pits. This subsurface leakage is in addition to the surface leakage that can take place if the pits are filled beyond capacity because the two leakage pathways operate on very different time-scales with the surface leakage being "prompt" and the subsurface leakage being slower and later (Lerche and Glaesser, 2006).

Figure 1, parts a and b, shows the river overflow from summer 2002 at Goitsche, in the vicinity of Halle (Saale), from the Mulde, one of the major tributaries to the Elbe River, indicating the widespread flooding that can take place in the neighborhood of the open pits.

What is of interest is to attempt to provide dynamically evolving models of such behavior so that one can estimate the consequent influence of the open pit waters on the surrounding region, both in terms of direct flooding due to overflow from the pits and also as the subsurface leakage waters mix with the groundwaters present prior to flooding, and transporting also residual pit waters that were present prior to flooding. The reason for such an interest is centered not only on possible contamination of surface and groundwater supplies, as well as the flood sediment deposits (mainly fine-grained shales, silts and muds) from the flooding onto the landscape that, in turn, change the richness of the soil used for crops, but also on the damage to infrastructures , such as houses and industrial sites, as well as for the possible preventative measures that could be taken to minimize future flooding and the attendant risks and hazards.

The simplest form of such quantitative flooding and overflow models proceeds as follows. One starts with an open pit that contains water from steady accumulation due to rains, river input, and groundwater seepage. The pit is usually very far from water filled. Then, in a season of torrential sustained rain the neighboring river swells so that there is a considerable excess of water discharge along the river. This excess water is either diverted to the open pit purposely, or naturally overflows and/or bursts the river banks and so finds its way automatically to the open pit. The pit then fills to capacity. Any waters in excess of the pit capacity then bypass the pit and so flood the surrounding land until the excess flow in the river is finished, when the river returns to normal- in the sense that the river then reoccupies its bed between the banks with no further overflow.

During the time the pit is filled, or partially filled by the excess river waters, the pit can develop "leaks" that allow drainage of the pit waters either to the surrounding landscape or to the underground, depending on the subsurface hydrogeological

Fig. 1 a. Aerial picture of the breakthrough of the Mulde River at high water stand into a former open pit mine area at Goitsche on 15 August 2002; **b.** Similar to Fig. 1a but taken from a different aerial perspective to show the larger region covered by the floodwater breakthrough

connections to the pit. Such leakage normally occurs after the pit is partially filled because the excess water pressure in the pit causes breakage of surface containing dykes that are already water-logged (in the case of leakage to the landscape) or of reverse flow into the subsurface caused by the increased hydrostatic pressure of the deepened waters in the pit. Both of these events lower the level of water in the pit, thereby allowing more river flood waters to enter the pit that, in turn, increase the leakage again. These processes carry on until the river flood waters are exhausted and the river re-occupies its bed.

Of considerable interest is: to estimate the total flow of the excess river waters that must bypass the pit, to estimate the total amount of water that can enter the pit and then leak from the pit, the total ground area that is flooded by such leakage and the depth of water, together with the length of time such flooding is likely to take before the waters recede.

Based on such ideas we have constructed a spreadsheet program that allows all of the above processes to take place and for which the time-dependent nature of the flooding, leakage and overflow are all handled simultaneously. In addition, the

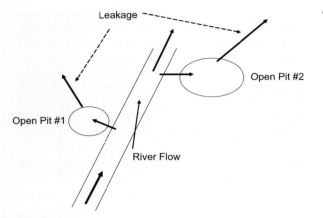

Fig. 2 Sketch of the river excess water flow into the two open pit areas

spreadsheet program allows for the presence of two open pits, one downstream from the other, so that waters by-passing the first pit can then be captured by the second pit, with the concomitant chances of leakage from the second pit as well as from the first. A pictorial representation of the flow situation is given in Fig. 2.

The ideas expressed above are best illustrated through simple numerical examples so that one can see immediately how the various components of the flooding system fit together. This point is now examined.

2 Numerical Illustrations

Consider the following illustrative situation. Due to heavy sustained rains a river contains extra waters in the amount of 7150 MMl that can overflow the banks of the river. This extra water component persists for a period of 140 h, corresponding to a "slug" of water over and above the river banks holding capability, and so is capable of causing serious flooding of the landscape. The flow rate is then 50 MMl/h. The two open pits are taken to have holding capacities of 250 MMl (the left open pit) and 350 MMl (the right open pit) respectively, and already contain 100 MMl (the left open pit) and 50 MMl (the right open pit). The right open pit is taken to be downstream from the left open pit. Not all the excess water in the river will be diverted to either the first or second open pits; precisely how much gets diverted to each pit depends on the flow pathways available to the pits from the river. For illustrative purposes only, the examples consider that a maximum of 50% of the river water excess can be diverted to the left open pit, and a maximum of 30% can be diverted to the right open pit, leaving 20% of the excess river waters to flow in the river bed for further downstream flooding possibilities.

Diversion of Flooding Rivers to Residual Mining Open Pits 539

As the pits fill, there can be leakage from the pits. Let the left open pit have a leakage rate of 10 MMl/h and the right open pit a leakage rate of 4 MMl/h. Further, there is a delay between the onset of fill of the pits and the onset of leakage from the pits. Take it that the left open pit starts to leak 30 h after fill commences and that the right open pit starts to leak 50 h after its fill commences. Note that with these numbers, the left open pit is already filled to capacity after 6 h, so that bypass of that open pit must then occur. The river waters then flow further downstream to the right open pit, filling it and so leading to further bypass once the right pit is full. Once leakage sets in for the right open pit then one can again divert river waters into the pit; the same is true for the left open pit.

Thus the dynamical problems of charging of the pits, bypass, recharging, leakage and landscape flooding are intimately interconnected in terms of the pit capacities, the river flow rate, and the onsets of leakage times and amounts that can be leaked.

Shown in Fig. 3 is the fill of the left open pit (called the left pit, in line with the sketch of Fig. 2) with time. Note that the fill is uniform in time until the pit reaches its capacity of 250 MMl, the pit then stays full until 30 h is reached when the leak commences, causing 10 MMl/h to be leaked. The fill of the left pit continues but there is also a steady drain so that the fill remains at 240 MMl until the river stops supplying the flood waters at time 140 h, when the residual waters in the left open pit are steadily drained out of the pit. We have set the drainage to remain open after the leak commences until all water in the pit is drained, although the program has been constructed so that one can also investigate situations in which the drainage stops after a prescribed time.

Figure 4 shows the water bypass for the left open pit (in MMl/h), indicating that, once the pit is filled, there is maximum bypass of the residual river excess flux until leakage commences, at time marker 30 h, from the pit. Thereafter the bypass is reduced by the leakage rate, of course. The cumulative bypass for the left pit is shown in Fig. 5, indicating that the total bypass is around 2200 MMl and is roughly, but not precisely, linear with time. Figure 6 shows the cumulative leakage from the left open pit. Until time 30 h, when the leak starts, there is no leakage, thereafter the leakage proceeds at the constant rate of 10 MMl/h so that by the time the excess river flux ends at time 150 h, almost 1400 MMl have leaked from the left pit.

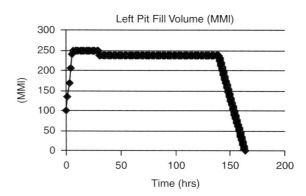

Fig. 3 Open pit fill volume (in MMl) as a function of time for the left open pit

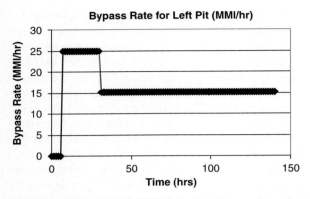

Fig. 4 Flow bypass rate (in MMl/h) for the left open pit as a function of time

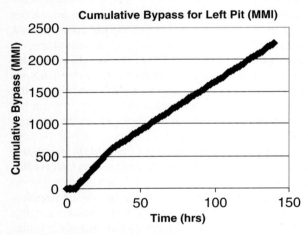

Fig. 5 Cumulative bypass (in MMl) for the left open pit as a function of time

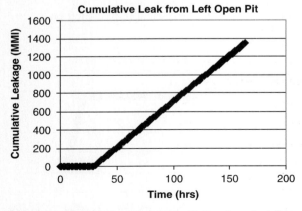

Fig. 6 Cumulative leakage (in MMl) with time from the left open pit

Diversion of Flooding Rivers to Residual Mining Open Pits   541

For the right open pit (referred to as the right pit, in line with the sketch of Fig. 2) the fill, bypass and leakage scenario is more complicated than that for the left open pit because the right open pit is further downstream. Thus not only does one have a different diversion fraction of available water from the river but the flux of river water is itself time dependent because of the bypass that occurred upstream at the left open pit. This bypass means that the flux of river waters available to the right open pit depends on the amount of water captured by the left open pit. The maximum divertible flux (at a fraction of 30% of intrinsic river flux) to the right open pit is then shown in Fig. 7 as a function of elapsed time. Note that, initially, the flux available to fill the right pit is at 25 MMl/h but this flux rises once the left pit is filled, at 6 h, to the full 50 MMl/h that is possible. However, once the left pit starts to leak (at 10 MMl/h) the flux available to the right pit drops, as also shown in Fig. 7, and stays lower by 10 MMl/h throughout the rest of the time of high river discharge.

Correspondingly, as exhibited in Fig. 8, the right pit fill volume rises to the maximum of 350 MMl and then holds at that value until the right pit starts to leak (at a rate of 4 MMl/h) when the fill of the right pit is dropped by this amount until time 150 h, when the river excess stops. Thereafter, the right pit steadily drains at the 4 MMl/h rate until the pit is empty at about time 225 h i.e. about 75 h after the river excess is contained in the river banks or has moved even further downstream. Thus the drainage causes flooding for a longer time that the high river excess in this illustration (Fig. 9).

The total bypass rate with time for both the left and right pits taken together then reflects this complicated interdependence of both pits, as shown in Fig. 10. The cumulative bypass with time for both pits is shown in Fig. 11, indicating that just over 5000 MMl (out of the initial 7150 MMl excess volume in the river) eventually bypasses both pits and flows further downstream.

Also of concern is to figure out the flooding that can be caused by the uncontained leakage. This aspect of the problem can be most easily addressed if one takes the total volume leaked from each open pit, lets the volume not be adsorbed by the ground, which is usually water-logged anyway, and spreads the total volume over

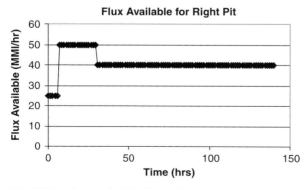

Fig. 7 Flux of water (in MMl/h) available to fill the right open pit that occurs further downstream than the left open pit

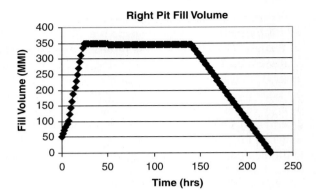

Fig. 8 Open pit fill volume (in MMl) as a function of time for the right open pit

the low lying land around the open pit. In East Germany, for instance, the land is extremely flat, almost to the Ural Mountains in Russia, so that widespread flooding is commonplace once an open pit disgorges its water containment volume.

In the specific illustration given above the total volume of water leaked from the left (right) open pit is 1336 MMl (704 MMl). If spread over an area of 900 km^2 (400 km^2), a not unusually extreme area, then the average depth of water in the affected zone is 1.48 m (1.76 m). Such a depth of water is capable of causing massive damage to infrastructure, to houses, to crop areas, and to storage areas. For example, the massive flooding in East Germany in 2002 was so severe in terms of

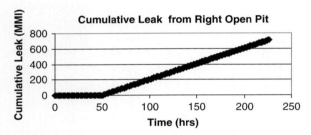

Fig. 9 Cumulative leakage (in MMl) as a function of time from the right open pit

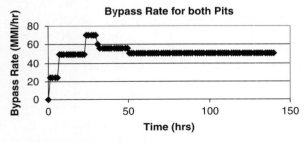

Fig. 10 Bypass rate (in MMl/h) for both the left and right open pits, representing waters still in the river that can cause damage and flooding further downstream

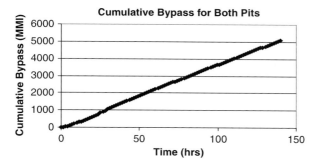

Fig. 11 Cumulative bypass (in MMl) for both open pits as a function of time

its impact that insurance companies are still recovering their outlay by attaching a premium surcharge of around 10–20% on their insurance policies over 3 years later (Allianz, 2003, 2004, 2005).

3 Discussion and Conclusion

The ability to provide quantitative model patterns for flood diversion, in this case into open pits that are the residual remnants of the lignite coal industry in East Germany, and to handle quantitatively the overflow out of the pits, as well as the bypass of river flood waters, goes a long way to allowing damage control scenarios to be investigated for their effectiveness for future potential flooding. In the case of East Germany such periodic flooding has a history going back to before the first recorded settlements, as measured by core information of prehistorical flood deposits and by historical records themselves. Given the flat nature of the land, and the significant feeder rivers to the major Elbe river running through southern Poland and Germany, presumably such periodic flooding events are likely to occur over and over again. The construction of holding dykes around the open pits to catch overflow is one method used for alleviation of the periodic flooding problem, but such dykes have to be placed so that they are high enough to contain the worst case flood leakage scenarios, they have to be constructed of materials that will not become water-logged during persistent rains, and they have to be strong enough that water pressure will not burst the dykes. Earthen dykes have been the solution that has been used until the present day but, judging from the water damage, as exhibited in Fig. 1, they are not capable of doing a sufficient retention job in times of massive persistent rains. But any other dyke solution is extremely expensive because of material costs and emplacement costs. It would appear that flooding will continue under conditions similar to those of summer 2002.

What this paper brings to the subject is the ability to quickly and accurately provide model scenarios that one can use to investigate potential leakage and flooding scenarios. In this way it is to be hoped that flood abatement control authorities can more readily understand the different possible patterns that can be caused, and that they can then provide optional solutions and fall-back positions as emergency management controls to minimize, if not entirely remove, the consequences of flooding by more carefully designed use of the open pit diversion possibilities and their attendant follow-on effects. Such is, arguably, the main message the quantitative procedure developed here has been designed to illustrate.

Acknowledgments This work has been supported by the DAAD through their award of a Visiting Professorship at the University of Leipzig. The University of Leipzig is also thanked for its contribution to this support and Professor Werner Ehrmann is particularly thanked for the courtesies and support he and his group at Leipzig have made available during the course of this work.

References

Allianz Versicherung (2003) Letter to the author informing on the perceived need to impose a surcharge on insurance premiums to cover catastrophic losses
Allianz Versicherung (2004) Letter to the author informing on the perceived need to impose a surcharge on insurance premiums to cover catastrophic losses
Allianz Versicherung (2005) Letter to the author informing on the perceived need to impose a surcharge on insurance premiums to cover catastrophic losses
Lerche I, Glaesser W (2006) Environmental risk assessment. Springer Verlag, Heidelberg, p 341
MZ – MittelDeutsche Zeitung (Central Germany Newspaper), Issue of 16 August 2002

A Christmas Parking Lot Problem

Ian Lerche

Abstract Based on much personal experience over many Christmas shopping campaigns, and personal dread at the prospect of further participating in another Christmas shopping expedition, I have designed simple models of such situations in order to see if merchants are indeed maximizing their total profit. Such an investigation seems to me to be a much better use of my time than to be forced to participate in the Christmas parking lot roulette. This brief paper shows how one can investigate the Christmas parking lot problem quantitatively from the merchant's perspective without the need to be involved personally, a major blessing from my viewpoint.

Keywords Simulation model · traffic · shopping

1 Introduction

In the USA it is generally estimated that around 60% of a retail merchant's profits are made during the Christmas season that extends from Thanksgiving until Christmas Eve. A retail merchant's store inevitably has an associated parking lot for customers, and also employees, so that the customers, who drive everywhere in the USA, can park their vehicles, buy products, and transport them home with ease.

In order to maximize his bulk sales, and so his profit, the retail merchant wants each customer to spend as much as possible but, equally, he wants to minimize the time each customer spends in the store so that the customer leaves as quickly as possible, thereby creating a parking space for the next customer. In this way the merchant thinks he can attain the greatest flow-through of customers and highest profit (reckoned as a fraction of bulk sales).

Personal experience, however, suggests that there is a further problem with parking. Assuming the merchant has set the prices of products attractively low in order to maximize the number of potentially interested customers, there will then be a

---

I. Lerche
Institut fuer Geophysik und Geologie, Universitaet Leipzig, Talstrasse 35, 04103 Leipzig, Germany, e-mail: lercheian@yahoo.com

large flux of cars in the road leading to his store. This flux, once sustained for an incredibly brief period of time during the day, means that the parking lot fills to capacity very quickly once the store is opened, and often before the store is opened too. Until at least one customer leaves the store and exits the parking lot there is no room for other cars, which then bypass the store and go elsewhere to shop. The retail merchant has then lost potential business.

Accordingly, most retail stores have two parking lots; the primary parking lot near the store and a secondary parking lot a bit further away to handle the overflow traffic. This second parking lot has to be near enough to the store so that potential customers are prepared to actually walk from the second lot to the store and, even more so, walk back to their cars with arms laden with Christmas goodies. Often a merchant will rent a field neighboring the original parking lot as a second parking lot during this season of frenzied buying. The ultimate aim, under all conditions, is for the merchant to maximize the number of customers that enter his store, to maximize the money each customer will spend, to minimize the time each customer spends in the store, and to minimize the potential loss of customers because the primary and secondary parking lots are full.

Based on much personal experience over many Christmas shopping campaigns, and personal dread at the prospect of further participating in another Christmas shopping expedition under the nightmarish conditions that generally prevail, I have often mused over the years if it is possible to design simple models of such situations in order to see if merchants are indeed maximizing their total profit. Such an investigation seems to me to be a much better use of my time than to be forced into personally participating in the Christmas parking lot roulette.

This brief paper shows how one can investigate the Christmas parking lot problem quantitatively from the merchant's perspective without the need to be involved personally, a major blessing from my viewpoint.

2 Statement of the Problem

The representation of Fig. 1 shows the basic flow pattern that happens. Parking lot #1 has a capacity of C cars and parking lot #2 a capacity of D cars. However, the store employees diminish the capacity of one or both parking lots. Let there be c employee cars parked during the store opening hours in parking lot #1 and d employee cars in parking lot #2. In real situations the number of employees allowed to park in parking lot #1 is usually zero because the store manager needs to keep the nearest spaces for customers. The sole exception is customarily for handicapped employees, who most often make up only a small percentage of the total employees. Let the store be open daily from time $t = 0$ until time T_{shut}.

Let the flux of cars along the road to the store be S cars/h. Most drivers would wish to park as close to the store as possible, thereby minimizing both their walking and also the time spent carrying all the goodies they bought at the store. However, a smaller number of drivers will opt for the more distant parking lot #2 because either they do not wish to spend a longer time looking for a parking space in lot #1

A Christmas Parking Lot Problem

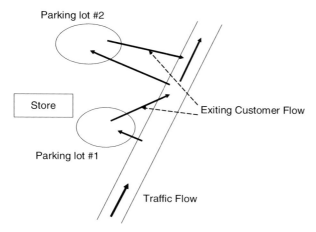

Fig. 1 Sketch of the parking lot problem

or because they do not wish to have their car undergo a higher chance of being scratched by the thoughtless customers, or because they are part of the minority who enjoy walking. Let the potential fraction of the flux of cars that opt for parking lot #1 be f, and the potential fraction of the flux of cars that bypass parking lot #1 opt for parking lot #2 be g (note that f + g can be greater than unity but must be less than 2). But not all such options can come true. First, if the first parking lot is full then cars have no choice but to continue to parking lot #2. Second, if the more remote parking lot #2 is also full then cars must continue along the road and so the merchant has lost potential customers.

Let the average amount spent by customers who do enter the store be $/car. Let the average time customers spend in the store buying their wonderful Christmas goodies be T_{buy}. Once the store has closed to new customers, the customers already in the store are allowed to stay and complete their purchases. The parking lots then empty after a time $t > T_{shut}$, and, on average, at time $t = T_{shut} + T_{buy}$. (Actually, personal observation shows that the time to get out of the parking lots is much longer because the exits are usually also the entrances. The less than friendly car drivers then tend to block the entrance as they try to enter so that other cars cannot leave. This problem causes a bottle-neck in the traffic flow from the parking lots and leads to major loss of potential customers as we will also show. In addition, the exiting cars must enter the road flow of traffic, which can be hideous, and this slow down effect is also a major blockade to parting potential customers from their money).The employees also leave the store and, because they too wish to complete their Christmas shopping as rapidly as possible, they also eventually exit the parking lots after cleaning up the store and restocking the shelves for the next day's mad frolic.

The question is: Can the merchant adjust the buying time and/or the product average price and/or the flux of consumers so that, over the opening hours of the store, he maximizes his profit? Clearly the answer to this question must also depend on the net flux S of cars per hour along the road and also on the holding capacity of each parking lot.

For a low flux of cars and a high holding capacity, the parking lots are never full and every potential customer can enter the store. However, as the flux of cars rises, for fixed parking lot capacities, there comes a point when the parking lots are full and only when the first customers who have bought goodies start leaving (at a time T_{buy} after opening of the store) can new cars enter the parking lots. The potential profit must also depend on the fraction of customers who are prepared to enter either parking lots #1 or #2, as well as on those who take one look at the mayhem that is going on in the parking lots and decide to drive further – either to a different store or to home, where they work on building numerical programs to evaluate the Christmas parking lot problem in the quiet of their studies. I count myself among the latter, naturally, or this paper would not exist.

The number of people who choose not to go to the stores for Christmas shopping is small, either because they are forced to go in order to buy that "must have" gift that is the demand of some small child, or because they feel it is a patriotic duty to spend their remaining hard won salaries before the end of the calendar year and so "max out" their plastic credit cards. For illustrative purposes we take this percentage of the populace to be effectively zero.

In general, there must be a bypass of cars throughout the opening hours of the store and the aim of the merchant is to minimize the total bypass so that customers enter his store. Once in the store they spend their average $/car and then exit. There can be bypass of cars from parking lot #1 but then the merchant must try to capture such a bypass with parking lot #2 so that no customers are lost. At a high enough flux of cars this desire will fail because the parking lots are full. Furthermore, only a fraction g of the cars that have bypassed parking lot # 1 (either because they wished to or because parking lot #1 was full) choose to enter parking lot #2.

Based on the above ideas, and choosing not to participate in the mindless fiscal frenzy of Bah Humbug time, in the relative peace and quiet far from the madding throng I have built a spreadsheet program that allows me to play peacefully with this Christmas parking lot problem. The next illustrations show how one can estimate the profit the merchant makes as parameters are adjusted to minimize the loss of potential customers.

3 Numerical Illustrations

3.1 Base Case

Consider the situation where a store has 50 employees, of whom 10% only are allowed to park in parking lot #1, with the remaining 45 relegated to the long trek from and to parking lot #2. Let parking lot #1 have a total capacity of 500 cars and let the more distant parking lot #2 have the same capacity. Let the lemming attitude of the public be such that 80% of the original traffic flow will try to enter parking lot #1

A Christmas Parking Lot Problem

in search of a space if they can when parking lot #1 is not full; the remaining cars will opt for parking lot #2 because they believe they will spend less time looking for a parking space and also waste less gas cruising around the parking lot in search of such elusive spaces. But when parking lot #1 is full then not even 80% of the original traffic flow can enter the parking lot and the overflow must carry on to parking lot #2 where only a fraction g of that residual flow choose to enter parking lot #2. We take this fraction to be 100% for the base case so that $g = 1$.

Let the store manager arrange to have his store open 20 h a day in this high potential profit time of year (actually many stores are open 24 h per day during this plastic credit card meltdown season so 20 h is somewhat conservative). Suppose further that those who do enter the store spend, on average, $100 per car (again this sum is probably somewhat conservative given the observations of damage to my bank account that has prevailed over the years). Let the average time for customers to shop, pay, and exit be about an hour (although judging, again personally, from shopping at Christmas with one or more of my daughters this one hour limit is undoubtedly a very gross underestimate). But based on this one hour shopping number the earliest shoppers start to leave the store, loaded with their goodies, an hour after it has opened.

Suppose that the store manager has priced his goods so that about 100 cars per hour flow down the street to his store, for a total car count over the 20 h opening time of some 2,000 cars. Remember this is the base case and is not likely to return a very good profit for the store because so few cars flow past per hour. But the base case will be used so that one can draw comparisons with the later cases to follow set in a familiar framework. Then, as shown in Fig. 2, there are no cars that bypass either parking lots #1 or #2; all customers can have their money replaced by trashy

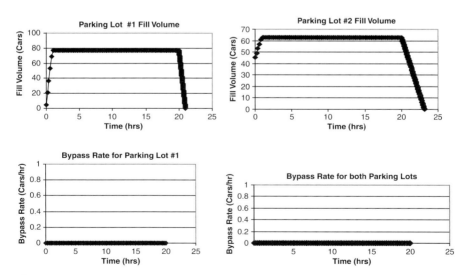

Fig. 2 Results of the Base Case calculations for fill of the parking lots with time and the bypass (zero) of cars

presents, and there is an orderly flow of traffic into and out of the parking lots. Basically this ease of shopping is due to the small volume of cars, only 100/h, and the large capacity of the two parking lots (500 cars each) so that there is always ample parking space. The exit rates out of both parking lots have been taken to be at the entrance rates of 80 cars/h for parking lot #1 and 20 cars/h for parking lot #2 because of this stress-free shopping. However, the situation is much different if the flow of cars is higher but the exit rates out of the parking lots are limited to 100 cars/h under all conditions, as we now show.

3.2 High Traffic Flow

Let there be a dominant toy that is a "must have" for almost every child and let the store have the sole rights of distribution (Claus et al., 2008). Now every car in a neighborhood from, apparently, millions of miles around the store has no choice except to try to purchase this ultimate toy at the one store. (In my case it was a "Dancerina" doll one year and a "Raggedy Ann and Raggedy Andy" alarm clock another year (Penney, 1999). I still twitch every time I think about the two situations). Suppose this dominance causes a traffic flow of 10 times the base case, i.e. 1,000 cars per hour for the 20 h period, implying 20,000 cars are all trying to squeeze into the now no longer capacious parking lots (I truly believe this sort of situation is a better description, possibly marginally underestimated a touch, of the reality that prevails at the season of "Spend everything you have ever earned and do it NOW").

Now the mayhem of the parking lots kicks into high gear. Shown in Fig. 2 are the volume fills of parking lots #1 and #2 with time, using the same parameters as in the base case except the exit rate from both parking lots is limited to 100 cars/h and the original flux of cars on the road is 1,000 per h for the 20 h opening time of the store, for a total of 20,000 cars in total. Also shown in Fig. 2 are the cumulative bypass of cars for parking lot #1 and the cumulative bypass for both parking lots together. The two upper panels of Fig. 2 show that both parking lots fill almost immediately and stay full until the store closes. The time to get out of the parking lot can extend up to 5 h after the store shuts, at an exit rate of 100 cars/h (How very true that is!). The two bottom panels of Fig. 2 show the cumulative bypass of cars, which therefore do not purchase anything from the store. Note that parking lot #1 fills in about 0.5 h so that most of the cars thereafter have to bypass the lot. Cumulatively, the total is about 14,000 cars over the 20 h opening period that do not get a space in parking lot #1. They add to the flow of traffic heading towards parking lot #2. However, as shown in the cumulative bypass panel of Fig. 2, just under 16,000 cars bypass both parking lots. The problem here is the bottleneck of exiting cars. At 100 cars/h one cannot refill the parking lots at better than 100 cars/h. so that one look at the massive traffic jams that exist causes only the foolhardy (of which there are about 4,000 for this example) to even attempt to enter the parking lots. The rest of the cars bypass both lots and, presumably, try to find the desired toy elsewhere; alternatively some 80% of the children demanding the "must have" toy are disappointed, leading to

A Christmas Parking Lot Problem

major fits, sulks, tears, etc. at opening time of the presents (and I know wherefrom I speak!). The merchant, in this case, despite having a monopoly on the "must-have" toy, can service only about 4,000 customers because of the insanity of the drivers attempting to both leave and enter the parking lots in terms of their driving and need to enter the high density steam of road traffic. A better design system for traffic entrance and exit is required to increase the rate at which cars can leave the lots and enter the main traffic stream and so maximize the merchant's profit.

3.3 A Better Traffic Flow

Let the merchant redesign his traffic flow for the parking lots so that he can raise the exit rate to 200 cars/h. from each parking lot. Now the problem of capturing money-bearing customers is somewhat better, as shown in Fig. 4, using the same parameter values as for the high flux of cars situation (1,000 cars/h for 20 h), and with the same 80% trying to get a space in parking lot #1 if available.

While the two parking lots are nearly always filled to capacity during the store opening hours, as shown in the upper two panels of Fig. 4, and while the cumulative bypass of cars for parking lot #1 is almost 12,000 as shown in the left lower panel of Fig. 4, some of that bypass is now captured by the second parking lot, so that only

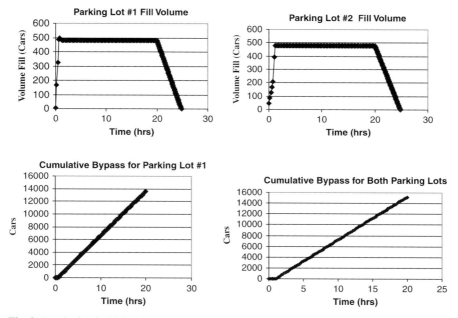

Fig. 3 Results for the high density of traffic flow (1,000 cars/h when the exit from the parking lots is limited to 100 cars/h for parking lot #1 and also 100 cars/h for parking lot #2)

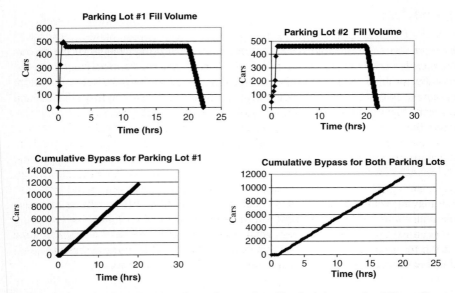

Fig. 4 Same as Fig. 3 but when the exit rate from each parking lot is increased to 200 cars/h

about 11,800 cars bypass the store completely; the store can now extract money from about 41% of the potential customers, a total customer base of 8,200 cars is serviced, leading to a doubling of the profit, and a halving of the small child tantrums at the opening time for presents.

3.4 A Corollary Effect

Once more, personal observation indicates that none of the suggestions above is ideal. Because of the high traffic density both on the road and entering and exiting the parking lots, and due to the idiocy of many drivers, there are car crashes at an increased rate during this season of self-reduction to poverty. The car crashes in the parking lots tend to do minor damage to the cars (although there is a huge number of such "fender-benders" because the Pauli Exclusion Principle dictates that two cars cannot occupy the same parking space simultaneously), while the crashes on the road, due to the slow-down of traffic trying to enter the lots and due to the drivers trying to exit the lots into the high speed road traffic, tend to be more serious per car involved, even if less in number-generally.

So as well as going bankrupt due to the store purchases, there is also a risk one will go deeply into debt due to car repairs (AAA, 1999), increased car insurance costs (Sylvan, 2000) and medical bills. The time-honored American tradition of "Sue anyone and everyone" can then be invoked to attempt to recover a portion, or all, of this risked outlay-including suing the store because it is not security conscious.

4 Conclusion

In short: there are no ways the store merchant can *maximize* his profit but he can increase his profit above the base case by suitably pricing his merchandise so that the flow of traffic to his store is increased, and by designing the parking lots for easier and faster entrance and exit. The disadvantage is that by so doing the merchant raises the accident rate of cars in the parking lots and on the access roads, and will undoubtedly be sued and have to pay, thereby lowering his potential profit margin.

Avoidance therapy for the potential risk of car crashes, for avoiding the long delay in getting a parking space and so also minimizing gas consumption, for saving the costs of buying trashy gifts for real money, for avoiding bankruptcy, and for peace and quiet is to stay home, build the Christmas Parking Lot program and have fun playing with it, and write this paper. The downside is that one then guarantees child tantrums on Christmas day because of the absence of the "must-have" toy.

I can hardly wait for summer vacation and the traffic flow to the beach and the search for accommodation; the makings of another program writing session are already causing my fingers to tremble with hope. Merry Christmas anyway.

Acknowledgments Many years of observations and personal involvement in the Christmas parking lot problem have been of great help. I thank my bank manager over the years for being so accommodating with my bank account overdrafts during the frenzied buying season. I thank Santa Claus and elves for detailed discussions of many of the points. My lawyer has also been most helpful on occasions of car damage caused during the so-called festive season.

This work has been supported by the DAAD through their award of a Visiting Professorship at the University of Leipzig. The University of Leipzig is also thanked for its contribution to this support and Professor Werner Ehrmann is particularly thanked for the courtesies and support he and his group at Leipzig have made available during the course of this work.

References

AAA, American Automobile Association (1999) Letter to author, dated 28 December 1999, confirming transport of damaged car to a garage
Claus S, Noel P, Weihnachtsmann D (2008) Personal communication
Penney JC (1999) Issuance of rain check for inability to provide "Raggedy Ann and Raggedy Andy" alarm clock prior to Christmas Day 1999
Sylvan G (2000) Insurance letter to author concerning traffic accident of 24 December 1999

Printing: Krips bv, Meppel, The Netherlands
Binding: Stürtz, Würzburg, Germany